Lecture Notes in Computer Science

Edited by G. Goos and J. Hartmanis

Dines Bjørner Manfred Broy
Igor V. Pottosin (Eds.)

Formal Methods in Programming and Their Applications

International Conference
Academgorodok, Novosibirsk, Russia
June 28 - July 2, 1993
Proceedings

Springer-Verlag
Berlin Heidelberg New York
London Paris Tokyo
Hong Kong Barcelona
Budapest

Series Editors

Gerhard Goos
Universität Karlsruhe
Postfach 69 80
Vincenz-Priessnitz-Straße 1
D-76131 Karlsruhe, Germany

Juris Hartmanis
Cornell University
Department of Computer Science
4130 Upson Hall
Ithaca, NY 14853, USA

Volume Editors

Dines Bjørner
United Nations University, International Institute for Software Technology
P. O. Box 3058, Macau

Manfred Broy
Institut für Informatik, Technische Universität München
Postfach 20 24 20, D-80290 München, Germany

Igor V. Pottosin
Institute of Informatics Systems
Av. Acad. Lavrentyev 6, Novosibirsk 630090, Russia

CR Subject Classification (1991): F.3-4, D.2-3, I.1-2

ISBN 3-540-57316-X Springer-Verlag Berlin Heidelberg New York
ISBN 0-387-57316-X Springer-Verlag New York Berlin Heidelberg

Typesetting: Camera-ready by author
Printing and binding: Druckhaus Beltz, Hemsbach/Bergstr.
45/3140-543210 - Printed on acid-free paper

Preface

The volume comprises the papers selected for presentation at the international conference "Formal Methods in Programming and Their Applications", held in Academgorodok (Novosibirsk, Russia), June 28 - July 2, 1993.

The conference was organized by the Institute of Informatics Systems of the Siberian Division of the Russian Academy of Sciences. The Institute is engaged in active research in the field of theoretical programming. The Institute has been an organizer of several international conferences related to programming and formal methods in programming but the latter have been considered together with other problems of programming. The current conference is the first forum organized by the Institute which is entirely dedicated to formal methods.

The main scientific tracks of the conference have been centered around the formal methods of program development and program construction. They include:

— specification, synthesis, transformation and verification of programs;

— parallel and distributed computations;

— semantics and logic of programs;

— theory of compilation and optimization;

— mixed computation, partial evaluation and abstract interpretation.

One of the main goals of the conference has been to promote formal methods in programming and to present and discuss the most interesting approaches to practical programming. A number of papers delivered at the conference are aimed at such a goal.

Scientists from eleven countries have been participants to the conference (Austria, Brazil, Canada, Denmark, England, France, Germany, the Netherlands, Russia, Turkey, and the USA as well as from the Territory of Macau)!

Also in the opinion of the participants the conference has been a success and similar conferences should be held in the future, perhaps, with a greater focus on the application of formal methods and the problems connected with it.

The organizers of the conference express their deep gratitude to the colleagues of Russian and international communities who supported actively the conference by rendering assistance and advice, review and participation.

We would also like to thank the Springer-Verlag for excellent co-operation.

July 1993 Dines Bjørner, Manfred Broy, Igor Pottosin

FMP&TA Organization

Programme Committee Co-Chairs:

Dines Bjørner United Nations University
Intl.Inst.f.Softw.Techn., Macau

Manfred Broy Technische Universität München
Institut für Informatik, Munich, Germany

Igor Pottosin Institute of Informatics Systems
Novosibirsk, Russia

Programme Committee:

Jan Barzdin, Riga
Mikhail Bulyonkov, Novosibirsk
Patrick Cousot, Paris
Pierre Deransart, Rocquencourt
Yuri Ershov, Novosibirsk
Yuri Gurevich, Ann Arbor
Victor Ivannikov, Moscow
Cliff Jones, Manchester
Philippe Jorrand, Grenoble
Vadim Kotov, Palo Alto
Reino Kurki-Suonio, Tampere
Alexander Letichevski, Kiev
Gregory Mints, Stanford
Valery Nepomniaschy, Novosibirsk
Anil Nerode, Ithaca
Amir Pnueli, Rehovot
Mikhail Taitslin, Tver
Boris Trachtenbrot, Tel-Aviv
Enn Tyugu, Stockholm

Invited Speakers

ChaoChen Zhou, Macau
Patrick Cousot, Paris
Yuri Ershov, Novosibirsk
Philippe Jorrand, Grenoble
Bernhard Möller, Augsburg
Peter Pepper, Berlin
Mikhail Taitslin, Tver
Martin Wirsing, Munich

Organizing Committee:

Valery Nepomniaschy, chairman
Victor Sabelfeld, publication of proceedings
Mikhail Bulyonkov, general organization
Alexander Bystrov, treasury
Tatiana Churina, scientific secretary

List of Referees

K.Andrew
J.Barzdin
Yu.Gurevich
P.Deransart
Yu.Ershov
N.Jones
C.Jones
P.Jorrand
R.Kurki-Suonio
A.Letichevsky
G.Mints
B.Möller
T.Mogensen
V.Nepomniaschy
V.Sazonov
V.Sabelfeld
N.Shilov
M.Taitslin
E.Tyugu
M.Valiev
I.Virbitskaite
C.Zhou

TABLE OF CONTENTS

Theory of Domains and Nearby *

Yu. L. Ershov

Novosibirsk State University, RIMIBE
Universitetsky pr. 4
Novosibirsk 630090
Russia

Abstract. The author presents a topological approach to the development of the theory of domains.

0. In the present paper we deal with historical, methodological, and mathematical aspects of the theory of domains. A recent book [1] by A. Jung demonstrates a noticeable progress in the theory.

Theory of domains arose in the late 60s in research that was carried out independently by Prof. D. Scott in Oxford and by the author in Novosibirsk. D. Scott was interested in a natural mathematical model for the type-free λ-calculus, whereas the author developed a theory of partial computable functionals of finite types. Both problems were solved quite satisfactorily. The corresponding results were reported by the author at the International Congress of Mathematicians in Nice, 1970 [2] and by D. Scott at the International Congress for Logic, Methodology, and Philosophy of Science in Bucharest, 1971 [3].

It turned out that there was a great resemblance between the mathematical models developed. The exact relation between these models was established in [4]. In particular, the notion of Scott's domain (S-domain) and the one of complete f_0-space were proved to be equivalent.

N. Bourbaki in "*L'Architecture des mathématiques*" distinguishes three basic mathematical structures: algebraic, topological, and that of partial order. All these structures are found in the theory of domains. The approach of D. Scott to the introduction of S-domain by means of (directed-complete) partial orders dominates in the current literature on computer science, though many basic concepts of the theory, e.g., the way-below relation, are rather difficult to comprehend. This was the reason why D. Scott repeatedly returned to the theory of domains, attempting to clarify the foundations. Thus, for this purpose he introduced information systems [5].

In the author opinion, topology should be the basic structure in the development of the theory. The author supposes that the topological approach of [4] is more preferable than the one based on a partial order, both for better reception and for potentially greater generality that is needed if one wants to study domains which contain only constructive points. In the present paper the author will try to substantiate this point of view following the ideas expressed in [4].

* Research supported in part by the Russian Foundation for Fundamental Research (93–011–16014).

1. Let $\langle X, T\rangle$ be a topological space (T is a topology on X, i.e., a family of all open sets). We define a preorder $\leqslant_T$ on X, related to the topology T, as follows: for $x, y \in X$

$$x \leqslant_T y \Leftrightarrow \text{for every open set } V \subseteq X (V \in T)(x \in V \rightarrow y \in V).$$

This relation is a partial order provided that $\langle X, T\rangle$ is a T_0-space, i.e., the weakest separation axiom holds: *for every $x, y \in X$, if $x \neq y$, then there exists an open set $V \subseteq X$ such that $x \in V$ and $y \notin V$, or $x \notin V$ and $y \in V$.*

The subscript T in the notation $\leqslant_T$ will usually be omitted. We introduce the following notation: $\hat{x} = \{y \mid y \in X,\, y \leqslant x\}$, $\overset{\vee}{x} = \{y \mid y \in X,\, x \leqslant y\}$.

If $\langle X, T\rangle$ is a T_1-space (i.e., $\forall x, y \in X\ (x \neq y \rightarrow \exists V \in T(x \in V \wedge y \notin V))$), then the preorder $\leqslant$ degenerates to the identity relation. In the sequel, we will consider only T_0-spaces.

We introduce one more relation, namely the *approximation relation* $\prec$ on elements of X as follows: for $x, y \in X$

$$x \prec y \Leftrightarrow \text{there exists an open set } V \subseteq X \text{ such that } (y \in V \text{ and } \forall z \in V (x \leqslant z)).$$

Remark. An equivalent definition may be given as follows: $x \prec y \Leftrightarrow y \in \operatorname{Int} \overset{\vee}{x}$, where $\operatorname{Int} Y$ is the interior of Y, i.e., the largest open subset of the set $Y \subseteq X$. Note that $x \prec y$ implies $x \leqslant y$.

We will use the following notation: $\overset{\curlywedge}{x} = \{y \mid y \in X,\, y \prec x\}$, $\overset{\curlyvee}{x} = \{y \mid y \in X,\, x \prec y\}$.

We call a topological space $\langle X, T\rangle$ *approximative* (or an *α-space*) if the following condition holds: for any open set $V \subseteq X$ and any element $x \in V$ there exists $y \in V$ such that $y \prec x$.

It is easy to see that the following holds:

1. If $\langle X, T\rangle$ is an α-space and a set $V \subseteq X$ is open, then

$$V = \bigcup_{x \in V} \overset{\curlyvee}{x}.$$

2. If $\langle X, T\rangle$ is an α-space and $x \in X$, then for every y, z such that $y \prec x$ and $z \prec x$ there exists $u \prec x$ such that $y \prec u$, $z \prec u$.
3. $x = \sup \overset{\curlywedge}{x}$, i.e., x is the least upper bound (relative to the order $\leqslant$) of the set $\overset{\curlywedge}{x}$.

Let $\langle X, T\rangle$ be an α-space. A set $X_0 \subseteq X$ is called a *base subset of* X if the following condition holds: for any open set $V \subseteq X$ and any $x \in V$ there exists $y \in V \cap X_0$ such that $y \prec x$.

Remark. X is a base subset of X.

Remark. If X_0 is a base subset of X, then $V = \bigcup_{x \in V \cap X_0} \overset{\curlyvee}{x}$ for every open set $V \subseteq X$.

Remark. If X_0 is a base subset of X, then for any $x \in X$ the set $\hat{x} \cap X_0$ is directed and $x = \sup(\hat{x} \cap X_0)$.

Now we proceed to the closure properties of α-spaces. (In the sequel, we will usually omit an explicit indication of the topology.)

Proposition 1. *If X and Y are α-spaces, then the Cartesian product $X \times Y$ is an α-space.*

Remark. The topology of the product $X \times Y$ is defined in a standard way.

Proposition 1 can be extended to products of arbitrary number of spaces, if we impose an additional quite natural restriction. An α-space X is called an *α_0-space* if the partially ordered set $\langle X, \leqslant_T \rangle$ has a least element.

Proposition 1′. *Let X_i, $i \in I$, be a family of α_0-spaces and $X = \prod_{i \in I} X_i$ be the Cartesian product of the family (equipped with the Tychonoff topology). Then X is an α_0-space.*

Many important constructions in the theory use the notions of retract and project. Remind that a continuous mapping $\rho : X \to X$ of a topological space X into itself is called a *retraction* if $\rho^2 = \rho$. The image $\rho(X)$ considered as a subspace of X is called a *retract of* X. A retraction $\rho : X \to X$ is called a *projection* if $\rho(x) \leqslant x$ for all $x \in X$. In this case $\rho(X)$ is called a *project of* X.

Proposition 2. *If X is an $\alpha(\alpha_0)$-space and $Y \subseteq X$ is a retract of X, then Y is an $\alpha(\alpha_0)$-space.*

Now we introduce an important notion of a complete α-space.

An α-space X is called *complete* if, given an α-space Y, its base subset Y_0, and a homeomorphism h of Y_0 into X such that $h(Y_0)$ is a base subset of X, there exists an extension of h to a continuous mapping of Y into X. (This extension will in fact be a homeomorphic embedding of Y into X.)

Proposition 3. *For every α-space Y there exists a complete α-space X and a homeomorphic embedding $\pi : Y \to X$ such that $\pi(Y)$ is a base subset of X.*

The α-space X in Proposition 3 is called *the completion of* Y. It is unique in a reasonable sense.

Now we establish a crucial connection between α-spaces and directed-complete partial orders.

Theorem 4. *If $\langle X, T \rangle$ is a complete α-space, then $\langle X, \leqslant_T \rangle$ is a continuous directed-complete partial order. If $\langle X, \leqslant \rangle$ is a continuous directed-complete partial order, then X, equipped with the Scott-topology, is a complete α-space and the approximation relation $\prec$ coincides with the way-below relation $\ll$.*

2. An important subclass of the class of approximative spaces is the class of finitary spaces. An element x in T_0-space X is called *finitary* if the relation $x \prec x$ holds or, equivalently, if the set $\check{x}$ is open. The set of all finitary elements of a space X will be denoted by $F(X)$. An approximative space X is called a *finitary space* (or a *φ-space*) if $F(X)$ is a base subset of X.

Remark. For an arbitrary base subset X_0 of X we have $F(X) \subseteq X_0$. If X_0 is a base subset of X and $x \in X_0 \setminus F(X)$, then $X_0 \setminus \{x\}$ is a base subset of X. Thus, an α-space X is finitary if and only if it has the least (under set inclusion) base subset.

Theorem 5. *If $\langle X, T\rangle$ is a complete φ-space, then $\langle X, \leqslant_T\rangle$ is an algebraic directed-complete partial order. If $\langle X, \leqslant\rangle$ is an algebraic directed-complete partial order, then X, equipped with the Scott-topology, is a complete φ-space.*

A φ-space X is called an *f-space* if $\langle F(X), \leqslant\rangle$ is a partial upper semilattice, i.e., a partial order such that, for any $x, y \in F(X)$, a consistency of x and y (i.e., $\exists z \in F(X)(x \leqslant z \wedge y \leqslant z)$) implies the existence of the least upper bound $x \sqcup y$ in $F(X)$. An f-space with a least element is called an *f_0-space* (cf. [4]).

Theorem 6. *If $\langle X, T\rangle$ is a complete f_0-space, then $\langle X, \leqslant_T\rangle$ is an S-domain. If $\langle X, \leqslant\rangle$ is an S-domain, then X, equipped with the Scott-topology, is a complete f_0-space.*

A φ-space X is called a *b-space* if $\langle F(X), \leqslant\rangle$ satisfies the condition: every finite subset $F \subseteq F(X)$ is contained in a finite subset $F_0 \subseteq F(X)$ such that

$$\forall F_1 \subseteq F_0 \ \forall x \in F(X) \ (\forall x_0 \in F_1(x_0 \leqslant x) \rightarrow$$
$$\rightarrow \exists x_1 \in F_0 \ (\forall x_0 \in F_1(x_0 \leqslant x_1) \wedge x_1 \leqslant x)).$$

Finite sets F_0 satisfying this condition are called *perfect*. A b-space with the least element is called a *b_0-space*.

Theorem 7. *If $\langle X, T\rangle$ is a complete b_0-space, then $\langle X, \leqslant_T\rangle$ is a B-domain. If $\langle X, \leqslant\rangle$ is a B-domain, then X, equipped with the Scott-topology, is a complete b_0-space.*

Remark. Every $f(f_0)$-space is a $b(b_0)$-space.

Proposition 8. *If X is a b-space and Y is a b_0-space, then the set $C(X,Y)$ of all continuous mappings of X into Y, equipped with the topology of pointwise convergence, is a b_0-space. Moreover, if Y is complete, then $C(X,Y)$ is complete.*

We point out the basic elements of the proof.

1. If F_0 is a finite perfect subset of $F(X)$ and $f_0 : F_0 \rightarrow F(Y)$ is monotone, then we can extend f_0 to a continuous mapping $f_0^* : X \rightarrow Y$ as follows. Notice that if $x \in X$, then $\hat{x} \cap F_0$ is empty or contains the greatest element c_x. In the first case we put $f_0^*(x)$ equals $\perp_Y$, the least element of Y; in the second we put $f_0^*(x) = f_0(c_x)$.

2. The finite elements of $C(X, Y)$ are exactly the functions of the form f_0^*.
3. Assume that $f_0^*, \ldots, f_n^*$ are obtained from the monotone mappings $f_0 : F_0 \to F(Y), \ldots, f_n : F_n \to F(Y)$. We put:
   $F_{n+1} \subseteq F(X)$ is a finite perfect subset of $F(X)$, containing $F_0 \cup F_1 \cup \ldots \cup F_n$;
   F_{n+2} is a finite perfect subset of $F(Y)$, containing $\{\perp_Y\} \cup f_0(F_0) \cup \ldots \cup f_n(F_n)$;
   $G = \{f \mid f$ is a monotone mapping of F_{n+1} into $F_{n+2}\}$;
   $G^* = \{f^* \mid f \in G\}$.

Then G^* is a finite perfect subset of $F(C(X, Y))$ and $\{f_0^*, \ldots, f_n^*\} \subseteq G^*$.

Proposition 9. *The category of b_0-spaces is Cartesian closed.*

Remark. A corresponding statement for f_0-spaces was proved in [6].

Remark. The category of b_0-spaces is closed under limits of bispectra, i.e., an analog of Theorem 1 [4, §5] holds.

Since retracts of a Cartesian closed category of topological spaces constitute a Cartesian closed category in themselves, it is useful to obtain a description for retracts of f_0-spaces. It turns out that the following generalization of Theorem 4.1 [1] holds. (We recall that, according to [1], a *deflation* of a topological space X is a continuous mapping $f : X \to X$ of X into itself such that $f(X)$ is finite and $f(x) \leqslant_T x$ for all $x \in X$.)

Proposition 10. *If an α_0-space X is a retract of a b_0-space, then there exists a directed family $f_i, i \in I$, of deflations of X such that* $\sup f_i = \mathrm{id}_X$. *If an α_0-space X possesses such a family of deflations, then X is a project of a b_0-space.*

The second part of the proposition is stronger than the corresponding assertion of Theorem 4.2 [1] even for complete b_0-spaces (= B-domains) and answers the question raised in [1, p. 92].

Remark. An explicit description of retracts (or projects) of complete f_0-spaces as complete A_0-spaces is given in [4].

3. As in [7], we give an effective version of b_0-spaces. Let X be a b_0-space. An enumeration $\nu : \omega \to F(X)$ is called a *constructivization of the base subset of* X if the following conditions hold:
1) the set $\{\langle n, m\rangle \mid n, m \in \omega, \nu n \leqslant \nu m\}$ is recursive;
2) there exists a recursive function $g : \omega \to \omega$ such that for every $n \in \omega$

$$\nu D_n (= \{\nu m \mid m \in D_n\}) \subseteq \nu D_{g(n)} (= \{\nu m \mid m \in D_{g(n)}\})$$

and $\nu D_{g(n)}$ is a perfect subset of $F(X)$. (Here D_n is a finite subset of ω with a canonical index n, cf. [6].)

A b_0-space X *has a constructivizible base* if there exists a constructivization of the base subset $F(X)$ of X.

Proposition 11. *The category of b_0-spaces with constructivizible base subsets is Cartesian closed.*

Let $\nu : \omega \to F(X)$ be a constructivization of the base subset of a b_0-space X. An element $x \in X$ is called *constructive*, if the set $\{n | \nu(n) \leqslant x\}$ is recursively enumerable.

A good theory of f_0-spaces which have constructivizible bases and such that all their elements are constructive is developed in [6]. In particular, the notions of computable enumeration of these spaces, completeness, and principal computable enumerations are defined there. This theory serves as a tool for the construction of partial computable functionals of finite types acting on partial continuous functionals (the model $\mathbb{C}$, [4, 7]). But the theory of b_0-spaces in which all points are constructive is not quite satisfactory as the following example shows.

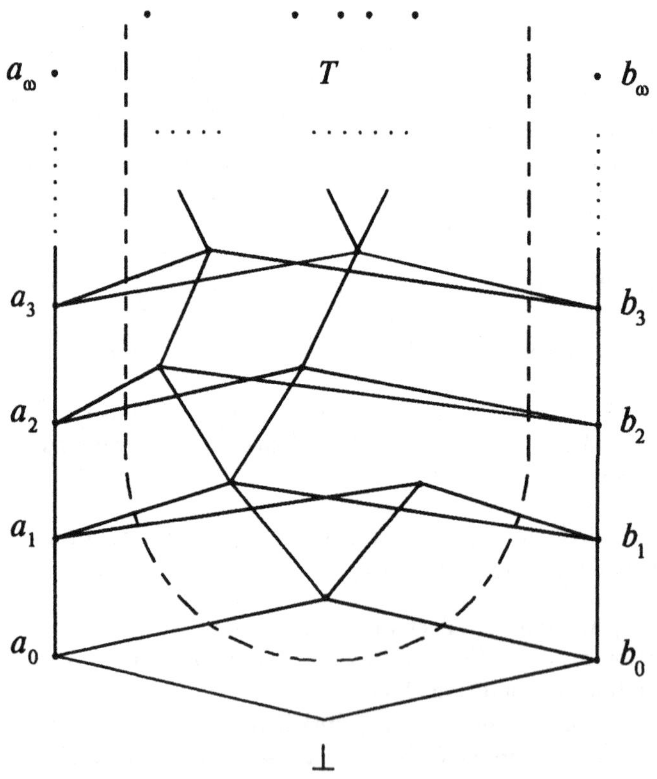

Here T is an infinite recursive binary tree without infinite recursive branches. Notice that every infinite recursively enumerable branch of a recursive tree is recursive. Hence, there is no infinite recursively enumerable branch in T. The existence of such trees is well known (cf. [8]).

Elements of T at a level n are (minimal) upper bounds of the pair a_n, b_n. We add limit points: a_ω for $a_0, a_1, \ldots$; b_ω for $b_0, b_1, \ldots$; and limit points corresponding to every infinite branch of the tree T. A topology on the obtained set X_T is defined by a subbasis constituted by open sets of the form $\check{c}$ where $c \in F(X_T) = T \bigcup \{a_0, b_0, \ldots\}$. Then X_T is a complete b_0-space (or a B-domain; moreover, a BL-domain in terms of [1]). Obviously, the base subset $F(X_T)$ is constructivizible. The points a_ω and b_ω are constructive, whereas all other limit points (which are the upper bounds of a_ω, b_ω) are not. Thus, a_ω and b_ω, being consistent in X_T, are inconsistent in the subspace $C(X_T)$ of all constructive points of X_T.

The example shows that from the "constructive" point of view f_0-spaces behave better than b_0-spaces.

To conclude, we mention that spaces with constructive points can be used to define an effective semantics which, in turn, can serve as a programming language (semantic programming, cf. [9]). Moreover, effective versions of the spaces enable one to obtain generalizations of the theory through the use of arbitrary admissible sets (instead of ω) as it was done in [10] in the case of f_0-spaces.

References

1. Jung, A.: Cartesian Closed Categories of Domains. CWI Tract **66** Stichting Math. Centrum Amsterdam (1989)
2. Ershov, Yu. L.: La théorie des énumération. Actes Congr. Intern. Math. **1** Gauthier–Villars Paris (1971) 223–227
3. Scott, D.: Models for various type-free calculi. LMPS IV. Studies in Logic No 74 North–Holland Amsterdam (1973) 157–188
4. Ershov, Yu. L.: Theory of A-spaces. Algebra i Logika **12** No 4 (1973) 369–416
5. Scott, D.: Domains for denotational semantics. ICALP 82. Springer LNCS No 140 (1982) 577–613
6. Ershov, Yu. L.: Theory of enumerations [in Russian]. Nauka Moscow (1977)
7. Ershov, Yu. L.: Model $\mathbb{C}$ of partial continuous functionals. Logic Colloq.'76. Studies in Logic No 87 North–Holland Amsterdam (1977) 455–467
8. Peretyat'kin, M. G.: Strongly constructive models and enumerations of the boolean algebra of recursive sets. Algebra i Logika **10** No 5 (1971) 535–557
9. Goncharov, S. S., Ershov, Yu. L., and Sviridenko, D. I.: Semantic programming. Information Processing 86. North–Holland Amsterdam (1986) 1993–1100
10. Ershov, Yu. L.: On f_A-spaces. Algebra i Logika **25** No 5 (1986) 533–543

Predicative Programming — A Survey

Jim Grundy*

University of Cambridge, Computer Laboratory
New Museums Site, Pembroke Street, Cambridge CB2 3QG, England

Abstract. The idea of using a predicate to specify behaviour is a compelling one, and leads to a desire to refine specifications into implementations in languages whose semantics have also been specified with predicates.
Many of us believe we share a common intuition about what a specification phrased as a predicate means. It may be surprising to learn that there are several ways to interpret a predicate as a specification. Under these interpretations, the same predicate can specify different behaviour. This paper examines three simple styles of specification using predicates.

1 Introduction

The idea of using a predicate to specify behaviour is a compelling one. The specification languages HP-SL [1] and Z [7] are based on it. Predicates in higher-order logic are used for the specification and verification of hardware [2]. The use of predicates to specify the required behaviour of programs leads to a desire to be able to refine specifications into implementations in programming languages whose semantics have also been specified with predicates. Various techniques for making such refinements have been proposed [4, 5, 8, 10, 12, 14, 18].

One reason for the popularity of specifying with predicates is clarity; predicates and the predicate calculus are well understood logical concepts. When a behaviour is specified with a predicate, those behaviours that satisfy the predicate meet the specification, all others do not.

Many of us believe we share a common intuition about what a specification phrased as a predicate means. It may be surprising to learn that there are several ways to interpret a predicate as a specification. Under these interpretations the same predicate can describe different behaviour. This paper examines three simple styles of specification using predicates. We compare these styles, highlighting their strengths and weaknesses. In particular, we examine how the semantics of an imperative programming language can be expressed in each style.

2 The Basic Concepts

Before we begin, we must introduce some terminology.

* The author is supported by DSTO Australia.

state: We use the term *state* to refer to the state of a computer. The state is a vector of program variables, for example (w, x, y, z), which we refer to as σ. The set of all states is denoted by the symbol 'Σ'.

initial state: The *initial state* of a computer is the state before a computation begins. We denote the value of program variable x in the initial state by $`x$, and the entire initial state by $`\sigma$.

final state: The *final state* of a computer is the state that it is left in after a computation finishes. We denote the value of program variable x in the final state by x', and the entire final state by σ'.

nontermination: When a computation finishes it is said to have *terminated.* The term *nontermination* is used to describe a computation that never finishes. This may happen either because the computation encounters an error and aborts, or because the computation enters an infinite loop. We do not distinguish between these two possibilities.

outcome: The *outcome* of a computation is either the final state in which it terminates, or nontermination. We introduce the symbol '$\perp$' to denote nontermination. The set of all outcomes, written '$\Sigma_\perp$', is the disjoint union of the set of all states and nontermination:

$$\Sigma_\perp \stackrel{\text{def}}{=} \Sigma + \{\perp\}$$

behaviour: For any computation, the *behaviour* of a computer can be characterised by the initial state in which the computation was begun and the outcome of the computation. Behaviours are therefore elements of $\Sigma \times \Sigma_\perp$.

specification: A *specification* is a description of the possible behaviours of a computation. A specification is a function from behaviours to booleans, that is of type $(\Sigma \times \Sigma_\perp) \to \mathbb{B}$. We say that a behaviour b *meets* a specification S, if S applied to b is true.

nondeterminism: If for any initial state only one behaviour meets a specification, then the specification is said to be *deterministic.* If, however, several behaviours would meet a specification, then that specification is said to be *nondeterministic.*

If we are attempting to prove a property of some system, we should assume that it behaves *demonically.* If a specification is nondeterministic and only some of the possible behaviours are ones that we want, then a demonic interpretation of the specification always chooses one of the behaviours that we *do not* want.

The opposite of demonic nondeterminism is *angelic* nondeterminism. If a specification is nondeterministic and only some of the possible behaviours are ones that we want, then an angelic interpretation of the specification always chooses one of the behaviours that we *do* want.

refinement: We say that a specification T is a *refinement* of another specification S, if any behaviour that meets T also meets S. The relationship 'S is refined by T' is written '$S \sqsubseteq T$'.

$$\vdash^{\text{def}} (S \sqsubseteq T) = (\forall b. (T\, b) \supset (S\, b)) \qquad [\sqsubseteq\text{\_DEF}]$$

From this definition we can see that refinement is the process of reducing nondeterminism in a specification.

2.1 Basic Specifications

Here are the descriptions of four basic specifications.

chaos

For some initial state, a specification allows *chaotic* (i.e. completely arbitrary) behaviour if the set of allowable outcomes is the set of all outcomes.

$$\overset{\text{def}}{\vdash} \mathsf{chaos}(i, o) = \mathbf{T} \qquad [\mathsf{chaos\_DEF}]$$

chance

For some initial state, a specification allows *chance* (i.e. arbitrary but terminating) behaviour if the set of allowable outcomes is precisely the set of all states.

$$\overset{\text{def}}{\vdash} \mathsf{chance}(i, o) = (o \neq \bot) \qquad [\mathsf{chance\_DEF}]$$

abort

For some initial state, a specification requires behaviour that never terminates, or *aborts*, if the set of allowable outcomes contains only the element $\bot$.

$$\overset{\text{def}}{\vdash} \mathsf{abort}(i, o) = (o = \bot) \qquad [\mathsf{abort\_DEF}]$$

miracle

For some initial state, a specification can never be satisfied (or requires *miraculous* behaviour) if the set of allowable outcomes is empty.

$$\overset{\text{def}}{\vdash} \mathsf{miracle}(i, o) = \mathbf{F} \qquad [\mathsf{miracle\_DEF}]$$

A miracle is a refinement of every specification. Note that it is not possible to implement miracle.

2.2 Predicates

Predicates are boolean-valued expressions which may contain free variables from the initial and final states $`\sigma$ and σ'. In this way predicates describe a relationship between these states. This relationship can be made explicit by regarding the predicate P as a shorthand for the function below:

$$\lambda(`\sigma, \sigma'). P$$

This function is of the type $(\Sigma \times \Sigma) \to \mathbb{B}$. Whenever it is not ambiguous, we will use predicates to stand for both their boolean values and functions from pairs of states to boolean values. For example, the predicate '**F**' stands for both the truth value false, and the constant function $\lambda(`\sigma, \sigma'). \mathbf{F}$.

The interpretation of a predicate as a specification is a function of type:

$$\underbrace{((\Sigma \times \Sigma) \to \mathbb{B})}_{\text{Predicates}} \to \underbrace{((\Sigma \times \Sigma_\bot) \to \mathbb{B})}_{\text{Specifications}}$$

There are many possible interpretations of predicates as specifications, but only a few preserve any of our intuitions about what a predicate should mean as a specification.

These few can be distinguished by the different ways in which nontermination is introduced into the interpretation.

The most obvious interpretation is the one that does not allow the description of nonterminating behaviour. Under this interpretation, any initial state i and outcome o will meet the specification described by P, if o is not $\perp$, and i and o satisfy P. Unfortunately since programs do not always terminate, this interpretation is unsuitable for describing programming language semantics. We are interested only in interpretations that can describe nonterminating behaviour.

2.3 A Simple Language

In this section we describe a small imperative programming language. In later sections we give this language a semantics using the various predicative models. By giving programming language constructs meanings as predicates, we are able to mix these constructs freely with other predicate notations. As a result we can treat predicates as a wide-spectrum language in which we can both phrase specifications and develop implementations.

Although any predicate can be interpreted as a specification, only a restricted syntactic category of expressions are considered executable. The restriction is syntactic so that executable specifications are easily recognised by both humans and machines. We do not give a complete definition of which expressions are executable. One might reasonably expect that expressions like '$1 + 2$' are executable, while expressions like '$\int_a^b f\ x\ dx$' are not. Note that in order to be considered executable, all variables in an expression must be drawn from the initial state.

We now describe the constructs of our programming language and the conditions under which they are executable.

$v' := e$

The assignment construct $x' := e$ assigns an expression e to a variable x', and leaves the other variables of the final state with the values they held initially. The assignment $x' := e$ is executable if the expression e is executable.

if b then P else Q

The conditional construct behaves according to P if the guard b is true. Otherwise, it behaves in accordance with Q. The conditional is executable if b, P and Q are executable. Note that although b is a boolean-valued expression it should not be interpreted as a specification.

$P \mid Q$

The choice construct makes an arbitrary choice between P and Q, and then behaves according to that specification. The choice $P \mid Q$ is executable if both P and Q are executable.

$P; Q$

The sequential composition $P; Q$ behaves according to the specification P and then (if this behaviour terminates) according to Q. The composition $P; Q$ is executable if both P and Q are executable.

var $x.\ P$

The var construct is used to introduce a local variable. The new variable x exists only within P. The block var $x.\,P$ is executable if P is executable. Note that within the scope of the var construct, the state space Σ should be extended to include the new variable.

label $l.\,P$
The label construct is used for creating recursively defined blocks. Within the body of the block, P, the label l should be interpreted as a recursive call to the start of the block. The block label $l.\,P$ is executable if assuming l is executable, then P is executable.

2.4 Notes

A possible misconception about specifications is that nontermination is never a desired behaviour. It is therefore sometimes assumed that the demonic choice between any specification and one requiring nontermination should always choose the one requiring nontermination. Suppose, however, we have a specification that requires nontermination, for example abort, and a potential implementation is proposed that allows a nondeterministic choice between some terminating behaviour and nontermination, for example $x' := \grave{}x$ | abort. Clearly, in this case the demonic choice is to choose the specification that allows terminating behaviour.

In general, a demonic choice between two specifications should choose the least refined of the two. However, not every set of specifications has a least refined element. We do know that the specification chaos is less refined that any other. Therefore the demonic choice between any specification and one that permits chaotic behaviour should choose the specification that permits chaotic behaviour. We also know that the specification miracle is more refined than any other. Therefore the demonic choice between a specification that requires miraculous behaviour and any other specification should choose the other specification.

3 The Relational Model

We call the first interpretation of predicates that we examine the *relational* model. The relational model is used by the HP-SL [1] and Z [7] specification languages. Mili's work on program development also uses the relational model [12]. More recently Gravell [4] and Sekerinski [18] have developed refinement methods based on this model. Nguyen has proposed a semantics for Dijkstra's guarded command language using the relational model [16].

3.1 Description

Consider a predicate P. The following is an interpretation of P as a specification using the relational model.

> Given any initial state, if there exists a final state satisfying P, then terminate in such a state. If there is no final state that satisfies P, any behaviour (including nontermination) is acceptable.

Let us denote the interpretation of the predicate P as a specification using the relational model by '$[\![P]\!]_r$'. When formalised, this interpretation is equivalent to the definition below:

$$\stackrel{\text{def}}{\vdash} [\![P]\!]_r(i, o) = ((\exists \sigma.\, P(i, \sigma)) \supset (o \neq \bot) \wedge P(i, o)) \qquad [[\![\text{-}]\!]_r\text{-DEF}]$$

A refinement relation between predicates in the relational model ($\sqsubseteq_r$) follows immediately from their interpretation.

$$\stackrel{\text{def}}{\vdash} (P \sqsubseteq_r Q) = ([\![P]\!]_r \sqsubseteq [\![Q]\!]_r) \qquad [\sqsubseteq_r\text{-DEF}]$$

A more direct definition of refinement would be easier to work with. Consider the choices available to us when attempting to satisfy the specification described by P. We have greatest choice when, for a particular initial state, there is no final state such that P holds—we can terminate in any state, or even fail to terminate. We can therefore refine a specification by increasing the number of initial states for which some final state will satisfy the predicate. When, for a particular initial state, there are several final states that satisfy P, we may terminate in any of those final states. Therefore, we can also refine a specification by decreasing the number of final states that will satisfy the predicate for any given initial state. Accordingly, the refinement relation between predicates in the relational model can be expressed as follows:

$$\begin{array}{l} \vdash (P \sqsubseteq_r Q) = \\ \quad \forall {}^{\backprime}\sigma.\, (\exists \sigma'.\, P({}^{\backprime}\sigma, \sigma')) \supset \\ \qquad (\exists \sigma'.\, Q({}^{\backprime}\sigma, \sigma')) \wedge (\forall \sigma'.\, Q({}^{\backprime}\sigma, \sigma') \supset P({}^{\backprime}\sigma, \sigma')) \end{array} \qquad [\sqsubseteq_r\text{-THM}]$$

3.2 Basic Specifications

Let us examine how the basic specifications of Sect. 2.1 are treated in the relational model:

chaos

The specification chaos is represented by the predicate **F** (false).

$$\vdash \mathsf{chaos} = [\![\mathbf{F}]\!]_r$$

chance

The specification chance is represented by the predicate **T** (true).

$$\vdash \mathsf{chance} = [\![\mathbf{T}]\!]_r$$

abort

There is no way to describe abort using the relational model.

$$\vdash \neg \exists P.\, \mathsf{abort} = [\![P]\!]_r$$

miracle

There is no way to describe miracle using the relational model.

$$\vdash \neg \exists P.\, \mathsf{miracle} = [\![P]\!]_r$$

3.3 The Simple Language

We now examine each of the programming language constructs described in Sect. 2.3 to see how they are represented in the relational model:

$x' := e$

Assignment can be expressed by stating that the final state is equal to the initial state with the appropriate substitutions.[2]

$$(x' := e) \stackrel{\text{def}}{=} (\sigma' = \grave{}\sigma)[e/\grave{}x]$$

if b then P else Q

The conditional construct is expressed as a choice between two mutually exclusive alternatives.

$$(\text{if } b \text{ then } P \text{ else } Q) \stackrel{\text{def}}{=} (b \wedge P) \vee (\neg b \wedge Q)$$

We could have given the meaning for this construct using a conditional expression, but this is the form traditionally used with the relational model.

$P \mid Q$

One might think that the nondeterministic choice between P and Q can be expressed by $P \vee Q$. Unfortunately, for any initial state if either P or Q would allow chaotic behaviour, this semantics would angelically chose the other specification. The meaning below forces a demonic interpretation of the choice between P and Q.

$$P \mid Q \stackrel{\text{def}}{=} (\exists \sigma' . P) \wedge (\exists \sigma' . Q) \wedge (P \vee Q)$$

$P; Q$

At first sight, it might appear that the composition of two specifications, P and Q, can be expressed as relational composition $\exists \sigma . P[\sigma/\sigma'] \wedge Q[\sigma/\grave{}\sigma]$. Unfortunately, this semantics requires P to angelically choose a final state from which Q will not allow chaotic behaviour. The meaning below gives a demonic interpretation to the composition by explicitly allowing chaotic behaviour whenever P may terminate in a state from which Q allows chaotic behaviour.

$$P; Q \stackrel{\text{def}}{=} (\forall \sigma . P[\sigma/\sigma'] \supset (\exists \sigma' . Q[\sigma/\grave{}\sigma])) \wedge (\exists \sigma . P[\sigma/\sigma'] \wedge Q[\sigma/\grave{}\sigma])$$

var $x . P$

To introduce a local variable, we must provide a quantified scope for its initial and final values. At first it might appear that the statement var $x . P$ should be defined as $\exists \grave{}x\, x' . P$. Unfortunately, this semantics is angelic. If there are some initial values of x such that P describes chaotic behaviour, then this meaning will (if possible) angelically avoid choosing one of those values. The correct meaning for var $x . P$ must allow chaotic behaviour whenever any initial value of x could result in chaotic behaviour.

$$(\text{var } x . P) \stackrel{\text{def}}{=} (\forall \grave{}x . \exists \sigma'\, x' . P) \wedge (\exists \grave{}x\, x' . P)$$

[2] The notation $E[e/x]$ denotes the substitution of e for the variable x in the expression E, while renaming bound variables to avoid capture.

label $l.\, P$

It is not always possible to give a meaning for a recursively defined block. However, if the body of the block represents a monotonic function of the label[3] (as is the case if the block is executable), then we can express the meaning as the least fixed-point of that function.[4]

$$\text{monotonic}(\lambda l.\, P) \vdash^{\text{def}} (\text{label}\, l.\, P) = \text{fix}(\lambda l.\, P)$$

4 The Partial Model

We call the next model we examine the *partial* model. The partial model can establish only the partial correctness of one specification with respect to another.[5] Despite this failing, the partial model has been extensively used for the specification and verification of hardware [2] (where the problem analogous to avoiding nontermination is avoiding short-circuits). The partial model also has been applied to software. Gordon has formalised Hoare logic using the partial model [3]. In our earlier work, the partial model was proposed as a basis for a refinement tool [5].

4.1 Description

Consider a predicate P. The following is an interpretation of P as a specification using the partial model.

> For any initial state, either choose a final state satisfying P and terminate in that state, or fail to terminate. If there is no final state that satisfies P, then nontermination is the only acceptable behaviour.

We denote the interpretation of the predicate P as a specification according to the partial model as '$[\![P]\!]_p$'. We can formalise this interpretation as follows.

$$\vdash^{\text{def}} [\![P]\!]_p(i, o) = P(i, o) \vee (o = \bot) \qquad [[\![\_]\!]_p\text{-DEF}]$$

Under this interpretation, termination in a state that does not satisfy the predicate is the only incorrect behaviour. Nontermination always meets the specification.

The definition of refinement between predicates in the partial model ($\sqsubseteq_p$) follows directly from their interpretation.

$$\vdash^{\text{def}} (P \sqsubseteq_p Q) = ([\![P]\!]_p \sqsubseteq [\![Q]\!]_p) \qquad [\sqsubseteq_p\text{-DEF}]$$

A direct definition of refinement would be of more practical use. Refinement is the process of reducing nondeterminism. A specification is refined by reducing the number of outcomes that satisfy the specification for any given initial state. Refinement of predicates in the partial model can therefore be expressed as follows:

$$\vdash (P \sqsubseteq_p Q) = (\forall \grave{}\sigma\, \sigma'.\, Q(\grave{}\sigma, \sigma') \supset P(\grave{}\sigma, \sigma')) \qquad [\sqsubseteq_p\text{-THM}]$$

---

[3] The function f is said to be *monotonic* if the following property is known to holds of it: $\forall x\, y.\, (x \sqsubseteq y) \supset ((f\, x) \sqsubseteq (f\, y))$.

[4] A *fixed point* of a function, f, is a value, X, such that $f\, X = X$. The *least fixed-point* of a function, denoted 'fix f', is the least of all such fixed points. Any monotonic function has a least fixed-point.

[5] Partial correctness is correctness under the assumption of termination.

4.2 Basic Specifications

Here we consider the treatment of the basic specifications of Sect. 2.1 in the partial model:

chaos
: The predicate **T** describes chaos. Termination in any final state satisfies **T** and, under the partial model, nontermination also satisfies **T**.

$$\vdash \mathsf{chaos} = [\![\mathbf{T}]\!]_p$$

chance
: The specification chance cannot be represented in the partial model.

$$\vdash \neg\exists P.\,\mathsf{chance} = [\![P]\!]_p$$

abort
: The specification abort requires nontermination, which is described by the predicate **F** in the partial model.

$$\vdash \mathsf{abort} = [\![\mathbf{F}]\!]_p$$

miracle
: In the partial model, **F** denotes nonterminating behaviour. While this is not the same thing as a miracle, it does fill an analagous role in the partial model where nontermination is a refinement of every specification, a property characteristic of miracles.

4.3 The Simple Language

We now examine the definition of the programming language constructs, described in Sect. 2.3 under the partial model.

$x' := e$
: Assignment requires termination and can therefore not be properly expressed in the partial model. If we ignore termination, we can capture the rest of the meaning as follows:

$$(x' := e) \stackrel{\text{def}}{=} (\sigma' = `\sigma)[e/`x]$$

if b then P else Q
: The conditional construct can be naturally expressed in the partial model by using implication:

$$(\text{if } b \text{ then } P \text{ else } Q) \stackrel{\text{def}}{=} ((b \supset P) \wedge (\neg b \supset Q))$$

Note that this definition is equivalent to the one used in the relational model. They are expressed differently because of the different conventions associated with the models.

$P \mid Q$
: Choice is simply expressed as disjunction:

$$P \mid Q \stackrel{\text{def}}{=} P \vee Q$$

Since nontermination is always acceptable in the partial model, there is no need to explicitly allow for it as we did when giving the relational semantics of choice.

$P; Q$

Sequential composition is expressed in the partial model as follows:

$$P; Q \stackrel{\text{def}}{=} (\exists \sigma.\, P[\sigma/\sigma'] \wedge Q[\sigma/\grave{}\sigma])$$

As with choice, the potential for nontermination does not complicate the definition of composition in the partial model.

$\mathsf{var}\, x.\, P$

We introduce a local variable simply by providing an existentially quantified scope for the initial and final values.

$$(\mathsf{var}\, x.\, P) \stackrel{\text{def}}{=} (\exists \grave{}x\, x'.\, P)$$

$\mathsf{label}\, l.\, P$

It is not always possible to capture the meaning of a recursive block. However, if the body of the block is a monotonic function of the label then the meaning can be expressed as a fixed point of that function. It is usual to associate the meaning of a recursive function with its least-fixed point. However, in the partial model, the meaning of a recursive function is more accurately captured by the greatest-fixed point.

Consider the following simple recursive block: $\mathsf{label}\, l.\, l$. The body of this block is the identity function. The least-fixed point of the identity function is **T**, the predicate that denotes chaotic behaviour. However, no invocation of this block can ever terminate, so **F**—the predicate that denotes nonterminating behaviour—is a more accurate meaning. The predicate **F** is the greatest fixed-point of the identity function.

Accordingly we define the meaning of a recursive block in the partial model as follows (we use the notation '$\overline{\text{fix}}f$' to denote the greatest fixed-point of the function f).

$$\text{monotonic}(\lambda l.\, P) \models^{\text{def}} (\mathsf{label}\, l.\, P) = \overline{\text{fix}}(\lambda l.\, P)$$

5 The Total Model

The next model to be examined we call the *total* model. The total model has been used by Hoare [10] and Hehner [8] to specify behaviour and program semantics.

5.1 Description

As with the previous models, we consider the interpretation of a predicate P as a specification, this time under the total model.

> For any initial state, if every final state would satisfy P, then we assume that the specifier cares so little about the behaviour that not only is termination in any final state acceptable, but so is nontermination. If, however, not every final state would satisfy P, then we must choose a state that does and terminate in it.

We denote the interpretation of the predicate P as a specification according to the total model as '$[\![P]\!]_t$'. When formalised, this interpretation is equivalent to the definition below:

$$\models^{\mathrm{def}} [\![P]\!]_t(i,o) = ((\neg\forall\sigma.\,P(i,\sigma)) \supset ((o \neq \bot) \wedge P(i,o))) \qquad [[\![\text{-}]\!]_t\text{-DEF}]$$

As before, the definition of refinement in this model ($\sqsubseteq_t$) follows directly from the interpretation of predicates as specifications.

$$\models^{\mathrm{def}} (P \sqsubseteq_t Q) = ([\![P]\!]_t \sqsubseteq [\![Q]\!]_t) \qquad [\sqsubseteq_t\text{ -DEF}]$$

Refinement can be seen as the process of reducing nondeterminism. A specification is refined by reducing the number of outcomes that satisfy the specification for any given initial state. Therefore, refinement in the total model may also be expressed as follows:

$$\vdash (P \sqsubseteq_t Q) = (\forall \grave{}\sigma\, \sigma'.\, Q(\grave{}\sigma,\sigma') \supset P(\grave{}\sigma,\sigma')) \qquad [\sqsubseteq_t\text{ -THM}]$$

5.2 Basic Specifications

We now examine the basic specifications of Sect. 2.1 using the total model.

chaos

The predicate **T** represents chaos.

$$\vdash \mathsf{chaos} = [\![\mathbf{T}]\!]_t$$

chance

It is not possible to represent chance in this model. This is because **T**, the only predicate that allows termination in any final state, also allows nontermination.

$$\vdash \neg\exists P.\,\mathsf{chance} = [\![P]\!]_t$$

abort

It is not possible to represent abort in this model. This is because **T**, the only predicate that allows nontermination, also allows termination in any final state.

$$\vdash \neg\exists P.\,\mathsf{abort} = [\![P]\!]_t$$

miracle

The predicate **F** represents the miracle in the total model, it is impossible to satisfy.

$$\vdash \mathsf{miracle} = [\![\mathbf{F}]\!]_t$$

5.3 The Simple Language

We now examine each of the programming constructs described in Sect. 2.3 to see how they are defined in the total model:

$x' := e$

Parallel assignment can be expressed by stating that the final state is equal to the initial state with the appropriate substitutions.

$$(x' := e) \stackrel{\text{def}}{=} (\sigma' = `\sigma)[e/`x]$$

if b then P else Q

The conditional construct can be naturally expressed in the total model by using implication.

$$(\text{if } b \text{ then } P \text{ else } Q) \stackrel{\text{def}}{=} ((b \supset P) \wedge (\neg b \supset Q))$$

$P \mid Q$

Choice is simply expressed as disjunction:

$$P \mid Q \stackrel{\text{def}}{=} P \vee Q$$

This semantics for choice is demonic because **T** represents chaotic behaviour in the total model.

Note that this semantics has one curious feature. Consider the following meanings for the predicates P and Q:

P: even x'

This requires behaviour that terminates in a state where x is an even number.

Q: odd x'

This requires behaviour that terminates in a state where x is an odd number.

Since we know that all numbers are either even or odd, the nondeterministic choice between P and Q is equivalent to the predicate **T** (which is equivalent to chaos in the total model). This means that the nondeterministic choice between two specifications both of which require terminating behaviour may allow nontermination.

$P; Q$

One might think that the sequential composition of two specifications should be represented in the total model as follows: $\exists\sigma.\, P[\sigma/\sigma'] \wedge Q[\sigma/`\sigma]$. Suppose, however, for some initial state P specifies chaotic, i.e. potentially nonterminating, behaviour. Suppose also that for all initial states Q, specifies some terminating behaviour. This meaning for composition incorrectly requires that the composition of these two specification always terminates. The following meaning for composition avoids this problem:

$$P; Q \stackrel{\text{def}}{=} ((\neg\forall\sigma'.\, P) \supset (\exists\sigma.\, P[\sigma/\sigma'] \wedge Q[\sigma/`\sigma]))$$

Note that this meaning for sequential composition has a flaw in that the sequential composition of two specifications, both of which require termination, may allow nonterminating behaviour. This problem means that sequential composition is not always associative in the total model. The example below illustrates the problem:

$$((x' \neq `x; x' \neq `x); x' = 0) = \mathbf{T} \qquad (x' \neq `x; (x' \neq `x; x' = 0)) = x' = 0$$

$\mathsf{var}\, x.\, P$

As in the partial model, we introduce a local variable by simply providing an existentially quantified scope for the initial and final values.

$$(\mathsf{var}\, x.\, P) \stackrel{\text{def}}{=} (\exists \grave{x}\, x'.\, P)$$

$\mathsf{label}\, l.\, P$

It is not always possible to give a meaning for a recursively defined block. However, if the body of the block represents a monotonic function of the label (as is the case if the block is executable), then we can express the meaning as the least fixed-point of that function.

$$\text{monotonic}(\lambda l.\, P) \stackrel{\text{def}}{\vdash} (\mathsf{label}\, l.\, P) = \text{fix}(\lambda l.\, P)$$

6 Comparison

In this section we make some comparisons between the various models. In particular we shall discuss the relative difficulty of using the various models for specifying behaviour and proving properties of programs. We shall also highlight areas where the models are unusual or run counter to our expectations.

6.1 The Refinement Ordering

The table below shows the definition of the refinement ordering for each of the models.

	$P \sqsubseteq Q$
Relational	$\forall \grave{\sigma}.\, (\exists \sigma'.\, P) \supset ((\exists \sigma'.\, Q) \wedge (\forall \sigma'.\, Q \supset P))$
Partial	$\forall \grave{\sigma}\, \sigma'.\, Q \supset P$
Total	$\forall \grave{\sigma}\, \sigma'.\, Q \supset P$

If we wish to use a predicative model as the basis of a refinement technique, and in particular a refinement technique where we are not restricted to using preproved refinement laws, then the simpler the definition of our refinement relation, the simpler our proof obligations will be. From this perspective the simplicity of the refinement relation as defined in the partial and total models gives them a clear advantage over the relational model. If we wish to make a refinement in the relational model that is not an instance of some preproved class of refinements, then we will have to perform an existence proof.

In addition to the simplicity of the partial and total models, these models support the intuitive association between correctness and logical implication held by many people.

6.2 The Basic Specifications

The table below gives the predicates used to describe each of the basic specifications in the various models.

	miracle	abort	chance	chaos
Relational			**T**	**F**
Partial	**F**	**F**		**T**
Total	**F**			**T**

The predicate **F** is used to describe miracle in both the partial and total models. The relational model cannot describe miracle. The inability to describe miracles is not a disadvantage in itself, since miracles cannot be implemented. However, various authors have argued that the presence of miracles in a calculus leads to a simpler and more powerful formalism [13, 15]. Note that by describing miracle with **F**, the total and partial models support the intuitive association between the specification which is impossible to meet and the predicate which is impossible to satisfy. Note that the partial model represents both miracle and abort with the predicate **F**. The resulting inability to distinguish between miraculous behaviour and nonterminating behaviour is why refinement in this model proves only partial correctness.

The relational model describes chance with the predicate **T**. Neither of the other models can describe arbitrary but terminating behaviour. In the partial model this is not possible because every specification is satisfied by nontermination. Users of the total model make the assumption that if a specifier does not care in which state a program terminates, then the specifier also does not care whether the program terminates at all.

All three models can describe chaos. By describing chaos with the predicate **T**, the partial and total models support the intuitive association between the specification which is trivial to meet and the predicate which is trivial to satisfy.

6.3 The Simple Language

In this section we examine those programming language constructs that are treated differently by the different models.

The Choice Construct. The table below sets out the meaning for the nondeterministic choice construct for each model.

	$P \mid Q$
Relational	$(\exists\sigma' . P) \wedge (\exists\sigma' . Q) \wedge (P \vee Q)$
Partial	$P \vee Q$
Total	$P \vee Q$

Because chaos is represented by **F** in the relational model we must explicitly ensure that nondeterministic choice behaves demonically. In the other models, where chaos is represented by **T**, demonic behaviour is the default. This results in a simpler meaning for choice in the partial and total models, which in turn should result in simpler proof obligations for programs involving choice. The partial and total models support the intuitive association held by many people between choice and logical or.

Unfortunately, in the total model, the nondeterministic choice between two specifications each requiring termination may sometimes allow nontermination. This problem was illustrated in Sect. 5.3.

Sequential Composition. The table below gives the meaning assigned to sequential composition in each of the models.

	$P;Q$
Relational	$(\forall \sigma.\, P[\sigma/\sigma'] \supset (\exists \sigma'.\, Q[\sigma/`\sigma])) \wedge (\exists \sigma.\, P[\sigma/\sigma'] \wedge Q[\sigma/`\sigma])$
Partial	$\exists \sigma.\, P[\sigma/\sigma'] \wedge Q[\sigma/`\sigma]$
Total	$(\neg \forall \sigma'.\, P) \supset (\exists \sigma.\, P[\sigma/\sigma'] \wedge Q[\sigma/`\sigma])$

Again the meaning in the relational model is complicated by the need to explicitly ensure that the construct behaves demonically. The meaning in the total model is slightly complicated by the fact that if a specification allows termination in any state, then it also allows nontermination. Also, this meaning is not associative, as we expect it should be.

The meaning of composition used in the partial model is the simplest, and supports the intuitive association between sequential composition and composition of relations. As with the other constructs, a simpler meaning for sequential composition leads to simpler proof obligations when proving properties of programs.

Local Variables. The table below sets out the meaning for a local variable declaration in each of the models.

	$\mathrm{var}\, x.\, P$
Relational	$(\forall `x.\, \exists \sigma'\ x'.\, P) \wedge (\exists `x\, x'.\, P)$
Partial	$\exists `x\, x'.\, P$
Total	$\exists `x\, x'.\, P$

The meaning in the relational model must explicitly ensure that the construct behaves demonically. For example, suppose the for some initial values of x the predicate P is always false. The expression $\exists `x\, x'.\, P$ avoids, if possible, choosing such values for $`x$. In the relational model where $\mathbf{F}$ corresponds to chaos, this would represent an angelic choice. In the other models where $\mathbf{F}$ represents miracle, this is the demonic choice.

General Recursion. In the table below, we set out the meaning given to recursive blocks in each of the models. These meanings are valid only if the bodies of the blocks are monotonic functions. This is the case if the blocks are executable, but may not be the case otherwise. Hence, the meanings given below are not a useful starting place for developing refinement techniques for general recursion. (Such a technique is outlined in the companion paper [6].)

	$\mathrm{label}\, l.\, P$
Relational	$\mathrm{fix}(\lambda l.\, P)$
Partial	$\overline{\mathrm{fix}}(\lambda l.\, P)$
Total	$\mathrm{fix}(\lambda l.\, P)$

Both the relational and total models define recursion as a least-fixed point. The Partial model uses the unusual construction of a greatest fixed-point because nontermination is the most defined specification that can be represented in this model. We could have modelled recursion with a least-fixed point in the partial model as well. This would equate infinite recursion with chaotic behaviour as is done in the other models. However, by using a greatest fixed-point we can be more accurate and equate infinite recursions with nonterminating behaviour.

6.4 Extremes

In order to highlight some curiosities, it is illustrative to examine what happens as we approach extreme specifications. Suppose that we have two integer-valued constants **min** and **max**, such that **min** is less than **max**.

Approaching the Impossible. Consider the following predicate:

$$\mathbf{min} \leq x' \leq \mathbf{max}$$

For any initial state this predicate describes a set of final states. In all three models this predicate represents a specification that allows termination in any state where the value of x lies between **min** and **max**. In the partial model, nontermination also meets the specification. If we move **min** and **max** closer together, we reduce the number of final states in which termination is acceptable. If we continue this process until the set described contains only one value for x then we must choose that value. Something curious happens if we move **min** and **max** further so that they pass each other and the predicate describes an empty set. In the total model no behaviour meets the specification. Similarly, in the partial model no terminating behaviour meets the specification. In the relational model, however, any behaviour including nontermination will now meet the specification. This sudden change represents a 'discontinuity' in the meaning of predicates in the relational model.

Approaching the Trivial. Consider the following predicate which is similar to the one used in the last section:

$$\neg(\mathbf{min} \leq x' \leq \mathbf{max})$$

For any initial state this predicate describes all those final states where the value of x does *not* lie between **min** and **max**. Under all three interpretations, termination in any of those states is acceptable. Again, the partial model also permits nontermination. If we move **min** and **max** closer together then we increase the number of final states in which termination is acceptable. This process continues smoothly up to and including the point at which there is only one unacceptable state. As before, something curious happens if we move **min** and **max** further so that they overlap and the specification describes the universal set of final states. In both the relational and partial models the meaning is simply that termination in any state is now acceptable. As always, the partial model also permits nontermination. In the total model, however, this last step alters the meaning of the specification so that for the first time nontermination is also an acceptable behaviour. This sudden change represents a discontinuity in the meaning of predicates in the total model.

There are no discontinuities in the partial model because the possibility of nontermination remains constant throughout the entire range of specifications.

7 Conclusion

We have examined three interpretations of predicates. (Readers interested in a comparison of these interpretations with approaches based on pairs of predicates are referred to [9].) In doing so we have shown there is no commonly held understanding of what a predicate means as a specification of behaviour. Some aspects of these models are simpler and support more of our intuitions about programs than others. We can see, however, that each of these models has its failings.

The relational model chooses to describe those primitive specifications that are of practical importance, namely chance and chaos. Perhaps this explains why the relational model has proved such a popular basis for specification languages. It was also the relational model that gave our programming language the most complex semantics and had the most complex refinement relation. This may explain why many formal developments begun with a specification phrased in the relational model are completed only after re-expressing the specification in another formalism.

One of the difficulties of specifying large systems is avoiding an inconsistent specification. However, the consequences of making this error are much more serious in the relational model where any behaviour will vacuously satisfy an inconsistent specification. In the total model, no behaviour will satisfy an inconsistent specification. Similarly, in the partial model no terminating behaviour will satisfy an inconsistent specification.

The partial model gave our programming language the simplest semantics. Unfortunately this model is unable to support proofs of total correctness. This may explain the unpopularity of this model as the basis of a refinement technique. It may also explain why this model is considered well suited to program verification [3] where the separation of the partial correctness proof and the termination proof is a welcome division of the task of proving total correctness.

The total model preserves much of the simplicity of the partial model while supporting proofs of total correctness, but this model is not without its difficulties. Its basic assumption that if a specifier does not care in which state a computation terminates, then they also does not care if the computation terminates at all is not universally shared [17]. Certainly the consequences can be counter to our intuitions; in particular, the nonassociativity of sequential composition, and the fact the nondeterministic choice between two terminating behaviours can allow nontermination are most unwelcome.

The various difficulties with the predicative models examined in this survey have meant that while predicates have proved popular for use in specification and even for verification, they have been all but abandoned in favour of specifications phrased as pairs of predicates for work in program refinement. The successor to this paper [6] elsewhere in these proceedings describes an alternative approach. There we use predicates from a three-valued logic to work around these problems while maintaining much of the original appeal of the predicative styles.

Acknowledgements

I am grateful for the support of the Australian Defence Science & Technology Organisation. I am also grateful to Mike Gordon for his guidance and supervision.

I would like to thank the following people who took time out from their normal schedule to read drafts of this paper: Emil Sekerinski of the Åbo Akademi; Richard Boulton, Mike Gordon and Tom Melham of the University of Cambridge; Andrew Gordon of the Chalmers University of Technology; Maris Ozols of the Australian Defence Science & Technology Organisation; and Ian Hayes of the University of Manchester. The paper has been much improved by their comments and corrections, any remaining flaws are, of course, my own doing.

References

1. Bear, S., Rush, T.: Rigorous software engineering: A method for preventing software defects. HP J. **42** (1991) 24–31
2. Gordon, M.: Why higher-order logic is a good formalism for specifying and verifying hardware. In Milne, G., Subrahmanyam, P. (eds), Proceedings of the 1985 Edinburgh Workshop on VLSI Design: Formal Aspects of VLSI Design, Edinburgh, Scotland (1985). North Holland, Amsterdam, The Netherlands 153–177
3. Gordon, M.: Mechanizing programming logics in higher-order logic. In Birtwistle, G., Subrahmanyam, P. (eds), Current Trends in Hardware Verification and Automated Theorem Proving (Proceedings of the 1988 Banff Workshop on Hardware Verification), Banff, Canada (1988). Springer-Verlag, Berlin, Germany 387–439
4. Gravell, A.: Constructive refinement of first-order specifications. In Jones et al. [11] 181–210
5. Grundy, J.: A window inference tool for refinement. In Jones et al. [11] 230–254
6. Grundy, J.: A three-valued logic for refinement. In Sabelfeld, V. (ed.), Proceedings of the International Conference on Formal Methods in Programming and Their Applications, Lecture Notes in Computer Science, Novosibirsk, Russia (1993). Springer-Verlag, Berlin, Germany
7. Hayes, I.: Specification Case Studies. Prentice Hall International Series in Computer Science. Prentice Hall International, London, England (1987)
8. Hehner, E.: Predicative programming — part 1. Commun. ACM **27** (1984) 134–143
9. Hehner, E.: Termination conventions and comparative semantics. Acta Inf. **25** (1988) 1–14
10. Hoare, C.: Programs are predicates. Trans. Royal Society — A **312** (1984) 475–490
11. Jones, C., et al. (eds), Proceedings of the 5th Refinement Workshop, Workshops in Computer Science, Lloyd's Register, London, England (1992). BCS-FACS, Springer-Verlag, London, England
12. Mili, A.: A relational approach to the design of deterministic programs. Acta Inf. **20** (1983) 315–328
13. Morgan, C.: Data refinement by miracles. Inf. Process. Lett. **26** (1988) 243–246
14. Neilson, D.: From Z to C: Illustration of a rigorous development method. Technical Monograph PRG-101, Oxford University Computing Laboratory, Programming Research Group, 8–11 Keble Road, Oxford OX1 3QD, England (1990)
15. Nelson, G.: A generalization of Dijkstra's calculus. ACM Trans. Programming Languages & Syst. **11** (1989) 517–561
16. Nguyen, T.: A relational model of demonic nondeterministic programs. J. Foundations Comput. Sci. **2** (1991) 101–131
17. Parnas, D.: Technical correspondence: Predicative programming. Commun. ACM **28** (1985) 354–356
18. Sekerinski, E.: A calculus for predicative programming. In Woodcock, J., et al. (eds), Proceedings of the 2nd Mathematics of Program Construction Conference, Lecture Notes in Computer Science, Oxford University, Oxford, England (1992). Springer-Verlag, Berlin, Germany

A Three-Valued Logic for Refinement

Jim Grundy*

University of Cambridge, Computer Laboratory
New Museums Site, Pembroke Street, Cambridge CB2 3QG, England

Abstract. Predicates are a popular tool for specification. Unfortunately, they are less well suited to defining the semantics of programming languages. As as result, program refinement techniques are usually based on predicate transformers. This paper proposes a three-valued logic with a richer kind of predicate that is suitable for both specification and defining programming language semantics.
The other contribution of this work is a method of refining specifications into recursive programs. It has been suggested that it is more difficult to develop recursive programs than iterative ones. We sidestep the difficulties by using an approximation to the meaning of recursive specifications. We prove that once the refinement process is complete, our approximation is refined by the true meaning.

1 Introduction

This paper introduces a three-valued logic and uses it to give a semantics to a simple imperative programming language. Predicates in this logic give clear specifications of behaviour. The strength of the technique is the ease with which it describes potentially nonterminating behaviour.

The idea of using a predicate to specify behaviour is a compelling one. Predicates of ordinary, two-valued, logic are best regarded as relations between pairs of states. Although such relations are well suited to modelling terminating behaviour, it is more difficult to find a suitable interpretation that allows them to specify both terminating and nonterminating behaviour. Many possible interpretations have been tried, several of which are examined in the companion paper [6] (elsewhere in these proceedings) which highlights the difficulties with those interpretations.

The other contribution of this work is to propose a method of refining specifications into recursive programs. Some authors [4, 11] have noted that general recursion requires a more elaborate semantics than that required for simple iteration. As a result, most of the work on program refinement has concentrated on methods of developing iterative programs. Much of the difficulty in proving the correctness of a derivation method for recursive programs lies in the fact that very little can be assumed about a program during the intermediate stages of its development. For example, we may not be able to assume that the body of a partly developed recursive block is a monotonic function of its label. We sidestep these difficulties by using an approximation to the meaning of recursive blocks. This approximation is accurate enough to support a natural style of refinement. We prove that our approximation is refined by the true meaning once the refinement

* The author is supported by DSTO Australia.

process is complete. We illustrate the technique with our three-valued predicates, but it could be applied with equal effectiveness to another semantics.

2 The Basic Concepts

This paper uses the same terminology and notation as its companion paper [6] elsewhere in these proceedings. The explanation of this terminology and notation is *not* repeated here. Readers are advised to read at least the section entitled 'Basic Concepts' from that paper before proceeding.

3 A Three-Valued Logic

We wish to define a three-valued logic for reasoning about computations. Various systems of three-valued logics were proposed in the 1920s and 1930s, the most notable of these being the systems of Bočvar [3], Kleene [7] and Łukasiewicz [8, 9]. More recently, logics in which every type—not just the booleans—is extended with an additional value have been developed for reasoning about partial functions. The most well know of these are Jones's 'Logic of Partial Functions' [1] and Scott's 'Logic of Computable Functions' [12]. In this section we propose a three-valued logic for reasoning about programs. We need to extend only the boolean type with a distinguished value. Our three-valued logic is similar to Bočvar's internal logic.

We begin by defining the three-valued booleans $\mathbb{B}_3$. The elements of $\mathbb{B}_3$ denote the truth values of the logic: false, true and undefined.

$$\mathbb{B}_3 \stackrel{\text{def}}{=} \mathbf{F}_3 \mid \mathbf{T}_3 \mid \bot_3$$

We define an injection, $\lfloor \_ \rfloor$, from $\mathbb{B}$ into $\mathbb{B}_3$.

$$\stackrel{\text{def}}{\vdash} (\lfloor \mathbf{T} \rfloor = \mathbf{T}_3) \wedge (\lfloor \mathbf{F} \rfloor = \mathbf{F}_3) \qquad [\lfloor \_ \rfloor\text{-DEF}]$$

We also partially define a function, $\lceil \_ \rceil$, mapping values from $\mathbb{B}_3$ into $\mathbb{B}$.

$$\stackrel{\text{def}}{\vdash} (\lceil \mathbf{T}_3 \rceil = \mathbf{T}) \wedge (\lceil \mathbf{F}_3 \rceil = \mathbf{F}) \qquad [\lceil \_ \rceil\text{-DEF}]$$

A 'strict' version of each logical connective is defined as follows. If any argument of the connective is $\bot_3$, then the result is $\bot_3$; otherwise, the result is as it would have been in two valued logic. We denote such a translated connective by underlining the symbol that denotes it in two valued logic. For example, here are the truth tables for the three-valued versions of disjunction, implication, equality, and negation:

$A \underline{\vee} B$		B: $\bot_3$	$\mathbf{T}_3$	$\mathbf{F}_3$
A	$\bot_3$	$\bot_3$	$\bot_3$	$\bot_3$
	$\mathbf{T}_3$	$\bot_3$	$\mathbf{T}_3$	$\mathbf{T}_3$
	$\mathbf{F}_3$	$\bot_3$	$\mathbf{T}_3$	$\mathbf{F}_3$

$A \underline{\supset} B$		B: $\bot_3$	$\mathbf{T}_3$	$\mathbf{F}_3$
A	$\bot_3$	$\bot_3$	$\bot_3$	$\bot_3$
	$\mathbf{T}_3$	$\bot_3$	$\mathbf{T}_3$	$\mathbf{F}_3$
	$\mathbf{F}_3$	$\bot_3$	$\mathbf{T}_3$	$\mathbf{T}_3$

$A \underline{\equiv} B$		B: $\bot_3$	$\mathbf{T}_3$	$\mathbf{F}_3$
A	$\bot_3$	$\bot_3$	$\bot_3$	$\bot_3$
	$\mathbf{T}_3$	$\bot_3$	$\mathbf{T}_3$	$\mathbf{F}_3$
	$\mathbf{F}_3$	$\bot_3$	$\mathbf{F}_3$	$\mathbf{T}_3$

A	$\underline{\neg} A$
$\bot_3$	$\bot_3$
$\mathbf{T}_3$	$\mathbf{F}_3$
$\mathbf{F}_3$	$\mathbf{T}_3$

We retain the traditional 'strong' meaning for the undecorated equality symbol, namely that it is true if and only if the arguments are identical.

$A = B$		B: $\bot_3$	B: $\mathbf{T}_3$	B: $\mathbf{F}_3$
A	$\bot_3$	**T**	**F**	**F**
	$\mathbf{T}_3$	**F**	**T**	**F**
	$\mathbf{F}_3$	**F**	**F**	**T**

We also translate quantifiers into our three-valued logic. If for any value of the bound variable the predicate is $\bot_3$, then the result is $\bot_3$; otherwise, the result is as it would have been in two-valued logic.[2]

$$\models^{def}(\underline{\exists} v.\, P) = ((\exists v.\, P = \bot_3) \Rightarrow \bot_3 \,|\, \lfloor \exists v.\, \lceil P \rceil \rfloor) \qquad [\underline{\exists}\text{\_DEF}]$$

$$\models^{def}(\underline{\forall} v.\, P) = ((\exists v.\, P = \bot_3) \Rightarrow \bot_3 \,|\, \lfloor \forall v.\, \lceil P \rceil \rfloor) \qquad [\underline{\forall}\text{\_DEF}]$$

We now define two connectives unique to our three-valued logic. The first of these, $\Supset$, is a kind of conditional that is usefull for phrasing specifications. The second, $\mathbin{\wedge\!\!\!\wedge}$, is a form of conjunction used in defining the semantics of a simple programming langauge.

$A \Supset B$		B: $\bot_3$	B: $\mathbf{T}_3$	B: $\mathbf{F}_3$
A	**T**	$\bot_3$	$\mathbf{T}_3$	$\mathbf{F}_3$
	F	$\bot_3$	$\bot_3$	$\bot_3$

$A \mathbin{\wedge\!\!\!\wedge} B$		B: $\bot_3$	B: $\mathbf{T}_3$	B: $\mathbf{F}_3$
A	$\bot_3$	$\bot_3$	$\bot_3$	$\mathbf{F}_3$
	$\mathbf{T}_3$	$\bot_3$	$\mathbf{T}_3$	$\mathbf{F}_3$
	$\mathbf{F}_3$	$\mathbf{F}_3$	$\mathbf{F}_3$	$\mathbf{F}_3$

4 Three-Valued Predicates as Specifications

Three-valued predicates are three-valued boolean expressions which may contain free variables from the initial and final states $`\sigma$ and σ'. In this way, three-valued predicates describe a relationship between these states. This relationship can be made explicit by regarding the three-valued predicate P as a short hand for the function below:

$$\lambda(`\sigma, \sigma').\, P$$

This function is of type $(\Sigma \times \Sigma) \rightarrow \mathbb{B}_3$. We call objects of this type *descriptions*. Whenever it is not ambiguous, we will use predicates to stand for both their boolean values and their corresponding descriptions. For example, the predicate '$\mathbf{F}_3$' stands for both the value $\mathbf{F}_3$ and the description $\lambda(`\sigma, \sigma').\, \mathbf{F}_3$.

The type of specifications, $(\Sigma \times \Sigma_\bot) \rightarrow \mathbb{B}$, is similar to the type of descriptions. Let us now consider an interpretation function, $[\![\_]\!]$, from descriptions to specifications:

$$[\![\_]\!] : \underbrace{((\Sigma \times \Sigma) \rightarrow \mathbb{B}_3)}_{\text{Descriptions}} \rightarrow \underbrace{((\Sigma \times \Sigma_\bot) \rightarrow \mathbb{B})}_{\text{Specifications}}$$

The following is an informal interpretation of a description D as a specification:

[2] The notation $B \Rightarrow X \,|\, Y$ denotes the conditional expression which equals X if B is true, and Y otherwise.

> Given any initial state, choose a final state such that D is not $\mathbf{F}_3$. (If there is no such state, then the specification requires a miracle.) If D is $\mathbf{T}_3$, then terminate in that final state; otherwise, if D is $\bot_3$, either terminate in that final state, or fail to terminate.

When formalised, this interpretation is equivalent to the following definition:

$$\models_{\mathrm{def}} [\![D]\!](i,o) = ((o \neq \bot) \Rightarrow (D(i,o) \neq \mathbf{F}_3) \mid (\exists \sigma.\, D(i,\sigma) = \bot_3)) \quad [[\![\_]\!]\text{\_DEF}]$$

Whenever it is not ambiguous we identify descriptions with the specifications they denote. For example, consider the predicate below:

$$\lfloor (\grave{}y > 0) \supset (z' = \grave{}x \div \grave{}y) \rfloor$$

We regard this predicate as denoting the informal specification:

> Given an initial state where $\grave{}y > 0$ holds, the computer should terminate in a state where $z' = \grave{}x \div \grave{}y$. If $\grave{}y > 0$ does not hold in the initial state, termination in any final state is acceptable.

As another example, consider the similar (but not identical) predicate:

$$(\grave{}y > 0) \sqsupset \lfloor z' = \grave{}x \div \grave{}y \rfloor$$

We regard this predicate as denoting this informal specification:

> Given an initial state where $\grave{}y > 0$ holds, the computer should terminate in a final state where $z' = \grave{}x \div \grave{}y$ holds. If $\grave{}y > 0$ does not hold in the initial state, either nontermination or termination in any final state is acceptable.

4.1 Basic Specifications

We can now examine how the basic specifications noted in the companion paper [6] are modelled by three-valued predicates:

chaos
: The three-valued predicate $\bot_3$ represents completely arbitrary (potentially nonterminating) behaviour.

$$\vdash \mathsf{chaos} = [\![\bot_3]\!]$$

chance
: The three-valued predicate $\mathbf{T}_3$ represents arbitrary but terminating behaviour.

$$\vdash \mathsf{chance} = [\![\mathbf{T}_3]\!]$$

abort
: It is not possible to describe abort, the specification that requires nontermination, with our interpretation of three-valued predicates. In practice, this is not a great loss, since specifications that require nontermination are rarely of use.

$$\vdash \neg \exists D.\, \mathsf{abort} = [\![D]\!]$$

miracle

The three-valued predicate $\mathbf{F}_3$ represents a miracle (the specification that is impossible to meet).

$$\vdash \mathsf{miracle} = [\![\mathbf{F}_3]\!]$$

This interpretation supports some pleasing intuitions. The predicate that is impossible to satisfy, $\mathbf{F}_3$, represents the specification that is impossible to meet, miracle. The predicate that is trivial to satisfy, $\mathbf{T}_3$, represents the specification that is met by any terminating behaviour, chance. The predicate $\bot_3$, traditionally associated with undefinedness, corresponds to the specification that admits completely arbitrary behaviour, chaos.

4.2 Redundancy

This model of specifications contains some redundancy. To illustrate this, we represent descriptions as sets of tuples. Suppose we wish to represent the description D by the set S. If D maps the pair $(\grave{}\sigma, \sigma')$ to X, then the tuple $((\grave{}\sigma, \sigma'), X)$ occurs in S. We assume that we wish to specify the behaviour of a simple machine with just two states s_1 and s_2. Consider the description represented by the set below:

$$\{((s_1, s_1), \bot_3), ((s_1, s_2), \mathbf{T}_3), ((s_2, s_1), \mathbf{T}_3), ((s_2, s_2), \mathbf{F}_3)\}$$

If we interpret this as a specification, then a computation commenced in the state s_1 may yield any of the following outcomes: termination in s_1, termination in s_2, or nontermination. Computations commenced in s_2 are required to terminate in s_1. The description below models the same specification:

$$\{((s_1, s_1), \bot_3), ((s_1, s_2), \bot_3), ((s_2, s_1), \mathbf{T}_3), ((s_2, s_2), \mathbf{F}_3)\}$$

While not desirable, this redundancy does not prove to be inconvenient.

5 Refining with Three-Valued Predicates

We wish to refine descriptions into programs in a language whose semantics is expressed with descriptions. It is easy to infer the refinement relation for descriptions.

$$\stackrel{\text{def}}{\vdash} (D \sqsubseteq_D D') = ([\![D]\!] \sqsubseteq [\![D']\!]) \qquad [\sqsubseteq_D\text{-DEF}]$$

Each step in the refinement process requires a proof that the new description is a refinement of its predecessor. To reduce the burden of these proofs, the refinement relation should be as simple as possible. Although $\sqsubseteq_D$ is the relationship we must preserve, its definition is a more complex than we would like (it appears simple, the complexity lies in the definition of $[\![\_]\!]$). In the following section we define a simpler refinement relation and show why it is acceptable to preserve this relationship instead.

5.1 A Simple Refinement Relation

We recall from our definition of refinement in the companion paper [6], that refinement can be viewed as a process of reducing nondeterminism. We now restate the informal interpretation of a description D as a specification in order to introduce a simplified concept of refinement for descriptions:

Given an initial state $`\sigma$:

1. Choose a state σ' such that $D(`\sigma, \sigma')$ is not $\mathbf{F}_3$.
2. If $D(`\sigma, \sigma')$ is $\mathbf{T}_3$, terminate in the state $\sigma'$; otherwise, if $D(`\sigma, \sigma')$ is $\bot_3$, either terminate in σ' or fail to terminate.

Let us consider refining the description D; in particular, refining the value of D at the point $(`\sigma, \sigma')$, leaving all other points unchanged. A value of $\mathbf{F}_3$ is the best since this would eliminate the possibility of choosing σ' at step 1. Failing that, a value of $\mathbf{T}_3$ is better than $\bot_3$, because while both $\mathbf{T}_3$ and $\bot_3$ allow termination in σ', $\bot_3$ also allows nontermination. Therefore, our refinement ordering over the three-valued booleans $\sqsubseteq_3$ is that $\mathbf{F}_3$ is better than $\mathbf{T}_3$, which is better than $\bot_3$. The truth table below gives a complete definition of the refinement relation which is summarised by the accompanying diagram:

$A \sqsubseteq_3 B$		B		
		$\bot_3$	$\mathbf{T}_3$	$\mathbf{F}_3$
A	$\bot_3$	**T**	**T**	**T**
	$\mathbf{T}_3$	**F**	**T**	**T**
	$\mathbf{F}_3$	**F**	**F**	**T**

$\mathbf{F}_3$
$\sqsubseteq_3 |$
$\mathbf{T}_3$
$\sqsubseteq_3 |$
$\bot_3$

The ordering is just the natural lifting of the reverse implication ordering over booleans. Implication and the refinement ordering are related as follows:

$$\vdash (P' \supset P) = (\lfloor P \rfloor \sqsubseteq_3 \lfloor P' \rfloor) \qquad [\supset\sqsubseteq_3\text{-THM}]$$

We define the simplified refinement relation for descriptions, $\sqsubseteq_d$, by stating that D' refines D if D' is a pointwise refinement of D.

$$\vdash_{def} (D \sqsubseteq_d D') = (\forall \sigma_1\, \sigma_2.\, D(\sigma_1, \sigma_2) \sqsubseteq_3 D'(\sigma_1, \sigma_2)) \qquad [\sqsubseteq_d\text{-DEF}]$$

Our simplified refinement relation for descriptions bears the following obviously desirable relationship to the true refinement relation:

$$\vdash (D \sqsubseteq_d D') \supset (D \sqsubseteq_D D') \qquad [\sqsubseteq_d\sqsubseteq_D\text{-THM}]$$

The following examples illustrate refinement with three-valued predicates:

- In this first refinement we make the specification more deterministic.

$$\vdash \lfloor x' > `x \rfloor \sqsubseteq_3 \lfloor x' = `x + 1 \rfloor$$

- In the second refinement we increase the domain of guaranteed termination.

$$\vdash (`x > 0 \supset \lfloor z' = `x \times `y \rfloor) \sqsubseteq_3 \lfloor z' = `x \times `y \rfloor$$

Note that from a theorem like $\Gamma \vdash P \sqsubseteq_3 P'$ where we use predicates to denote descriptions, we can deduce a theorem about the descriptions themselves, provided no components of $\grave{}\sigma$ or σ' occur free in Γ.

$$\frac{\Gamma \vdash P \sqsubseteq_3 P'}{\Gamma \vdash \big(\lambda(\grave{}\sigma,\sigma').\,P\big) \sqsubseteq_d \big(\lambda(\grave{}\sigma,\sigma').\,P'\big)}[\text{no part of } \grave{}\sigma \text{ or } \sigma' \text{ free in } \Gamma]$$

5.2 A Potential Problem

In the previous section we noted that it is possible to refine a description by refining it with respect to the simplified refinement relation $\sqsubseteq_d$.

$$\vdash (D \sqsubseteq_d D') \supset (D \sqsubseteq_D D')$$

The implication does not hold in the other direction. Consider the following descriptions:

$$\begin{aligned} D_1 &= \big\{\big((s_1,s_1),\bot_3\big), \big((s_1,s_2),\mathbf{T}_3\big), \big((s_2,s_1),\mathbf{T}_3\big), \big((s_2,s_2),\mathbf{F}_3\big)\big\} \\ D_2 &= \big\{\big((s_1,s_1),\mathbf{F}_3\big), \big((s_1,s_2),\bot_3\big), \big((s_2,s_1),\mathbf{T}_3\big), \big((s_2,s_2),\mathbf{F}_3\big)\big\} \end{aligned}$$

The meanings of these descriptions as specifications are as follows:

$$\begin{aligned} [\![D_1]\!] &= \{(s_1,\bot),(s_1,s_1),(s_1,s_2),(s_2,s_1)\} \\ [\![D_2]\!] &= \{(s_1,\bot),(s_1,s_2),(s_2,s_1)\} \end{aligned}$$

From this we can see that D_2 is clearly a refinement of D_1, and the relationship $D_1 \sqsubseteq_D D_2$ holds as expected. Unfortunately D_1 is not refined by D_2 under the relationship $\sqsubseteq_d$. This suggests that if we were to use $\sqsubseteq_d$ as our refinement relation, we might be prevented from making some valid refinement of a description. This would be a most undesirable situation.

5.3 No Problem

Consider another description, D_1'.

$$D_1' = \big\{\big((s_1,s_1),\bot_3\big), \big((s_1,s_2),\bot_3\big), \big((s_2,s_1),\mathbf{T}_3\big), \big((s_2,s_2),\mathbf{F}_3\big)\big\}$$

This description represents the same specification as was represented by D_1 in the previous section; that is $[\![D_1']\!] = [\![D_1]\!]$. Note, however, that D_1' *is* refined by D_2.

It is easy to identify a class of descriptions whose members are refined, according to $\sqsubseteq_d$, by any genuine refinement. We call these the *clear* descriptions.

$$\begin{aligned} &\overset{\text{def}}{\vdash} \mathit{clear}\, D = \\ &\quad \forall \sigma_0.\, \big(\exists \sigma_1.\, D(\sigma_0,\sigma_1) = \bot_3\big) \supset \big(\forall \sigma_1.\, D(\sigma_0,\sigma_1) \neq \mathbf{T}_3\big) \end{aligned} \qquad [\mathit{clear}\text{\_DEF}]$$

$$\vdash \mathit{clear}\, D = \big(\forall D'.\, (D \sqsubseteq_d D') = (D \sqsubseteq_D D')\big) \qquad [\mathit{clear}\text{\_THM}]$$

If we partition the set of descriptions into sets whose members denote the same specification, then each set will contain precisely one *clear* description.

Fortunately, when interpreted as descriptions, the three-valued predicates which most naturally capture our intended specification correspond to *clear* descriptions. In fact, any predicate that does not explicitly relate particular final states to the value $\bot_3$ denotes a *clear* description. For example, if we wish to specify a behaviour which is always required to terminate, then that specification can be captured by a description of the form:

$$\lfloor Rel \rfloor$$

If for some initial states nontermination is allowed, then the specification can be captured by a description of the form (where Pre does not contain variables from σ'):

$$Pre \sqsupset \lfloor Rel \rfloor$$

Descriptions of this form are always *clear*. Specification phrased in this precondition and relation style can be found in various guises in Hoare Logic, VDM and the various refinement calculi. They are generally regarded as providing a natural method of specification.

In the next section we give a semantics to a nondeterministic imperative programming language. The meaning of each command in the programming language is a *clear* description. The constructs of the programming language which combine descriptions to form new descriptions produce *clear* descriptions if their arguments are *clear* descriptions. A refinement process that begins with a *clear* description and progresses to an implementation via steps that combine program constructs and *clear* descriptions, will never encounter the potential problem highlighted in Sect. 5.2.

6 Programs as Three-Valued Predicates

In this section we describe a small imperative programming language. We give each construct in the language a meaning with a three-valued predicate. We are thereby able to mix these constructs freely with other predicate notations. As a result, we can treat predicates as a wide-spectrum language in which we can both phrase specifications and develop implementations.

Although any three-valued predicate can be interpreted as a specification, only some expressions are considered executable. The restriction is syntactic so that executable specifications are easily recognised by both humans and machines. We do not give a complete definition of which expressions are executable. One might reasonably expect that expressions like '$1 + 2$' are executable, while expressions like '$\int_a^b f\ x\ dx$' are not. Note that in order to be considered executable, all variables in an expression must be drawn from the initial state.

We now describe the constructs of our programming language and the conditions under which they are considered executable.

$x' := e$

The assignment statement $x' := e$ assigns the expression e to the variable x' and leaves the rest of the variables in the final state with the values they had in the initial state. The meaning of an assignment statement can be expressed as follows:[3]

$$(x' := e) \stackrel{\text{def}}{=} \lfloor `\sigma = \sigma' \rfloor [e/`x]$$

[3] The notation '$E[e/x]$' denotes the substitution of e for the variable x in the expression E, while renaming bound variables to avoid capture.

An assignment statement is executable if the expression e is executable. An assignment statement can never fail to terminate.

if b then P else Q

The conditional statement behaves according to P if the guard b holds; otherwise it behaves according to Q. This meaning is captured in the following definition, which uses a conditional expression:

$$(\text{if } b \text{ then } P \text{ else } Q) \stackrel{\text{def}}{=} (b \Rightarrow P \mid Q)$$

A conditional is executable if the guard expression is executable and both the then and else clauses are executable. A conditional statement may lead to nontermination if the chosen clause may lead to nontermination.

$P \mid Q$

The nondeterministic choice between P and Q can be expressed with three-valued disjunction.

$$P \mid Q \stackrel{\text{def}}{=} P \veebar Q$$

The choice statement $P \mid Q$ is executable if both P and Q are executable. A choice statement may fail to terminate if either clause may fail to terminate.

$P; Q$

The sequential composition $P; Q$ behaves according to P and then (if this behaviour terminates) according to Q. It relates an initial and final state if there exists some intermediate state such that P relates the initial and the intermediate states and Q relates the intermediate and final states.

$$(P; Q) \stackrel{\text{def}}{=} (\underline{\exists}\sigma.\, P[\sigma/\sigma'] \mathbin{\wedge\!\!\!\wedge} Q[\sigma/\grave{}\sigma])$$

A sequential composition $P; Q$ is executable if both P and Q are executable. It may fail to terminate if either P may fail to terminate when started in the initial state or Q may fail to terminate when started in any of the possible intermediate states.

var $x.\, D$

We introduce a local variable x simply by providing an existentially quantified scope for its initial and final values.

$$(\text{var } x.\, P) \stackrel{\text{def}}{=} (\underline{\exists}\grave{}x\, x'.\, P)$$

Note that within the scope of the var construct, the state space Σ should be extended to include the new variable.

label $l.\, D$

The label statement is used for creating recursive blocks. Within the body of a block, the label should be interpreted as a description representing a recursive call to the block. For example, suppose x is a natural number; the following program decrements x until it reaches 0:

$$\text{label } l.\ \text{if } \grave{}x = 0 \text{ then skip else } (x' := \grave{}x - 1; l)$$

The body of a labelled block is a function which takes a description and returns a description. In the example just given, the body was the following function:

$$\lambda l.\ \text{if } \grave{}x = 0 \text{ then skip else } (x' := \grave{}x - 1; l)$$

It is not always possible to give a meaning for a recursive block. However, if the body is a monotonic function of the label[4] (as is the case if the block is executable), then we can express the meaning as the least fixed-point of that function.[5]

$$\mathrm{monotonic}(\lambda l.\, D) \vdash_{\mathrm{def}} (\mathsf{label}\; l.\, D) = \mathrm{fix}(\lambda l.\, D)$$

A labelled block is executable if, assuming that the label of the block is executable, then the body of the block is executable.

6.1 Nonstandard predicates

The notation proposed for use with program refinement is not quite that of standard logic. Fortunately, the difference is small and our notation can be regarded as a syntactic sugaring of the standard notation. Consider the predicate below:

$$\lfloor x' = \grave{}x + 1 \rfloor ; \lfloor x' = \grave{}x + 1 \rfloor$$

The intended meaning of this partial program is that x should be incremented, and then x should be incremented again. Assuming σ_0 and σ_1 are the initial and final states, our original predicate is syntactic sugar for a predicate of the following form (x_0 is a variable from the initial state, x is a variable from the intermediate state, and x_1 is a variable from the final state):

$$\underline{\exists}\sigma.\, \lfloor x = x_0 + 1 \rfloor \mathrel{\underline{\wedge}} \lfloor x_1 = x + 1 \rfloor$$

Although in keeping with standard programming practice, the program-like notation is unusual because the term contains two instances of x' (and $\grave{}x$) that have the same name, but denotes different variables. Similarly, the first x' and the second $\grave{}x$ denotes the same variable (namely x), despite having different names. This means that sequential compositions are not standard logical terms when expressed in a program-like notation. However, the meaning of a sequential composition, as defined earlier in this section, is a standard logical term. Since we wish to express our programs in a familiar notation, and reason about them using standard logic, we regard sequential compositions as syntactic sugar for their meanings.

Our notation is more general than previously described. In addition to writing $\grave{}x$ to denote the value of x before the current statement, we also write $\grave{}\grave{}x$ to denote the value of x before that, and $\grave{}\grave{}\grave{}x$ for the value of x before that, and so on. Similarly, while x' denotes the value of x after the current statement, x'' denotes the value of x after that, and so on. Consider the following predicate which uses multiple priming:

$$\lfloor x' = \grave{}x + 1 \rfloor ; \lfloor x' = \grave{}\grave{}x + 1 \rfloor$$

The meaning of this specification is that x should be incremented once. The form of the specification suggests that x should be incremented twice, but that the result of the first

---

[4] The function f is said to be *monotonic* if the following property is known to hold of it: $\forall x\, y.\, (x \sqsubseteq y) \supset ((f\ x) \sqsubseteq (f\ y))$.

[5] A *fixed point* of a function f is a value, X, such that $f\ X = X$. The *least fixed-point* of a function, denoted fix f, is the least of all such fixed points. Any monotonic function has a least fixed-point.

increment is ignored when calculating the second. Again, assuming σ_0 and σ_1 are the initial and final states, this predicate is syntactic sugar for the standard notation below:

$$\underline{\exists}\sigma.\, \lfloor x = x_0 + 1 \rfloor \mathbin{\wedge\!\!\wedge} \lfloor x_1 = x_0 + 1 \rfloor$$

Note that while expressions involving multiple primes, like $``x$ and x'', can be valuable in the intermediate stages of a refinement, they can never be executable. Note also that sequential compositions that do not involve multiple priming are associative, as we expect them to be.

Although the translation between the sugared and standard notations is straightforward, it is more complex than the kinds of translation usually associated with the phrase 'syntactic sugaring'. This is because the translation of terms depends upon their context. The translation begins in an initial context and builds upon the context as it descends the structure of the term. Consider the following predicate:

$$x_1 = x_0 + 1$$

If σ_0 and σ_1 are our initial and final states, then this predicate would correspond to the following syntactic sugar:

$$x' = `x + 1$$

However, if this predicate were a subterm in a larger predicate, it might appear differently. In our previous example where this term was the second part of a sequential composition, it was represented by the following syntactic sugar:

$$x' = ``x + 1$$

7 Refining Specifications by Component

Conditional statements, nondeterministic choice, sequential compositions, local variable blocks, and labelled blocks form new predicates from existing ones. In the process of refining a predicate into executable form, we decompose the predicate into smaller predicates combined with these constructs. We may be able to refine some of these components directly into an executable form, others we will have to decompose further. The results of this section are a collection of theorems which show that refining an individual component of a specification results in a refinement of the specification as a whole. These theorems also allow us to make assumptions based on the context of a component when refining it. In the case of recursive blocks this amounts to a method for developing recursive programs. These results are intended to form the basis of a transformational refinement system like the one described in [5].

7.1 Refining Conditional Statements

A conditional statement can be refined by refining the then clause or the else clause.

Theorem. *The refinement of the* then *clause of a conditional statement may be made under the assumption that the guard of the conditional is true.*

$$\vdash (b \supset (P \sqsubseteq_3 P')) \supset ((\mathsf{if}\, b\, \mathsf{then}\, P\, \mathsf{else}\, Q) \sqsubseteq_3 (\mathsf{if}\, b\, \mathsf{then}\, P'\, \mathsf{else}\, Q))$$

A similar theorem states that the refinement of the else clause may be made under the assumption that the guard is false. The validity of these theorems follows directly from the definition of a conditional statement and the monotonicity of conditional expressions.

7.2 Refining Nondeterministic Choices

A nondeterministic choice may be refined by refining either of its clauses. When refining one of the clauses it is possible to assume that the other clause could not lead to nontermination.

Theorem. *The second clause of a choice may be refined under the assumption that the first clause could not lead to nontermination.*

$$\vdash ((P \neq \bot_3) \supset (Q \sqsubseteq_3 Q')) \supset (P \mid Q \sqsubseteq_3 P \mid Q')$$

A similar theorem asserts that the first clause of a choice may be refined under the assumption that the second clause could not lead to nontermination. These theorems follow directly from the definition of choice and the monotonicity of three-valued disjunction.

7.3 Refining Sequential Compositions

A sequential composition may be refined by refining either of its clauses.

Theorem. *The second clause of a sequential composition may be refined under the assumption that the first clause has already been executed.*

$$\vdash (\forall \sigma . (P[\sigma/\sigma'] \neq \mathbf{F}_3) \supset (Q[\sigma/\grave{}\sigma] \sqsubseteq_3 Q'[\sigma/\grave{}\sigma])) \supset ((P;Q) \sqsubseteq_3 (P;Q'))$$

A similar theorem states that the first clause of a composition may be refined under the assumption that the second clause is about to be executed. Both these theorems follows from the definition of sequential composition and the monotonicity of $\underline{\wedge}$ and $\underline{\exists}$.

7.4 Refining Variable Blocks

A local variable block may be refined by refining its body.

Theorem. *We may refine a local block by refining the body of the block.*

$$\vdash (\forall \grave{}v\, v' . P \sqsubseteq_3 P') \supset ((\mathsf{var}\, v . P) \sqsubseteq_3 (\mathsf{var}\, v . P'))$$

This theorem follows from the definition of var and the monotonicity of $\underline{\exists}$.

7.5 Refining Labelled Blocks

Refining the body of the labelled block is more interesting. Not only do we wish to be able to refine a block by refining its body, but we want to be able to make recursive calls to the label within the body. We would like to support a style of refinement with the following properties:

- We can refine a labelled block by refining its body.
- If we refine the body of the block into some function of the original description that the block is being used to implement, then the block can be further refined by replacing the original description with a recursive call to the label.

Refinement to Terminating Code. If we examine the requirements of our refinement style, we see that they are not valid in general. We could give any description a label, and then, realising that the body of the block is the original specification, we could further refine the body into a recursive call to the label. This is, of course, not a valid refinement. The problem is caused by attempting to refine specifications into nonterminating implementations. Perhaps if we could force refinement to terminating implementations, then this style of refinement would be valid.

Let us contrive a new meaning for labelled blocks that forces us to make refinements to terminating solutions and which is demonstrably weaker than the true meaning for blocks. Each block has associated with it a function, v, from the state space to some well-ordered domain like the natural numbers. (For the examples in this paper we assume, without loss of generality, that the domain *is* the natural numbers.) This function is called the *variant* of the block. The syntax of the new blocks is the same as for the old blocks, only annotated with a subscripts, v and O, denoting the variant and the original description that the block was introduced to implement. The meaning associated with such a block is as follows:

$$\begin{array}{l} \models^{\mathrm{def}} (\mathsf{label}_{(v,O)}\, l.\, D)(\sigma_0, \sigma_1) = \\ \quad (\lambda l.\, D) \\ \quad\quad (\lambda(\sigma_a, \sigma_b).\, (v\, \sigma_a) < (v\, \sigma_0) \supset O(\sigma_a, \sigma_b)) \\ \quad\quad (\sigma_0, \sigma_1) \end{array} \qquad [\mathsf{label}_{(v,O)}\text{-DEF}]$$

From this definition we are able to formulate theorems that allow us to introduce and refine labelled blocks.

Theorem. *We can always refine a description by giving it a label (as long as that label does not occur free within the description).*

$$\vdash O(\grave{}\sigma, \sigma') \sqsubseteq_3 (\mathsf{label}_{(v,O)}\, l.\, O)(\grave{}\sigma, \sigma')$$

The validity of this theorem follows directly from our definition of $\mathsf{label}_{(v,O)}$.

Theorem. *If you want to refine a block* $(\mathsf{label}_{(v,O)}\, l.\, D)$, *you can do so by refining* D *into* D' *under the assumption that* l *is a description that refines the original description* O *in states where the variant has decreased.*

$$\begin{array}{l} \vdash \left(\begin{array}{l} \forall \sigma_0\, \sigma_1\, l. \\ \quad (\forall \sigma_a\, \sigma_b.\, ((v\, \sigma_a) < (v\, \sigma_0)) \supset (O(\sigma_a, \sigma_b) \sqsubseteq_3 l(\sigma_a, \sigma_b))) \supset \\ \quad\quad D(\sigma_0, \sigma_1) \sqsubseteq_3 D'(\sigma_0, \sigma_1) \end{array} \right) \supset \\ \quad (\mathsf{label}_{(v,O)}\, l.\, D)(\grave{}\sigma, \sigma') \sqsubseteq_3 (\mathsf{label}_{(v,O)}\, l.\, D')(\grave{}\sigma, \sigma') \end{array}$$

This theorem follows directly from the definition of $\mathsf{label}_{(v,O)}$.

At the end of a refinement process begun by giving a description a label and advanced by repeatedly refining the body of the description, we will have a block of the form $(\mathsf{label}_{(v,O)}\, l.\, D)$, where D is executable. The product of this process is a theorem of the form:

$$\vdash O(\grave{}\sigma, \sigma') \sqsubseteq_3 (\mathsf{label}_{(v,O)}\, l.\, D)(\grave{}\sigma, \sigma')$$

The next step in the refinement is to drop the subscripts from the labelled block by refining it into its true meaning.

Theorem. *Our new contrived meaning for a labelled block is weaker than the true meaning whenever the block is executable (and therefore monotonic).*

$$\forall \sigma_a\, \sigma_b.\, O(\sigma_a, \sigma_b) \sqsubseteq_3 (\mathsf{label}_{(v,O)}\, l.\, D)(\sigma_a, \sigma_b), \text{monotonic}(\lambda l.\, D) \vdash$$
$$(\mathsf{label}_{(v,O)}\, l.\, D)(\sigma_0, \sigma_1) \sqsubseteq_3 (\mathsf{label}\, l.\, D)(\sigma_0, \sigma_1)$$

Proof. The proof takes the form of a transfinite induction over the value of the variant in the initial state. Note that the proof assumes that $(\lambda l.\, D)$ is monotonic, this follows from our assumption that the block is executable.

Inductive Hypothesis: $v\, \sigma_a < v\, \sigma_0, O \sqsubseteq_d (\mathsf{label}_{(v,O)}\, l.\, D) \vdash$
$$(\mathsf{label}_{(v,O)}\, l.\, D)(\sigma_a, \sigma_b) \sqsubseteq_3 \text{fix}(\lambda l.\, D)(\sigma_a, \sigma_b)$$

For the induction we hypothesise that we have proved our contrived meaning is weaker than the real meaning for any initial state where the variant is less than it is in σ_0. We then show that our hypothesis also holds for the state σ_0.

$(\mathsf{label}_{(v,O)}\, l.\, D)(\sigma_0, \sigma_1)$

We replace the block with its definition.

$= \quad (\lambda l.\, D)$ $\quad$ [$\mathsf{label}_{(v,O)}$ _DEF]

$\quad (\lambda(\sigma_a, \sigma_b).\, (v\, \sigma_a) < (v\, \sigma_0) \supset O(\sigma_a, \sigma_b))$

$\quad (\sigma_0, \sigma_1)$

We use the assumption that O is refined by the block, together with the fact that $(\lambda l.\, D)$ is monotonic, to refine O into the block.

$\sqsubseteq_3 \quad (\lambda l.\, D)$ $\quad$ [D_MONO]

$$\left(\begin{array}{l}\lambda(\sigma_a, \sigma_b).\\ \quad (v\, \sigma_a) < (v\, \sigma_0) \supset (\mathsf{label}_{(v,O)}\, l.\, D)(\sigma_a, \sigma_b)\end{array}\right)$$

$\quad (\sigma_0, \sigma_1)$

In the context of the term '$(\mathsf{label}_{(v,O)}\, l.\, D)(\sigma_a, \sigma_b)$', we can assume that $v\, \sigma_a$ is less than $v\, \sigma_0$. Therefore we can use the inductive hypothesis and the monotonicity of $(\lambda l.\, D)$ to further refine the block into its true meaning.

$\sqsubseteq_3 \quad (\lambda l.\, D)$ $\quad$ [INDUCT_HYP]

$$\left(\begin{array}{l}\lambda(\sigma_a, \sigma_b).\, (v\, \sigma_a) < (v\, \sigma_0) \supset\\ \qquad\qquad \text{fix}(\lambda l.\, D)(\sigma_a, \sigma_b)\end{array}\right)$$

$\quad (\sigma_0, \sigma_1)$

Since $(\lambda l.\, D)$ is monotonic, the conditional clause can be refined simply to $\text{fix}(\lambda l.\, D)(\sigma_a, \sigma_b)$.

$\sqsubseteq_3 \quad (\lambda l.\, D)\big(\lambda(\sigma_a, \sigma_b).\, \text{fix}(\lambda l.\, D)(\sigma_a, \sigma_b)\big)(\sigma_0, \sigma_1)$ $\quad$ [D_MONO]

The abstraction over (σ_a, σ_b) can be removed through η-conversion.

$= \quad (\lambda l.\, D)\big(\text{fix}(\lambda l.\, D)\big)(\sigma_0, \sigma_1)$ $\quad$ [η_CONV]

Using the definition of a fixed point and the assumption that $(\lambda l.\, D)$ is monotonic, we can remove the application of $(\lambda l.\, D)$ to its fixed point.

$= \quad \text{fix}(\lambda l.\, D)(\sigma_0, \sigma_1)$ $\quad$ [fix _DEF]

We have now proved the following theorem:

$$\forall \sigma_a\, \sigma_b.\, O(\sigma_a, \sigma_b) \sqsubseteq_3 (\mathsf{label}_{(v,O)}\, l.\, D)(\sigma_a, \sigma_b), \text{monotonic}(\lambda l.\, D) \vdash$$
$$(\mathsf{label}_{(v,O)}\, l.\, D)(\sigma_0, \sigma_1) \sqsubseteq_3 \text{fix}(\lambda l.\, D)(\sigma_0, \sigma_1)$$

We complete the proof using the definition of $(\mathsf{label}\, l.\, D)$.

$$\forall \sigma_a\, \sigma_b.\, O(\sigma_a, \sigma_b) \sqsubseteq_3 (\mathsf{label}_{(v,O)}\, l.\, D)(\sigma_a, \sigma_b), \text{monotonic}(\lambda l.\, D) \vdash$$
$$(\mathsf{label}_{(v,O)}\, l.\, D)(\sigma_0, \sigma_1) \sqsubseteq_3 (\mathsf{label}\, l.\, D)(\sigma_0, \sigma_1)$$

□

Analogous proofs can be performed for nested recursive blocks by using Bekič's theorem [2], and multiple inductions. Note also that with a little extra effort, a similar proof can be produced for recursive blocks whose meanings have been defined as a least upper-bound rather than as a least fixed-point.

Invariants. Invariants are an important aid for reasoning about recursive programs. We would like to associate an invariant with each labelled block. The meaning of a block extended with an invariant should be such that in order to introduce a block or a recursive call to a label, we would be required to prove that the invariant is satisfied. In return for these additional restrictions, we would be able to assume that the invariant held in the context of the body of the block.

Here is another contrived meaning for blocks which has the properties we require. The subscripts v and O denote, as before, the variant function of the block and the original description that the block was used to refine. The new subscript I is the invariant of the block and is a function from the program state to boolean.

$$\models^{\mathrm{def}} (\mathsf{label}_{(v,O,I)}\, l.\, D)(\sigma_0, \sigma_1) = \left(\begin{array}{l} (I\,\sigma_0) \supset \\ \quad (\lambda l.\, D) \\ \qquad \left(\begin{array}{c} \lambda(\sigma_a, \sigma_b).\, ((I\,\sigma_a) \wedge (v\,\sigma_a) < (v\,\sigma_0)) \supset \\ O(\sigma_a, \sigma_b) \end{array} \right) \\ \quad (\sigma_0, \sigma_1) \end{array} \right) \qquad [\mathsf{label}_{(v,O,I)}\text{\_DEF}]$$

The following theorems allow us to introduce and refine labelled blocks with invariants.

Theorem. *We can refine any description by giving it a label, provided that we can show that the chosen invariant holds (and that the label is not free in the description).*

$$\vdash I\,\sigma_0 \supset \big(O(\sigma_0, \sigma_1) \sqsubseteq_3 (\mathsf{label}_{(v,O,I)}\, l.\, O)(\sigma_0, \sigma_1)\big)$$

This theorem follows directly from the definition of $\mathsf{label}_{(v,O,I)}$.

Theorem. *In the initial state inside the body of the block we can assume that the invariant holds. We can refine a block by refining its body. If we wish to refine an instance of O inside the body of the block into a recursive call to the label, we must show that the invariant has been established and that the variant of the block has decreased.*

$$\vdash \left(\begin{array}{l} \forall \sigma_0\, \sigma_1\, l. \\ \quad \left(\begin{array}{l} I\,\sigma_0 \\ \wedge \left(\begin{array}{c} \forall \sigma_a\, \sigma_b.\, ((I\,\sigma_a) \wedge ((v\,\sigma_a) < (v\,\sigma_0))) \supset \\ O(\sigma_a, \sigma_b) \sqsubseteq_3 l(\sigma_a, \sigma_b) \end{array} \right) \end{array} \right) \supset \\ \quad D(\sigma_0, \sigma_1) \sqsubseteq_3 D'(\sigma_0, \sigma_1) \end{array} \right) \supset$$
$$(\mathsf{label}_{(v,O,I)}\, l.\, D)(`\sigma, \sigma') \sqsubseteq_3 (\mathsf{label}_{(v,O,I)}\, l.\, D')(`\sigma, \sigma')$$

This theorem also follows directly from the definition of $\mathsf{label}_{(v,O,I)}$.

This last theorem states that once you have refined a block into an executable form using the contrived meaning, you can replace the contrived meaning with the real meaning.

Theorem. *The contrived meaning is weaker than the true meaning whenever the block is executable.*

$$\forall \sigma_1\, \sigma_b.\, I\, \sigma_a \supset \left(O(\sigma_a, \sigma_b) \sqsubseteq_3 (\mathsf{label}_{(v,O,I)}\, l.\, D)(\sigma_a, \sigma_b)\right), \mathrm{monotonic}(\lambda l.\, D) \vdash$$
$$(\mathsf{label}_{(v,O,I)}\, l.\, D)(\sigma_0, \sigma_1) \sqsubseteq_3 (\mathsf{label}\, l.\, D)(\sigma_0, \sigma_1)$$

The proof of this theorem is almost identical to that given for blocks without an invariant, and is therefore omitted.

8 Conclusion

The work described here has two main features. The first is the use of a three-valued logic to describe behaviour. We claim that our three-valued logic allows us to specify potentially nonterminating behaviour more elegantly than other approaches based on two-valued logic. It is common to assume that nonterminating behaviour is undesirable. However, if we wish to be able to mix specifications and program constructs in a unified framework, then the specification language must be able to describe all behaviours permitted by the programming language. Our language of three-valued predicates retains the clarity of specification that motivates the treatment of programs as predicates. Indeed, if a specification does not call for potential nontermination, it can be expressed as a two-valued predicate.

The other contribution of this paper is an approach to refining specifications into recursive implementations. We begin by describing a natural approach to the refinement of specifications into recursive implementations. We then posit a contrived meaning for recursive blocks that trivially supports the proposed refinement process, and which can be regarded as a approximation to the true meaning. Finally we prove that once the refinement has been completed and the block is executable, then the approximation can be replaced by the true meaning. This approach is outlined in the context of our three-valued predicative semantics, but the ideas could be translated into other formalisms.

Acknowledgements

I am grateful for the support of the Australian Defence Science & Technology Organisation. I am also grateful to Mike Gordon for his guidance and supervision.

I would like to thank the following people who read drafts of this paper: Richard Boulton, Mike Gordon, Tom Melham and Joakim von Wright of the University of Cambridge; Maris Ozols of the Australian Defence Science & Technology Organisation; and Ian Hayes of the University of Manchester. The paper has been much improved by their comments and corrections, any remaining flaws are, of course, my own doing.

References

1. Barringer, H., et al.: A logic covering undefinedness in program proofs. Acta Inf. **21** (1984) 251–269
2. Bekič, H.: Definable operations in general algebras, and the theory of automata and flowcharts. In Jones, C. (ed.), Programming Languages and Their Definition, volume 177 of Lecture Notes in Computer Science. Springer-Verlag, Berlin, Germany (1984) 30–55
3. Bočvar, D.: Ob odnom trexznačnom isčislenii i ego primenenii k analizu paradoksov klassičeskogo rasširennogo funkcional'nogo isčislenija [on a three-valued calculus and its application to the analysis of contraditions of expanded functional calculus]. Mathematičeskij sbornik **4** (1938) 287–308
4. Dijkstra, E.: A Discipline of Programming. Prentice Hall Series in Automatic Computation. Prentice Hall, Englewood Cliffs, United States (1976)
5. Grundy, J.: A window inference tool for refinement. In Jones, C., et al. (eds), Proceedings of the 5th Refinement Workshop, Workshops in Computer Science, Lloyd's Register, London, England (1992). BCS-FACS, Springer-Verlag, London, England 230–254
6. Grundy, J.: Predicative programming — a survey. In Sabelfeld, V. (ed.), Proceedings of the International Conference on Formal Methods in Programming and Their Applications, Lecture Notes in Computer Science, Novosibirsk, Russia (1993). Springer-Verlag, Berlin, Germany
7. Kleene, S.: On notation for ordinal numbers. J. Symbolic Logic **3** (1938) 150–155
8. Łukasiewicz, J.: On the notion of possibility. In McCall [10] 15–16
9. Łukasiewicz, J.: On three-valued logic. In McCall [10] 16–18
10. McCall, S. (ed.) Polish Logic 1920–1939. Oxford University at the Clarendon Press, Oxford, England (1967)
11. Nelson, G.: A generalization of Dijkstra's calculus. ACM Trans. Programming Languages & Syst. **11** (1989) 517–561
12. Scott, D., Strachey, C.: Towards a mathematical semantics for computer languages. In Fox, J. (ed.), Proceeding of the Symposium on Computers and Automata, volume 21 of Microwave Research Institute Symposia Series, Brooklyn, United States (1971). Polytechnic Institute of Brooklyn, Microwave Research Institute, Wiley-Interscience 19–46

A Compositional Semantics of Combining Forms for Gamma Programs

David Sands[1]

Department of Computer Science, University of Copenhagen
Universitetsparken 1, DK-2100 Copenhagen Ø, DENMARK. (e-mail: dave@diku.dk)

Abstract. The Gamma model is a minimal programming language based on local multiset rewriting (with an elegant chemical reaction metaphor); Hankin *et al* derived a calculus of Gamma programs built from basic reactions and two composition operators, and applied it to the study of relationships between parallel and sequential program composition, and related program transformations. The main shortcoming of the "calculus of Gamma programs" is that the refinement and equivalence laws described are not compositional, so that a refinement of a sub-program does not necessarily imply a refinement of the program.

In this paper we address this problem by defining a compositional (denotational) semantics for Gamma, based on the *transition trace* method of Brookes, and by showing how this can be used to verify substitutive refinement laws, potentially widening the applicability and scalability of program transformations previously described.

The compositional semantics is also useful in the study of relationships between alternative combining forms at a deeper semantic level. We consider the semantics and properties of a number of new combining forms for the Gamma model.

1 Background

The Gamma Model The Gamma formalism was proposed by Banâtre and Le Métayer [BM93] as a means for the high level description of parallel programs with a minimum of explicit control. Gamma is a minimal language based on local rewriting of a finite multiset (or *bag*), with an appealing analogy with the chemical reaction process. As an example, a program that sorts an array $a_0, \ldots, a_n$ (of integers, say) could be defined as follows. Represent the array as a multiset of pairs $\{(0, a_0), \ldots, (n, a_n)\}$, then just specify a single rule: "exchange ill-ordered values"

$$sort_A : ((i,x),(j,y) \rightarrow (i,y),(j,x) \Leftarrow i < j \;\&\; x > y).$$

$i < j \;\&\; x > y$ specifies a property (a *reaction condition*) to be satisfied by the selected elements (i, x) and (j, y); these elements are replaced in the multiset (the *chemical solution*) by the elements $(i, y), (j, x)$ (the product of the reaction). Nothing is said in this definition about the order of evaluation of the comparisons; if several disjoint pairs of elements satisfy the reaction condition the comparisons and replacements can even be performed in parallel.

The computation terminates when a stable state is reached, that is to say when no elements of the multiset satisfy the reaction condition (or in general,

[1] Research supported by the Danish Research Council, DART Project (5.21.08.03)

any of a number of reaction conditions). The interested reader may find a long series of examples illustrating the Gamma style of programming in [BM93]. The benefit of using Gamma in systematic program construction in the Dijkstra-Gries style are illustrated in [BM90].

A Calculus of Gamma Programs For the sake of modularity it is desirable that a language offers a rich set of operators for combining programs. It is also fundamental that these operators enjoy a useful collection of algebraic laws in order to make it possible to reason about composed programs. Hankin, Le Métayer and Sands [HMS92] defined two composition operators for the construction of Gamma programs from basic reactions, namely sequential composition $P_1 \circ P_2$ and parallel composition $P_1 + P_2$. The intuition behind $P_1 \circ P_2$ is that the stable multiset reached after the execution of P_2 is given as argument to P_1. On the other hand, the result of $P_1 + P_2$ is obtained (roughly speaking) by executing the reactions of P_1 and P_2 (in any order, possibly in parallel), terminating only when neither can proceed further—in the spirit of the original Gamma programs.

As a further simple example, consider the following program which sorts a multiset of integers:

$$\begin{array}{l} sort_B : match \circ init \\ \text{where} \quad init : (x \rightarrow (0, x) \Leftarrow integer(x)) \\ \qquad match : ((i, x), (i, y) \rightarrow (i, x), (i+1, y) \Leftarrow x \leq y) \end{array}$$

The reaction *init* gives each integer an initial rank of zero. When this has been completed, *match* takes any two elements of the same rank and increases the rank of the larger.

In fact we get a sort program for a multiset of integers by placing $sort_A$ in parallel with $sort_B$ as $sort_A + sort_B$.

Hankin *et al.* [HMS92] derived a number of program refinement and equivalence laws for parallel and sequential composition, by considering the input-output behaviour induced by an operational semantics. So, for example, the program $sort_A + sort_B$ which is by definition $sort_A + (match \circ init)$ is refined by the program

$$(sort_A + match) \circ init.$$

This refinement is an instance of a general refinement law:

$$P + (Q \circ R) \geq (P + Q) \circ R.$$

Amongst other things, Hankin *et. al.* went on to describe how the laws were used in the design of a "pipelining" program transformation method.

Overview

The main shortcoming of the "calculus of Gamma programs" described in [HMS92] is that the refinement and equivalence laws described are not compositional, so that a refinement of a sub-program does not necessarily imply a refinement of the program. This limits the applicability and scalability of program transformations and inhibits the study of the relationships between alternative combining forms.

In this paper we directly address this problem by defining a compositional (denotational) semantics for Gamma, show that this can be used to verify substitutive refinement laws, and study properties of the language and relationships between alternative combining forms at a deeper semantic level.

The remainder of the paper is organised as follows. In **Section 2** we define the operational semantics of Gamma programs. In **Section 3** we describe the simple partial correctness refinement ordering and show that it is not substitutive. We then define *operational approximation* as the largest substitutive preordering contained in this.

In **Section 4** we motivate and adapt a method of Brookes [Bro93] for giving a (fully abstract) denotational semantics for a parallel shared variable *while* language originally studied by Hennessy and Plotkin [HP79]. The key technique is to give the meaning of a program as a set of *transition traces* which represent both the computation steps that the program can perform, and the ways that the program can interact with its environment.

The induced ordering from the compositional model implies operational approximation, and so in **Section 5** it is shown that a number of refinement laws can be proved. We then show that the semantics is not fully abstract (it does not completely characterise operational approximation) by proving an interesting property of the language: there is no analogy of the "Big Bang" for Gamma programs. In other words, there is no program which can add some elements to the empty set and then terminate.

Section 6 shows how the approach is useful for studying alternative combining forms for Gamma programs and the relationships between them, and **Section 7** concludes.

2 Operational Semantics

In this section we consider the operational semantics of programs consisting of basic reactions (written $(A \Leftarrow R)$, where R is the reaction condition, and A is the associated action, both assumed to have the same arity), together with two *combining forms*: sequential composition, $P_1 \circ P_2$, and parallel combination, $P_1 + P_2$ as introduced in [HMS92].

$$P \in \mathbf{P} ::= (A \Leftarrow R) \mid P \circ P \mid P + P$$

To define the semantics for these programs we define a single step transition relation between *configurations*. The *terminal* configurations are just multisets, and the *intermediate* configurations are program, multiset pairs written $\langle P, M\rangle$, where $M \in \mathbf{M}$ is the set of finite multisets of elements. The domain of the elements is left unspecified, but is expected to include integers, booleans and tuples.

We define the single step transitions first for the individual reactions $(A \Leftarrow R)$. A reaction terminates on some multiset exactly when there is no sub-multiset of elements which (when viewed as a tuple) satisfy the reaction condition. We will not specify the details of the reaction conditions and actions, but assume, as previously, that they are total functions over tuples of integers, truth values, tuples etc.

$$\langle (A \Leftarrow R), M\rangle \rightarrow M \quad \text{if } \neg\exists\vec{a} \subseteq M. R\vec{a}$$

Otherwise, if we can form a tuple from elements of the multiset, and this tuple satisfies the reaction condition, the selected elements can be replaced by the elements produced by the associated action function:

$$\langle (A \Leftarrow R), M \uplus \vec{a} \rangle \rightarrow \langle (A \Leftarrow R), M \uplus A\vec{a} \rangle \quad \text{if } R\vec{a}$$

The *terminal transitions* have the form $\langle P, M \rangle \rightarrow M$, and are only defined for programs not containing sequential composition, so the remaining *terminal* transitions are defined by the following synchronised-termination rule:

$$\frac{\langle P, M \rangle \rightarrow M \quad \langle Q, M \rangle \rightarrow M}{\langle P + Q, M \rangle \rightarrow M}$$

The remaining intermediate transitions are given by first defining what we will call *active contexts.*

DEFINITION **2.1** *An* active context, **A** *is a term containing a single hole* [ ]*:*

$$\mathbf{A} ::= [\,] \mid P + \mathbf{A} \mid \mathbf{A} + P \mid P \circ \mathbf{A}$$

Let $\mathbf{A}[P]$ denote active context $\mathbf{A}$ with program P in place of the hole.

The idea of active contexts is that they isolate parts of a program that can affect the next transition, so that for example, the left-hand side of a sequential composition is not active (but it can become active once the right-hand side has terminated). The remaining transitions are defined by the following two rules:

$$\frac{\langle P, M \rangle \rightarrow \langle P', M' \rangle}{\langle \mathbf{A}[P], M \rangle \rightarrow \langle \mathbf{A}[P'], M' \rangle} \qquad \frac{\langle Q, M \rangle \rightarrow M}{\langle \mathbf{A}[P \circ Q], M \rangle \rightarrow \langle \mathbf{A}[P], M \rangle}$$

The first says that if there is a possible reaction in an active context then it can proceed, while the second says that if the right hand side of a sequential composition can terminate, then it can be erased.

Notice that since a program does not uniquely factor into a sub-program in an active context, so this is another source of nondeterminism. We make the assumption that the one step evaluation relation is total, ie. that for all nonterminal configurations $\langle P, M \rangle$ there is at least one configuration U such that $\langle P, M \rangle \rightarrow U$. This amounts to an assumption that the transition relation on single $(A \Leftarrow R)$ pairs is total.

In [HMS92] an equivalent definition is given in terms of the more traditional SOS style, and active contexts are given as a derived construction to facilitate proofs of certain properties.

3 Operational Program Orderings

In [HMS92] defined a variety of orderings on programs based on their termination and nontermination behaviour. In this study we only consider the simplest of these, the partial correctness ordering (the $\leq_L$ order of [HMS92]).

We define $P_1 \leq P_2$ whenever, for each possible input M, if P_1 can terminate producing some multiset N then so can P_2.

This partial correctness preorder and associated equivalence $\sim$, satisfy the following properties (a subset of those given in [HMS92], where they are stated for a stronger ordering which also treats nontermination as a possible "result"). The reaction $(A \Leftarrow \mathit{False})$ stands for a reaction whose reaction-condition is not satisfiable, hence the action A and its arity are irrelevant.

1. $P \circ (Q \circ R) \sim (P \circ Q) \circ R$
2. $P + (Q + R) \sim (P + Q) + R$
3. $P + Q \sim Q + P$
4. $(A \Leftarrow \mathit{False}) + P \sim P + (A \Leftarrow \mathit{False}) \sim P$
5. $(A \Leftarrow \mathit{False}) \circ P \sim P \circ (A \Leftarrow \mathit{False}) \sim P$
6. $P \leq P + P$
7. $(P_1 + P_3) \circ P_2 \leq (P_1 \circ P_2) + P_3$

Furthermore, the sequential composition operator is monotonic in both operands, ie. if $P_i \leq Q_i$ then $P_1 \circ P_2 \leq Q_1 \circ Q_2$. Unfortunately it is not monotonic with respect to parallel composition[1] as the following example shows:

Consider the following two atomic programs:

$$P_1 : x \to 0 \Leftarrow x = 1$$
$$P_2 : x \to 1 \Leftarrow x = 0$$

By property 5 we know that $(A \Leftarrow \mathit{False}) \circ P_1 \leq P_1$. Now $\langle P_1 + P_2, \{1\}\rangle$ can never terminate, but $\langle((A \Leftarrow \mathit{False}) \circ P_1) + P_2, \{1\}\rangle \to^* \{1\}$, and so $((A \Leftarrow \mathit{False}) \circ P_1) + P_2 \not\leq P_1 + P_2$.

Because of the lack of a general substitutivity property, the use of the partial correctness ordering in reasoning about programs is limited. As is standard, we are therefore interested in the largest contextual (pre) congruence relation contained in $\leq$.

DEFINITION **3.1** *Let* **C** *range over general program contexts, then we define* operational approximation *($\sqsubseteq_o$) and* operational congruence*($\equiv_o$) respectively by:*

$$P_1 \sqsubseteq_o P_2 \iff \forall \mathbf{C}.\, \mathbf{C}[P_1] \leq \mathbf{C}[P_2]$$
$$P_1 \equiv_o P_2 \iff P_1 \sqsubseteq_o P_2 \;\&\; P_2 \sqsubseteq_o P_1$$

The task now is to find some characterisation of operational approximation.

4 Transition Traces

In the previous section we showed that the partial correctness preorder was not substitutive with respect to parallel composition (and hence arbitrary program contexts). In other words, to characterise the behaviour of programs in all possible program contexts it is not enough to distinguish between them on the basis of their input-output behaviour alone.

[1] As was also observed for the stronger "relational" ordering in [HMS92].

Traces It is clear that we must consider intermediate behaviour of programs to obtain a substitutive equivalence. Consider the ***strict traces***, $\mathcal{ST}[\![P]\!]$ of a program defined as all sequences of multisets which describe the single steps leading to a terminal configuration:

$$\mathcal{ST}[\![P]\!] = \{M_1, \ldots, M_k \mid \langle P, M_1\rangle \rightarrow \langle P_2, M_2\rangle \rightarrow \cdots \rightarrow \langle P_{k-1}, M_{k-1}\rangle \rightarrow M_k\}$$

Now clearly if $\mathcal{ST}[\![P_1]\!] \subseteq \mathcal{ST}[\![P_2]\!]$ then $P_1 \leq P_2$. As a possible account of program meaning, the strict traces are in some cases rather strong since they unnecessarily distinguish between $(A \Leftarrow \mathit{False})$ and $(A \Leftarrow \mathit{False}) \circ (A \Leftarrow \mathit{False})$. However, in other respects they are still not strong enough to provide a denotation for programs which is compositional, as the following example shows:

Consider the following three atomic programs:

$$\begin{aligned} P_1 &: x \rightarrow 0 \Leftarrow x = 1 \\ P_2 &: x \rightarrow 2 \Leftarrow x = 1 \\ P_3 &: x \rightarrow 1 \Leftarrow x = 2 \end{aligned}$$

Since P_1 and P_2 have the same reaction condition, it should not be too difficult to see that

$$\mathcal{ST}[\![P_1 \circ P_2]\!] = \mathcal{ST}[\![(A \Leftarrow \mathit{False}) \circ P_2]\!],$$

but that in the context $[\,] + P_3$ they give different traces; in particular,

$$\begin{aligned} &\{1\}\{2\}\{2\}\{1\}\{0\}\{0\} \\ &\in \mathcal{ST}[\![(P_1 \circ P_2) + P_3]\!] \setminus \mathcal{ST}[\![((A \Leftarrow \mathit{False}) \circ P_2) + P_3]\!]. \end{aligned}$$

The reason why a semantics based on traces as opposed to just input-output behaviour is still not compositional is that they do not take into account the possible interference (from the program's surrounding context) that can occur during execution. We can think of the computational model as a form of shared-variable language, and adapt standard techniques developed in that context as a possible solution.

Resumptions Hennessy and Plotkin [HP79] described a denotational semantics for a simple *while*-language with a parallel composition operator. The basis of their method is to define the meaning of a command as a ***resumption***.

A resumption models the meaning of a command as a function from the current state (the store, or in our case the multiset) to a set containing all possible terminal states reachable in one step, together with a set of state-resumption pairs, each of which represents a possible next nonterminal state together with a resumption for the rest of the program. This construction can model the fact that after each atomic step in the computation of the command it may be interrupted by the context in which it is being executed. Since this interruption may change the state, the resumption component of the pair describes how the command may then resume execution.

Move Traces At about the same time, Abrahamson [Abr79], in a study of a modal logic for parallel programs, sketched a semantics for a parallel shared-state language based on sequences of "moves". A move is a pair of states representing an atomic computation step of the program. Adjacent moves model a possible interference by some other process executing in parallel with the program. A set of possible move-sequences of a program can be thought of as an "unraveling" of the program's resumption semantics, but with the advantage of being mathematically simpler to work with, since it does not involve any powerdomain constructions.

Transition Traces Brookes [Bro93] independently came up with the idea of using sequences of state-pairs to give a compositional semantics for the Hennessy-Plotkin *while*-language, but with an important improvement: by considering only those traces with certain closure properties (most importantly an ***absorption*** property which we will describe below) he was able to give a fully abstract semantics for the *while* language without having to add the rather unnatural co-routine operator of [HP79].

We adapt Brooke's transition trace model for our language. We define the meaning of a command as a set of transition traces, and show that the definition can be given compositionally.

4.1 Transition Trace Semantics

The idea is to define the meaning of a program P as a set of nonempty finite sequences of multiset pairs.

DEFINITION **4.1** *The transition trace function* $\mathcal{TT}[\![\_]\!] : \mathbf{P} \to \wp((\mathbf{M} \times \mathbf{M})^{+})$ *is given by*

$$\begin{aligned}\mathcal{TT}[\![P]\!] = \{&(M_0, N_0)(M_1, N_1)\ldots(M_k, N_k) \mid \\ &\langle P, M_0\rangle \to^* \langle P_1, N_0\rangle \; \& \\ &\langle P_1, M_1\rangle \to^* \langle P_2, N_1\rangle \;\&\; \ldots \;\&\; \langle P_k, M_k\rangle \to^* N_k\}\end{aligned}$$

The intuition behind the use of transition traces is that each transition trace

$$(M_0, N_0)(M_1, N_1)\ldots(M_k, N_k) \in \mathcal{TT}[\![P]\!]$$

represents a terminating execution of program P in some context, starting with multiset M_0, and in which each of the pairs (M_i, N_i) represents computation steps performed by (derivatives of) P and the adjacent multisets N_{i-1}, M_i represent possible interfering computation steps performed by the context. The partial correctness ordering is derivable from the transition traces by considering traces of length one:

PROPOSITION **4.2**

$$P_1 \le P_2 \iff \{(M,N) \mid (M,N) \in \mathcal{TT}[\![P_1]\!]\} \subseteq \{(M,N) \mid (M,N) \in \mathcal{TT}[\![P_2]\!]\}$$

A key feature of the definition of transition traces is that they use the reflexive-transitive closure of the one-step evaluation relation, $\to^*$. An important consequence of this reflexivity and transitivity is that they are closed under "stuttering" and "absorption" properties described below. In the following let ϵ denote the empty sequence. Let α and β range over elements of $(\mathbf{M} \times \mathbf{M})^*$.

DEFINITION 4.3 *A set* $T \subseteq \wp((\mathbf{M} \times \mathbf{M})^{+})$ *is closed under left-stuttering and absorption if it satisfies the following two conditions*

$$\text{left-stuttering}\,\frac{\alpha\beta \in T, \beta \neq \epsilon}{\alpha(M,M)\beta \in T} \qquad \text{absorption}\,\frac{\alpha(M,N)(N,M')\beta \in T}{\alpha(M,M')\beta \in T}$$

Let $\ddagger T$ denote the left-stuttering and absorption closure (henceforth just closure) of a set T.

PROPOSITION 4.4 *For all programs* P, $\mathcal{TT}[\![P]\!] = \ddagger\mathcal{TT}[\![P]\!]$.

"Stuttering" represents the fact that we can have an arbitrary interference by the context without any visible steps performed by the program, The concept and terminology are well-known from Lamport's work on temporal logics for concurrent systems [Lam91]. Notice that we say *left*-stuttering to reflect that the context is not permitted to change the state after the termination of the program. In this way each transition trace of a program only charts interactions with its context up to the point of the programs termination. The "absorption" property is important because prevents "idle" computation steps from becoming semantically significant, which is a problem with the resumption approach.

4.2 A Compositional Definition of Transition Traces

We now show that the transition traces of a command can be given a denotational definition, which shows that transition trace semantics can be used to prove operational equivalence.

For the basic reaction-action pairs $(A \Leftarrow R)$, we build transition traces by simply considering all sequences of mediating transitions, followed by a terminal transition. Define the following sets[2]

$$\begin{aligned}\mathsf{mediators}_{(A\Leftarrow R)} &= \{(M,N) \mid \langle (A \Leftarrow R), M\rangle \rightarrow^{*} \langle (A \Leftarrow R), N\rangle\}\\ \mathsf{terminals}_{(A\Leftarrow R)} &= \{(M,N) \mid \langle (A \Leftarrow R), M\rangle \rightarrow^{*} N\}\end{aligned}$$

Sequential composition has an easy definition. We just concatenate the transition traces from the transition traces of the components, and take their closure. Define the following sequencing operation for transition trace sets:

$$T_1 \mathbin{;} T_2 = \{\alpha\beta \mid \alpha \in T_1, \beta \in T_2\}$$

Not surprisingly parallel composition is built with the use of an interleaving combinator. For α and β in $(\mathbf{M},\mathbf{M})^{*}$ let $\alpha \between \beta$ be the set of all their interleavings, given inductively by

$$\begin{aligned}\epsilon \between \beta = \beta \between \epsilon &= \{\beta\}\\ (M,M')\alpha \between (N,N')\beta &= \{(M,M')\gamma \mid \gamma \in \alpha \between (N,N')\beta\}\\ &\quad \cup \{(N,N')\gamma \mid \gamma \in (M,M')\alpha \between \beta\}.\end{aligned}$$

Now to define the transition traces of $P_1 + P_2$ we must ensure that the traces of P_1 and P_2 are interleaved, but not arbitrarily. The termination step of a parallel composition requires an agreement at the point of their termination. For this purpose, we define the following interleaving operation on transition traces:

$$T_1 \oplus T_2 = \{\alpha(M,M) \mid \alpha_1(M,M) \in T_1, \alpha_2(M,M) \in T_2, \alpha \in \alpha_1 \between \alpha_2\}$$

[2] Note that the definitions extend to all simple programs, that is, programs not containing sequential composition

DEFINITION 4.5 *The transition trace mapping* $\mathsf{T}[\![\_]\!] : \mathbf{P} \to \wp((\mathbf{M}\times\mathbf{M})^+)$ *is given by induction on the syntax as:*

$$\mathsf{T}[\![(A \Leftarrow R)]\!] = (\mathsf{mediators}_{(A\Leftarrow R)})^* \mathbin{;} \mathsf{terminals}_{(A\Leftarrow R)}$$

$$\mathsf{T}[\![P_1 \circ P_2]\!] = \ddagger(\mathsf{T}[\![P_2]\!] \mathbin{;} \mathsf{T}[\![P_1]\!])$$

$$\mathsf{T}[\![P_1 + P_2]\!] = \ddagger(\mathsf{T}[\![P_1]\!] \oplus \mathsf{T}[\![P_2]\!])$$

In Appendix A we sketch the correctness proof of the compositional definition leading to the following result:

THEOREM 4.6 *For all programs* P, $\mathsf{T}[\![P]\!] = \mathcal{TT}[\![P]\!]$

Define the transition trace ordering on programs $\sqsubseteq_t$ as

$$P_1 \sqsubseteq_t P_2 \iff \mathcal{TT}[\![P]\!] \subseteq \mathcal{TT}[\![P_2]\!].$$

The operations used to build the compositional definition are all monotone with respect to subset inclusion, and so a simple induction on contexts is sufficient to give

$$P_1 \sqsubseteq_t P_2 \Rightarrow \forall \mathrm{C}.\, \mathrm{C}[P_1] \sqsubseteq_t \mathrm{C}[P_2]$$

and since $P \sqsubseteq_t Q$ implies $P \leq Q$, we have the desired characterisation of operational approximation:

$$P_1 \sqsubseteq_t P_2 \Rightarrow P_1 \sqsubseteq_o P_2$$

However, as we shall show in the next section, we cannot reverse the implication—the transition trace semantics is not (inequationally) *fully abstract*.

5 Laws and Full Abstraction

Many laws of operational approximation and equivalence can now be proved using the transition trace model. In this section we state that (almost all) the partial correctness laws presented in section 2 can be shown to hold for operational approximation and equivalence, using the transition traces. However, in spite of the pedigree of Brookes' transition trace method, we show that not all laws can be verified, ie. that the semantics is not fully abstract.

5.1 Laws

We can now state a number of laws for transition trace approximation and its associated equivalence, $\equiv_t$.

PROPOSITION 5.1

1. $P \circ (Q \circ R) \equiv_t (P \circ Q) \circ R$
2. $P + (Q + R) \equiv_t (P + Q) + R$
3. $P + Q \equiv_t Q + P$
4. $(A \Leftarrow False) + P \equiv_t P + (A \Leftarrow False) \equiv_t P$
5. $(A \Leftarrow False) \circ P \sqsubseteq_t P$
6. $P \circ (A \Leftarrow False) \equiv_t P$

7. $P \sqsubseteq_t P + P$

8. $(P_1 + P_3) \circ P_2 \sqsubseteq_t (P_1 \circ P_2) + P_3$

Note that none of the inequalities above can be strengthened to equalities. In particular, consider law 5. The transition traces of $(A \Leftarrow False)$ is the set $\ddagger\{(M, M) \mid M \in \mathbf{M}\}$, and so for all multisets N there exists a trace $\alpha(N, N) \in \mathcal{TT}[\![(A \Leftarrow False) \circ P]\!]$, which is clearly not true in general for $\mathcal{TT}[\![P]\!]$ (for a concrete example, take the one from section 3).

Many of these properties follow directly from properties of the transition trace operations. The other are proved from either the operational or compositional definitions of transition traces. A more thorough consideration of the laws is presented in [San93]

5.2 On Full Abstraction and the Big Bang

The transition trace semantics is fully abstract if transition trace approximation coincides with operational approximation. Unfortunately this does not hold, but the counterexample is interesting, and suggests modifications to either the language or the semantics for which the full abstraction question is still open.

In the construction of the transition traces of a program, when we have adjacent pairs $\cdots(M, M')(N, N')\cdots$, the change from M' to N represents changes made by the programs environment. If these changes are not realisable by an actual program then we are in danger of distinguishing between programs because of unfeasible context behaviour. First we show that when we consider the context's termination behaviour there are impossible interruptions in the transition traces. Then we give an example which shows that this leads to distinctions between operationally equivalent programs.

There are programs that always diverge. Consider any nullary reaction $(() \rightarrow A \Leftarrow True)$ which says: if the empty set is a subset of the multiset, replace that subset by A. Clearly such a reaction is always applicable, and as a consequence for all contexts $\mathbf{C}$ and programs P,

$$(() \rightarrow A \Leftarrow True) \equiv_t \mathbf{C}[(() \rightarrow A \Leftarrow True)] \sqsubseteq_t P.$$

We use this property to show that there is no "Big Bang" program which can add elements to the empty multiset and still terminate:

PROPOSITION **5.2** $\langle P, \emptyset\rangle \rightarrow^* M \Rightarrow M = \emptyset$

PROOF Suppose that $\langle P, \emptyset\rangle \rightarrow^* M \neq \emptyset$ then there is some P' and nonempty N such that $\langle P, \emptyset\rangle \rightarrow^* \langle P', \emptyset\rangle \rightarrow \langle P', N\rangle \rightarrow^* M$. From the operational semantics we must have $P' = \mathbf{A}[(A \Leftarrow R)]$ for some active context $\mathbf{A}$ and reaction $(A \Leftarrow R)$, such that $\langle (A \Leftarrow R), \emptyset\rangle \rightarrow \langle (A \Leftarrow R), N\rangle$. Therefore the reaction must be nullary, with $R() = True$ and $A() = N$, which contradicts the assumption that the program terminates. □

Now we can see that the interruptions modeled by the transition traces of a program are potentially too liberal, since they allow sequences of the form $\cdots(M_1, \emptyset)(M_2, N_2)\cdots$ for nonempty M_2, which could never happen in an execution of the program in any context for which the *combined* system terminates. We use this fact to build a counterexample to full abstraction.

Consider the unary reaction $Await_\emptyset : (x \rightarrow x \Leftarrow True)$. This program terminates only when the multiset is empty. Now consider composition with some arbitrary program P:

PROPOSITION 5.3 $P \circ Await_{\emptyset} \sqsubseteq_o Await_{\emptyset}$

The idea is that in any context, if either of these programs terminates it must be the case that at some intermediate point $Await_{\emptyset}$ terminates. Since this can only happen when the multiset becomes empty, it follows from proposition 5.2 that any computation following this point cannot change the multiset and still terminate. Therefore, the additional composition with P can never alter the result of a terminating computation, which must always be the empty multiset, but *can* make the program terminate less often.

PROPOSITION 5.4 *$\mathcal{TT}[\![\_]\!]$ is not fully abstract.*

PROOF Take $P = (A \Leftarrow False)$. Then

$$\mathcal{TT}[\![(A \Leftarrow False) \circ Await_{\emptyset}]\!] = \ddagger \{(\emptyset, \emptyset)(M, M) \mid M \in \mathbf{M}\}$$

but this is not a subset of $\mathcal{TT}[\![Await_{\emptyset}]\!] = \ddagger \{(\emptyset, \emptyset)\}$. □

We postpone further discussion of the full abstraction problem until the concluding section.

6 New Combining Forms

In this section we consider the addition of other combining forms for programs that provide a conservative extension of transition trace equivalence, and relationships between them.

Vanilla Parallel Composition The more usual form of parallel program composition does not require that the two programs terminate synchronously. Extend the syntax of the language with $P ::= P_1 || P_2$, and the active contexts with $\mathbf{A} ::= \mathbf{A}||P \mid P||\mathbf{A}$. Now we can give the rules for one step evaluation:

$$\frac{\langle Q, M\rangle \to M}{\langle P||Q, M\rangle \to \langle P, M\rangle} \qquad \frac{\langle P, M\rangle \to M}{\langle P||Q, M\rangle \to \langle Q, M\rangle}$$

The transition traces for $P||Q$ can be given compositionally just by (taking the closure of) interleaving those of P with those of Q, and the proof extends easily. The expected associativity and commutativity properties also hold for $||$. Some relationships with $\circ$ and $+$ can be summarised in the following diagram where $\mathsf{FA} = (A \Leftarrow False)$ and the arrows ($\to$) depict the ordering $\sqsubseteq_t$.

$$(\mathsf{FA} \circ P) + (\mathsf{FA} \circ Q)$$
$$\uparrow$$
$$P||Q$$
$$\nearrow \qquad \nwarrow$$
$$(\mathsf{FA} \circ P)+Q \qquad P+(\mathsf{FA} \circ Q)$$
$$\nearrow \quad \nwarrow \qquad \nearrow \quad \nwarrow$$
$$Q \circ P \qquad P + Q \qquad P \circ Q$$

Nondeterministic Choice A nondeterministic choice operator is also easily added, just as in [Bro93]. Extend $\mathbf{P} ::= P_1 \vee P_2$. Evaluation is specified by insisting that the choice is committed before any evaluation can occur. Therefore we do not extend the active contexts, but have the two rules:

$$\langle P \vee Q, M\rangle \to \langle P, M\rangle \qquad \langle P \vee Q, M\rangle \to \langle Q, M\rangle$$

Again the extension of the semantics is straightforward. The transition traces are modeled compositionally using set union, and $+$ and $\circ$ distribute over $\vee$. Perhaps more interestingly we get an exact characterisation of the above form of parallel composition:

$$P \| Q \equiv_t ((\mathsf{FA} \circ P) + Q) \vee (P + (\mathsf{FA} \circ Q))$$

Critical Region There are problems regarding language extensions when we attempt to add operators which allow the atomic steps to have a larger granularity. The problem is that the definition of $+$ requires, for termination, that both operands must agree. This demands that terminal transitions must be of the form $\langle P, M\rangle \to M$, otherwise implementation of parallel composition would require multiset duplication and possible backtracking of multiset computations.

However, some control over atomicity seems desirable to guarantee interference freedom. With the above remarks in mind we give the definition of a form of guarded *critical region* $P \lhd Q$, which allows for the uninterrupted execution of sub-program P as soon as Q terminates. To do this we extend active contexts to also include contexts of the form $P \lhd \mathbf{A}$, and define the following intermediate transition as follows:

$$\frac{\langle Q, M\rangle \to M \quad \langle P, M\rangle \to^* N}{\langle P \lhd Q, M\rangle \to \langle \mathsf{FA}, N\rangle}$$

The use of the "dummy" program FA in the right hand side of this transition means that a critical region construct fits with parallel composition in a computationally reasonable way. The price that is paid for this is that $\lhd$ is not associative (the compositional definition of its transition traces is left as an exercise).

7 Conclusions

We have presented a compositional semantics for a number of combining forms for the Gamma model of multiset programming, beginning with those defined in [HMS92]. The semantics is useful for verifying substitutional refinement laws, and thus extend the applicability of some of the transformation techniques discussed in previous work. It is also a useful tool for studying the impact of language extensions. In the remainder of the paper we consider areas of future work.

In section 5, the fact that a program cannot add elements to an empty multiset and still terminate was used to show that the semantics is not fully abstract. This suggests a modification to the semantics to only allow feasible transition traces for which adjacent pairs $(M_i, N_i)(M_{i+1}, N_{i+1})$ satisfy the property

$N_i = \emptyset \Rightarrow M_{i+1} = \emptyset$. It is an open question whether this modification leads to a fully abstract model.

Alternatively, we might view the above property as an indication of a lack of expressive power (since, for example, it implies that the constant programs $K_M : \forall N. \langle K_M, N \rangle \rightarrow^* M$ are not definable) and look for suitable extensions. One such extension involves the addition of reactions which are parameterised by a *critical mass* limiting the global size of the multisets in which they are applicable[3]. The critical mass reactions are sufficient to define the family of programs $\{Await_M\}$ which terminate if and only if the multiset becomes equal to M (cf. Hennessy and Plotkin's *await* commands). If we also include the guarded critical region construct suggested in the previous section, they can be used to give a full abstraction proof, directly adapting the method used by Brookes.

In spite of the absence of full abstraction, the transition traces are sufficiently abstract to prove the expected laws. A more detailed study of the laws of the parallel and sequential composition is given in [San93], in which we verify or refute all of the laws for the simple relational operational ordering given in [HMS92], including the "residual program" laws. In addition it is shown that every program is refined by a product ($\circ$) of sums ($+$) of basic reactions.

Acknowledgements Thanks to Chris Hankin and Daniel Le Métayer for numerous discussions and ideas. Thanks to Steve Brookes, Samson Abramsky and Geoff Burn for helpful discussions and advice.

A Proof of the compositional definition of the transition traces

In this appendix the proof of correctness for the compositional definition of transition traces is sketched.

LEMMA **A.1** $\mathsf{T}[\![P]\!]$ *is closed.*

LEMMA **A.2**

1. $\langle P, M \rangle \rightarrow^* N \Rightarrow (M, N) \in \mathsf{T}[\![P]\!]$
2. $\langle P, M \rangle \rightarrow^* \langle P', M' \rangle \;\&\; \alpha \in \mathsf{T}[\![P']\!] \Longrightarrow (M, N)\alpha \in \mathsf{T}[\![P]\!]$

PROOF Prove that the lemma holds for one-step evaluation (inductions in the structure of P), and extend to $\rightarrow^*$ by induction on the length of the reflexive-transitive closure of $\rightarrow$ using lemma A.1. □

LEMMA **A.3**

1. $\beta_1 \in \mathcal{TT}[\![P_1]\!] \;\&\; \beta_2 \in \mathcal{TT}[\![P_2]\!] \Longrightarrow \beta_2\beta_1 \in \mathcal{TT}[\![P_1 \circ P_2]\!]$
2. $\alpha(M, M) \in \mathcal{TT}[\![P_1]\!] \;\&\; \beta(M, M) \in \mathcal{TT}[\![P_2]\!] \;\&\; \gamma \in \alpha \between \beta \Longrightarrow \gamma(M, M) \in \mathcal{TT}[\![P_1 + P_2]\!]$

PROOF Inductions on the lengths of β_2 and γ respectively. □

---

[3] Although this would violate the "locality principal" [BM93] of the Gamma model.

THEOREM **A.4** *For all programs P*, $\mathcal{TT}[\![P]\!] = \mathsf{T}[\![P]\!]$.

PROOF We prove it in two halves:

- $\mathcal{TT}[\![P]\!] \subseteq \mathsf{T}[\![P]\!]$: The (implicitly inductive) definition of $\mathcal{TT}[\![\_]\!]$ can be given as the least fixed point of the monotonic functional $\mathrm{F} : (\mathbf{P} \to \wp((\mathbf{M} \times \mathbf{M})^*)) \to \mathbf{P} \to \wp((\mathbf{M} \times \mathbf{M})^*)$

$$\begin{aligned}\mathrm{F} = \lambda R.\lambda P. &\{(M,N) \mid \langle P,M\rangle \to^* N\}\\ &\cup \{(M,N)\alpha \mid \langle P,M\rangle \to^* \langle P',N\rangle, \alpha \in R(P')\}\end{aligned}$$

  From lemma A.2 we immediately have that $\mathsf{T}[\![\_]\!]$ is a post fixed point of F and so by the Knaster-Tarski fixed point theorem contains the least fixed point, $\mathcal{TT}[\![\_]\!]$.

- $\mathsf{T}[\![P]\!] \subseteq \mathcal{TT}[\![P]\!]$: By induction on the structure of P we show that $\delta \in \mathsf{T}[\![P]\!] \Rightarrow \delta \in \mathcal{TT}[\![P]\!]$. The base case ($P = (A \Leftarrow R)$) is obvious. In both the inductive cases we assume without loss of generality that δ comes from

$$\{\alpha_2\alpha_1 \mid \alpha_1 \in \mathsf{T}[\![P_1]\!], \alpha_2 \in \mathsf{T}[\![P_2]\!]\}, \text{ and}$$
$$\{\alpha(M,M) \mid \alpha_1(M,M) \in \mathsf{T}[\![P_1]\!], \alpha_2(M,M) \in \mathsf{T}[\![P_2]\!], \alpha \in \alpha_1 \between \alpha_2\}$$

  respectively (ie. it is sufficient to ignore the closure of these sets in the compositional definitions).
  Case ($P = P_1 \circ P_2$): $\delta = \delta_2\delta_1$ for some $\delta_1 \in \mathsf{T}[\![P_1]\!], \delta_2 \in \mathsf{T}[\![P_2]\!]$. The induction hypothesis gives that $\delta_1 \in \mathcal{TT}[\![P_1]\!], \delta_2 \in \mathcal{TT}[\![P_2]\!]$, and the result follows from lemma A.3.
  Case ($P = P_1 + P_2$): $\delta = \gamma(N,N)$ for some $\gamma \in \alpha \between \beta$, $\alpha(N,N) \in \mathsf{T}[\![P_1]\!]$ and $\beta(N,N) \in \mathsf{T}[\![P_2]\!]$. From the induction hypothesis we know that $\alpha(N,N) \in \mathcal{TT}[\![P_1]\!]$, $\beta(N,N) \in \mathcal{TT}[\![P_2]\!]$ and the result follows from lemma A.3.

□

References

[Abr79] K. Abrahamson. Modal logic of concurrent nondeterministic programs. In *Proceedings of the International Symposium on Semantics of Concurrent Computation*, volume 70, pages 21–33. Springer-Verlag, 1979.

[BM90] J.-P. Banâtre and D. Le Métayer. The Gamma model and its discipline of programming. *Science of Computer Programming*, 15:55–77, 1990.

[BM93] J.-P. Banâtre and D. Le Métayer. Programming by multiset transformation. *CACM*, January 1993. (INRIA research report 1205, April 1990).

[Bro93] S. Brookes. Full abstraction for a shared variable parallel language. In *Logic In Computer Science*, 1993. (to appear).

[HMS92] C. Hankin, D. Le Métayer, and D. Sands. A calculus of Gamma programs. Research Report DOC 92/22 (28 pages), Department of Computing, Imperial College, 1992. (short version to appear in the Proceedings of the Fifth Annual Workshop on Languages and Compilers for Parallelism, Aug 1992, Springer-Verlag).

[HP79] M. Hennessy and G. D. Plotkin. Full abstraction for a simple parallel programming lanuage. In *Mathematical Foundations of Computer Science*, volume 74 of *LNCS*, pages 108–120. Springer-Verlag, 1979.

[Lam91] L. Lamport. The Temporal Logic of Actions. Technical Report 79, DEC Systems Research Center, Palo Alto, CA, 1991.

[San93] D. Sands. Laws of parallel synchronised termination. In *Theory and Formal Methods 1993: Proceedings of the First Imperial College, Department of Computing, Workshop on Theory and Formal Methods*, Isle of Thorns, UK, 1993. Springer-Verlag Workshops in Computer Science.

Algebraic Properties of Loop Invariants

Gerald Futschek

Vienna University of Technology
Institut für Softwaretechnik, Technische Universität Wien
A-1040 Vienna, Austria

Abstract. A set P(DO, R) of all invariants that ensure termination and where the postcondition R is true after termination is defined for every loop DO and for every postcondition R. Complying with the corresponding properties required, these sets P(DO, R) induce a topology on wp(DO, R). The weakest precondition wp(DO, R) is the weakest invariant of DO with respect to R. The topology P(DO, R) has a non-trivial structure and contains arbitrary conjunctions of invariants.

1 Introduction

Finding invariants is the most important step when developing loops. The standard approach is to develop a loop invariant based on a given specification with precondition and postcondition by generalizing the postcondition. This technique is discussed in [Dij76] and [Gries81]. We want to define a set of invariants that ensure termination and where a postcondition R is true upon it for a given loop and postcondition R. We then proceed to discuss the algebraic properties of this set of invariants.

We consider a **do** loop that has the following form

DO: **do** B $\rightarrow$ S **od**

This loop is a simple form of Dijkstras guarded command loop [Dij75]. The loop body S can be any deterministic or nondeterministic program. A guarded command loop with more than one guard as described in [Dij75] can easily be transformed into this simple form. Hence the results that we obtain in this discussion can also be applied to guarded command loops.

In our proof outlines we will use the notation proposed by Dijkstra in [Dij90] that comprises the use of square brackets to denote universal quantification.

2 Loop Invariants with Respect to a Postcondition R

Definition: A predicate P is defined as an *invariant of a loop* DO if

$$P \wedge B \Rightarrow wp(S, P) \qquad \text{(Inv.0)}$$

is true. This condition corresponds to the predicate $\{P \wedge B\}$ S $\{P\}$ (total correctness) defined by C.A.R. Hoare in [Hoare69]. If P is true initially then P is true before and after each execution of the loop body S, hence P is a loop invariant. *If* B becomes false P remains invariantly true upon the termination of the loop.

We have to consider that by just establishing an invariant, termination cannot be ensured. The following example shows that while there may exist several valid invariants for a loop not all of them ensure termination.

Example: DO: $\{P\}$ **do** $i \neq 0 \rightarrow \{P \wedge i \neq 0\}$ $i := i - 2$ $\{P\}$ **od**

This loop has many invariants, some of them are

P: i **mod** 2 = 0	termination only if $i \geq 0$ is true initially
P: i **mod** 2 ≠ 0	no termination
P: i **mod** 2 = 0 ∧ i≥ 0	termination is ensured
P: i = 0	termination without execution of the loop body

Please note that only the last two invariants ensure the termination of the loop.

Invariants that ensure termination and imply a specific postcondition are of special interest when developing a program.

Definition: P is defined as a *loop invariant of DO with respect to R*, if in addition to Inv.0

$$P \Rightarrow wp(DO, R) \qquad \text{(Inv.1)}$$

is also true.

This definition means that P is a precondition of DO with respect to R, in other words $\{P\}$ DO $\{R\}$ is true. If P is initially true it not only remains invariantly true but termination is ensured and R is true upon it.

Notation: We denote the *set of all invariants of DO with respect to R* as

P(DO, R)

Thus a set of invariants is now defined for every loop and for any postcondition R. In the following we will use **P**(R) instead of **P**(DO, R) for reasons of simplicity and clearity, whenever we discuss a fixed loop DO.

3 Properties of P(R)

Considering the set of all loop invariants of DO with respect to R, we can derive a number of algebraic properties.

3.1 False is an Element of P(R)

The strongest assertion **false** is an element of every set **P**(R).

$$[\mathbf{false} \in \mathbf{P}(R)] \qquad \text{(P.0)}$$

This property implies that **false** is an invariant of all loops with respect to any postcondition R. Hence the sets **P**(R) are not empty.

Proof of P.0: **false** is an invariant, because $(\mathbf{false} \wedge B) \Rightarrow wp(S, \mathbf{false})$ is true (Inv.0). **false** is an invariant with respect to R, because $\mathbf{false} \Rightarrow wp(DO, R)$ is true (Inv.1).

3.2 Any Conjunction of a Positive Number of Invariants is an Element of P(R)

We consider an arbitrary non-empty set of invariants $V \subseteq \mathbf{P}(R)$. Then

$$[(\forall P\colon\ P \in V\colon\ P) \in \mathbf{P}(R)] \qquad \text{(P.1)}$$

All arbitrary but non-empty conjunctions of invariants are an element of **P**(R).

Using only two invariants P_1 and P_2 yields the specific property:

$$[P_1, P_2 \in \mathbf{P}(R) \ \Rightarrow\ (P_1 \wedge P_2) \in \mathbf{P}(R)] \qquad \text{(P.1')}$$

Proof of P.1: Since all $P \in V$ are invariants of DO (Inv.0), we have

$$[(\forall P\colon\ P \in V\colon\ P \wedge B \Rightarrow wp(S, P))]$$

$\Rightarrow$ $\{(\forall P\colon\ P \in V\colon\ P) \Rightarrow P$ for all $P \in V\}$

$$[(\forall P\colon\ P \in V\colon\ (\forall P\colon\ P \in V\colon\ P) \wedge B \ \Rightarrow\ wp(S, P))]$$

$\Rightarrow$ {predicate calculus}

$$[(\forall P\colon\ P \in V\colon\ P) \wedge B \ \Rightarrow\ (\forall P\colon\ P \in V\colon\ wp(S, P))]$$

$=$ {wp is positively conjunctive (wp.1)}

$$[(\forall P\colon\ P \in V\colon\ P) \wedge B \ \Rightarrow\ wp(S, (\forall P\colon\ P \in V\colon\ P))]$$

Hence $(\forall P:\ P \in V:\ P)$ is an invariant of DO.

$(\forall P:\ P \in V:\ P)$ is an invariant of DO with respect to R can be easily concluded from predicate calculus:

$[(\forall P:\ P \in V:\ P \Rightarrow wp(DO, R))]$

$\Rightarrow$ {predicate calculus}

$[(\forall P:\ P \in V:\ P) \Rightarrow wp(DO, R)]$

3.3 Any Disjunction of Invariants is an Element of P(R)

We proceed with the discussion of this property in the same manner as before. We consider an arbitrary set of invariants $V \subseteq \mathbf{P}(R)$. Then

$$[(\exists P:\ P \in V:\ P) \in \mathbf{P}(R)] \qquad (P.2)$$

The set $\mathbf{P}(R)$ contains an arbitrary number of disjunctions of its elements.

Using only two invariants P_1 and P_2 yields the specific property:

$$[P_1, P_2 \in \mathbf{P}(R) \;\Rightarrow\; (P_1 \vee P_2) \in \mathbf{P}(R)] \qquad (P.2')$$

Proof of P.2:

$[(\forall P:\ P \in V:\ P \wedge B \Rightarrow wp(S, P))]$

$\Rightarrow$ $\{P \Rightarrow (\exists P:\ P \in V:\ P)$ for all $P \in V$, wp is monotonous (wp.2)$\}$

$[(\forall P:\ P \in V:\ P \wedge B \Rightarrow wp(S, (\exists P:\ P \in V:\ P)))]$

$\Rightarrow$ {predicate calculus}

$[(\exists P:\ P \in V:\ P) \wedge B \;\Rightarrow\; wp(S, (\exists P:\ P \in V:\ P))]$

Hence $(\exists P:\ P \in V:\ P)$ is an invariant of DO.

$(\exists P:\ P \in V:\ P)$ is an invariant of DO with respect to R can be concluded from predicate calculus:

$[(\forall P:\ P \in V:\ P \Rightarrow wp(DO, R))]$

$=$ {predicate calculus}

$[(\exists P:\ P \in V:\ P) \Rightarrow wp(DO, R)]$

3.4 wp(DO, R) is an Invariant with Respect to R

The weakest precondition wp(DO, R) of a loop is always an invariant of this loop with respect to R.

$$[wp(DO, R) \in \mathbf{P}(R)] \tag{P.3}$$

Proof of P.3:

wp(DO, R) is defined as $(\exists i:\ 0 \le i:\ H_i(R))$ (see wp.DO in appendix).

We begin by proving that $H_i(R) \in \mathbf{P}(R)$ is true for all $i \ge 0$ because according to P.2 this implies that $(\exists i:\ 0 \le i:\ H_i(R)) \in \mathbf{P}(R)$ is true. Hence we have $wp(DO, R) \in \mathbf{P}(R)$.

a) all $H_i(R)$ are invariants of DO

$$
\begin{array}{ll}
 & H_i(R) \wedge B \\
\Rightarrow & \quad \{Hi.2\} \\
 & H_{i+1}(R) \wedge B \\
= & \quad \{Hi.1\} \\
 & (H_0(R) \ \vee \ (B \wedge wp(S, H_i(R)))) \ \wedge B \\
= & \quad \{Hi.0, \text{calculus logic}\} \\
 & wp(S, H_i(R)) \wedge B \\
\Rightarrow & \quad \{\text{calculus logic}\} \\
 & wp(S, H_i(R))
\end{array}
$$

b) all $H_i(R)$ are invariants of DO with respect to R:

$$
\begin{array}{ll}
 & H_i(R) \\
\Rightarrow & \quad \{\text{predicate logic}\} \\
 & (\exists i:\ 0 \le i:\ H_i(R)) \\
= & \quad \{wp.DO\} \\
 & wp(DO, R)
\end{array}
$$

a) and b) shows that all predicates $H_i(R)$ comply with the conditions Inv.0 and Inv.1 thus yielding

$$H_i(R) \ \in \ \mathbf{P}(R) \ \text{ for all } \ i \ge 0. \tag{Hi.3}$$

According to P.2 wp(DO, R) is an invariant of DO with respect to R.

3.5 All Predicates that are Stronger than $H_0(R)$ are Elements of P(R)

$$[(\forall P:\ P \Rightarrow H_0(R):\ P \in \mathbf{P}(R))] \tag{P.4}$$

Proof of P.4: for all P with $P \Rightarrow H_0(R)$ we observe

$$\begin{array}{ll} & P \wedge B \\ \Rightarrow & \quad \{P \Rightarrow H_0(R)\} \\ & H_0(R) \wedge B \\ = & \quad \{Hi.0\} \\ & R \wedge \neg B \wedge B \\ = & \quad \{logic\} \\ & \textbf{false} \\ \Rightarrow & \quad \{logic\} \\ & wp(S, P) \end{array}$$

Hence P is an invariant with respect to R because

$$\begin{array}{ll} & P \Rightarrow H_0(R) \\ \Rightarrow & \quad \{H_0(R) \Rightarrow wp(DO, R), wp.DO\} \\ & P \Rightarrow wp(DO, R) \end{array}$$

According to our definition $H_0(R)$ and all stronger predicates are invariants of DO but the loop body is never executed when we use this invariant. In this specific case the loop has the same effect as a skip statement. Hence these invariants are not really interesting for the development of a program.

Notice: P.4 implies theorem P.0 that says that false is always an invariant.

3.6 wp(DO, R) is the Weakest of all Invariants of P(R)

We can draw this conclusion from Inv.1 and P.3. According to P.3 wp(DO, R) $\in$ **P**(R) is true and $P \Rightarrow wp(DO, R)$ is true for all invariants $P \in$ **P**(R) (because of Inv.1). Hence wp(DO, R) is the weakest of all invariants of **P**(R).

3.7 P(R) has a Non-trivial Structure

According to the algebraic properties discussed so far **P**(R) could easily consist of all predicates that are stronger than wp(DO, R) thus making the structure of **P**(R) rather uninteresting. The following example shows that **P**(R) does not contain any predicates that are stronger than wp(DO, R).

DO: **do** $n \neq N \rightarrow n := n + 1$ **od** $\{R: n = N\}$

The predicate P: $n \leq N$ is the weakest precondition of DO with respect to R. The predicate $n+1 = N$ is stronger than P but is not an invariant of DO.

4 The Topology of Invariants

4.1 The Axioms of the Topological Space of Invariants

Every invariant P corresponds to a set of states {s: P} in the state space. Identifying the predicate P with its set of states and denoting this set with P yields the following properties of sets of states for the properties P.0, P.1', P.2 and P.3.

$$\emptyset \in \mathbf{P}(R) \tag{T.0}$$

$$P_1, P_2 \in \mathbf{P}(R) \Rightarrow P_1 \cap P_2 \in \mathbf{P}(R) \tag{T.1'}$$

$$(V \subseteq \mathbf{P}(R)) \Rightarrow (\cup P:\ P \in V) \in \mathbf{P}(R) \tag{T.2}$$

$$wp(DO, R) \in \mathbf{P}(R) \tag{T.3}$$

These are exactly the axioms of a topological space defined on the set wp(DO, R). P $\subseteq$ wp(DO, R) for every invariant P $\in$ **P**(R) is concluded from Inv.1. The invariants of DO with respect to R therefore define a topological space on the set wp(DO, R) with the invariants as open sets. We conclude from property P.1 the more general property of the corresponding topological space

$$(V \subseteq \mathbf{P}(R), V \neq \emptyset) \Rightarrow (\cap P:\ P \in V) \in \mathbf{P}(R) \tag{T.1}$$

4.2 P(R) is Monotonous

$$[R_1 \subseteq R_2 \Rightarrow \mathbf{P}(R_1) \subseteq \mathbf{P}(R_2)] \tag{T.4}$$

P(R_1) is a topology on wp(DO, R_1) and **P**(R_2) is a topology on wp(DO, R_2). According to the monotonicity of wp (wp.2) wp(DO, R_1) $\subseteq$ wp(DO, R_2) is true for $R_1 \subseteq R_2$.

Proof of T.4:

All $P_1 \in \mathbf{P}(R_1)$ are invariants according to Inv.0. Inv.1 can also be applied with respect to R_2:

$P_1 \in \mathbf{P}(R_1)$
$\Rightarrow$ {Inv.1}
$P_1 \subseteq wp(DO, R_1)$
$\Rightarrow$ {wp.2}
$P_1 \subseteq wp(DO, R_2)$ (Inv.1)

The order relation ($\subseteq$) on the state sets R_i therefore induces an order on the sets **P**(R_i).

P(true) is the most general topology on DO since it includes all invariants that ensure termination. **P(false)** is the strongest topology and it only contains the empty set since wp(DO, **false**) = **false.**

4.3 Denotation of P(R)

Lemma: For every $P \in$ **P**(R) the sets **P**(R) $\cap$ P define a topology of invariants on wp(DO, R) $\cap$ P.

Proof of Lemma: proof of the axioms T.0 to T.3

Theorem: $\quad$ **P**(R) = **P(true)** $\cap$ wp(DO, R) $\quad$ (T.5)

Each of the topologies **P**(R) can be derived from the maximum topology **P(true)**.

Proof of theorem:

1. "$\Rightarrow$"

$$
\begin{array}{ll}
 & P \in \mathbf{P}(R) \\
\Rightarrow & \quad \{T.4, Inv.1\} \\
 & P \in \mathbf{P}(\mathbf{true}),\ P \subseteq wp(DO, R) \\
\Rightarrow & \quad \{\text{set theory, Lemma}\} \\
 & P \in \mathbf{P}(\mathbf{true}) \cap wp(DO, R)
\end{array}
$$

2. "$\Leftarrow$"

$$
\begin{array}{ll}
 & P \in \mathbf{P}(\mathbf{true}) \cap wp(DO, R) \\
\Rightarrow & \quad \{Inv.0, \text{Lemma}\} \\
 & P \subseteq wp(DO, R),\ P \text{ is an invariant} \\
\Rightarrow & \quad \{Inv.0, Inv.1\} \\
 & P \in \mathbf{P}(R)
\end{array}
$$

4.4 Pot(H_0(R)) is a Complete Subtopology of P(R).

We conclude from P.4 that the power set of H_0(R) forms a subtopology of P(R).

4.5 The Topological Basis of P(R)

For every state $x \in$ wp(DO, R) we can easily construct the minimal invariant P_x that contains x.

$$P_x = (\bigcap P:\ x \in P, P \in \mathbf{P}(R))$$

$P_x \in$ **P**(R) follows from T.1.

The sets

$$B = \{P_x:\ x \in wp(DO, R)\} \cup \emptyset$$

are a basis of **P**(R). Every invariant P ∈ **P**(R), P ≠ ∅ can be denotated as

$$P = (\bigcup P_x:\ x \in P)$$

Example: For the following loop DO we give all invariants with respect to R

DO: {P} **do** $i \neq 0 \rightarrow$ $\{P \wedge i \neq 0\}$ $i := i - 2$ {P} **od** {R: $i = 0$}

	predicate	corresponding set of states
wp(DO, R)	$i \,\mathbf{mod}\, 2 = 0 \wedge 0 \leq i$	$\{i:\ i \,\mathbf{mod}\, 2 = 0 \wedge 0 \leq i\}$
P_x	$i \,\mathbf{mod}\, 2 = 0 \wedge 0 \leq i \leq x$	$\{i:\ i \,\mathbf{mod}\, 2 = 0 \wedge 0 \leq i \leq x\}$

$$\mathbf{P}(R) = \emptyset \cup \{P_x:\ x \in wp(DO, R)\} \cup \{wp(DO, R)\}$$

5 Conclusion

The properties of loop invariants that were presented in this paper induce a topological space on the maximal invariant wp(DO,R). This motivates topological reasoning about loop invariants. The knowledge of these properties may give more insight in program construction.

References

[Dij75] E.W.Dijkstra: *Guarded commands, nondeterminacy and formal derivation of programs*, Comm.ACM 18 (1975) 453-457.

[Dij76] E.W.Dijkstra: *A Discipline of Programming*, Prentice Hall, Englewood Cliffs, NJ, 1976.

[Dij90] E.W.Dijkstra, C.S.Scholten: *Predicate Calculus and Program Semantics*, Springer Verlag New York Inc., 1990.

[Gries81] D. Gries: *The Science of Programming*, Springer Verlag New York Inc. , 1981.

[Hoare69] C.A.R. Hoare: *An axiomatic basis for computer programming*, Comm. ACM 12 (1969) 576-580.

[Hoare87] C.A.R. Hoare, I.J. Hayes et al.: *Laws of Programming*, Comm. ACM 30 (1987) 672-686.

Appendix

In the following the prerequisites used in the proofs are presented.

Properties of wp (see [Dij76, Dij90])

0. Law of excluded miracle

$$[\neg wp(S, \mathbf{false})] \qquad \text{(wp.0)}$$

1. Distributivity of conjunction

for any non-empty bag V of predicates:

$$[(\forall R:\ R \in V:\ wp(S, R)) \ = \ wp(S, (\forall R:\ R \in V:\ R))] \qquad \text{(wp.1)}$$

$$[wp(S, R_1) \wedge wp(S, R_2) \ = \ wp(S, R_1 \wedge R_2)] \qquad \text{(wp.1')}$$

2. Monotonicity

$$[(R_1 \Rightarrow R_2) \ \Rightarrow \ (wp(S, R_1) \Rightarrow wp(S, R_2))] \qquad \text{(wp.2)}$$

Equivalency of Hoare Logic and wp Calculus (see [Gries81])

$$\{Q\}\ S\ \{R\} \quad \text{is equivalent to} \quad Q \Rightarrow wp(S, R). \qquad \text{(wp.3)}$$

Definition of wp(DO, R) (see [Gries81])

$$wp(DO, R) \ := \ (\exists i:\ 0 \le i:\ H_i(R)) \qquad \text{(wp.DO)}$$

$$H_0(R) \ := \ R \wedge \neg B \qquad \text{(Hi.0)}$$

$$H_{i+1}(R) \ := \ H_0(R) \ \vee \ (B \wedge wp(S, H_i(R))) \quad \text{for } i \ge 0 \qquad \text{(Hi.1)}$$

$$[H_i(R) \ \Rightarrow \ H_{i+1}(R)] \quad \text{for } i \ge 0. \qquad \text{(Hi.2)}$$

An Approach to Parameterized First-order Specifications: Semantics, Correctness, Parameter Passing

Wolfgang Reif *
University of Karlsruhe
reif@ira.uka.de

Abstract

This paper presents an alternative approach to loose specifications of parameterized data types. The specification language is full first-order logic, and the semantics is a particular class of functions, mapping parameter algebras to (parameter generated) target algebras. We investigate monomorphicity and correctness of parameterized first-order specifications, and present simple syntactic criteria for these notions. Furthermore, correctness of standard parameter passing is studied. We give a characterization as well as a simple sufficient criterion for specifications with correct parameter passing. Finally, the interaction between parameter passing and correctness of specifications is investigated: We present conditions under which actualization of parameterized specifications preserves the correctness of the constituent specifications.

1 Introduction

Since the pioneering work of Liskov, Zilles, the ADJ-group and many others in the early seventies, a number of different approaches to algebraic specification have emerged and gained a strong influence on software engineering. In combination with logics of programs, algebraic specifications have become an indispensable tool for the formal description and construction of verified software. Usually, three main stream approaches are distinguished according to their semantics: initial algebra semantics ([LZ 74], [Gu 76], [GTW 78], [TWW82], [Ga 83], [Pa 85], [EM 85], [Pa 90], [EM 90]), terminal algebra semantics ([Wa 79], [HR 80], [Ka 83]) and loose semantics ([GGM 76], [BDPPW 79], [WB 82], [WPPDB 83], [CIP 87], [Wi 90]).

In the first two approaches a specification is meant to define a specific abstract data type precisely up to isomorphism, and therefore the semantics of a specification is a particular

* Author's address: Institut für Logik, Komplexität und Deduktionssysteme, Universität Karlsruhe, Postfach 6980, W-7500 Karlsruhe, FRG, Tel. +721-608-4245. This research was partly sponsored by the BMFT-project KORSO.

class of isomorphic algebras (initial or final algebras resp.). In the case of initial semantics the specification language is in general restricted to universal equations or Horn clauses, in order to guarantee the existence of initial models (for the extension by constraints see [Pa 90], [EM 90]). In the loose approach, however, a specification is rather thought of as a collection of requirements, open to a number of different interpretations. The semantics of a specification is therefore a class of possibly non-isomorphic interpretations. This view of a specification is an appropriate basis for software specification and development, and is adopted in this paper. The loose approach has some practical advantages: The specification language is full first-order logic and is not restricted to Horn clauses. This increases the flexibility in practical applications considerably. Furthermore, first-order specifications exhibit better "deductive properties" than Horn clause specifications, in the sense that in general more consequences can be deduced from the explicit axioms of the specification (see [Re 92]). This is a good basis for verification systems.

In this paper we present a parameterization mechanism for loose first-order specifications. It combines a pushout-style parameterization on the syntactic level (e.g. [EKTWW 80], [BG 80], [Eh 82], [Ga 83], [EM 85]) with a semantics based on classes of functions (cf. [SST 92]). A parameterized specification PSPEC = ($SPEC_P$, SPEC) is a pair of two first-order specifications. $SPEC_P$ is the *parameter specification* and SPEC is the *target specification*, the former beeing a subspecification of the latter. A single parameterized data type satisfying PSPEC is modelled by a partial function $\mathcal{F}$, that takes a parameter algebra $\mathcal{M}$ and returns a ("parameter generated") target algebra $\mathcal{F}(\mathcal{M})$. Roughly, the semantics SEM(PSPEC) is the class of all functions satisfying PSPEC. This loose semantics differs from the ones given in ASL ([SW 83], [Wi 86]) or Extended ML (for functor specifications, [ST 88]), but is similar to the one given in [SST 92] for specifications of parametric algebras. There are also similarities to PLUSS [Gau 92].

In general, loose specifications admit non-isomorphic interpretations, and therefore are ambiguous. In the software development process this is exploited in abstract requirement specifications. Only the crucial properties of the system are specified, leaving out implementational or other details. In later phases of the development however, the specifications tend to restrict the classes of described models more and more, ending up with design specifications describing an interpretation up to isomorphism. Specifications with this property are called *monomorphic*. If a data type $\mathcal{F}$ satisfies a monomorphic specification PSPEC, then PSPEC is called *correct* for $\mathcal{F}$. In this paper the main stress is put on how to detect monomorphicity, and how monomorphic and correct specifications can be developed. We investigate both notions, and present a simple proof theoretic criterion for monomorphic specifications. Furthermore, we investigate the actualization of parameterized specifications. We define the notion of *strict persistency* which characterizes the correctness of *standard parameter passing* (actualization with an unparameterized specification). We present a simple sufficient criterion for strict persistency, and show that for strictly persistent parameterized specifications, actualization is a correctness preserving operation.

The paper is organized as follows: In the next section we recall some basic definitions, and define simple and parameterized specifications. In section 3 we investigate the semantics of parameterized specifications. Section 4 is devoted to monomorphic and correct specifications, and section 5 investigates standard parameter passing.

2 Basic Definitions

2.1 Simple and parameterized specifications

A *signature* SIG = (S, OP) consists of a set of sorts S and a set OP of function- and predicate symbols over S. S* are the finite words over S with empty word λ. For $w = s_1 .. s_n$ OP_w is the set of n-ary predicate symbols with the indicated typing, and $OP_{w,s}$ is the set of function symbols with range sorts w and target sort s. For every sort $s \in S$ there is a countably infinite reservoir X_s of individual variables. X is the system of all these variable sets. For SIG = (S, OP) and X, $T_{OP,s}(X)$ denotes the *terms* of sort s over OP and X. $T_{OP,s}$ is the corresponding set of ground terms (= $T_{OP,s}(\emptyset)$). $T_{OP}(X)$ and T_{OP} are the sets of all terms and ground terms, respectively. A signature SIG = (S, OP) is a *subsignature* of SIG' = (S', OP') if $S' \supseteq S$ and $OP' \supseteq OP$. L(SIG, X) and $\forall$L(SIG, X) denote the sets of first-order and universal fomulas over SIG and X, respectively . A *simple* (unparameterized) specification SPEC = (S, OP, T) consists of a signature SIG = (S, OP) and a set T of first-order formulas. SPEC is a *subspecification* of SPEC' = (S', OP', T') if (S, OP) is a subsignature of (S', OP') and $T \subseteq T'$.

Parameterized specifications are based on *parameterized signatures* SIG = $(S \cup S_p, OP \cup OP_p)$. We distinguish between *parameter sorts* S_p, *target sorts* S, *parameter operations* OP_p, and *target operations* OP. The sets are assumed to be disjoint, and $SIG_p = (S_p, OP_p)$ is called the *parameter signature*. Sometimes we use the notation $SIG_p \cup (S, OP)$ for SIG. A parameterized specification PSPEC = ($SPEC_p$, SPEC) consists of two simple specifications: the *parameter specification* $SPEC_p$ and the *target specification* SPEC, the former being a subspecification of the latter.

IND(SIG) denotes the set of first-order instances of a schema for *structural induction* over the target sorts S of SIG = $(S \cup S_p, OP \cup OP_p)$. IND(SIG) is defined as follows: Let $\phi = \langle \varphi_s : s \in S \rangle$ a vector of formulas φ_s and $\mathbf{x} = \langle x_s : s \in S \rangle$ a vector of induction variables x_s of sort $s \in S$. The variable x_s occurs free in φ_s, and $\varphi_s(c)$ stands for substituting c for x_s in φ_s. We define IND(SIG) = $\{ (\mathrm{Basis}(\mathbf{x}, \phi) \wedge \mathrm{Hyp}(\mathbf{x}, \phi)) \rightarrow \mathrm{Concl}(\mathbf{x}, \phi) : \phi = \langle \varphi_s : s \in S \rangle, \mathbf{x} = \langle x_s : s \in S \rangle \}$. We set $\mathrm{Basis}(\mathbf{x}, \phi) := \bigwedge_{c \in OP_{\lambda,s}} \varphi_s(c)$, $\mathrm{Concl}(\mathbf{x}, \phi) := \bigwedge_{s \in S} \forall x_s . \varphi_s$ and $\mathrm{Hyp}(\mathbf{x}, \phi) := \bigwedge \{ \mathrm{Step}(f, \mathbf{x}, \phi) : f \in OP_{w,s}, s \in S \}$. For $f \in OP_{w,s}$, $s \in S$, with $w = s_1..s_r$, and new variables $x_1, ..., x_r$ Step(f, x, ϕ) is the following formula:

$$\forall x_1, ..., x_r. ((\bigwedge \{ \varphi_{s'}(x') : \langle s', x' \rangle \in \{ \langle s_1, x_1 \rangle, ..., \langle s_r, x_r \rangle \}, s' \in S \}) \rightarrow \varphi_s(f(x_1, ..., x_r))).$$

2.2 Semantic concepts

For SIG = (S, OP) a *SIG-algebra* $\mathcal{A} = ((A_s)_{s \in S}, (f_{\mathcal{A}})_{f \in OP})$ consists of non-empty carrier sets A_s and the interpretations $f_{\mathcal{A}}$ for the symbols from OP. The interpretation $f_{\mathcal{A}}$ is a total function or a predicate depending on what kind of symbol f is. We use a relaxed notation for algebras by just enumerating the components, for example, ($\mathbb{N}$, 0, succ, <) for the natural numbers with zero, successor and the less-than predicate. The class of SIG-algebras is abbreviated by Alg(SIG). We write $\mathcal{A} \cong \mathcal{B}$, if $\mathcal{A}$ and $\mathcal{B}$ are isomorphic. If SIG = (S, OP) is a subsignature of SIG', and $\mathcal{A}$ a SIG'-algebra, then the *reduct* of $\mathcal{A}$ wrt. SIG,

$\mathcal{A}|_{SIG}$, is the algebra $\mathcal{B}$ with $B_s = A_s$ for all $s \in S$ and $f_{\mathcal{B}} = f_{\mathcal{A}}$ for all $f \in OP$. Let SIG a subsignature of SIG', $\mathcal{A}$ a SIG'-algebra, and g an isomorphism from $\mathcal{A}|_{SIG}$ to $\mathcal{M}$. Then a SIG'-algebra $(\mathcal{A} \nabla_{SIG,g} \mathcal{M})$ can be defined, where roughly $\mathcal{A}|_{SIG}$ is replaced in $\mathcal{A}$ by $\mathcal{M}$ according to g. We call this algebra the *isomorphic alteration* of $\mathcal{A}$ by $\mathcal{M}$ wrt. g. We get the properties $(\mathcal{A} \nabla_{SIG,g} \mathcal{M}) \equiv \mathcal{A}$ and $(\mathcal{A} \nabla_{SIG,g} \mathcal{M})|_{SIG} = \mathcal{M}$. Since in the sequel the actual isomorphism g is not important, we use the relaxed notation $(\mathcal{A} \nabla_{SIG} \mathcal{M})$.

The evaluation of a term $t \in T_{OP}(X)$ in $\mathcal{A}$ under the *valuation* $s: X \to \mathcal{A}$ is denoted by $t_{s,\mathcal{A}}$. We write $\mathcal{A}, s \models \varphi$ if the formula φ is true in $\mathcal{A}$ under s, $\mathcal{A} \models \varphi$ if $\mathcal{A}$ is a model of φ, and $T \models \varphi$ if every model of the formula set T is a model of φ. Alg(T) is the class of all models of T. An algebra is generated if each carrier element can be denoted by a ground term. Gen(T) is the class of generated models of T. For SPEC = (S, OP, T) Alg(SPEC) := Alg(T) and Gen(SPEC) := Gen(T).

Signature morphisms , *forgetful operations*, *specification morphisms* and *free functors* for equational parameterized specifications are defined as usual (see [EM 85] for details). A signature morphism h: SIG → SIG', maps sorts and operations of SIG to sorts and operations of SIG', a specification morphism h : SPEC → SPEC' is a signature morphism, which additionally guarantees that SPEC' entails the *translated axioms* of SPEC: SPEC' ⊨ h(SPEC). For a given signature morphism h: SIG → SIG' the forgetful operation V_h turns a SIG'-Algebra $\mathcal{A}$ into a SIG-algebra $\mathcal{B} := V_h(\mathcal{A})$ with $B_s := A_{h(s)}$ and $f_{\mathcal{B}} := h(f)_{\mathcal{A}}$.

3 Semantics of Parameterized First-order Specifications

We illustrate the syntax of parameterized specifications by the following description of finite sets over an arbitrary partially ordered domain: Sets_over_Partial_order is described in a compact notation for parameterized specifications. It refers to the parameter specification Partial_order and lists the target sorts, functions, predicates, variables and axioms. The sort *data*, and the partial ordering< are the only parameter sorts and operations. The target sort is *set*, the target operations are the empty set ∅, the insertion of a data element (insert), the membership predicate (∈), and an ordering (<<) on sets, a variant of the multiset ordering.

parameterized specification Sets_over_Partial_order
- **parameter** Partial_order
- **sorts** set
- **functions:** $\emptyset : \to set$
  - $insert : data \times set \to set$
- **predicates** infix $\in : data \times set$
  - infix $<< : set \times set$
- **vars** $n, m : set$
- **axioms** $\neg\ x \in \emptyset$
  - $x \in insert(y, m) \leftrightarrow (x = y \lor x \in m)$
  - $n = m \leftrightarrow \forall x. (x \in n \leftrightarrow x \in m)$
  - $n << m \leftrightarrow \forall x. (x \in n \to \exists y. (y \in m \land x < y))$

end

parameter specification Partial_order
  sorts data
  predicates infix $<$: data $\times$ data
  vars x, y, z : data
  axioms $\neg\, x < x$
  $x < y \wedge y < z \rightarrow x < z$

end

For simple specifications we adopt the loose semantics approach of Giarratana et al. ([GGM 76]) and the Munich CIP-group ([BDPPW 79]). The semantics of a simple specification SPEC is the class of its generated models Gen(SPEC). Our semantics of a parameterized specification PSPEC = ($SPEC_p$, SPEC) with signature SIG = ($S \cup S_p$, $OP \cup OP_p$) is based on the class of *parameter generated* or *S_p-generated* algebras, PGen(SIG). A SIG-algebra $\mathcal{A}$ is S_p-generated if each element of any target carrier (A_s for $s \in S$) can be constructed from the parameter carriers (A_s for $s \in S_p$) by finitely many applications of the target functions from OP. S_p-generated algebras are models of IND(SIG). We define the *target models* of PSPEC, Tar(PSPEC), to be the class of S_p-generated models of SPEC.

The interpretations of PSPEC are functions from Alg($SPEC_p$) to Tar(PSPEC). For example, the function $\mathcal{Set}$ mapping each model $\mathcal{M}$ of Partial_order to the algebra of finite sets over $\mathcal{M}$ (with the corresponding operations) is an interpretation of the above sample specification. However, we are not interested in arbitrary functions. We require that they preserve their parameters -like $\mathcal{Set}$ does-, or at least preserve them up to isomorphism. In general we cannot expect that the functions are defined for each parameter algebra, like $\mathcal{Set}$, since the target specification might be inconsistent. Therefore, we require that they are defined whenever possible. This means that the functions are defined for a parameter algebra $\mathcal{M}$ iff there exists a target model $\mathcal{A}$ with $\mathcal{A}|_{SIG_p} \equiv \mathcal{M}$. Finally, we require that the functions are compatible with isomorphism: The images of isomorphic arguments are again isomorphic.

3.1 Definition (Semantics of parameterized specifications)

Let PSPEC = ($SPEC_p$, SPEC) with signature SIG = ($S \cup S_p$, $OP \cup OP_p$). Then SEM(PSPEC) is the class of all partial functions $\mathcal{F} \in Alg(SIG_p) \rightarrow Alg(SIG)$ (with domain dom($\mathcal{F}$) and range rg($\mathcal{F}$)) satisfying the following properties:

1) $rg(\mathcal{F}) \subseteq Tar(PSPEC)$
2) $dom(\mathcal{F}) = Tar(PSPEC)|_{SIG_p}$
3) for all $\mathcal{M} \in dom(\mathcal{F})$: $\mathcal{F}(\mathcal{M})|_{SIG_p} \equiv \mathcal{M}$
4) for all $\mathcal{M}, \mathcal{M}' \in dom(\mathcal{F})$: $\mathcal{M} \equiv \mathcal{M}' \Rightarrow \mathcal{F}(\mathcal{M}) \equiv \mathcal{F}(\mathcal{M}')$ ■

Properties 1) and 2) guarantee that the images of a function $\mathcal{F}$ are target models, and that $\mathcal{F}$ is defined whenever possible. We call a function $\mathcal{F}$ with property 3) *persistent*. Property

4) states that $\mathcal{F}$ is compatible with isomorphism. An algebra $\mathcal{A}$ *satisfies* a simple specification SPEC if $\mathcal{A} \in$ SEM(SPEC). A function $\mathcal{F}$ *satisfies* PSPEC if $\mathcal{F} \in$ SEM(PSPEC). We call two functions $\mathcal{F}_1$, $\mathcal{F}_2$ *isomorphic* ($\mathcal{F}_1 \equiv \mathcal{F}_2$), if dom($\mathcal{F}_1$) = dom($\mathcal{F}_2$) and $\mathcal{F}_1(\mathcal{M}) \cong \mathcal{F}_2(\mathcal{M})$ for all $\mathcal{M} \in$ dom($\mathcal{F}_1$). The functions of SEM(PSPEC) are defined for Alg(SPEC_p) if each $\mathcal{M} \in$ Alg(SPEC_p) can be extended to a parameter generated model of SPEC, i.e. $\text{Tar(PSPEC)}|_{\text{SIG}_p}$ = Alg(SPEC_p). A specification PSPEC with this property is called *freely extendible.*

3.2 Remark (Elementary properties of SEM(PSPEC) and related work)

SEM(PSPEC) is closed under isomorphism: If $\mathcal{F}_1 \in$ SEM(PSPEC) and $\mathcal{F}_1 \equiv \mathcal{F}_2$ then $\mathcal{F}_2 \in$ SEM(PSPEC). However, since we allow full first-order specifications, it might happen that the target specification SPEC is inconsistent or has no parameter generated models (Tar(PSPEC) = $\emptyset$). Then (and only then) SEM(PSPEC) consists of the undefined function. If Tar(PSPEC) is not empty, then SEM(PSPEC) in general contains different non-isomorphic functions. In our example, however, SEM(Sets_over_Partial_order) is a single class of isomorphic functions. How this special case can be detected from the axioms is shown in section 4.

With the initial semantics approach to equational parameterized specifications (see f.e. [EM 85]), the above semantics shares the view of parameterized data types as mappings from parameter algebras to target algebras. In the initial semantics approach, the free functors are always defined for all parameter algebras, but may fail to be persistent. In our approach, however, the functions are persistent per definition but possibly not defined for every parameter algebra. The functions satisfying freely extendible specifications are persistent and defined for Alg(SPEC_p). In practice we are interested in freely extendible specifications.

In ASL ([SW 83], [Wi 86]) the semantics of a parameterized specification is a mapping that takes a class of (parameter) algebras and returns a class of (result) algebras. In Extended ML ([ST 88]) the semantics of a functor specification is a function mapping an algebra to a class of algebras. Both are different from the semantics given above. In [SST 92] Sanella, Sokolowski and Tarlecki introduce a parameterization mechanism for *parametric algebras*, using a semantics similar to one given above.

4 Monomorphicity and Correctness

In this section we investigate the notions of monomorphicity and correctness, give a syntactic criterion how to detect momonorphicity from the axioms, and illustrate it by an example. We concentrate on parameterized specifications. For simple specifications we refer to [Wi 90], [Re 92].

4.1 Definition (Monomorphic specifications)

A parameterized specification PSPEC is *monomorphic* if any two functions $\mathcal{F}_1$ and $\mathcal{F}_2 \in$ SEM(PSPEC) are isomorphic. ■

The following analysis of monomorphic specifications is based on the fact that a PSPEC is monomorphic iff any two target models $\mathcal{A}$ and $\mathcal{B}$ with identical parameter reduct ($\mathcal{A}|_{SIG_p} = \mathcal{B}|_{SIG_p}$) are isomorphic. The latter criterion can be equivalently reformulated in terms of quasi-ground literals.

4.2 Definition (Quasi-ground literals)

Let SIG = ($S \cup S_p$, $OP \cup OP_p$) be a parameterized signature with parameter subsignature $SIG_p = (S_p, OP_p)$. Let X_p be a set of variables for S_p. *Quasi-ground literals* are formulas of the shape: $t_1 = t_2$, $\neg t_1 = t_2$, $q(t_1,..,t_n)$ and $\neg q(t_1,...,t_n)$, where $t_1, t_2,.., t_n \in T_{OP}(X_p)$ and q is a predicate symbol from OP. Non-negated literals are called *positive*. ■

The term *quasi-ground literals* stems from the fact that the terms occuring in the literals are ground with respect to the target sorts, but may contain variables of the parameter sorts. Henceforth, a term $t \in T_{OP}(X_p)$ is called a *quasi-ground* term.

4.3 Fact (Monomorphicity and quasi-ground literals)

PSPEC is monomorphic $\Leftrightarrow$

for all $\mathcal{A}, \mathcal{B} \in$ Tar(PSPEC) with $\mathcal{A}|_{SIG_p} = \mathcal{B}|_{SIG_p}$, all positive quasi-ground literals φ, and for all valuations $s: X_p \rightarrow \mathcal{A}|_{SIG_p}$, the following holds: $\mathcal{A}, s \models \varphi \Leftrightarrow \mathcal{B}, s \models \varphi$. ■

Roughly speaking, a specification is monomorphic if it determines the truth of the positive quasi-ground literals for target models with identical parameter reducts. Monomorphic specifications arising in practice, often meet an even stronger condition: They determine the truth of the positive quasi-ground literals for all models with identical parameter reducts, that satisfy the target specification and the structural induction principle. This situation is studied in [Re 91] and leads to the following criterion for monomorphic specifications.

4.4 Theorem (Sufficient criterion for monomorphicity)

Let PSPEC = ($SPEC_p$, SPEC) with signature SIG = $SIG_p \cup$ (S, OP). PSPEC is monomorphic, if for each positive quasi-ground literal $\varphi(x_1, .., x_n)$ there exists a formula $\varphi'(x_1, .., x_n) \in L(SIG_p, X_p)$ such that:

SPEC $\cup$ IND(SIG) $\models \varphi(x_1, .., x_n) \leftrightarrow \varphi'(x_1, .., x_n)$. ■

Now we turn to the correctness of specifications. A simple specification SPEC is *correct* for an algebra $\mathcal{A}$, if $\mathcal{A}$ is the only generated model of SPEC up to isomorphism. Correspondingly, PSPEC is correct for a function $\mathcal{F}$, if PSPEC describes $\mathcal{F}$ uniquely up to isomorphism.

4.5 Definition (Correctness of specifications)

A parameterized specification PSPEC = ($SPEC_p$, SPEC) is *correct* for $\mathcal{F} \in Alg(SIG_p) \to Alg(SIG)$, if $\mathcal{F} \in$ SEM(PSPEC) and PSPEC is monomorphic. ■

Note that if Tar(PSPEC) = ∅, PSPEC is correct for the undefined function. In order to prove that PSPEC is correct for $\mathcal{F} \in Alg(SIG_p) \to Alg(SIG)$, the following criterion based on 4.4 and 4.5 can be used.

4.6 Corollary (Correctness of specifications)

A parameterized specification PSPEC = ($SPEC_p$, SPEC) is correct for $\mathcal{F} \in Alg(SIG_p) \to Alg(SIG)$, if the following conditions hold:

1) for all $\mathcal{M} \in dom(\mathcal{F})$: $\mathcal{F}(\mathcal{M})$ is S_p-generated
2) for all $\mathcal{M} \in dom(\mathcal{F})$: $\mathcal{F}(\mathcal{M}) \models$ SPEC
3) $dom(\mathcal{F}) \supseteq Tar(PSPEC)|_{SIG_p}$
4) for all $\mathcal{M} \in dom(\mathcal{F})$: $\mathcal{F}(\mathcal{M})|_{SIG_p} \cong \mathcal{M}$
5) for all positive quasi-ground literals $\varphi(x_1, ..., x_n)$ there exists $\varphi'(x_1, ..., x_n) \in L(SIG_p, X_p)$ such that SPEC ∪ IND(SIG) $\models \varphi(x_1, ..., x_n) \leftrightarrow \varphi'(x_1, ..., x_n)$ ■

4.7 Example (Sets_over_Partial_order revisited)

With the help of 4.4 we show that Sets_over_Partial_order is monomorphic. To this purpose we classify the positive quasi ground literals: set equations $t = t'$, predicate expressions $x \in t$ and $t << t'$. The terms t and t' take the form insert(x_1, insert(x_2, ..insert(x_k, ∅)..)) and insert(y_1, insert(y_2, ..insert(y_u, ∅)..)), respectively, and x_i and y_i are variables of the sort data. If the quasi-ground literal φ is $x \in t$ then φ' is $\bigvee_{i=1..k} x = x_i$ (*false* if k = 0). If φ is the equation $t = t'$ then φ' is the formula $\forall z. ((\bigvee_{i=1..k} z = x_i) \leftrightarrow (\bigvee_{i=1..u} z = y_i))$, and if φ is $t << t'$, we set φ' $\forall z. ((\bigvee_{i=1..k} z = x_i) \rightarrow \exists z'. ((\bigvee_{i=1..u} z' = y_i) \wedge z < z'))$. In all cases SPEC ∪ IND(SIG) $\models \varphi \leftrightarrow \varphi'$ holds.

Furthermore, by inspection of the conditions 1) to 4) of 4.6, we see that Sets_over_Partial_order, is correct for the function *Set* with dom(*Set*) = Alg(Partial_order) and *Set*($\mathcal{M}$) := (Pot_ω(M), M, ∅, insert, ∈, <<, <) for all $\mathcal{M} \in$ dom(*Set*). Pot_ω(M) denotes the finite subsets of M (carrier of $\mathcal{M}$) and the operations have the standard meaning. ■

The criteria 4.4 and 4.6 are useful to prove monomorphicity or correctness of a given specification. Example 4.7 illustrates the general technique to prove that the equality and the predicates are fixed by the axioms with respect to the parameter. However, for large specifications, this proof is very extensive. The higher the number of symbols in the signature, the more quasi-ground literals exist. Furthermore, from a methodological point of view, it is unrealistic to develop large specifications as a monolithic block. What is needed are structuring operations enabling the incremental design of correct specifications from smaller ones, preserving the correctness of the constituent specifications. Two operations that are available in almost all specification languages (CLEAR [BG 80], ASL [SW 83], [Wi 86], ACT ONE [EM 85], Extended ML [ST 88], SPECTRUM [Br 91]) are the disjoint union of specifications, and the enrichment of a specification with new axioms for new symbols. A large correct specification can be developed by starting out with a small fragment, proving it correct using the above criteria, and then extending it incrementally to the intended specification. Due to space limitations, we refer to [Re 91] for the criteria for union and enrichment.

5 Parameter Passing

In this section we study two questions: First, the correctness of standard parameter passing for parameterized full first-order specifications, and, second, the interaction between parameter passing and the correctness of the involved specifications. With standard parameter passing we mean actualization with simple specifications ([EM 85]).

Roughly, the actualization of a parameterized data type $\mathcal{F}$ with an actual parameter $\mathcal{M}$ is the function application of $\mathcal{F}$ to $\mathcal{M}$. $\mathcal{M}$ needs not necessarily be formulated over the parameter signature of $\mathcal{F}$. It suffices if there is a signature morphism h between the parameter signature of $\mathcal{F}$ and the signature of $\mathcal{M}$. The result of the actualization of $\mathcal{F}$ by $\mathcal{M}$ and h is denoted by $\mathcal{F}_h(\mathcal{M})$. Syntactic actualization of PSPEC = ($SPEC_p$, SPEC) with an actual parameter specification SPEC' replaces $SPEC_p$ in SPEC by SPEC'. The connection between $SPEC_p$ and SPEC' is described by a specification morphism h. The resulting simple specification is denoted by $PSPEC_h$(SPEC').

Correctness of parameter passing is the following problem : Does $\mathcal{F}_h(\mathcal{M})$ satisfy $PSPEC_h$(SPEC') if $\mathcal{M}$ and $\mathcal{F}$ satisfy SPEC' and PSPEC respectively ? And conversely, can every $\mathcal{A}'$ satisfying $PSPEC_h$(SPEC') be obtained by actualization of a function $\mathcal{F}$ satisfying PSPEC with a parameter $\mathcal{M}$ satisfying SPEC'? If both answers are positive for all specification morphisms then parameter passing is correct for PSPEC. We provide a solution to this problem by a characterization of the specifications with correct parameter passing and give a syntactic criterion for this class.

The second problem solved in this section is the interaction between parameter passing and the correctness of the involved specifications: We present sufficient conditions under which for all specification morphisms h, $PSPEC_h$(SPEC') is correct for $\mathcal{F}_h(\mathcal{M})$, provided PSPEC and SPEC' are correct for $\mathcal{F}$ and $\mathcal{M}$, respectively .

5.1 Definition (Syntactic actualization)

Standard parameter passing requires three things: A parameterized specification PSPEC = ($SPEC_p$, SPEC) with signature SIG = ($S \cup S_p$, $OP \cup OP_p$), a simple specification SPEC' = (S', OP', T') with signature SIG', being the *actual parameter*, and a specification morphism h : $SPEC_p \rightarrow$ SPEC' which indicates which sorts and operations of $SPEC_p$ are replaced by which sorts and operations of SPEC'. Then the *value specification* $PSPEC_h$(SPEC') with signature ($S' \cup S$, $OP' \cup OP^*$) results from replacing SPEC' for $SPEC_p$ in SPEC, thereby replacing the symbols of $SPEC_p$ that also may occur in SPEC according to h. OP* contains the same symbols like OP, but the typing of the symbols may have changed according to h. In the literature this syntactic parameter passing is illustrated with the following (pushout) parameter passing diagram (see [EM 85] for formal details):

$$
\begin{array}{ccc}
SPEC_p & \xrightarrow{\;u\;} & SPEC \\
{\scriptstyle h}\downarrow & & \downarrow{\scriptstyle h_1} \\
SPEC' & \xrightarrow[\;u'\;]{} & ASPEC = PSPEC_h(SPEC')
\end{array}
$$

The specification morphism h is called the *passing morphism*, u and u' are the *formal* and *actual parameter inclusions*. h_1 is the *induced passing morphism* which acts like h for symbols from $SPEC_p$ and is the identity for the target symbols of SPEC. Additionally, it is required that S and S' are disjoint as well as OP and OP'. ■

In the sequel we assume such a standard parameter passing diagram and refer to the above notation. The semantic actualization is based on *amalgamated sums* : If $\mathcal{M}' \in$ Alg(SPEC') and $\mathcal{A} \in$ Alg(SPEC) and $V_h(\mathcal{M}') = \mathcal{A}|_{SIG_p} =: \mathcal{M}$, then the amalgamated sum $\mathcal{B} := \mathcal{M}' +_{\mathcal{M}} \mathcal{A}$ is uniquely defined in Alg(ASPEC) with $\mathcal{B}|_{SIG'} = \mathcal{M}'$ and $V_{h_1}(\mathcal{B}) = \mathcal{A}$.

5.2 Definition (Semantic actualization)

Let PSPEC = ($SPEC_p$, SPEC), SPEC' and the specification morphism h be given like above. For $\mathcal{F} \in$ SEM(PSPEC) and $\mathcal{M}' \in$ Alg(SPEC') with $\mathcal{M} := V_h(\mathcal{M}') \in dom(\mathcal{F})$ we define the *actualization* of $\mathcal{F}$ by $\mathcal{M}'$ and h: $\mathcal{F}_h(\mathcal{M}') := \mathcal{M}' +_{\mathcal{M}} (\mathcal{F}(\mathcal{M}) \nabla_{SIG_p} \mathcal{M})$. ■

Since $\mathcal{F} \in$ SEM(PSPEC) is persistent per definition, $\mathcal{F}(\mathcal{M})|_{SIG_p}$ and $\mathcal{M}$ are isomorphic, and hence $\mathcal{F}(\mathcal{M}) \nabla_{SIG_p} \mathcal{M}$ is always defined. Now we give a formal definition of the correctness of parameter passing.

5.3 Definition (Correctness of parameter passing)

We assume the following parameter passing diagram.

$$\begin{array}{ccc} SPEC_p & \xrightarrow{u} & SPEC \\ h \downarrow & & \downarrow h_1 \\ SPEC' & \xrightarrow[u']{} & ASPEC = PSPEC_h(SPEC') \end{array}$$

Standard parameter passing is *correct* if PSPEC is freely extendible and if for every passing morphism h: $SPEC_p \rightarrow SPEC'$ the following holds:

$$SEM(PSPEC_h(SPEC')) = \bigcup_{\mathcal{F} \in SEM(PSPEC)} \mathcal{F}_h(SEM(SPEC')) \qquad (*)$$

where $\mathcal{F}_h(SEM(SPEC')) = \{ \mathcal{F}_h(\mathcal{M}') : \mathcal{M}' \in SEM(SPEC') \text{ and } V_h(\mathcal{M}') \in dom(\mathcal{F}) \}$. ■

A careful look at the definitions reveals, that the right to left inclusion of the set equation (*) in 5.3 is generally true. However, the reverse does not hold. There exist freely extendible specifications that do not have correct parameter passing. Essentially, the reason why a parameterized specification PSPEC might fail to have correct parameter passing is the following phenomenon: It might happen that for a certain h: $SPEC_p \rightarrow SPEC'$ there exists a generated model $\mathcal{B}$ of ASPEC, the SIG' reduct of which is not generated (see [Re 91] for examples). Fortunately, this is actually the only phenomenon hindering the correctness of parameter passing. The following definition is designed to prevent it.

5.4 Definition (Strict persistency)

Let PSPEC be a parameterized specification with signature SIG = $SIG_p \cup (S, OP)$ and parameter signature $SIG_p = (S_p, OP_p)$. PSPEC is *strictly persistent*, if for all target models $\mathcal{A} \in Tar(PSPEC)$, for all $s \in S_p$, all terms $t(x_1, .., x_n) \in T_{OP \cup OP_p, s}(X_p)$ and all valuations $s : X_p \rightarrow \mathcal{A}$ there exists a term $t' \in T_{OP_p, s}(\{x_1, .., x_n\})$ with: $t_{s, \mathcal{A}} = t'_{s, \mathcal{A}}$. ■

Under weak restrictions of SIG_p, strict persistency plus free extendibility characterize the specifications with correct parameter passing.

5.5 Theorem (Characterization of correct parameter passing)

Let PSPEC = ($SPEC_p$, SPEC) be a parameterized specification with signature SIG = $SIG_p \cup (S, OP)$. The parameter signature $SIG_p = (S_p, OP_p)$ is required to provide at least one term $t \in T_{OP_p, s}$ for all $s \in S_p$. Then the following holds:

Standard parameter passing is *correct* for PSPEC iff PSPEC is *strictly persistent* and *freely extendible.* ■

5.6 Remark (Strict persistency and related work)

In theorem 5.5 the "if-part" of the equivalence is relevant for practical applications. However, the restriction on SIG_p in 5.5 is only needed for the "only-if-part". Therefore, the restriction is irrelevant for practical applications. It can be dropped completely, if SIG_p has only a single parameter sort.

In order to compare result 5.5 with the corresponding one for initial semantics ([EM 85]), we assume (E_p, E) to be a freely extendible, parameterized equational specification. Then all possibly non-isomorphic interpretations of SEM((E_p, E)) are defined for all parameter algebras and are persistent. In the initial approach, persistency of the free functor already implies the correctness of parameter passing, whereas we need the additional requirement that (E_p, E) is strictly persistent. This is because persistent free functors additionally guarantee, that the reduct of the image of a generated actual parameter algebra is always generated. This is exactly what strict persistency guarantees in our setting. The definitions of strict persistency and free extendibility have also some similarity to the definitions of *sufficient completeness* and *consistency* in [Ga 83] for equational specifications. Actually, the corresponding result for (the standard version of) parameter passing in [Ga 83] for equational specifications can be obtained as a corollary of ours. The approach in [NO 87] is based on functorial semantics and considers Horn specifications with boolean constraints. Here we consider loose semantics and full first-order logic.

The following syntactic criterion is useful to establish strict persistency. It states that PSPEC is strictly persistent if the target operations can be eliminated from terms of a parameter sort.

5.7 Theorem (Criterion for strict persistency)

PSPEC = $(SPEC_p, SPEC)$ with signature SIG = $(S \cup S_p, OP \cup OP_p)$ is *strictly persistent*, if for all $s \in S_p$, all terms $t(x_1, ..., x_n) \in T_{OP \cup OP_p, s}(X_p)$ there exist terms t_1', .., $t_k' \in T_{OP_p, s}(\{x_1, ..., x_n\})$ such that SPEC $\cup$ IND(SIG) $\models \bigvee_{i=1,..,k} t = t_i'$. ■

In 5.7 it is not necessary that SPEC $\cup$ IND(SIG) determines a single t' for t. It may leave unspecified which of the terms t_i' is equal to t. In our sample specification Sets_over_Partial_order, there are no terms of a parameter sort at all. Therefore, 5.7 is vacuously true, and hence Sets_over_Partial_order is strictly persistent.

Now we turn our attention to the interaction of parameter passing and the correctness of the involved specifications. Correct parameter passing, or equivalently strict persistency plus free extendibility, and correctness of a specification are independent notions: There are strictly persistent and freely extendible specifications that are not correct, and conversely, there exist freely extendible specifications that are correct for an interpretation but not strictly persistent. However, strict persistency and free extendibility are important to ensure that actualization preserves the correctness of the involved specifications.

5.8 Theorem (Actualization and correct specifications)

If PSPEC = ($SPEC_p$, SPEC) is strictly persistent and freely extendible, then for all actual parameter specifications SPEC', and all passing morphisms h: $SPEC_p \rightarrow$ SPEC' the following holds:

PSPEC correct for $\mathcal{F}$, and SPEC' correct for $\mathcal{M}'$ $\Rightarrow$ $PSPEC_h$(SPEC') correct for $\mathcal{F}_h(\mathcal{M}')$ ■

Theorem 5.8 suggests to look at strictly persistent and freely extendible specifications, in order to guarantee correctness preserving actualization. Strict persistency can be checked by criterion 5.7. A specification PSPEC can be shown to be freely extendible by giving a concrete function $\mathcal{F}$ with dom($\mathcal{F}$) = Alg($SPEC_p$) which satisfies PSPEC. This can be done with our sample specification Sets_over_Partial_order. Consequently, actualization of Sets_over_Partial_order is correctness preserving. Under certain restrictions there is also a syntactic criterion for free extendibility.

5.9 Theorem (Criterion for free extendibility)

PSPEC with signature SIG and parameter signature SIG_p is freely extendible if the following conditions hold:

1) PSPEC strictly persistent
2) SPEC $\subseteq \forall L$(SIG, X)
3) SIG provides a quasi-ground term for each target sort
4) for all $\varphi \in \forall L(SIG_p, X_p)$: SPEC $\vdash \varphi \Rightarrow SPEC_p \vdash \varphi$.

($\vdash$ denotes derivability in first-order logic) ■

References

[BDPPW 79] Broy, M., Dosch, H., Partsch, H., Pepper, P., Wirsing, M., Existential Quantifiers in Abstract Data Types, Proc. 6th ICALP, Graz 1979, Springer LNCS 71

[BG 80] Burstall, R.M., Goguen, J. A., The semantics of CLEAR, a specification language, in: D. Bjorner, ed., Proc. Advanced Course on Abstract Software Specifications, Springer LNCS 86, 1980

[Br 91] Broy, Facchi, Grosu, Hettler, Hussmann, Nazareth, Regensburger, Stolen, The Requirement and Design Specification Language SPECTRUM, technical report TUM-I9140, TU Munich, 1991

[CIP 87] CIP-Group, The Munich Project CIP, Vol. 2, The Program Transformation System CIP-S, Springer LNCS 292 (1987)

[Eh 82] Ehrich, H.-D., On the theory of specification, implementation, and parameterization, of abstract data types, JACM 29(1), 1982

[EKTWW 80] Ehrig, Kreowski, Thatcher, Wagner, Wright, Parameterized data types in algebraic specification languages, Proc. Int. Coll. on Automata, Languages and Programming, Springer LNCS 85, 1980

[EM 85] Ehrig, H., Mahr, B., Fundamentals of Algebraic Specification 1, Equations and Initial Semantics, EATCS Monographs on Theoretical Computer Science, Vol. 6, Springer 1985

[EM 90] Ehrig, H., Mahr, B., Fundamentals of Algebraic Specification 2, Module Specifications and Constraints, EATCS Monographs on Theoretical Computer Science, Vol. 21, Springer 1990

[Ga 83] Ganzinger, H., Parameterized Specifications: Parameter Passing and Implementation with Respect to Observability, ACM TOPLAS, Vol. 5, No. 3, July 1983

[Gau 92] Gaudel, M.-C., Structuring and Modularizing Algebraic Specifications: the PLUSS specification language, evolutions and perspectives, Finkel, Jantzen (eds.), STACS 92, Springer LNCS 577, 1992

[GGM 76] Giarratana, V., Gimona, F., Montanari, U., Observability Concepts in Abstract Data Type Specifications, 5th Symposium Math. Foundations of Computer Science (1976), Springer LNCS 45

[GTW 78] Goguen, J., Thatcher, J., Wagner, E., An Initial Algebra Approach to the Specification, Correctness and Implementation of Abstract Data Types, Current Trends in Programming Methodology IV, Yeh, R. (Ed.), Prentice-Hall, Englewood Cliffs, 1978

[Gu 76] Guttag, J.V., Abstract Data Types and The Development of Data Structures, Supplement to Proc. Conference on Data Abstraction, Definition, and Structure, SIGPLAN Notices 8 (1976)

[HR 80] Hornung, G., Raulefs, P., Terminal Algebra Semantics and Retractions for Abstract Data Types, DeBakker, van Leuwen, 7th ICALP, Springer LNCS 85, pp. 310-323 (1980)

[Ka 83] Kamin, S., Final Data Types and Their Specification, ACM TOPLAS 5,1 1983

[LZ 74] Liskov, B.H., Zilles, S.N., Programming with Abstract Data Types, SIGPLAN Notices 6 (1974)

[NO 87] Navarro M., Orejas F., Parameterized Horn Clause Specifications: Proof Theory and Correctness, TAPSOFT 87, Springer LNCS 249, 1987

[Pa 85] Padawitz, P., Parameter Preserving Data Type Specifications, J. of Computer and System Sciences 34

[Pa 90] Padawitz, P., Horn Logic and Rewriting for Functional and Logic Program Design, Universität Passau, Fakultät für Mathematik und Informatik, MIP-9002, März 1990

[Re 91] Reif, W., Correctness of Specifications and Generic Modules, Dissertation, Univ. of Karlsruhe, 1991

[Re 92] Reif, W., Correctness of Full First-order Specifications, Proc. International Conference on Software Engineering and Knowledge Engineering, Capri, Italy, IEEE press, 1992

[SST 92] Sanella, D., Sokolowski, S., Tarlecki, A., Toward formal development of programs from algebraic specifi cations: parametrisation revisited, report ECS-LFCS-92-222, Laboratory for Foundations of Computer Science, University of Edinburgh, 1992

[ST 88] Sanella, D.,, Tarlecki, A., Specifications in an arbitrary institution, Inform. a. Comp. 76, 1988

[SW 83] Sanella D., Wirsing M., A kernel language for algebraic specification and implementation, Proc. Int. Conf. on Foundations of Conputation Theory, Borgholm, Springer LNCS 158, 1983

[TWW 82] Thatcher, J., Wagner, E., Wright, J., Data Type Specification: Parametrization and the Power of Specification Techniques, ACM TOPLAS 4,4 1982

[Wa 79] Wand, M., Final Algebra Semantics and Data Type Extensions, Journal of Computer and System Sciences 19, 1, 1979

[WB 82] Wirsing, M., Broy, M., An Analysis of Semantic Models for Algebraic Specifications, Theoretical Foundations of Programming Methodology, Broy, Schmidt (eds.), Nato ASIS, C91, Reidel, 1982

[Wi 86] Wirsing, M., Structured algebraic specifications: a kernel language, Theoretical Computer Science 42, 1986

[Wi 90] Wirsing, M., Algebraic Specification, Handbook of Theoretical Computer Science, J. van Leeuwen (ed.), volume B, Elsevier, 1990

[WPPDB 83] Wirsing, M., Pepper, P., Partsch, H., Dosch, W., Broy, M., On Hierarchies of Abstract Data Types, Acta Informatica 20 (1983)

Algebraic Modelling of Imperative Languages with Pointers

A. V. Zamulin

Institute of Informatics Systems
Siberian Division of the Russian Academy of Sciences
630090, Novosibirsk, Russia
e-mail: zam@isi.itfs.nsk.su

Summary. A modelling technique for an imperative programming language is presented in the paper. The technique is based on the algebraic approach to the data type specification and introduces such basic data types as memory, variable, record, function, and procedure and such basic kinds of expressions as linear and conditional expressions. It is shown that statements are a partial case of expressions, pointers are a partial case of variables, and procedures are a partial case of functions. The technique is illustrated by a simple Pascal program including variables, pointers, and assignment and conditional statements.

1 Introduction

All the problems of program verification, transformation, partial evaluation, and abstract interpretation require a formal representation of a program as a mathematical entity. The task is easier with respect to functional languages based on the conventional mathematical notions. It is not so in the case of imperative languages. The reason is that the computation model of such a language is based on a notion of *memory*. It means that both input data and intermediate data of a program are stored in memory for some time and used when needed. Therefore, we need to formally specify the memory and associated objects and tie these notions in with other notions (statement, function, procedure, etc.) of an imperative programming language.

One of the most promising approaches to the definition of programming languages is an algebraic one which is studied for a quite a long time [Let68, GTY74, GTW78, BWP87]. It was successfully applied to functional languages [BW82]. The attempts taken in the algebraic modelling of imperative programming languages are, to our mind, far from success. It is simply proposed that "programming languages should be studied in terms of algebraic theories" in [GTW78], for example. A many-sorted algebra is used in [Let68] to construct a general theory of a formal language. A special case of the many-sorted algebra, a multialgebra, is used in [GTY74] for the formal definition of several specific data structures.

* Author's present address: School of Computer Science, University Sains Malaysia, 11800 Penang, Malaysia; e-mail: zam@cs.usm.my

A substantial effort to algebraically define an imperative programming language was made in [BWP87] where it is regarded as a hierarchy of abstract data types (ADT) constituting a theory of many-sorted algebras. Such data types, as "expression", "statement", "loop", etc., are introduced in the paper to represent the most popular syntactic constructs of an imperative language. The paper does not give, however, a specification of memory and related concepts.

An advanced work in this field is represented by [Wag84]. It gives the definitions of the most important data types of an imperative programming language in terms of category theory. However, this author, reasonably criticising "the data theorists" for introducing models which are very far from the usual programmers' reality, offers such tricky things which are also far away from the usual programming. The model becomes even more complex when pointers are introduced [Wag87].

A simple model disregarding pointers and locations was proposed in [Don77]. Unnamed locations are necessary however when complex objects with computed addresses of their components are considered. Such a computed address is often used as the left hand side of an assignment statement or procedure argument called by reference. Therefore, a model of an imperative programming language is needed which, being abstract as much as possible, could simulate the conventional language notions in terms close to the practice of a usual programmer. One of the main tasks of the model is incorporation of pointers and unnamed locations so that they become a natural constituent of a programming language. An attempt to define pointers by means of the algebra of partial maps is undertaken in [Moe93]. Another version of pointer definition is given in [BBB85] where they are associated with a special structure "plexus". Our approach differs from these two by considering pointers as a special case of variables, thus avoiding the necessity of using an extra algebra or an extra data structure.

The classical algebraic theory is always a triple "set of sort names, set of sort-indexed operators, set of axioms". In programming languages, this triple is usually divided into several groups each consisting of a sort name and corresponding operators and axioms. Such a group is called an *abstract data type (ADT)*. The theory is regarded then as a family of abstract data types. This approach is followed in this paper which is organised in the following way.

The definition conventions are discussed in Section 2. The major concepts of the paper, the memory and variable data types, are specified in Section 3; they are illustrated by a sample Pascal program. Several forms of expressions are introduced in Section 4. A record type is specified in Section 5 and a function type is specified in Section 6. Some conclusions are drawn in Section 7.

2 Definition conventions

The main algebraic notions are given in [BG81] in the following way.

Defn. A signature, Σ, is a pair, $< S, \Omega >$, where S is a set (of sorts) and Ω is a family of sets (of operators) indexed by $S^* \times S$.

Defn. Let Σ be a signature. A Σ-algebra A is an S-indexed family of sets, $|A|$, called the carrier of A, together with an $S^* \times S$-indexed family of functions $\alpha_{us} : \omega_{us} \longrightarrow (|A|_u \longrightarrow |A|_s)$ where $\omega_{us} \in \Omega, u \in S^*, s \in S$, and $|A|_{u1...un} = |A|_{u1} \times ... \times |A|_{un}$.

Defn. A presentation is a pair $< \Sigma, E >$, where Σ is a signature and E is a set of equations.

Defn. A theory is a presentation $< \Sigma, E >$, such that E is closed (under usual derivation rules for equations).

According to these definitions, a theory is a set of sort names, a set of indexed operators, and a set of equations describing the functionality and semantics of the introduced operations. All these components (with natural language sentences defining the semantics of operations, instead of equations) can be found in a programming language. However, modern programming languages usually use a *data type* notion, meaning by this a set of values *together* with a set of operations manipulating these values. Therefore, the overall signature of the language is subdivided into groups containing one sort per group and a set of corresponding operators. Moreover, the modern languages permit the definition and subsequent instantiation of *polymorphic* or *generic data types* each being a kind of *type constructor* using one or more *type parameters* (set type, array type, etc.). *Sort constructors* were introduced in [BFG91] to model this language feature. This idea can be used to develop the type constructors which are more appropriate in the programming language.

In addition to the operations defined in data types (*basic operations*), a language can have a number of "detached" operations which can be expressed by basic operations. Both the set of detached operations and set of data type operations can have *generic operations*, each being supplied with a type parameter defining the type of some operands and/or result of the operation.

To build a formal model of a modern programming language, we need to modify the above definitions, taking into account these new language features. Note that after we have decided to create groups of operations building a data type, we do not need use the name of the corresponding sort inside the group and can invent a special symbol, for example "*", meaning "myself". A special function is used then to bind a type name with a group. To make the definitions simpler, we consider in this paper that each generic construction has only one type parameter. The reader is encouraged to extend the definitions for the case of arbitrary number of type parameters.

Defn. Assume that

S is a set of names (of concrete data types);
R is a set of names (of generic data types);
TT is a set of the following type terms:

1. $s \in S$ is a type term,
2. if T is a type term and $r \in R$, then $r(T)$ is a type term;

OT is a set of the following operation types:

1. a type term is an operation type,
2. if $T_1, ..., T_n, T_{n+1}$ are type terms, then $T_1, ..., T_n \longrightarrow T_{n+1}$ is an operation type;

D is a set of (detached) operators indexed by operation types from OT;

CS is a family of sets of (concrete data type) operators indexed by operation types from OT', where OT' is a set of operation types constructed as above by extending S by a distinguished name "*", and "*" occurs at least once in each operation type $oT \in OT'$; a member of CS is called a *concrete data type signature*;

GS is a family of pairs $< p, G >$, where p is a name (of a type parameter) and G is a set of (generic data type) operators indexed by operation types from OT'', where OT'' is a set of operation types constructed as above by extending S by a distinguished name "*" and p, and "*" occurs at least once in each $oT \in OT''$; a pair $< p, G >$ is called a *generic data type signature*.

Then a polymorphic signature ΣP is a quintuple $< S, R, CS, GS, D >$.
Notation. If op is an operator and oT is an operation type, then $op : oT$ is an operator indexed by oT.

The polymorphic signature as defined above does not bind names from S and R to data type signatures and therefore does not provide for a correspondence between data type signatures and type terms met in operation types. To meet this purpose, we introduce a notion of *signature expansion*.

Defn. Let ΣP be a polymorphic signature as defined above. A ΣP signature expansion is triplet of functions $< def^s, def^r, inst^r >$, where

$def^s : S \longrightarrow CS$ is a function binding each $s \in S$ to a data type signature from CS (concrete data type definition);

$def^r : R \longrightarrow GS$ is a function binding each $r \in R$ to a data type signature from GS (generic data type definition);

$inst^r : GS \times TT \longrightarrow CS$ is an instantiation function mapping a generic data type signature $< p, G > \in GS$ into a concrete data type signature C by replacing the type parameter p by the type term T in each $(op : oT) \in G$;

After these bindings are done, each type term T, where $T \in S$, denotes a signature resulting from $def^s(T)$ and each type term $r(T)$ denotes a signature resulting from $inst^r(def^r(r), T)$. In this paper, talking of a polymorphic signature, we mean a signature expansion as well. We can now introduce a notion of ΣP-algebra.

Defn. Let ΣP be a polymorphic signature as defined above. A ΣP-algebra A is a triplet of functions, $< map^c, map^g, map^d >$, such that:

map^c is a function binding each concrete data type signature $C \in CS$ to a *data type implementation*, Cim, consisting of a set of elements, denoted by

$|*|$ and called the *carrier* of the type in question, an element of the set $|*|$ per each nullary operator of this signature, and an implementation function $|op^c| : |T_1| \times ... \times |T_n| \longrightarrow |T_{n+1}|$, where $|T_i|, i = 1, ..., n+1$, is the carrier of the type denoted by the type term T_i, per each n-nary operator $op^c : T_1, ..., T_n \longrightarrow T_{n+1}$ of this signature; the set of all possible data type implementations is denoted by CIM, a function using CIM as its domain is called a *generic* function; we say that a data type implementation Cim corresponds to a data type signature C if it provides an implementation function for each operator from C;

map^g is a function binding each generic data type signature $< p, G > \in GS$ to a generic function $gim^g : CIM \longrightarrow CIM$, where CIM is the set of data type implementations as defined above, such that an instantiation $gim^g(Cim^T)$, where Cim^T is an implementation of a data type T, gives a data type implementation corresponding to the data type signature resulting from the instantiation $inst^r(< p, G >, T)$;

map^d is a function mapping each $(op^d : T) \in D$, where T is a type term, into an element of the set $|T|$ and each $(op^d : T_1, ..., T_n \longrightarrow T_{n+1}) \in D$ into an implementation function $|op^d| : |T_1| \times ... \times |T_n| \longrightarrow |T_{n+1}|$, where $|T_i|, i = 1, ..., n+1$, is a set of values of the type T_i.

Let us now allow the signature to have a set of pairs $< q, op^q >$ in both the detached operation specification D and each data type signature from CS, where q is a name (of a type parameter) and op^q is a (generic) operator indexed by an operation type $oT^q \in OT'''$, where OT''' is respectively an extension of either OT or OT' provided that q is a type term. Then we add:

to the signature expansion, an instantiation function $inst^q$, producing a concrete operator indexed by an operation type created by replacing the type parameter q in oT^q by the type term T;

to an algebra, a function map^q which binds each $< q, op^q >$ to a generic implementation function $gop^q : CIM \longrightarrow IMPL$, where $IMPL$ is the set of implementation functions as defined above, such that $gop^q(Cim^T)$, where Cim^T is an implementation of a data type T, gives an implementation function for the operator produced by the instantiation $inst^q(< q, op^q >, T)$.

Notation: an instantiation $inst^q(< q, op^q >, T)$ is usually denoted by $op^q(T)$.

For a given signature ΣP, any number of algebras implementing its sorts and operators can be constructed. One special algebra, $W_{\Sigma P}(X)$, called the *word algebra*, is distinguished among them. Let X_T be a set of variables of type T (a name which can be associated with any value of the given type is meant here by a variable). Then the elements of this algebra, which are called *terms* or *expressions*, are the following:

a nullary operator of type T is a term of type T;
a variable $x \in X_T$ is a term of type T;
for an operator $op : T_1, ..., T_n \longrightarrow T$ and all terms $t_1, ..., t_n$ of types $T_1, ..., T_n$ respectively, $op(t_1, ..., t_n)$ is a term of type T,

for an instantiated operator $op^q(T_p) : T_1, ..., T_n \longrightarrow T$ and all terms $t_1, ..., t_n$ of types $T_1, ..., T_n$ respectively, $op^q(T_p)(t_1, ..., t_n)$ is a term of type T.

Notation. Instead of $op^q(T_p)(t_1, ..., t_n)$, we write in this paper $op^q(T_p, t_1, ..., t_n)$ or simply $op^q(t_1, ..., t_n)$ when it seems appropriate.

A term without variables is called a *ground term*. We denote the set of all terms of type T by $|W(X)|_T$ and the set of ground terms of type T by $|W|_T$. Taking into account the uniqueness of the word algebra, we mean a value of type T to be an element of the set $|T|$ of a particular algebra of the given signature and we mean a value of type **term** T to be a corresponding term of the word algebra. So, for an argument of type **term** T, a value from the set $|W(X)|_T$ instead of a value from the set $|T|$ is used in any algebra of the signature.

For any algebra A of the given signature, there exist an evaluation function $eval = |W(X)|_T \longrightarrow |T|$, defined in the following way:

for a variable $x \in X_T$ and binding function $v : X_T \longrightarrow |W|_T$, $eval(x) = v(x)$;
for a nullary operator c, $eval(c) = map(c)$;
for an operator $op : T_1, ..., T_n \longrightarrow T$,
$eval(op(t_1, ..., t_n)) = map(op)(eval(t_1), ..., eval(t_n))$.
for an instantiated operator $op^q(T_p) : T_1, ..., T_n \longrightarrow T$,
$eval(op^q(T_p)(t_1, ..., t_n)) = map(op^q(T_p))(eval(t_1), ..., eval(t_n))$.

The function *map* here is either map^c or map^d, depending on the set the operator belongs to.

Notation. The functions *eval* and *map* are usually used implicitly, i.e. we simply write $op(t_1, ..., t_n)$ instead of $map(op)(eval(t_1), ..., eval(t_n))$. The function *eval* is explicitly used only in respect to explicit term arguments.

Note that the argument of type **term** T is evaluated not *before* but *after* its substitution when it becomes an object of computation. This method of parameter substitution corresponds to *name substitution* in Algol 60 [Nau63]; it is also called *normal-order evaluation* [Wat90].

Let ΣP be a polymorphic signature. Let us associate sets of equations E^c, E^g, and E^d with each $C \in CS, G \in GS$, and D respectively, so that each equation in E^c has at least one operator from C, each equation in E^g has at least one operator from G, and each equation in E^d has at least one operator from D. Then we get a presentation. A pair $< C, E^c >$ is usually called a *concrete data type specification*, a pair $< G, E^g >$ is called a *generic data type specification*, and a pair $< D, E^d >$ is called a *detached operation specification*. We usually say "a theory" when actually we have a presentation.

Thus, a theory is generally a set of data type specifications and a detached operation specification. Note that at the level of presentation the functions def^s and def^r bind names to data type specifications and the function $inst^r$ instantiate the corresponding equations in addition to the data type signature.

We consider that the language theory contains the data type Boolean with two truth values *true* and *false*. Therefore we can introduce a type equivalence operation

$eqv : TT \times TT \longrightarrow Boolean$ with the following specification:

$$\{\forall s_1, s_2 : S, r : R, T_1, T_2, T_3 : TT,$$
$$T_1 \ eqv \ T_1 = true;$$
$$T_1 \ eqv \ T_2 = T_2 \ eqv \ T_1;$$
$$T_1 \ eqv \ T_2 \ and \ T_2 \ eqv \ T_3 = T_1 \ eqv \ T_3;$$
$$s_1 \ eqv \ r(T_1) \ and \ s_2 \ eqv \ r(T_2) = s_1 \ eqv \ s_2 \ and \ T_1 \ eqv \ T_2\}$$

This type equivalence scheme is known in programming languages under the name *structural type equivalence*.

The theory is introduced in this paper by the word **theory**, the set S is introduced by the word **ctypes**, and the set R is introduced by the word **gtypes**. We may then assume the following predefinitions to be given:

theory imperative_language =
  ctypes {Boolean, nat, memory, ...};
  gtypes {var, const, func, ...};

Each concrete data type specification is introduced by the **module** operation giving a set of operators (in square brackets) and a set of axioms (in curly brackets). A generic data type specification is introduced by the **gen** operation consisting of the **module** operation with type parameters. Both def^s and def^r are denoted by the word **type**. Type parameters of a generic operation in a concrete data type are indexed by the word TYPE and enclosed in parentheses. The axioms are defined as quantified equations according to [WPP83]. However, since the developed theory adopts conditional terms, some axioms have the form of conditional expressions. Type arguments are not indicated in axioms when they seem unnecessary. Each data type is supplied with a special value *error* which indicates an exception. A universal equality predicate denoted by "=" is considered avalaible for all types.

We regard a data type specification as an abstract data type in the sense that any implementation satisfying the equations is allowed. This is known as *loose semantics* [BBB85]. According to this, a specification methodology known as *behavioural approach*, first advocated in [GH78], is used. In this case, the set of data type operations is divided into two groups: *constructors* and *observers*. The constructor yields a value of the specified data type and the observer yields a value of another data type. The minimal set of constructors which can be used to create any value of the specified type is called *basic constructors*, the others are called *secondary constructors*. So, each observer in a data type specification is applied to each value generated by basic constructors, and each secondary constructor is defined in terms of basic constructors. Moreover, since our model allows procedure operations (ie. operations dealing with memory), each of such operations is defined in terms of normal function operations.

3 Memory and variables

Let us consider a simple Pascal program:

```
begin
    var a,b : integer;
    var p, q : ↑ integer;
    new(p);
    new(q);
    a := 1;
    p↑ := a;
    q↑ := p↑;
    b := q↑;
    if p = q then print(1) else print(2);
    if p↑ = q↑ then print(2) else print(2)
end.
```

Four variables are declared in the program, two of them are integer variables and the other two are pointer variables. Two memories are used in this program, one for storing static variables, *a, b, p, q*, and one for storing dynamic variables referenced by pointers *p* and *q*. We have also several statements which will be discussed in the sequel.

So, we need a notion of *memory* consisting of *locations* marked by unique identifiers,say, *addresses*. Some addresses can be bound to identifiers, others can be unnamed. Each location can hold a constant or variable value of a definite type. Accordingly, two kinds of addresses (locations) can be imagined: *constant* and *variable* ones defined by *constant types* and *variable types* respectively. We say "a constant (variable) address of type T" if its location holds a constant (variable) value of type T. We say that the memory type defines a set of values which are called *memory states*. Each memory operation uses one memory state and produces another state.

Generally, it is possible to imagine a computation model which permits the naming of each memory state and its explicit use in computations. One example is taking backup memory copies in certain checkpoints of a program or database transaction and subsequent usage in the case of program failure. However, imperative languages use a model which assumes that a new memory state always replaces the previous one in all memory producing operations. For this reason, the memory argument is usually substituted implicitly in the corresponding operations. When several coexisting memories are possible in a program, special conventions on their usage are elaborated.

To specify the memory type, we firstly need operations constructing memory from separate locations and estimating the number of them. For simplicity, we consider the memory consisting only of variable locations and do not regard constant locations whose specification is very much like that of the variables.

type memory = **module**
    [create_mem: *;
                    — *creates an empty memory containing no locations*
      add_var: (T : TYPE), *, T ⟶ *;
                    — *constructs a new memory by adding a new variable*

```
            — location, with a value in it, to the existing memory
    delete_last: * ⟶ *; — deletes the last address in the memory
    numb_of_addr : * ⟶ nat
            — gives the number of created addresses]
  {∀ T: TYPE, x: T, m: *,
    numb_of_addr(create_mem) = 0;
    numb_of_addr(add_var(m, x)) = numb_of_addr(m) + 1
    delete_last(create_mem) = error;
    delete_last(add_var(m, x)) = m};
```

The axioms devide the memory operations into one basic constructor, one secondary constructor, and one observer; they show that the operation *add_var* increases the number of memory locations and *delete_last* deletes the last created address. This specification is completed by a specification of variables.

A *variable type* defines a set of location addresses capable to hold mutable values of a definite type. The pair "variable address, mutable value" is usually called a *variable object* or, simply, *variable*. All variable types have the same behaviour and therefore a generic variable type can be specified.

```
type var = gen(value_type: TYPE) module
  [ nil: *; — a distinguished address value
    create_var: memory ⟶ *; — creates a new variable address
    var_value: memory, * ⟶ value_type
            — fetches the value of a variable]
  {∀ m: memory, a: value_type, vad: *,
    var_value(m, nil) = error;
    var_value(m, create_var(m)) = error;
            — the address is not yet allocated in the memory
    var_value(create_mem, vad) = error; — the memory is empty
    var_value(add_var (m, a), vad) =
        if vad "=" create_var(m) — if this is a proper address
        then a — fetch the value
        else var_value(m, vad)
            — otherwise go to the previous memory}
```

Having definied the memory and variable types, we can specify a variable update operation:

```
[mod_var: (T: TYPE), memory, var(T), T ⟶ memory]
{∀ m: memory, T: TYPE, vad: var(T), a, a1: T,
  mod_var(create_mem, vad, a) = error;
  mod_var(add_var(m, a), vad, a1) =
      if vad "=" create_var(m) — the needed address is found
      then add_var(m, a1) — replace the value
      else add_var(mod_var(m, vad, a1), a)
          — update in the previous memory};
```

The variable declarations of our program

var a,b : integer;

can now be considered as a shorthand notation for

var a: integer := 0, b: integer := 0;

invoking the following operations for each of the variables:

1. creation of a new variable address in the current memory,
2. creation of a new memory state by attaching a new variable address, with an associated value, to the current memory.

Let us denote the subsequent states of the static memory by *sm1, sm2,* ...; the subsequent states of the dynamic memory by *dm1, dm2,* ... ; and the locations in the dynamic memory by *rp* and *rq*. Then we have for our program (it is assumed that an integer variable is initialised with 0):

a = create_var(integer, create_mem);
sm1 = add_var(integer, create_mem, 0);
b = create_var(integer, sm1);
sm2 = add_var(integer, sm1, 0);

Later in the program, each time when a variable identifier is indicated, the denoted address is used and the dereferencing operation *var_value* is implicitly applied if needed.

A variable declaration

var prtv: var(T) or
var prtv: ↑T, in another notation,

introduces a variable *prtv* which can accept variable addresses as its values. Such a variable is usually called a *pointer* and its type is called a *pointer type*. We can call it a *pointer to a variable* and its type a *variable pointer type* respectively. Thus, the variable types of the presented model serve as pointer types as well. So, the declarations

var p, q : ↑ integer;

can be represented in our model by the following (it is assumed a pointer variable is initialised with "nil"):

p = create_var(var(integer), sm2);
sm3 = add_var(var(integer), sm2, nil);
q = create_var(var(integer), sm3);
sm4 = add_var(var(integer), sm3, nil);

The operation *create_var* is used implicitly when a static variable is declared. However, it is used in Pascal explicitly, as a procedure *new*, when a dynamic variable is created. So, the following lines of our program:

new(p);
new(q);

are represented by:

rp = create_var(integer, create_mem);
  — *an integer location in the dynamic memory*
dm1 = add_var(integer, create_mem, rp);
sm5 = mod_var(var(integer), sm4, p, rp);
  — *a value of the pointer p is established*
rq = create_var(integer, dm1);
  — *an integer location in the dynamic memory*
dm2 = add_var(integer, dm1, rq);
sm6 = mod_var(var(integer), sm5, q, rq)
  — *a value of the pointer q is established*

In this way static variables, dynamic variables, and pointers are created in the model of a Pascal program.

The operations deleting locations in a heap-like memory (garbage collectors) can also be specified. Let us firstly introduce an operation deleting a location without updating the values of possible pointers.

[delete_simple: (T: TYPE), memory, var(T) ⟶ memory]
{∀ T: TYPE, a: T, vad: var(T), m: memory,
  delete_simple(create_mem, vad) = create_mem;
  delete_simple(add_var(m, a), vad) =
    if vad "=" create_var(T, m); — *the last address in memory*
    then m — *the current memory is the result*
    else add_var(delete_simple (m, vad), a)}
      — *delete from the previous memory.*

We now introduce two operations for comparing and substracting of addresses.

[" > ": (T,T1: TYPE), var(T), var(T1) ⟶ Boolean;
  " − ": (T, T1: TYPE), var(T), var(T1) ⟶ var(T)]
{∀ T, T1: TYPE, vad: var(T), m, m1: memory,
  nil > vad = false;
  create_var (T, m) > nil = true;
  create_var (T, m) > create_var(T1, m1) =
    number_of_addr(m) > number_of_addr(m1);
  nil − vad = nil;
  create_var(T, m) − vad = create_var(T, delete_simple(m, vad))
    — *delete the substracted address from the current memory*
    — *and create an address in the new memory*}

And finally a general operation updating the values of possible pointers is specified.

[delete_var: (T: TYPE), memory, var(T) ⟶ memory]

```
{∀ T,T1: TYPE, a: T1, vad: var(T), m: memory, ∃ T2: TYPE,
  delete_var(create_mem, vad) = create_mem;
  delete_var(add_var(T1, m, a), vad) =
        if vad "=" create_var(m) — the last address in memory
        then m
        else if T1 eqv var(T2) — if the last location is a pointer location
             then if a "=" vad — the location points
                        — to the address to be deleted
                  then error — a dangling pointer
                  else if a > vad
                        — the pointed address is greater then "vad"
                        then add_var(delete_var(m, vad), a−vad)
                             — the pointer value is updated
                        else add_var(delete_var(m, vad), a)
                             — the pointer value is not updated
             else add_var(delete_var(m, vad), a)}
```

4 Expressions and statements

The form of terms introduced in Section 2 is the most general one for it does not involve the peculiarities of particular theories and constituting data types. A term of this form could be called *linear* because each of its operators is always evaluated once and only once. Since our theory contains the Boolean type, we can include the following generic operation into it:

```
[ if: (T : TYPE), Boolean, term T, term T ⟶ T]
{∀ T: TYPE, t1, t2: term T,
  if(true, t1, t2) = eval(t1);
  if(false, t1, t2) = eval(t2)};
```

that permits the construction of *conditional terms*. Using a conventional notation, it is possible to give the following definition of such a term:

> if *p* is a term of type Boolean and *t1, t2* are terms of type *s*, then *if p then t1 else t2* is a term of type *s*.

Terms of type memory are called *statements*. One of the most popular of them in programming languages is an *assignment statement* of the form

v := e,

where *e* is an expression and *v* is a variable. In our model, this statement is regarded as a syntactic form for the memory type operation

mod_var (T, m, v, e),

where *T* is the type of *e*, and *m* is the memory state after the evaluation of *v* and *e*. So, for our program, the statement:

a := 1;

stands for:

sm7 = mod_var(integer, sm6, a, 1);

The operation *var_value* is used in programs both implicitly (a variable value is fetched) and explicitly (a pointer is dereferenced). Therefore the following statements of our program:

```
p↑ := a;
q↑ := p↑;
b := q↑;
```

can be represented in our model by:

```
dm3 = mod_var(integer, dm2, var_value(var(integer), sm7, p),
                    var_value(integer, sm7, a));
    — a general rule in Pascal: each variable in the right hand side of
    — the assignment statement is always dereferenced once implicitly;
    — in the left hand side only explicit dereferencings are counted;
    — thus, p↑ := a has been translated;
dm4 = mod_var(integer, dm3, var_value( var(integer), sm7, q),
                    var_value(integer, dm3, var_value(var(integer), sm7, p)));
    — the statement q↑ := p↑ has been translated;
    — note one dereferencing for q and two dereferencings for p;
sm8 = mod_var(integer, sm7, b, var_value(integer, dm4,
                    var_value(var(integer), sm7, q)));
    — the statement b := q↑ has been translated;
```

Conditional expressions were already used in the above specifications. Those of them which produce a result of type *memory* are *conditional statements.* We can now represent in our model the last two statements of the sample program:

```
if(var_value(var(integer), sm8, p) = var_value(var(integer), sm8, q),
   print(1), print(2)); — Pascal always dereferences once the both
         — sides of equality; in this case the value 2 will be printed
         — because p and q point to different locations
if(var_value(integer, dm4, var_value(var(integer), sm8, p)) =
                    var_value(integer, dm4, var_value(var(integer), sm8, q)),
   print(1), print(2));
         — this time 1 will be printed because the referenced variables
         — have the same value.
```

Let now *s1, s2, ... , sn* be statements, and *ts* be a term of type *s* , then the term *ts(sn(... s2(s1(m)))*, where *m* is an initial memory state, is called a *compound expression* and have a special notation, for example:

begin s1; s2; ... ; sn; ts **end**;

Now, if *def1, def2, ... , defm* are constant and variable definitions, a term

ts(sn(... s2(s1(defm(... (def2(def1(m))))))))

can be called a *block expression* and have a special notation, for example:

block def1; def2; ... ; defm **begin** s1; s2; ... ; sn; ts **end**.

In the cases when *ts* is a statement, we have a *compound statement* and *block* accordingly. However we should note that block and compound expressions always implicitly produce a new memory which is used when needed.

5 Record types

Let us apply the developed technique to specify a *record type*. When a concrete record type is defined, such as

record p1 : t1; p2 : t2; . . . ; pn : tn **end**,

the following data type specification is constructed:

```
module [ create_rec : t1, t2, . . . , tn ⟶ * ;
         p1 : * ⟶ t1;
         updatep1 : *, t1 ⟶ * ;
         p1 : var(*) ⟶ var(t1);
         p2 : * ⟶ t2;
         updatep2 : *, t2 ⟶ * ;
         p2 : var(*) ⟶ var(t2);
                  .
                  .
         pn : * ⟶ tn;
         updatepn : *, tn ⟶ * ;
         pn : var(*) ⟶ var(tn)]
        {∀ i: 1 . . n, ai, bi: ti, r: *, rv: var(*), m: memory,
         pi(create_rec(a1, a2, . . . , an)) = ai;
         updatepi(create_rec(a1, a2, ... , ai, ..., an), bi) =
                  create_rec(a1, a2, ... , bi, ...an);
         var_value(m, pi(rv)) = pi(var_value(m, rv));
         mod_var(m, pi(rv), bi) =
                  mod_var(m, rv, updatepi(var_val(m, rv), bi))}.
```

The operation *create_rec* creates a new record as a n-tuple of component (field) values. Three operations exist for each declared field, one is for fetching a field's value, the other is for producing a new record by modifying a field's value and the third one is for producing a field variable for a record variable (the overloading of operation names is allowed here). Note that the availability of a field updating operation gives us a possibility to create complex mutable objects. However, it is possible to imagine a constant field in a record type (see Napier 88 [MBC89] as an example). Such a field is then not supplied with the updating operation.

The axioms define formally the above semantics. The second axiom shows that any updating operation can be expressed by a corresponding creating operation and therefore any complex updating term can be reduced to a simple record creation operation. Following the last axiom, an assignment statement

pi(rv) := a or rv.pi := a in the conventional notation

is equivalent to the statement

rv := updatepi (rv, a)

and this gives us a right to use field variables as arguments for function (procedure) parameters called by reference.

6 Function types

The usage of variable addresses in terms makes the evaluation of the terms dependent on the values associated with these addresses. Therefore, the set of terms of type T containing variable addresses of types $T1, T2, ..., Tn$ could be considered as a set of functions mapping $T1, T2, ..., Tn$ into T. Since all these functions have the same domain and codomain, we can consider them as values of a same function type. Then the specification of a function type:

func(T1, T2, ..., Tn)T

can be represented in the following way:

module [create_func: var(T1), var(T2), ... var(Tn), **term** T ⟶ * ;
apply : *, T1, T2, ..., Tn ⟶ T];
{∀ p1: var(T1), p2: var(T2), ..., pn: var(Tn),
t: **term** T, a1: T1, a2: T2, ..., an: Tn,
apply(create_func(p1, p2, ... , pn, t), a1, a2, ... , an) =
block var p1: T1 := a1, ..., pn: Tn := an
begin t **end**}.

The operation *create_func* creates a function from a sequence of addresses usually represented by their identifiers (formal parameter names), and a term (function body) of a result type. The operation *apply* applies a function to a set of arguments and produces a result of the indicated type. In programming languages the definition of a function type and the operation *create_func* are usually combined in a so called *function denotation*, which might be defined as a functional term in the following way:

if $p1, p2, ..., pn$ are identifiers, $T1, T2, ..., Tn, T$ are data types,
and t is a term of type T, then
func(p1: T1, p2: T2, ... , pn: Tn) T :: t
is a term of type $T1, T2, ..., Tn \longrightarrow T$.

The operation *apply* is always used in programming languages implicitly in the construction *function call* which consists of a name of a function followed by a sequence of arguments in parentheses. A function producing a result of type "memory" is called a *procedure* and denoted in a special way, for example:

proc(p1: T1, p2: T2, ... , pn: Tn) :: s

where s is a statement. The operations of procedure creation and application are called *procedure denotation* and *procedure call* respectively. Note that a result of type "memory" can be implicitly produced by a function whose body is a block or compound expression; in this way functions with side effects are possible.

The model introduces formally only one kind of parameter substitution, *call by value*. However, variable type parameters accept addresses as their values, and therefore they model *call by reference* substitution. Functions as parameters give another kind of parameter substitution.

Example. Suppose the following record variable and procedure are defined in a Pascal program:

var r: record a, b: integer end;
proc aproc (var p1: integer; p2: integer); begin p1 := p2 end;

The procedure will be represented in the following way in our model:

aproc = **proc**(p1: var(integer); p2: integer) :: p1↑ := p2↑;

(note that variable parameters in Pascal are actually pointers dereferenced implicitly in the procedure body, in a formal model they are dereferenced explicitly; here and below we use the symbol "↑" for the operation "var_value"). The following procedure call is possible:

aproc(r.a, r.b)

which will be represented by the following simplified expression in the model:

apply(aproc, a(r), b(r)↑) = — *note that we do not need*
— *dereference a(r) because it has a proper type*
apply(create(p1, p2, (p1↑ := p2↑)), a(r), b(r)↑) =
block var p1: var(integer) := a(r), p2: integer := b(r)↑
begin p1↑ := p2↑ **end** =
a(r) := b(r)↑ =
r := updatea(r, b(r)↑).

Note that the availability of a field update operation in the record type gives us a possibility to substitute record fields for variable type parameters.

7 Conclusion

An algebraic specification technique is used to introduce a formal model of an imperative programming language which incorporates such basic constructions as memory, variables, pointers, expressions, statements, functions, and procedures. These are represented either as abstract data types with all the necessary operations or as high level term composition forms. It is shown that statements are a special case of expressions, pointers are a special case of variables, and procedures are a special case of functions.

An imperative program represented as a word algebra term can be formally analised, transformed into another term, reduced to a shorter term when part of input data is supplied, and fully evaluated when all input data are supplied. The model can serve as a basis of a database specification language requiring the notions of memory and persistent variable. It can also serve as a high-level denotation [Sch86] of an imperative programming language.

8 Acknowledgements

I should like to thank Gregory Kucherov and Bernhard Moeller for suggestions and comments on earlier drafts of the paper, and Kasem Lellahi, Nicolas Spyratos, and Stefan Sokolowski for the stimulating discussions.

References

[BBB85] F.L. Bauer, R. Berghammer, M. Broy, W. Dosch, F. Geiselbrechtinger, R. Gnatz, E. Hangel, W. Hesse, B. Krieg-Brückner, A. Laut, T.A. Matzner, B. Möller, F. Nickl, H. Partsch, P. Pepper, K. Samelson, M. Wirsing, H. Wössner. The Munich project CIP. Volume I: The wide spectrum language CIP-L. Lecture Notes in Computer Science, v.183, Berlin: Springer 1985

[BW82] Broy M. and Wirsing M. Algebraic definition of functional programming languages. RAIRO Inf. Theor., v.17, 1982, No.2, p. 137-162.

[BWP87] Broy M. and Wirsing M. and Pepper P. On the algebraic definition of programming languages. ACM TOPLAS, v.9, 1987, No.1, p.54-99.

[BFG91] Broy M., Facchi C., Grosu R., ea. The Requirement and Design Specification, Language Spectrum, An Informal Introduction, Version 0.3. Technische Universitaet Muenchen, Institut fur Informatik, October 1991.

[BG81] Burstall R.M., Goguen J.A. Algebras, Theories, and Freeness: An Introduction for Computer Scientists // Proc. Marktoberdorf Summer School on Theoretical Foundations of Programming Methodology, August, 1981.

[Don77] Donahue J.E. Locations considered unnecessary. Acta Informatica, v.8, pp. 221-242 (1977).

[GTY74] Glushkov V.M., Tseitlin G. E., Yutshenko E.L. Algebra, Languages, Programming. Kiev, Naukova dumka, 1974.

[GTW78] Gougen J.E., Thatcher J.W. , and Wagner E. G. An initial algebra approach to the specification, correctness and implementation of abstract data types. Current Trends in Programming Methodology IV: Data Structuring. Pre Cliffs, N.J., 1978, p. 80-144.

[GH78] Guttag J., Hornung J.J. The algebraic specification of abstract data types. Acta Informatika, 1978, v.10, No.1, pp. 27-52.

[Let68] Letichevski A.A. Syntax and semantics of formal languages. Kibernetika, 1968, No.4, p. 1-9.

[Moe93] Moeller B. Towards Pointer Algebra. Science of Computer Programming, 1993 (to appear).

[MBC89] Morrison R., Brown A.L., Connor R.C.H., Dearle A. The Napier 88 Reference manual. PPRR- -77-89, University of St. Andrews and Glasgow, 1989.

[Nau63] Naur P. (Editor). Revised Report on the Algorithmic Language ALGOL 60. Comm. ACM, 1963, v.6, No.1, pp. 1-17.

[Sch86] Schmidt D. A. Denotational semantics. Allyn and Bacon, 1986.

[Wag84] Wagner E.G. Categorical Semantics, Or Extending Data types to Include Memory. Recent Trends in Data Type Specification: Informatik Fachberichte, v. 116, 1984, p. 1-21.

[Wag87] Wagner E.G. Semantics of Block Structured Languages with Pointers. Mathematical Foundations of Programming Language Semantics, LNCS, 1987, No 298, p. 57-84.

[Wat90] Watt D.A. Programming Language Concepts and Paradigms. Prentice Hall, 1990.

[WPP83] Wirsing M., Pepper P., Partsch H., e.a. On Hierarchies of Abstract Data Types. Acta Informatica, 1983, v.20, p. 1-33.

Galois Connection Based Abstract Interpretations for Strictness Analysis

Patrick Cousot[1] and Radhia Cousot[2]

[1] LIENS, École Normale Supérieure
45, rue d'Ulm
75230 Paris cedex 05 (France)
cousot@dmi.ens.fr

[2] LIX, École Polytechnique
91128 Palaiseau cedex (France)
radhia@polytechnique.fr

Abstract. The abstract interpretation framework based upon the approximation of a fixpoint collecting semantics using Galois connections and widening/narrowing operators on complete lattices [CC77a, CC79b] has been considered difficult to apply to Mycroft's strictness analysis [Myc80, Myc81] for which denotational semantics was though to be more adequate (because non-termination has to be taken into account), see e.g. [AH87], page 25.

Considering a non-deterministic first-order language, we show, contrary to expectation, and using the classical Galois connection-based framework, that Mycroft strictness analysis algorithm is the abstract interpretation of a relational semantics (a big-steps operational semantics including non-termination which can be defined in $G^{\infty}SOS$ either in rule-based or fixpoint style by induction on the syntax of programs [CC92])

An improved version of Johnsson's algorithm [Joh81] is obtained by a subsequent dependence-free abstraction of Mycroft's dependence-sensitive method.

Finally, a compromise between the precision of dependence-sensitive algorithms and the efficiency of dependence-free algorithms is suggested using widening operators.

Keywords: Abstract interpretation; Relational semantics; Strictness analysis; Galois connection; Dependence-free and dependence-sensitive analysis; Widening.

Abstract interpretation [CC77a, CC79b] is a method for designing approximate semantics of programs which can be used to gather information about programs in order to provide safe answers to questions about their run-time behaviours. These semantics can be used to design manual proof methods or to specify automatic program analyzers. When the semantic analysis of programs is to be automated, the answers can only be partial or approximate (that is correct/safe/sound but incomplete) since questions such as termination for all input data are undecidable.

By considering that non-terminating and erroneous behaviors are equivalent, *call-by-need* can be replaced by *call-by-value* in functional programs either whenever the actual argument is always evaluated at least once in the function body or upon the later recursive calls (so that if the evaluation of the actual argument does not terminate or is erroneous then so does the function call) or whenever the function call does not terminate or is erroneous (whether the actual argument is evaluated or not). Alan Mycroft's *strictness analysis* [Myc80, Myc81] is an abstract interpretation designed to recognize these situations. Observe the importance of taking non-termination $\perp$ into account: left-to-right addition is strict in its second argument since it does not terminate either when evaluation of its first argument is guaranteed not to terminate (in which case the second argument is not needed) or the evaluation of its first argument does terminate but that of the second does not. In the classical definition of strictness, errors Ω must also be identified with non-termination $\perp$. Otherwise left-to-right addition would not be strict in its second argument since $\mathtt{true} + \perp = (1/0) + \perp = \Omega$.

Given a recursive function declaration $f(x, y) \equiv e$, Alan Mycroft has designed an abstract function $f^!(x, y) \equiv e^!$ on the abstract domain $\{0, 1\}$ such that:

- if $f^!(0, b) = 0$ for $b \in \{0, 1\}$ then f is strict in x;
- if $f^!(a, 0) = 0$ for $a \in \{0, 1\}$ then f is strict in y;
- if $f^!(1, 1) = 0$ then f never terminates;
- if $f^!(a, b) = 1$ for $a, b \in \{0, 1\}$ then the abstraction is too approximate and no conclusion can be drawn.

The abstract value $e^!$ of an expression e is determined as follows:

- If the expression e is reduced to a constant k then its abstract value $e^!$ is 1 which is the best possible approximation to the fact that it is an integer.
- If the expression e is reduced to a formal parameter x then its evaluation is erroneous or does not terminate if and only if the evaluation of the actual argument is erroneous or does not terminate so that its abstract value $e^!$ is x which denotes the value of the abstract actual argument.
- If the expression e is a basic operation such as $e_1 + e_2$ (always needing its two arguments e_1 and e_2) and the abstract values $e_1^!$ and $e_2^!$ of expressions e_1 and e_2 are obtained recursively, then $(e_1 + e_2)^! = (e_1^! \wedge e_2^!)$ where $(0 \wedge 0) = (0 \wedge 1) = (1 \wedge 0) = 0$ and $(1 \wedge 1) = 1$ since the evaluation of $e_1 + e_2$ is erroneous or does not terminate when that of e_1 or that of e_2 is erroneous or does not terminate, which we conclude from $e_1^! = 0$ or $e_2^! = 0$.
- The evaluation of a conditional $e \equiv (e_1 \rightarrow e_2, e_3)$ is erroneous or does not terminate when the evaluation of the condition e_1 is erroneous (for example non-boolean) or does not terminate so that $e_1^! = 0$ implies $e^! = 0$. If the evaluation of the condition e_1 terminates and the returned boolean value is unknown (so that $e_1^! = 1$), then erroneous termination or non-termination of e is guaranteed only if those of both e_2 and e_3 are guaranteed so that $e^! = 0$ only if $e_2^! = 0$ and $e_3^! = 0$. Therefore $e^!$ can be defined as $(e_1^! \wedge (e_2^! \vee e_3^!))$ where $(0 \vee 0) = 0$ and $(0 \vee 1) = (1 \vee 0) = (1 \vee 1) = 1$.

For the traditional addition example:

$$\mathtt{f(x,\ y)} \equiv \mathtt{((x\ =\ 0)} \rightarrow \mathtt{y,\ (1\ +\ f(x\ -\ 1,\ y)))}$$

we have:

$$f^!(x, y) = (x \wedge 1) \wedge (y \vee (1 \wedge f^!(x \wedge 1, y)))$$

After simplifications (such as $(x \wedge 1) = (1 \wedge x) = x$, $(x \wedge 0) = (0 \wedge x) = 0$, etc.), we get:

$$f^!(x, y) = x \wedge (y \vee f^!(x, y))$$

so that:

$$f^{\natural}(0,0) = 0 \qquad f^{\natural}(1,0) = f^{\natural}(1,0) \qquad f^{\natural}(0,1) = 0 \qquad f^{\natural}(1,1) = 1$$

Constraint $f^{\natural}(0,1) = 0$ shows that f is strict in its first parameter. Constraint $f^{\natural}(1,0) = f^{\natural}(1,0)$ is satisfied by $f^{\natural}(1,0) = 0$ proving that f is also strict in its second parameter. It is also satisfied by $f^{\natural}(1,0) = 1$ which states that the call may terminate or not, which is compatible with but less precise than $f^{\natural}(1,0) = 0$ stating that the call is erroneous or does not terminate.

Mycroft's strictness analysis is dependence-sensitive[3] in that for example we might have $f^{\natural}(0,0) = 0$ whereas $f^{\natural}(0,1) = 1$ and $f^{\natural}(1,0) = 1$ so that f is jointly strict in its two parameters without being strict in any of them. Thomas Johnsson [Joh81], followed by John Hugues [Hug88], proposed a dependence-free strictness analysis method that would not discover this property. However it is more efficient and can discover useful information when no dependence relationship between parameters is needed. For the above addition example, the equations:

$$f^{\times}[\mathrm{x}](x) = x \qquad f^{\times}[\mathrm{y}](y) = y \vee f^{\times}[\mathrm{y}](y)$$

are such that $f^{\times}[\mathrm{x}](0) = 0$ and $f^{\times}[\mathrm{y}](0) = 0$ proving that f is strict in its two parameters.

The purpose of abstract interpretation is to prove the safeness of such program analysis methods with respect to a semantics, or better to formally design them by approximation of the semantics of programs.

1 Principles of Abstract Interpretation

Abstract interpretation, when reduced to its basic principles introduced in [CC77a], can be understood as the approximation of a *collecting semantics*[4].

– The collecting semantics collects a class of properties of interest about a program[5]. It is assumed to be defined as the least fixpoint $\mathrm{lfp}_{\perp}^{\sqsubseteq} F$ of a monotonic operator F where the set of *concrete properties*:

$$\mathcal{F}(\sqsubseteq, \perp, \top, \sqcup, \sqcap) \text{ is a complete lattice.} \qquad (1)$$

$\sqsubseteq$ is called the *computational ordering*. The definition of the operator F is by structural induction on the syntax of the program. F is assumed to be monotonic[6] (but,

---

[3] In order to later avoid the confusion between "relational semantics" and "relational analysis" in the sense of S.S. Muchnick and N.D. Jones [JM81], we use "dependence-sensitive" for "relational" and "dependence-free" for "independent attribute" analysis.

[4] introduced in [CC77a] under the name *static semantics*, and renamed "collecting" by F. Nielson.

[5] invariance properties in [CC77a].

[6] $S \overset{m}{\underset{\sqsubseteq}{\longmapsto}} T$ denotes the set of total functions $\varphi \in S \mapsto T$ on posets $S(\sqsubseteq)$ and $T(\preceq)$ which are *monotonic*, that is, $\forall x, y \in S : x \sqsubseteq y \Rightarrow \varphi x \preceq \varphi y$.

for simplicity, could be assumed to be continuous[7,8]):

$$F \in \mathcal{F} \stackrel{\text{m}\sqsubseteq}{\longmapsto} \mathcal{F} \tag{2}$$

Proposition 1 Tarski's fixpoint theorem [CC79a]. *Let $\mathbb{O}$ be the class of ordinals*[9]. *Define $F^0(X) = X$, $F^{\lambda+1}(X) = F(F^{\lambda}(X))$ for successor ordinals $\lambda + 1$ and $F^{\lambda}(X) = \bigsqcup_{\beta<\lambda} F^{\beta}(X)$ for limit ordinals λ. (1) and (2) imply that the transfinite iteration sequence $F^{\lambda}(\bot)$, $\lambda \in \mathbb{O}$ is well-defined, increasing for $\sqsubseteq$ and ultimately stationary $(\exists \epsilon \in \mathbb{O} : \forall \lambda \geq \epsilon : F^{\lambda}(\bot) = F^{\epsilon}(\bot))$. Its limit:*

$$\text{lfp}_{\bot}^{\sqsubseteq} F \;=\; \bigsqcup_{\lambda\in\mathbb{O}} F^{\lambda}(\bot) \tag{3}$$

is the least fixpoint of F for $\sqsubseteq$. In particular if $F \in \mathcal{F} \stackrel{\text{c}\sqcup}{\longmapsto} \mathcal{F}$ is continuous then:

$$\text{lfp}_{\bot}^{\sqsubseteq} F \;=\; \bigsqcup_{n\in\mathbb{N}} F^{n}(\bot) \tag{4}$$

– Since the collecting semantics is not effectively computable and sometimes not even computer representable, approximations must be considered. The notion of approximation is formalized using an *approximation relation*: $\varphi \leq \phi$ means that property ϕ safely approximates φ. To effectively compute such an approximation of $\text{lfp}_{\bot}^{\sqsubseteq} F$, we replace concrete properties in $\mathcal{F}$ by more crude abstract properties in a well-chosen set $\mathcal{F}^{!}$. Concrete properties $\varphi \in \mathcal{F}$ and abstract properties $\varphi^{!} \in \mathcal{F}^{!}$ should be related by a "semantic function" $\gamma \in \mathcal{F}^{!} \longmapsto \mathcal{F}$. The idea is then to mimic the computation of $\text{lfp}_{\bot}^{\sqsubseteq} F = \sqcup_{\lambda\in\mathbb{O}} F^{\lambda}(\bot)$ within $\mathcal{F}^{!}$ as $\sqcup^{!}_{\lambda\in\mathbb{O}} F^{!\lambda}(\bot^{!})$ by defining $\bot^{!}$, $F^{!}$ and $\sqcup^{!}$, such that:

$$\bot \leq \gamma(\bot^{!}) \tag{5}$$

for all $\varphi \in \mathcal{F}$ and $\phi^{!} \in \mathcal{F}^{!}$:

$$\varphi \leq \gamma(\phi^{!}) \;\Rightarrow\; F(\varphi) \leq \gamma(F^{!}(\phi^{!})) \tag{6}$$

and for all $\{\varphi_{\beta} \mid \beta \in \mathbb{O}\} \subseteq \mathcal{F}$, $\{\phi^{!}_{\beta} \mid \beta \in \mathbb{O}\} \subseteq \mathcal{F}^{!}$ and all limit ordinals λ (ω only if F is continuous):

$$\left(\forall \beta \leq \beta' < \lambda : \varphi_{\beta} \sqsubseteq \varphi_{\beta'} \wedge \varphi_{\beta} \leq \gamma(\phi^{!}_{\beta})\right) \;\Rightarrow\; \bigsqcup_{\beta<\lambda} \varphi_{\beta} \leq \gamma\left(\bigsqcup^{!}_{\beta<\lambda} \phi^{!}_{\beta}\right) \tag{7}$$

We get a fixpoint approximation method:

$$\text{lfp}_{\bot}^{\sqsubseteq} F \leq \gamma\left(\bigsqcup^{!}_{\lambda\in\mathbb{O}} F^{!\lambda}(\bot^{!})\right) \tag{8}$$

---

[7] If $\lim \in \wp(P) \rightarrowtail P$ then a mapping $\varphi \in P \longmapsto P$ is said to be a *complete* lim-*morphism*, written $\varphi \in P \stackrel{\lim}{\longmapsto} P$, (respectively lim-*upper-continuous*, lim-*lower-continuous*) if $\lim\{\varphi(x) \mid x \in X\} = \varphi(\lim\{x \mid x \in X\})$ for all $X \subseteq P$ (respectively increasing and decreasing chains $X \subseteq P$) when the limits are well-defined. For short, it is *continuous* if $\sqcup$-upper-continuous.

[8] Continuity would require the non-determinism of the primitives of the functional language considered in section 2 to be bounded.

[9] Ordinals constitute an extension into the infinite of the order properties of the naturals numbers: $0, 1, 2, \ldots, \omega, \omega+1, \omega+2, \ldots, \omega.2, \omega.2+1, \omega.2+2, \ldots, \omega^2, \ldots, \omega^{\omega}, \ldots, \omega^{\omega^{\cdots}}, \ldots$ where successor ordinals have the form $\lambda + 1$, $\lambda \in \mathbb{O}$ and limit ordinals are of the form $\omega.\lambda$, $\lambda \in \mathbb{O} - \{0\}$ and ω is the least upper bound of the naturals.

Proposition 2 Fixpoint approximation. *(1), (2), (5), (6) and (7) imply (8).*

– It is often the case that the approximation relation is a partial ordering:

$$\leq \text{ is a partial order on } \mathcal{F} \tag{9}$$

Then, we can consider an abstract version $\leq^{!}$ on $\mathcal{F}^{!}$ of the concrete approximation ordering $\leq$ on $\mathcal{F}$:[10]

$$\leq^{!} \text{ is a partial order on } \mathcal{F}^{!} \tag{10}$$

The abstract approximation ordering approximates the concrete one, hence the semantic function γ is monotonic:

$$\gamma \in \mathcal{F}^{!} \overset{m\leq^{!}}{\longmapsto} \mathcal{F} \tag{11}$$

An interesting situation is when all concrete properties $\varphi \in \mathcal{F}$ have a best corresponding approximation $\alpha(\varphi) \in \mathcal{F}^{!}$ where $\alpha \in \mathcal{F} \mapsto \mathcal{F}^{!}$. Since $\alpha(\varphi)$ is an approximation that correctly describes φ, we have:

$$\forall \varphi \in \mathcal{F} : \varphi \leq \gamma(\alpha(\varphi)) \tag{12}$$

and since it is the most precise one, we also have:

$$\forall \varphi \in \mathcal{F} : \forall \phi^{!} \in \mathcal{F}^{!} : \varphi \leq \gamma(\phi^{!}) \Rightarrow \alpha(\varphi) \leq^{!} \phi^{!} \tag{13}$$

(9) and (10) imply that (11), (12) and (13) are equivalent to the characteristic property of *Galois connections* written:

$$\mathcal{F}(\leq) \underset{\alpha}{\overset{\gamma}{\leftrightarrows}} \mathcal{F}^{!}(\leq^{!}) \overset{\text{def}}{=} \forall \varphi \in \mathcal{F} : \forall \phi^{!} \in \mathcal{F}^{!} : \varphi \leq \gamma(\phi^{!}) \Leftrightarrow \alpha(\varphi) \leq^{!} \phi^{!} \tag{14}$$

As suggested in [CC77c], a Galois connection can be lifted to higher-order functional spaces for the pointwise ordering:

$$\varphi \leq \phi \overset{\text{def}}{=} \forall x : \varphi(x) \leq \phi(x) \tag{15}$$

as follows:

Proposition 3 Functional Galois connection. *(9), (10), (14) for $\langle \alpha_0, \gamma_0 \rangle$ and $\langle \alpha_1, \gamma_1 \rangle$ and (15) for $\leq$ and $\leq^{!}$ imply:*

$$\mathcal{F}_0 \overset{m\leq}{\longmapsto} \mathcal{F}_1(\leq) \underset{\lambda\Phi\cdot\alpha_1\circ\Phi\circ\gamma_0}{\overset{\lambda\Psi\cdot\gamma_1\circ\Psi\circ\alpha_0}{\leftrightarrows}} \mathcal{F}_0^{!} \overset{m\leq^{!}}{\longmapsto} \mathcal{F}_1^{!}(\leq^{!}) \tag{16}$$

– When no abstract property is useless, that is the abstraction function α is surjective or equivalently $\alpha \circ \gamma$ is the identity, we have a *Galois surjection* written:

$$\mathcal{F}(\leq) \underset{\alpha}{\overset{\gamma}{\leftrightarrows}} \mathcal{F}^{!}(\leq^{!}) \overset{\text{def}}{=} \mathcal{F}(\leq) \underset{\alpha}{\overset{\gamma}{\leftrightarrows}} \mathcal{F}^{!}(\leq^{!}) \wedge \forall \phi \in \mathcal{F}^{!} : \alpha \circ \gamma(\phi) = \phi \tag{17}$$

Corollary 4 Functional Galois connection. *(9), (10), (14) for $\langle \alpha_1, \gamma_1 \rangle$, (17) for $\langle \alpha_0, \gamma_0 \rangle$ and (15) for $\leq$ and $\leq^{!}$ imply:*

$$\mathcal{F}_0 \overset{m\leq}{\longmapsto} \mathcal{F}_1(\leq) \underset{\lambda\Phi\cdot\alpha_1\circ\Phi\circ\gamma_0}{\overset{\lambda\Psi\cdot\gamma_1\circ\Psi\circ\alpha_0}{\leftrightarrows}} \mathcal{F}_0^{!} \longmapsto \mathcal{F}_1^{!}(\leq^{!}) \tag{18}$$

Corollary 5 Functional Galois surjection. *(9), (10), (17) for $\langle \alpha_0, \gamma_0 \rangle$ and $\langle \alpha_1, \gamma_1 \rangle$ and (15) for $\leq$ and $\leq^{!}$ imply:*

$$\mathcal{F}_0 \overset{m\leq}{\longmapsto} \mathcal{F}_1(\leq) \underset{\lambda\Phi\cdot\alpha_1\circ\Phi\circ\gamma_0}{\overset{\lambda\Psi\cdot\gamma_1\circ\Psi\circ\alpha_0}{\leftrightarrows}} \mathcal{F}_0^{!} \overset{m\leq^{!}}{\longmapsto} \mathcal{F}_1^{!}(\leq^{!}) \tag{19}$$

---

[10] We can always define $\varphi^{!} \leq^{!} \phi^{!} \overset{\text{def}}{=} \gamma(\varphi^{!}) \leq \gamma(\phi^{!})$ and reason on the equivalence classes $\mathcal{F}^{!}/\equiv^{!}$ where $\varphi^{!} \equiv^{!} \phi^{!} \overset{\text{def}}{=} \varphi^{!} \leq^{!} \phi^{!} \wedge \phi^{!} \leq^{!} \varphi^{!}$.

– Proposition 3 and corollary 5 lead to the hypothesis that approximations are preserved by the semantic function F:

$$F \in \mathcal{F} \overset{m\leq}{\longmapsto} \mathcal{F} \tag{20}$$

Then, by defining $\bot^{\mathsf{I}}$, F^{I} and $\sqcup^{\mathsf{I}}$ such that:

$$\bot^{\mathsf{I}} \geq^{\mathsf{I}} \alpha(\bot) \tag{21}$$

$$F^{\mathsf{I}} \geq^{\mathsf{I}} \alpha \circ F \circ \gamma \tag{22}$$

and for all $\{\varphi_\beta \mid \beta \in \mathbb{O}\} \subseteq \mathcal{F}$, $\{\phi^{\mathsf{I}}_\beta \mid \beta \in \mathbb{O}\} \subseteq \mathcal{F}^{\mathsf{I}}$ and all limit ordinals λ (ω only if F is continuous):

$$\left(\forall \beta \leq \beta' < \lambda : \varphi_\beta \sqsubseteq \varphi_{\beta'} \wedge \alpha(\varphi_\beta) \leq^{\mathsf{I}} \phi^{\mathsf{I}}_\beta\right) \Rightarrow \alpha\Big(\bigsqcup_{\beta<\lambda} \varphi_\beta\Big) \leq^{\mathsf{I}} \bigsqcup^{\mathsf{I}}_{\beta<\lambda} \phi^{\mathsf{I}}_\beta \tag{23}$$

we obtain the following fixpoint approximation method:

Proposition 6 Fixpoint approximation. *(1), (2), (9), (10), (14), (15), (20), (21), (22) and (23) imply (8).*

– If the computational ordering has an abstract correspondent:

$$\mathcal{F}^{\mathsf{I}}(\sqsubseteq^{\mathsf{I}}, \bot^{\mathsf{I}}, \top^{\mathsf{I}}, \sqcup^{\mathsf{I}}, \sqcap^{\mathsf{I}}) \text{ is a complete lattice.} \tag{24}$$

such that:

$$\mathcal{F}(\sqsubseteq) \underset{\alpha}{\overset{\gamma}{\leftrightarrows}} \mathcal{F}^{\mathsf{I}}(\sqsubseteq^{\mathsf{I}}) \tag{25}$$

then we can characterize $\bot^{\mathsf{I}}$ and $\sqcup^{\mathsf{I}}$, as follows:

$$\bot^{\mathsf{I}} = \alpha(\bot) \tag{26}$$

$$\bigsqcup^{\mathsf{I}}{}_{\lambda\in\mathbb{O}} \varphi^{\mathsf{I}}_\lambda = \alpha\Big(\bigsqcup_{\lambda\in\mathbb{O}} \gamma(\varphi^{\mathsf{I}}_\lambda)\Big) \tag{27}$$

Lemma 7. *(1), (24) and (25) imply (26) and (27).*

If the abstract computational and approximation orderings coincide then we can simplify hypothesis (23) as follows:

Lemma 8. *(1), (24), (25) and* $\leq^{\mathsf{I}} = \sqsubseteq^{\mathsf{I}}$ *imply (23).*

Corollary 9 Fixpoint approximation. *(1), (2), (9), (10), (14), (15), (20), (22), (24), (25) and* $\leq^{\mathsf{I}} = \sqsubseteq^{\mathsf{I}}$ *imply (8).*

2 Syntax and Relational Semantics of a First-order Functional Language

2.1 Syntax

In order to simplify the manipulation of tuples of parameters, we understand a tuple $(e_1, \ldots, e_i, \ldots, e_n)$ of actual arguments corresponding to formal parameters $(v_1, \ldots, v_i, \ldots, v_n)$ as a vector $\vec{e}$. This vector $\vec{e}$ is a notation for a special kind of functions from its domain $\mathrm{Dom}\,\vec{e} = \{v_1, \ldots, v_i, \ldots, v_n\}$ into the set of expressions such that $\vec{e}[v_i] = e_i$. The domain of $\vec{e}$ is left implicit in the notation $(e_1, \ldots, e_i, \ldots, e_n)$ but can be obtained from the syntactic context. The correspondence between v_i and e_i is given as usual by a positional notation so that each v_i and e_i has a rank $\mathrm{rk}(v_i) = \mathrm{rk}(e_i) = i$. We write $v \in \vec{e}$ for $v \in \mathrm{Dom}\,\vec{e}$ so that $\vec{e} = \prod_{v\in\vec{e}} \vec{e}[v]$. For uniformity, we also use a vector notation $\vec{v}$ for the tuple of formal parameters

$(v_1, \ldots, v_i, \ldots, v_n)$ with the convention that $v_i = \vec{v}[v_i]$. Finally, the same notation is used for functions, so that a program consists of function declarations $\vec{f}\vec{v} \equiv \vec{F}$, also written:

$$\prod_{f \in \vec{f}} f\vec{v} \equiv \vec{F}[f]$$

where the body $\vec{F}[f]$ of function f is an expression e depending upon the formal parameters $v \in \vec{v}$ and containing recursive calls to the functions $f \in \vec{f}$. This expression e is written according to the following syntax[11]:

$e ::= k$	constant
$\mid v$	variable
$\mid b\vec{e}$	basic operation
$\mid f\vec{e}$	function call
$\mid (e_1 \rightarrow e_2, e_3)$	conditional
$\vec{v} ::= (v_1, \ldots, v_n)$	tuple of formal parameters
$\vec{e} ::= (e_1, \ldots, e_n)$	tuple of actual arguments

2.2 Semantic Domains

Computed values belong to a given set Υ containing the booleans $\{tt, ff\}$ and other basic values such as natural numbers $\mathbb{N}$, integers $\mathbb{Z}$, reals $\mathbb{R}$, etc. We use $\Omega \notin \Upsilon$ to denote run-time errors and $\bot \notin \Upsilon$ to denote non-termination. Therefore saying that an expression has value Ω means that its evaluation terminates by the production of an error message. Saying that its value is $\bot$ means that its evaluation does not terminate. We define $\Upsilon^{\Omega} \stackrel{\text{def}}{=} \Upsilon \cup \{\Omega\}$, $\Upsilon_{\bot} \stackrel{\text{def}}{=} \Upsilon \cup \{\bot\}$ and $\Upsilon_{\bot}^{\Omega} \stackrel{\text{def}}{=} \Upsilon \cup \{\Omega, \bot\}$. Values ν taken by a variable v belong to $\mathcal{D}^{*}$ and values $\vec{\nu}$ taken by a tuple $\vec{v}$ of variables belong to $\vec{\mathcal{D}}^{*}$[12]:

$$\mathcal{D}^{*} \stackrel{\text{def}}{=} \Upsilon_{\bot}^{\Omega} \qquad \vec{\mathcal{D}}^{*} \stackrel{\text{def}}{=} \prod_{v \in \vec{v}} \mathcal{D}^{*}$$

Using Ω to denote run-time errors and $\bot$ to denote non-termination, we can always consider total functions over $\Upsilon_{\bot}^{\Omega}$ corresponding to partial functions over Υ. Contrary to a common practice in denotational semantics, which is explained and followed by [Mos90] (see page 599), we avoid representing errors by $\bot$ since we want to discriminate finite against infinite executions[13]. The *relational semantics* of a function declaration $f\vec{v} \equiv e$ is a relation $f^{*} \in \mathcal{F}^{*}$ where:

$$\mathcal{F}^{*} \stackrel{\text{def}}{=} \wp(\vec{\mathcal{D}}^{*} \times \mathcal{D}^{*})$$

We write $f(\vec{\nu}) \rightsquigarrow \nu$ for $\langle \vec{\nu}, \nu \rangle \in f^{*}$ which means that the call of function f with actual arguments having values $\vec{\nu}$ may return the result ν. If this call may terminate

---

[11] Expressions e and tuples $\vec{e}$ of expressions have an attribute $e.\vec{v}$ (resp. $\vec{e}.\vec{v}$) which is the list of visible parameters and an attribute $e.\partial$ (resp. $\vec{e}.\partial$) which is the set of headings $f\vec{v}$ of the visible functions f which can occur free in the expression. Tuples of variables $\vec{v}$ (respectively tuples of expressions $\vec{e}$) have a domain attribute Dom $\vec{v}$ (resp. Dom $\vec{e}$) which is the set of variables occurring in the tuple (resp. the set of formal parameters corresponding to each expression in the tuple).

[12] $\vec{\mathcal{D}}^{*}$ is a shorthand for $\vec{\mathcal{D}}_{\vec{v}}^{*} = \prod_{v \in \text{Dom } \vec{v}} \mathcal{D}^{*}$ where $\vec{v}$ is left context-dependent.

[13] More refinement of Ω might be useful. For example we could differentiate the class of run-time failures due to a type error (using "wrong" as in [Mil78] for the value $1/tt$) from those due to partially defined well-typed functions (such as $1/0$). Each kind of error might even be described by a different error value (corresponding to different error messages).

erroneously then $f(\vec{\nu}) \rightsquigarrow \Omega$ whereas $f(\vec{\nu}) \rightsquigarrow \bot$ indicates that nontermination is possible. The relational semantics $\vec{f}^{*}$ of a program $\prod_{f \in \vec{f}} f\vec{v} \equiv \vec{F}[f]$ (where the expression $\vec{F}[f]$ is the body of function f) is a tuple of functions belonging to $\vec{\mathcal{F}}^{*}$:

$$\vec{\mathcal{F}}^{*} \stackrel{\text{def}}{=} \prod_{f \in \vec{f}} \mathcal{F}^{*}$$

An expression e can be evaluated when the relational semantics $\vec{\varphi}$ of the functions $\vec{f}$ and the value $\vec{\nu}$ of the variables $\vec{v}$ which may occur free in e are known. We write $\vec{\varphi}, \vec{\nu} \vdash e \rightsquigarrow \nu$ to mean that evaluation of e in environment $\vec{\varphi}$, $\vec{\nu}$ may return ν. In particular this evaluation returns an error if $\nu = \Omega$ and does not terminate when $\nu = \bot$. Therefore the relational semantics e^{*} of the expression e belongs to $\mathcal{E}^{*}$[14]:

$$\mathcal{E}^{*} \stackrel{\text{def}}{=} \vec{\mathcal{F}}^{*} \stackrel{\text{m}\subseteq}{\longmapsto} \wp(\vec{\mathcal{D}}^{*} \times \mathcal{D}^{*})$$

whereas the relational semantics $\vec{e}^{*}$ of a tuple of expressions $\vec{e}$ belongs to $\vec{\mathcal{E}}^{*}$[14]:

$$\vec{\mathcal{E}}^{*} \stackrel{\text{def}}{=} \vec{\mathcal{F}}^{*} \stackrel{\text{m}\subseteq}{\longmapsto} \wp(\vec{\mathcal{D}}^{*} \times \vec{\mathcal{D}}^{*})$$

2.3 Rule-based Presentation of the Relational Semantics

We now present the relational semantics of expressions, by induction on their syntax and then the semantics of programs. We use the technique of bi-inductive definitions introduced in [CC92], but resort only to the intuitive understanding of the reader. $\vec{f}^{*}, \vec{\nu} \vdash e \rightsquigarrow \nu$ means that evaluation of expression e may return value ν in the evaluation environment specified by $\vec{f}^{*}$, giving the relational semantics $\vec{f}^{*}[f]$ of the functions f used in e, and $\vec{\nu}$ giving the value $\vec{\nu}[v]$ of the variables v which can occur free in e.

– We assume that the value $\underline{k} \in \Upsilon$ of the constant k is given. Therefore:

$$\vec{f}^{*}, \vec{\nu} \vdash k \rightsquigarrow \underline{k}\ + \tag{28}$$

Such a positive axiom schema $\vec{f}^{*}, \vec{\nu} \vdash e \rightsquigarrow \nu\ +$, marked "+", describes finite behaviors in $\vec{\mathcal{F}}^{*} \mapsto \wp(\vec{\mathcal{D}}^{*} \times \Upsilon^{\Omega})$ that is terminating or erroneous evaluations of expression e for all possible values of $\vec{f}^{*}$, $\vec{\nu}$ and ν.

– A formal parameter v denotes the value of the corresponding actual parameter which is stored in $\vec{\nu}[v]$:

$$\vec{f}^{*}, \vec{\nu} \vdash v \rightsquigarrow \vec{\nu}[v]\ + \quad \text{if } \vec{\nu}[v] \in \Upsilon^{\Omega} \tag{29}$$

If the evaluation of the actual parameter does not terminate, which is formally represented by the fact that $\vec{\nu}[v] = \bot$, and the corresponding formal parameter v is used, then this use leads to a non-terminating evaluation:

$$\vec{f}^{*}, \vec{\nu} \vdash v \rightsquigarrow \bot\ - \quad \text{if } \vec{\nu}[v] = \bot \tag{30}$$

Such a negative axiom schema, marked "-", describes infinite behaviors in $\vec{\mathcal{F}}^{*} \mapsto \wp(\vec{\mathcal{D}}^{*} \times \{\bot\})$ that is non-terminating computations for all possible values of $\vec{f}^{*}$, $\vec{\nu}$ and v satisfying the side condition $\vec{\nu}[v] = \bot$[15].

---

[14] Once again, e and $\vec{e}$ is left context-dependent so as to avoid the precise but heavy notation $\vec{\mathcal{E}}^{*}_{\vec{e}} = \left(\prod_{f\vec{v} \in \vec{e}.\vec{\theta}} \wp\left(\left(\prod_{v \in \text{Dom}\,\vec{v}} \mathcal{D}^{*}\right) \times \mathcal{D}^{*}\right)\right) \stackrel{\text{m}\subseteq}{\longmapsto} \wp\left(\left(\prod_{v \in \vec{e}.\vec{v}} \mathcal{D}^{*}\right) \times \left(\prod_{v \in \text{Dom}\,\vec{e}} \mathcal{D}^{*}\right)\right)$.

[15] here all non-terminating computations deliver $\bot$ but in lazy languages they might also result in infinite data structures.

– The evaluation of a tuple $\vec{e}$ of actual parameters consists in evaluating each parameter $\vec{e}[v]$, $v \in \vec{e}$. This evaluation does not terminate if the evaluation of at least one component does not terminate. Else the evaluation of the tuple $\vec{e}$ of actual parameters is terminating or erroneous:

$$\frac{\forall v \in \vec{e} : \vec{f}^{*}, \vec{\nu} \vdash \vec{e}[v] \rightsquigarrow \vec{\nu}'[v]}{\vec{f}^{*}, \vec{\nu} \vdash \vec{e} \rightsquigarrow \vec{\nu}'}^{+} \quad \text{if } \forall v \in \vec{e} : \vec{\nu}'[v] \in \Upsilon^{\Omega}$$

$$\frac{\forall v \in \vec{e} : \vec{f}^{*}, \vec{\nu} \vdash \vec{e}[v] \rightsquigarrow \vec{\nu}'[v]}{\vec{f}^{*}, \vec{\nu} \vdash \vec{e} \rightsquigarrow \vec{\nu}'}^{-} \quad \text{if } \exists v \in \vec{e} : \vec{\nu}'[v] = \bot$$

– We assume that the relational semantics of each basic operation b is specified by a total relation $b^{*} \in \mathbb{P}(\vec{\Upsilon}_{\bot}^{\Omega} \times \Upsilon_{\bot}^{\Omega})$ [16] (or a function $\underline{b} \in \vec{\Upsilon}_{\bot}^{\Omega} \mapsto \Upsilon_{\bot}^{\Omega}$ if the language is deterministic, in which case $b^{*} \stackrel{\text{def}}{=} \{\langle \vec{\nu}, \underline{b}(\vec{\nu})\rangle \mid \vec{\nu} \in \vec{\Upsilon}_{\bot}^{\Omega}\}$). We write $b(\vec{\nu}) \rightsquigarrow \nu$ for $\langle \vec{\nu}, \nu\rangle \in b^{*}$.

Example 1. • The integer addition is strict in its two parameters and is defined only for integers so that $\underline{\text{plus}}(\nu, \nu') \stackrel{\text{def}}{=} ((\nu = \bot \vee \nu' = \bot) \rightarrow \bot, ((\nu \notin \mathbb{Z} \vee \nu' \notin \mathbb{Z}) \rightarrow \Omega, (\nu + \nu')))$ where $+$ denotes the usual mathematical addition.

• For McCarthy's conjunction, we would have $\underline{\text{cand}}(\nu, \nu') \stackrel{\text{def}}{=} (\nu = \bot \rightarrow \bot, (\nu = \text{ff} \rightarrow \text{ff}, (\nu = \text{tt} \rightarrow ((\nu' \in \{\bot, \text{tt}, \text{ff}\} \rightarrow \nu', \Omega)), \Omega)))$. Observe that the first parameter is passed by value since evaluation of the conjunction does not terminate as soon as this first parameter does not terminate whereas the second is passed by need, and used only if the first is tt.

• For the unbounded random assignment '?', which does not need its argument, we have $?(\nu) \rightsquigarrow n$ if and only if $n \in \mathbb{N}$.

• For the bounded random assignment, we have:

$$\begin{aligned}
?(\langle \nu, \nu'\rangle) \rightsquigarrow \bot \quad &\text{iff} \quad (\nu = \bot) \vee (\nu' = \bot),\\
?(\langle \nu, \nu'\rangle) \rightsquigarrow \Omega \quad &\text{iff} \quad (\nu \in \Upsilon^{\Omega} - \mathbb{Z}) \vee (\nu' \in \Upsilon^{\Omega} - \mathbb{Z}) \vee\\
&\qquad (\nu \in \mathbb{Z} \wedge \nu' \in \mathbb{Z} \wedge \nu > \nu'),\\
?(\langle \nu, \nu'\rangle) \rightsquigarrow z \quad &\text{iff} \quad (\nu \in \mathbb{Z} \wedge \nu' \in \mathbb{Z} \wedge \nu \leq z \leq \nu').
\end{aligned}$$ □

The value of $b\vec{e}$ is obtained by applying b to the value of the arguments $\vec{e}$ (which may be Ω or $\bot$ without preventing the result to be in Υ when the argument is not needed):

$$\frac{\vec{f}^{*}, \vec{\nu} \vdash \vec{e} \rightsquigarrow \vec{\nu}' \ \wedge\ b(\vec{\nu}') \rightsquigarrow \bot}{\vec{f}^{*}, \vec{\nu} \vdash b\vec{e} \rightsquigarrow \bot}^{-}$$

$$\frac{\vec{f}^{*}, \vec{\nu} \vdash \vec{e} \rightsquigarrow \vec{\nu}' \ \wedge\ b(\vec{\nu}') \rightsquigarrow \nu}{\vec{f}^{*}, \vec{\nu} \vdash b\vec{e} \rightsquigarrow \nu}^{+} \quad \text{if } \nu \in \Upsilon^{\Omega}$$

– The evaluation of the conditional $(e_1 \rightarrow e_2, e_3)$ does not terminate when the evaluation of the condition e_1 does not terminate:

$$\frac{\vec{f}^{*}, \vec{\nu} \vdash e_1 \rightsquigarrow \bot}{\vec{f}^{*}, \vec{\nu} \vdash (e_1 \rightarrow e_2, e_3) \rightsquigarrow \bot}^{-}$$

[16] We define the set $\mathbb{P}(S \times T)$ of *total binary relations* on $S \times T$ as $\{\rho \in \wp(S \times T) \mid \forall s \in S : \exists t \in T : \langle s, t\rangle \in \rho\}$ so that $\rho \in \mathbb{P}(S \times T)$ implies that for every $s \in S$ there exists at least one $t \in T$ such that the pair $\langle s, t\rangle$ belongs to ρ.

The evaluation of the conditional $(e_1 \rightarrow e_2, e_3)$ terminates with an error when the evaluation of the condition e_1 terminates without returning a boolean:

$$\frac{\vec{f}^{*}, \vec{\nu} \vdash e_1 \rightsquigarrow \nu \ \wedge\ \nu \in \Upsilon^{\Omega} - \{\mathrm{tt}, \mathrm{ff}\}}{\vec{f}^{*}, \vec{\nu} \vdash (e_1 \rightarrow e_2, e_3) \rightsquigarrow \Omega}^{+}$$

If the evaluation of the condition e_1 returns a boolean b, then the value of the conditional $(e_1 \rightarrow e_2, e_3)$ is that of e_2 if b is true and that of e_3 if b is false, including the run-time error and non-termination cases:

$$\frac{\vec{f}^{*}, \vec{\nu} \vdash e_1 \rightsquigarrow \mathrm{tt} \ \wedge\ \vec{f}^{*}, \vec{\nu} \vdash e_2 \rightsquigarrow \nu}{\vec{f}^{*}, \vec{\nu} \vdash (e_1 \rightarrow e_2, e_3) \rightsquigarrow \nu}^{+} \quad \text{if } \nu \in \Upsilon^{\Omega}$$

$$\frac{\vec{f}^{*}, \vec{\nu} \vdash e_1 \rightsquigarrow \mathrm{tt} \ \wedge\ \vec{f}^{*}, \vec{\nu} \vdash e_2 \rightsquigarrow \bot}{\vec{f}^{*}, \vec{\nu} \vdash (e_1 \rightarrow e_2, e_3) \rightsquigarrow \bot}^{-}$$

$$\frac{\vec{f}^{*}, \vec{\nu} \vdash e_1 \rightsquigarrow \mathrm{ff} \ \wedge\ \vec{f}^{*}, \vec{\nu} \vdash e_3 \rightsquigarrow \nu}{\vec{f}^{*}, \vec{\nu} \vdash (e_1 \rightarrow e_2, e_3) \rightsquigarrow \nu}^{+} \quad \text{if } \nu \in \Upsilon^{\Omega}$$

$$\frac{\vec{f}^{*}, \vec{\nu} \vdash e_1 \rightsquigarrow \mathrm{ff} \ \wedge\ \vec{f}^{*}, \vec{\nu} \vdash e_3 \rightsquigarrow \bot}{\vec{f}^{*}, \vec{\nu} \vdash (e_1 \rightarrow e_2, e_3) \rightsquigarrow \bot}^{-}$$

– The relational semantics of the function call $f\vec{e}$ can be defined knowing the value $\vec{\nu}$ of the free variables $\vec{v}$ which may appear within $\vec{e}$ and the relational semantics $f^{*} = \vec{f}^{*}[f]$ of the function $f \in \vec{f}$ of the program $\prod_{f \in \vec{f}} f\vec{v} \equiv \vec{F}[f]$. f^{*} specifies a set of arguments-result pairs $f(\vec{\nu}') \rightsquigarrow \nu$. The call $f\vec{e}$ may return value ν if, given the value $\vec{\nu}'$ of the actual parameters $\vec{e}$, the function f may return that value ν, that is $f(\vec{\nu}') \rightsquigarrow \nu$. Therefore, we have:

$$\frac{\vec{f}^{*}, \vec{\nu} \vdash \vec{e} \rightsquigarrow \vec{\nu}' \ \wedge\ f(\vec{\nu}') \rightsquigarrow \nu}{\vec{f}^{*}, \vec{\nu} \vdash f\vec{e} \rightsquigarrow \nu}^{+} \quad \text{if } \nu \in \Upsilon^{\Omega} \tag{31}$$

$$\frac{\vec{f}^{*}, \vec{\nu} \vdash \vec{e} \rightsquigarrow \vec{\nu}' \ \wedge\ f(\vec{\nu}') \rightsquigarrow \bot}{\vec{f}^{*}, \vec{\nu} \vdash f\vec{e} \rightsquigarrow \bot}^{-} \tag{32}$$

Observe that in an effective implementation of the above rules (31) and (32), parameters must be passed by need since a non-terminating actual parameter $\vec{e}[v]$, such that $\vec{\nu}'[v] = \bot$, will actually affects the computation only if the corresponding formal parameter v is needed in the body $\vec{F}[f]$ of function f according to rules (29) and (30). Moreover call-by-name would be inadequate since the effective parameters are evaluated only once. However this kind of implementation detail is omitted in the mathematical description of the relational semantics where the behaviours of the actual parameters, including the non-terminating ones, can be described prior to describing the behaviour of the function call.

– The relational semantics $\vec{f}^{*}$ of a program $\prod_{f \in \vec{f}} f\vec{v} \equiv \vec{F}[f]$ specifies that a call of function f consists in evaluating the function body $\vec{F}[f]$ with formal parameters bound to their actual values:

$$\frac{\vec{f}^{*}, \vec{\nu} \vdash \vec{F}[f] \rightsquigarrow \nu}{f(\vec{\nu}) \rightsquigarrow \nu}^{+} \quad \text{if } \nu \in \Upsilon^{\Omega} \tag{33}$$

$$\frac{\vec{f}^{*}, \vec{\nu} \vdash \vec{F}[f] \rightsquigarrow \bot}{f(\vec{\nu}) \rightsquigarrow \bot}^{-} \tag{34}$$

These rules are inductive since recursive function calls may occur in function bodies. For the positive rule (33), this is an induction of the length of the computations: the computations for the recursive calls, are included in the computation of the main call, hence shorter. The basis of this induction is given by the axioms (28) and (29) which involve a single step computation. This reasoning is no longer valid for non-terminating calls, that is in the negative rule (34), since the main and recursive sub-calls all involve infinite computations. Hence the induction involved in the negative rules should be understood quite differently. One must imagine that the basis is given by all infinite computations $f(\vec{\nu}) \rightsquigarrow \bot$ and that the rule eliminates those which cannot be obtained by an evaluation of the function body $\vec{F}[f]$ where recursive calls must correspond to non-terminating calls which are not yet eliminated. Repeating this process ad infinitum, only the non-terminating behaviours will be left out.

Example 2. Let us consider $\mathtt{f(x)} \equiv (\mathtt{x = 0 \rightarrow 1, f(x-1)})$. Starting with the basis $f^{\star} = \{f(n) \rightsquigarrow \bot \mid n \in \mathbb{Z}_{\bot}^{\Omega}\}$ and applying the above rules to the function body we derive that $f(\Omega) \rightsquigarrow \Omega$, $f(0) \rightsquigarrow 1$ and $f(n) \rightsquigarrow \bot$ if $n \in \mathbb{Z}_{\bot} - \{0\}$ so that, by elimination of the impossible infinite behaviour, we get $f^{\star} = \{\langle\Omega,\ \Omega\rangle, \langle 0,\ 1\rangle\} \cup \{\langle n,\ \bot\rangle \mid n \in \mathbb{Z}_{\bot} - \{0\}\}$. Repeating this process, we discover that $f(1)$ calls $f(0)$ which has no infinite behaviour so that the non-termination of $f(1)$ is impossible. The second approximation of $f^{\star}$ is now $\{\langle\Omega,\ \Omega\rangle\} \cup \{\langle n,\ 1\rangle \mid 0 \leq n \leq 1\} \cup \{\langle n,\ \bot\rangle \mid \mathbb{Z}_{\bot} - \{0,1\}\}$. After i repetitions of this elimination process we get $f^{\star} = \{\langle\Omega,\ \Omega\rangle\} \cup \{\langle n,\ 1\rangle \mid 0 \leq n < i\} \cup \{\langle n,\ \bot\rangle \mid \mathbb{Z}_{\bot} - \{0, \ldots, i-1\}\}$ so that passing to the limit we conclude that $f^{\star} = \{\langle\Omega,\ \Omega\rangle\} \cup \{\langle n,\ 1\rangle \mid 0 \leq n\} \cup \{\langle n,\ \bot\rangle \mid n < 0 \vee n = \bot\}$. □

More generally, such bi-inductive definitions can be given a precise meaning as specifications of fixpoints of monotone operators on complete lattices [CC92] which we now illustrate for the relational semantics.

2.4 Fixpoint Presentation of the Relational Semantics

In order to capture the idea that terminating evaluations of recursive functions are obtained inductively by construction from the terminating evaluation of recursive calls while non-terminating evaluations are obtained destructively by elimination of inaccessible recursive calls, let us introduce the partial order $\sqsubseteq^{\star}$ on $\mathcal{F}^{\star}$ together with the corresponding least upper bound and greatest lower bound defined by:

$$\bot^{\star} \stackrel{\text{def}}{=} \vec{\Upsilon}_{\bot}^{\Omega} \times \{\bot\} \tag{35}$$

$$\top^{\star} \stackrel{\text{def}}{=} \vec{\Upsilon}_{\bot}^{\Omega} \times \Upsilon^{\Omega} \tag{36}$$

$$\varphi \sqsubseteq^{\star} \varphi' \stackrel{\text{def}}{=} (\varphi \cap \top^{\star}) \subseteq (\varphi' \cap \top^{\star}) \wedge (\varphi \cap \bot^{\star}) \supseteq (\varphi' \cap \bot^{\star}) \tag{37}$$

$$\sqcup^{\star}_{i \in \Delta} \varphi_i \stackrel{\text{def}}{=} \cup_{i \in \Delta}(\varphi_i \cap \top^{\star}) \cup \cap_{i \in \Delta}(\varphi_i \cap \bot^{\star}) \tag{38}$$

$$\sqcap^{\star}_{i \in \Delta} \varphi_i \stackrel{\text{def}}{=} \cap_{i \in \Delta}(\varphi_i \cap \top^{\star}) \cup \cup_{i \in \Delta}(\varphi_i \cap \bot^{\star})^{17}$$

$\mathcal{F}^{\star}(\sqsubseteq^{\star}, \bot^{\star}, \top^{\star}, \sqcup^{\star}, \sqcap^{\star})$ is a complete lattice. The pointwise extension to tuples of relations $\vec{\mathcal{F}}^{\star}(\vec{\sqsubseteq}^{\star}, \vec{\bot}^{\star}, \vec{\top}^{\star}, \vec{\sqcup}^{\star}, \vec{\sqcap}^{\star})$ is also a complete lattice. Therefore we can define the relational semantics $\vec{f}^{\star}$ of the program $\prod_{f \in \vec{f}} f\vec{v} \equiv \vec{F}[f]$ (where the body $\vec{F}[f]$ of function $f \in \vec{f}$ is an expression) as the least fixpoint of a monotonic operator

[17] This is for $\Delta \neq \emptyset$. As usual, $\sqcup^{\star}\emptyset = \bot^{\star}$ and $\sqcap^{\star}\emptyset = \top^{\star}$.

$\vec{F}^{\sharp}$ on this complete lattice $\vec{\mathcal{F}}^{\sharp}$ (thus avoiding the difficulties resulting from the non-existence of arbitrary upper bounds in cpos). This operator $\vec{F}^{\sharp} \in \vec{\mathcal{F}}^{\sharp} \stackrel{m\sqsubseteq^{\sharp}}{\longmapsto} \vec{\mathcal{F}}^{\sharp}$ corresponds to rule schemata (33) and (34) [18]:

$$\vec{f}^{\sharp} \stackrel{\text{def}}{=} \text{lfp}_{\perp^{\sharp}}^{\sqsubseteq^{\sharp}} \vec{F}^{\sharp} \qquad \vec{F}^{\sharp} \stackrel{\text{def}}{=} \lambda\vec{\varphi}\cdot\prod_{f\in\vec{f}} \vec{F}[f]^{\sharp}[\vec{\varphi}] \tag{39}$$

The operator $\vec{F}[f]^{\sharp} \in \vec{\mathcal{F}}^{\sharp} \stackrel{m\sqsubseteq^{\sharp}}{\longmapsto} \mathcal{F}^{\sharp}$ is defined by induction on the syntax of the expression $\vec{F}[f]$ constituting the body of function f according to the axiom and rule schemata (28) to (34)[19]:

$$k^{\sharp}[\vec{\varphi}] \stackrel{\text{def}}{=} \{\langle\vec{\nu}, \underline{k}\rangle \mid \vec{\nu} \in \vec{\mathcal{D}}^{\sharp}\} \tag{40}$$

$$v^{\sharp}[\vec{\varphi}] \stackrel{\text{def}}{=} \{\langle\vec{\nu}, \vec{\nu}[v]\rangle \mid \vec{\nu} \in \vec{\mathcal{D}}^{\sharp}\} \tag{41}$$

$$\vec{e}^{\sharp}[\vec{\varphi}] \stackrel{\text{def}}{=} \{\langle\vec{\nu}, \vec{\nu}'\rangle \mid \forall v \in \vec{e} : \langle\vec{\nu}, \vec{\nu}'[v]\rangle \in \vec{e}[v]^{\sharp}[\vec{\varphi}]\} \tag{42}$$

$$b\vec{e}^{\sharp}[\vec{\varphi}] \stackrel{\text{def}}{=} \vec{e}^{\sharp}[\vec{\varphi}] \circ b^{\sharp} \tag{43}$$

$$\begin{aligned}(e_1 \rightarrow e_2, e_3)^{\sharp}[\vec{\varphi}] \stackrel{\text{def}}{=}\ & \{\langle\vec{\nu}, \perp\rangle \mid \langle\vec{\nu}, \perp\rangle \in e_1^{\sharp}[\vec{\varphi}]\} \\ \cup\ & \{\langle\vec{\nu}, \Omega\rangle \mid \exists\nu \in \Upsilon^{\Omega} - \{\text{tt}, \text{ff}\} : \langle\vec{\nu}, \nu\rangle \in e_1^{\sharp}[\vec{\varphi}]\} \\ \cup\ & \{\langle\vec{\nu}, \nu\rangle \mid \langle\vec{\nu}, \text{tt}\rangle \in e_1^{\sharp}[\vec{\varphi}] \wedge \langle\vec{\nu}, \nu\rangle \in e_2^{\sharp}[\vec{\varphi}]\} \\ \cup\ & \{\langle\vec{\nu}, \nu\rangle \mid \langle\vec{\nu}, \text{ff}\rangle \in e_1^{\sharp}[\vec{\varphi}] \wedge \langle\vec{\nu}, \nu\rangle \in e_3^{\sharp}[\vec{\varphi}]\}\end{aligned} \tag{44}$$

$$f\vec{e}^{\sharp}[\vec{\varphi}] \stackrel{\text{def}}{=} \vec{e}^{\sharp}[\vec{\varphi}] \circ \vec{\varphi}[f] \tag{45}$$

Example 9. For the program:

```
f(x) ≡ (x = 0 → 0, (x < 0 → f(?(())), f(x - 1)))
```

where {()} is the unit type and ?(()) returns any nonnegative integer (so that the nondeterminism is unbounded) the equation is:

$$\begin{aligned}\vec{F}^{\sharp}[f][\vec{\varphi}] = \ & \{\langle\perp, \perp\rangle, \langle\Omega, \Omega\rangle, \langle 0, 0\rangle\} \cup \{\langle x, y\rangle \mid x < 0 \wedge \exists n \geq 0 : \langle n, y\rangle \in \vec{\varphi}[f]\} \\ & \cup \{\langle x, y\rangle \mid x > 0 \wedge \langle x - 1, y\rangle \in \vec{\varphi}[f]\}\end{aligned}$$

The transfinite iterates $\varphi^{\lambda} = \vec{F}^{\sharp}[f]^{\lambda}(\perp^{\sharp})$ are:

$$\begin{aligned}\varphi^{0} &= \perp^{\sharp} = \mathbb{Z}_{\perp}^{\Omega} \times \{\perp\} \\ \varphi^{1} &= \{\langle\perp, \perp\rangle, \langle\Omega, \Omega\rangle, \langle 0, 0\rangle\} \cup \{\langle x, \perp\rangle \mid x \neq 0\} \\ \varphi^{2} &= \{\langle\perp, \perp\rangle, \langle\Omega, \Omega\rangle\} \cup \{\langle x, 0\rangle \mid x < 2\} \cup \{\langle x, \perp\rangle \mid x < 0 \vee x \geq 2\} \\ &\dots \\ \varphi^{n} &= \{\langle\perp, \perp\rangle, \langle\Omega, \Omega\rangle\} \cup \{\langle x, 0\rangle \mid x < n\} \cup \{\langle x, \perp\rangle \mid x < 0 \vee x \geq n\} \\ &\dots \\ \varphi^{\omega} &= \{\langle\perp, \perp\rangle, \langle\Omega, \Omega\rangle\} \cup \{\langle x, 0\rangle \mid x \in \mathbb{Z}\} \cup \{\langle x, \perp\rangle \mid x < 0\} \\ \varphi^{\omega+1} &= \{\langle\perp, \perp\rangle, \langle\Omega, \Omega\rangle\} \cup \{\langle x, 0\rangle \mid x \in \mathbb{Z}\} \\ \varphi^{\omega+2} &= \varphi^{\omega+1}\end{aligned}$$

proving that the program returns 0 for all integer parameters. □

---

[18] $\text{lfp}_{x}^{\leq}\varphi$ is the least fixpoint of φ greater than or equal to x for partial ordering $\leq$.

[19] The *composition* $\rho \circ \rho'$ of two binary relations $\rho \in \wp(S \times T)$ and $\rho' \in \wp(T \times U)$ is the relation $\{\langle s, u\rangle \in S \times U \mid \exists t \in T : \langle s, t\rangle \in \rho \wedge \langle t, u\rangle \in \rho'\}$.

We observe that the semantics is well-defined:

Lemma 10.

$$
\begin{array}{c}
\vec{\varphi} \,\vec{\subseteq}\, \vec{\varphi}' \Rightarrow e[\vec{\varphi}] \subseteq e[\vec{\varphi}'] \wedge \vec{e}[\vec{\varphi}] \subseteq \vec{e}[\vec{\varphi}'] \\
\vec{\varphi} \,\vec{\sqsubseteq}^{\natural}\, \vec{\varphi}' \Rightarrow e[\vec{\varphi}] \sqsubseteq^{\natural} e[\vec{\varphi}'] \wedge \vec{e}[\vec{\varphi}] \sqsubseteq^{\natural} \vec{e}[\vec{\varphi}'] \\
\vec{F}^{\natural} \in \vec{\mathcal{F}}^{\natural} \stackrel{\mathrm{m}\vec{\subseteq}}{\longmapsto} \vec{F}^{\natural} \;\wedge\; \vec{F}^{\natural} \in \vec{\mathcal{F}}^{\natural} \stackrel{\mathrm{m}\vec{\sqsubseteq}^{\natural}}{\longmapsto} \vec{F}^{\natural}
\end{array}
\tag{46}
$$

Proposition 11. *If $b^{\natural} \in \mathbb{P}(\vec{\mathcal{D}}^{\natural} \times \mathcal{D}^{\natural})$ is a total binary relation for all basic operations b then a program defines a total binary relation*[20]*, that is $\forall f \in \vec{f} : \vec{f}^{\natural}[f] \in \mathbb{P}(\vec{\mathcal{D}}^{\natural} \times \mathcal{D}^{\natural})$.*

$$
\begin{array}{ll}
\nu & : \mathcal{D}^{\natural} \stackrel{\text{def}}{=} \Upsilon_{\perp}^{\Omega} \\
\vec{\nu} & : \vec{\mathcal{D}}^{\natural} \stackrel{\text{def}}{=} \prod_{v \in \vec{v}} \mathcal{D}^{\natural} \\
\varphi, f^{\natural} & : \mathcal{F}^{\natural} \stackrel{\text{def}}{=} \wp(\vec{\mathcal{D}}^{\natural} \times \mathcal{D}^{\natural}) \\
\vec{\varphi}, \vec{f}^{\natural} & : \vec{\mathcal{F}}^{\natural} \stackrel{\text{def}}{=} \prod_{f \in \vec{f}} \mathcal{F}^{\natural}
\end{array}
\qquad
\begin{array}{ll}
e^{\natural} & : \mathcal{E}^{\natural} \stackrel{\text{def}}{=} \vec{\mathcal{F}}^{\natural} \stackrel{\mathrm{m}\vec{\sqsubseteq}^{\natural}}{\longmapsto} \wp(\vec{\mathcal{D}}^{\natural} \times \mathcal{D}^{\natural}) \\
\vec{e}^{\natural} & : \vec{\mathcal{E}}^{\natural} \stackrel{\text{def}}{=} \vec{\mathcal{F}}^{\natural} \stackrel{\mathrm{m}\vec{\sqsubseteq}^{\natural}}{\longmapsto} \wp(\vec{\mathcal{D}}^{\natural} \times \vec{\mathcal{D}}^{\natural}) \\
\vec{F}^{\natural} & : \vec{\mathcal{F}}^{\natural} \stackrel{\mathrm{m}\vec{\sqsubseteq}^{\natural}}{\longmapsto} \vec{\mathcal{F}}^{\natural}
\end{array}
$$

$$
\begin{array}{ll}
k^{\natural}[\vec{\varphi}] & \stackrel{\text{def}}{=} \{\langle \vec{\nu}, \underline{k} \rangle \mid \vec{\nu} \in \vec{\mathcal{D}}^{\natural}\} \\
v^{\natural}[\vec{\varphi}] & \stackrel{\text{def}}{=} \{\langle \vec{\nu}, \vec{\nu}[v] \rangle \mid \vec{\nu} \in \vec{\mathcal{D}}^{\natural}\} \\
b\vec{e}^{\natural}[\vec{\varphi}] & \stackrel{\text{def}}{=} \vec{e}^{\natural}[\![\vec{\varphi}]\!] \circ b^{\natural} \\
(e_1 \rightarrow e_2, e_3)^{\natural}[\![\vec{\varphi}]\!] & \stackrel{\text{def}}{=} \quad \{\langle \vec{\nu}, \perp \rangle \mid \langle \vec{\nu}, \perp \rangle \in e_1^{\natural}[\vec{\varphi}]\} \\
& \quad \cup \{\langle \vec{\nu}, \Omega \rangle \mid \exists \nu \in \Upsilon^{\Omega} - \{\mathrm{tt}, \mathrm{ff}\} : \langle \vec{\nu}, \nu \rangle \in e_1^{\natural}[\vec{\varphi}]\} \\
& \quad \cup \{\langle \vec{\nu}, \nu \rangle \mid \langle \vec{\nu}, \mathrm{tt} \rangle \in e_1^{\natural}[\vec{\varphi}] \wedge \langle \vec{\nu}, \nu \rangle \in e_2^{\natural}[\vec{\varphi}]\} \\
& \quad \cup \{\langle \vec{\nu}, \nu \rangle \mid \langle \vec{\nu}, \mathrm{ff} \rangle \in e_1^{\natural}[\vec{\varphi}] \wedge \langle \vec{\nu}, \nu \rangle \in e_3^{\natural}[\vec{\varphi}]\} \\
f\vec{e}^{\natural}[\vec{\varphi}] & \stackrel{\text{def}}{=} \vec{e}^{\natural}[\![\vec{\varphi}]\!] \circ \vec{\varphi}[f] \\
\vec{e}^{\natural}[\vec{\varphi}] & \stackrel{\text{def}}{=} \{\langle \vec{\nu}, \vec{\nu}' \rangle \mid \forall v \in \vec{e} : \langle \vec{\nu}, \vec{\nu}'[v] \rangle \in \vec{e}[v]^{\natural}[\vec{\varphi}]\}
\end{array}
$$

$$
\begin{array}{ll}
\perp^{\natural} & \stackrel{\text{def}}{=} \vec{\mathcal{D}}^{\natural} \times \{\perp\} \\
\top^{\natural} & \stackrel{\text{def}}{=} \vec{\mathcal{D}}^{\natural} \times \Upsilon^{\Omega} \\
\varphi \sqsubseteq^{\natural} \varphi' & \stackrel{\text{def}}{=} (\varphi \cap \top^{\natural}) \subseteq (\varphi' \cap \top^{\natural}) \;\wedge\; (\varphi \cap \perp^{\natural}) \supseteq (\varphi' \cap \perp^{\natural}) \\
\bigsqcup^{\natural}_{i \in \Delta} \varphi_i & \stackrel{\text{def}}{=} \bigcup_{i \in \Delta}(\varphi_i \cap \top^{\natural}) \;\cup\; \bigcap_{i \in \Delta}(\varphi_i \cap \perp^{\natural}) \\
\vec{F}^{\natural} & \stackrel{\text{def}}{=} \lambda \vec{\varphi} \cdot \prod_{f \in \vec{f}} \vec{F}[f]^{\natural}[\vec{\varphi}] \\
\vec{f}^{\natural} & \stackrel{\text{def}}{=} \mathrm{lfp}^{\vec{\sqsubseteq}^{\natural}}_{\vec{\perp}^{\natural}} \vec{F}^{\natural} \;=\; \vec{\bigsqcup}^{\natural}_{\lambda \in \mathbb{O}} \vec{F}^{\natural\,\lambda}(\vec{\perp}^{\natural})
\end{array}
$$

$$
\begin{array}{ll}
b(\vec{\nu}) \leadsto \nu & \stackrel{\text{def}}{=} \langle \vec{\nu}, \nu \rangle \in b^{\natural} \text{ given such that } b^{\natural} \in \mathcal{F}^{\natural} \\
f(\vec{\nu}) \leadsto \nu & \stackrel{\text{def}}{=} \langle \vec{\nu}, \nu \rangle \in \vec{f}^{\natural}[f] \\
\vec{\varphi}, \vec{\nu} \vdash e \leadsto \nu & \stackrel{\text{def}}{=} \langle \vec{\nu}, \nu \rangle \in e^{\natural}[\vec{\varphi}] \\
\vec{\varphi}, \vec{\nu} \vdash \vec{e} \leadsto \vec{\nu}' & \stackrel{\text{def}}{=} \langle \vec{\nu}, \vec{\nu}' \rangle \in \vec{e}^{\natural}[\vec{\varphi}]
\end{array}
$$

Fig. 1. Synopsis of the relational semantics of $\prod_{f \in \vec{f}} f\vec{v} \equiv \vec{F}[f]$

3 Forward Strictness Analysis by Abstract Interpretation

By applying Alan Mycroft's approximation to the relational semantics, we find again his dependence-sensitive strictness analysis method by mere calculus. In order to

[20] Partial relations r are represented by a total relation R such that $\langle \vec{\nu}, \Omega \rangle \in R$ when $\forall \nu \in \Upsilon_{\perp} : \langle \vec{\nu}, \Omega \rangle \notin r$.

avoid combinatorial explosion, we shortly explore dependence-free strictness analysis methods and finally suggest a compromise using widenings.

3.1 Definition of Strictness

A function φ is *strict* in its parameters $v \in I$ if the evaluation of φ cannot terminate whenever that of its parameters $v \in I$ does not terminate, that is $\varphi(\perp) = \perp$ when φ has only one parameter. If we define $f(x) \equiv (0 \rightarrow f(x), f(x))$ then $f(x) = \Omega$ since 0 is not a boolean so that f is not strict. However, using Mycroft's equations, we conclude that f is strict. Hence differentiating between terminating errors and non-termination is possible but would lead to a different strictness analysis algorithm. Since we want to rediscover Mycroft's algorithm, we can assume, by type checking, that Ω never appears in positions where the abstract equations would differ (i.e. essentially in boolean tests). This is not convincing, e.g. if we want to infer simultaneously type and strictness information. Another solution, as chosen by Mycroft, consists in assimilating Ω and $\perp$ in the semantics. Although commonly accepted, we find that this reasoning is an approximation which should be made explicit. This can be explained by assimilating Ω to $\perp$ in the strictness analysis only, as follows:

Definition 12. $\varphi \in \mathcal{F}^{*} = \mathbb{P}(\vec{\mathcal{D}}^{*} \times \mathcal{D}^{*})$ is *strict* in its parameters $v \in I$ if and only if for all $\vec{\nu} \in \vec{\mathcal{D}}^{*}$ and $\nu \in \mathcal{D}^{*}$:

$$(\forall v \in I : \vec{\nu}[v] \in \{\perp, \Omega\} \wedge \varphi(\vec{\nu}) \rightsquigarrow \nu) \Rightarrow \nu \in \{\perp, \Omega\}$$

Since relation $\varphi(\vec{\nu}) \rightsquigarrow \nu$ is not effectively computable, Mycroft [Myc80, Myc81] uses an approximation φ' of φ for the *approximation ordering* $\subseteq$:

Proposition 13. *For all φ, $\varphi' \in \mathcal{F}^{*}$, if φ' is strict in its parameters $v \in I$ and $\varphi \subseteq \varphi'$ then φ is strict in its parameters $v \in I$.*

Moreover this approximation φ' of φ can be effectively computed by abstract interpretation.

3.2 Forward Dependence-Sensitive Strictness Analysis

À la Mycroft Abstraction Mycroft's original idea [Myc80, Myc81] is to approximate functions $\varphi^{\circ} \in \mathcal{F}^{\circ} = \vec{\mathcal{D}}^{\circ} \mapsto \mathcal{D}^{\circ}$ (where $\mathcal{D}^{\circ}$ is Scott's flat domain) by functions $\varphi^{!} \in \mathcal{F}^{!} = \vec{\mathcal{D}}^{!} \mapsto \mathcal{D}^{!}$. $\mathcal{D}^{!}$ is the complete lattice $\{0, 1\}(\leq, 0, 1, \vee, \wedge)$ with $0 \leq 0 < 1 \leq 1$. 0 is the abstraction of $\perp$ (and Ω) and 1 that of any other value $\nu \in \Upsilon$:

$$\begin{array}{lll} \mathcal{D}^{!} \stackrel{\text{def}}{=} \{0,1\} & \alpha^{!}_{\Upsilon}(\nu) \stackrel{\text{def}}{=} 1 & \text{if } \nu \neq \perp \text{ and } \nu \neq \Omega \\ \alpha^{!}_{\Upsilon} \in \mathcal{D}^{*} \mapsto \mathcal{D}^{!} & \alpha^{!}_{\Upsilon}(\perp) \stackrel{\text{def}}{=} 0 & \\ & \alpha^{!}_{\Upsilon}(\Omega) \stackrel{\text{def}}{=} 0 & \end{array} \tag{47}$$

We only slightly deviate from his judicious choice in that instead of considering functions we use a similar abstraction for relations $\varphi^{*} \in \mathcal{F}^{*}$, using Galois connections.

A set is approximated by the most imprecise approximation of its elements:

$$\alpha^{!}_{\mathcal{D}}(V) \stackrel{\text{def}}{=} \bigvee\{\alpha^{!}_{\Upsilon}(\nu) \mid \nu \in V\} \qquad \gamma^{!}_{\mathcal{D}}(\nu) \stackrel{\text{def}}{=} \{\nu' \mid \alpha^{!}_{\Upsilon}(\nu') \leq \nu\} \tag{48}$$

Proposition 14. *If $\mathcal{D}^{!}(\leq, \vee)$ is a complete lattice and $\alpha^{!}_{\Upsilon} \in \mathcal{D}^{*} \mapsto \mathcal{D}^{!}$ then (48) defines a Galois connection; If $\alpha^{!}_{\Upsilon}$ is surjective then it is a Galois surjection:*

$$\wp(\mathcal{D}^{*})(\subseteq) \underset{\alpha^{!}_{\mathcal{D}}}{\overset{\gamma^{!}_{\mathcal{D}}}{\leftrightarrows}} \mathcal{D}^{!}(\leq) \tag{49}$$

A set of vectors in $\vec{\mathcal{D}}^{\natural}$ is approximated componentwise ($\vec{\mathcal{D}}^{!} \stackrel{\text{def}}{=} \prod_{v \in \vec{v}} \mathcal{D}^{!}$ is a complete lattice for the componentwise ordering $\vec{\nu} \vec{\leq} \vec{\nu}'$ if and only if $\forall v \in \vec{v} : \vec{\nu}[v] \leq \vec{\nu}'[v]$):

$$\begin{aligned} \alpha^{!}_{\vec{\mathcal{D}}}(\vec{V}) &\stackrel{\text{def}}{=} \prod_{v \in \vec{v}} \alpha^{!}_{\mathcal{D}}(\{\vec{\nu}[v] \mid \vec{\nu} \in \vec{V}\}) \\ \gamma^{!}_{\vec{\mathcal{D}}}(\vec{\nu}) &\stackrel{\text{def}}{=} \{\vec{\nu}' \in \vec{\mathcal{D}}^{\natural} \mid \forall v \in \vec{v} : \vec{\nu}'[v] \in \gamma^{!}_{\mathcal{D}}(\vec{\nu}[v])\} \end{aligned} \tag{50}$$

Proposition 15. *(49) and (50) imply:*

$$\wp(\vec{\mathcal{D}}^{\natural})(\subseteq) \underset{\alpha^{!}_{\vec{\mathcal{D}}}}{\overset{\gamma^{!}_{\vec{\mathcal{D}}}}{\leftrightarrows}} \vec{\mathcal{D}}^{!}(\vec{\leq}) \tag{51}$$

Lemma 16. *If $\alpha^{!}_{\Upsilon}$ is surjective then (48) and (50) imply that:*

$$\forall \vec{\nu}^{!} \in \vec{\mathcal{D}}^{!} : \exists \vec{\nu} \in \gamma^{!}_{\vec{\mathcal{D}}}(\vec{\nu}^{!}) : \alpha^{!}_{\vec{\mathcal{D}}}(\{\vec{\nu}\}) = \vec{\nu}^{!} \tag{52}$$

The lifting (16) of a Galois connection to higher-order functional spaces for the pointwise ordering can be generalized to relations as follows[21, 22, 23]:

$$\begin{aligned} \alpha^{!}(\phi) &\stackrel{\text{def}}{=} \lambda \vec{\nu}^{!} \cdot \alpha^{!}_{\mathcal{D}}(\{\nu \mid \exists \vec{\nu} \in \gamma^{!}_{\vec{\mathcal{D}}}(\vec{\nu}^{!}) : \langle \vec{\nu}, \nu \rangle \in \phi\}) = \alpha^{!}_{\mathcal{D}} \circ \phi^{\star} \circ \gamma^{!}_{\vec{\mathcal{D}}} \\ \gamma^{!}(\varphi) &\stackrel{\text{def}}{=} \{\langle \vec{\nu}, \nu \rangle \mid \nu \in \gamma^{!}_{\mathcal{D}} \circ \varphi \circ \alpha^{!}_{\vec{\mathcal{D}}}(\{\vec{\nu}\})\} \qquad = {}^{\varrho}\gamma^{!}_{\mathcal{D}} \circ \varphi \circ \alpha^{!}_{\vec{\mathcal{D}}}{}' \end{aligned} \tag{53}$$

We observe that $\alpha^{!\star}(\mathcal{F}^{\natural}) \subseteq \vec{\mathcal{D}}^{!} \stackrel{m\leq}{\longmapsto} \mathcal{D}^{!}$ since $\alpha^{!}(\phi) = \alpha^{!}_{\mathcal{D}} \circ \phi^{\star} \circ \gamma^{!}_{\vec{\mathcal{D}}}$ is the composition of monotonic functions and complete $\cup$-morphisms, hence monotonic. Therefore, we define:

$$\mathcal{F}^{!} \stackrel{\text{def}}{=} \vec{\mathcal{D}}^{!} \stackrel{m\leq}{\longmapsto} \mathcal{D}^{!} \tag{54}$$

which is a complete lattice $\mathcal{F}^{!}(\leq, \lambda\vec{\nu}\cdot 0, \lambda\vec{\nu}\cdot 1, \vee, \wedge)$ for the pointwise ordering $\varphi \leq \varphi'$ if and only if $\forall \vec{\nu} \in \vec{\mathcal{D}}^{!} : \varphi(\vec{\nu}) \leq \varphi'(\vec{\nu})$. We have:

Proposition 17. *(49), (51), (53) and (54) imply:*

$$\mathcal{F}^{\natural}(\subseteq) \underset{\alpha^{!}}{\overset{\gamma^{!}}{\leftrightarrows}} \mathcal{F}^{!}(\leq) \tag{55}$$

Prop. 17 is independent of the particular definition of $\alpha^{!}_{\Upsilon}$. Taking (47) into account, we have:

Proposition 18 Connection between the relational and strictness semantics. *If $\alpha^{!}_{\Upsilon}$ is surjective (hence, in particular, if (47) holds) then:*

$$\mathcal{F}^{\natural}(\subseteq) \underset{\alpha^{!}}{\overset{\gamma^{!}}{\leftrightarrows}} \mathcal{F}^{!}(\leq) \tag{56}$$

A vector of relations in $\vec{\mathcal{F}}^{\natural} = \prod_{f \in \vec{f}} \mathcal{F}^{\natural}$ is approximated componentwise in $\vec{\mathcal{F}}^{!} \stackrel{\text{def}}{=} \prod_{f \in \vec{f}} \mathcal{F}^{!}$ as follows:

$$\vec{\alpha}^{!}(\vec{\phi}) \stackrel{\text{def}}{=} \prod_{f \in \vec{f}} \alpha^{!}(\vec{\phi}[f]) \qquad \vec{\gamma}^{!}(\vec{\varphi}) \stackrel{\text{def}}{=} \prod_{f \in \vec{f}} \gamma^{!}(\vec{\varphi}[f]) \tag{57}$$

---

[21] The *image* of $X \subseteq S$ by relation $\rho \in \wp(S \times T)$ is $\rho^{\star}(X)$ where $\rho^{\star} \in \wp(S) \mapsto \wp(T)$ is defined by $\rho^{\star}(X) \stackrel{\text{def}}{=} \{t \mid \exists s \in X : \langle s, t \rangle \in \rho\}$. In particular for a total function $\varphi \in S \mapsto T$, $\varphi^{\star}(X) = \{\varphi(s) \mid s \in X\}$.

[22] The *projection* $\varphi' \in S \mapsto T$ of $\varphi \in \wp(S) \mapsto T$ is defined by $\varphi'(s) \stackrel{\text{def}}{=} \varphi(\{s\})$.

[23] If $f \in S \mapsto \wp(T)$ then ${}^{\varrho}f$ is the relation $\{\langle s, t \rangle \mid t \in f(s)\}$.

Proposition 19. *(56) and (57) imply:*

$$\vec{\mathcal{F}}^{\star}(\vec{\subseteq}) \underset{\vec{\alpha}^{!}}{\overset{\vec{\gamma}^{!}}{\leftrightarrows}} \vec{\mathcal{F}}^{!}(\vec{\lesssim}) \tag{58}$$

Using Prop. 5, we lift this approximation to function spaces as follows:

$$\alpha^{!}_{f}(\vec{\phi}) \overset{\text{def}}{=} \vec{\alpha}^{!} \circ \vec{\phi} \circ \vec{\gamma}^{!} \qquad \gamma^{!}_{f}(\vec{\varphi}) \overset{\text{def}}{=} \vec{\gamma}^{!} \circ \vec{\varphi} \circ \vec{\alpha}^{!} \tag{59}$$

Proposition 20. *(58) implies:*

$$\vec{\mathcal{F}}^{\star} \overset{m\subseteq}{\longmapsto} \vec{\mathcal{F}}^{\star}(\vec{\subseteq}) \underset{\alpha^{!}_{f}}{\overset{\gamma^{!}_{f}}{\leftrightarrows}} \vec{\mathcal{F}}^{!} \overset{m\lesssim}{\longmapsto} \vec{\mathcal{F}}^{!}(\vec{\lesssim}) \tag{60}$$

The same idea (19) can be applied to the approximation $e^{!} \in \mathcal{E}^{!}$ of the semantics $e^{\star} \in \mathcal{E}^{\star}$ of expressions e by defining $\mathcal{E}^{!} \overset{\text{def}}{=} \vec{\mathcal{F}}^{!} \overset{m\lesssim}{\longmapsto} \vec{\mathcal{D}}^{!} \overset{m\lesssim}{\longmapsto} \mathcal{D}^{!}$ and:

$$\alpha^{!}_{\varepsilon}(\varepsilon) \overset{\text{def}}{=} \alpha^{!} \circ \varepsilon \circ \vec{\gamma}^{!} \qquad \gamma^{!}_{\varepsilon}(\epsilon) \overset{\text{def}}{=} \gamma^{!} \circ \epsilon \circ \vec{\alpha}^{!} \tag{61}$$

Proposition 21. *(61), (56) and (58) imply:*

$$\mathcal{E}^{\star}(\vec{\subseteq}) \underset{\alpha^{!}_{\varepsilon}}{\overset{\gamma^{!}_{\varepsilon}}{\leftrightarrows}} \mathcal{E}^{!}(\vec{\lesssim}) \tag{62}$$

In order to approximate the semantics $\vec{e}^{\star}$ of vectors $\vec{e}$ of expressions, we start by approximating elements of $\wp(\vec{\mathcal{D}}^{\star} \times \vec{\mathcal{D}}^{\star})$ by elements of the complete lattice $\vec{\mathcal{D}}^{!} \overset{m\lesssim}{\longmapsto} \vec{\mathcal{D}}^{!}$ $(\vec{\lesssim}, \lambda\vec{v}\cdot\vec{0}, \lambda\vec{v}\cdot\vec{1}, \vec{\vee}, \vec{\wedge})$ for the componentwise ordering $\vec{\varphi} \vec{\lesssim} \vec{\varphi}'$ if and only if $\forall v \in \vec{v} : \vec{\varphi}[v] \leq \vec{\varphi}'[v]$ so that $\vec{0} = \prod_{v \in \vec{v}} 0$. We define:

$$\begin{aligned} \alpha^{!}_{\pi}(\vec{\phi}) &\overset{\text{def}}{=} \lambda\vec{v}^{!}\cdot\alpha^{!}_{\vec{\mathcal{D}}}(\{\vec{v}' \mid \exists\vec{v} \in \gamma^{!}_{\vec{\mathcal{D}}}(\vec{v}^{!}) : \langle\vec{v}, \vec{v}'\rangle \in \vec{\phi}\}) = \alpha^{!}_{\vec{\mathcal{D}}} \circ \vec{\phi}^{\star} \circ \gamma^{!}_{\vec{\mathcal{D}}} \\ \gamma^{!}_{\pi}(\vec{\varphi}) &\overset{\text{def}}{=} \{\langle\vec{v}, \vec{v}'\rangle \mid \vec{v}' \in \gamma^{!}_{\vec{\mathcal{D}}} \circ \vec{\varphi} \circ \alpha^{!}_{\vec{\mathcal{D}}}(\{\vec{v}\})\} \qquad = {}^{\varrho}\gamma^{!}_{\vec{\mathcal{D}}} \circ \vec{\varphi} \circ \alpha^{!\,'}_{\vec{\mathcal{D}}} \end{aligned} \tag{63}$$

Proposition 22.

$$\wp(\vec{\mathcal{D}}^{\star} \times \vec{\mathcal{D}}^{\star})(\subseteq) \underset{\alpha^{!}_{\pi}}{\overset{\gamma^{!}_{\pi}}{\leftrightarrows}} \vec{\mathcal{D}}^{!} \overset{m\lesssim}{\longmapsto} \vec{\mathcal{D}}^{!}(\vec{\lesssim}) \tag{64}$$

Finally the approximation $\vec{e}^{!} \in \vec{\mathcal{E}}^{!}$ of the semantics $\vec{e}^{\star} \in \vec{\mathcal{E}}^{\star}$ of a vector $\vec{e}$ of expressions is defined in $\vec{\mathcal{E}}^{!} \overset{\text{def}}{=} \vec{\mathcal{F}}^{!} \overset{m\lesssim}{\longmapsto} \vec{\mathcal{D}}^{!} \overset{m\lesssim}{\longmapsto} \vec{\mathcal{D}}^{!}$ by:

$$\alpha^{!}_{\vec{\varepsilon}}(\varepsilon) \overset{\text{def}}{=} \alpha^{!}_{\pi} \circ \varepsilon \circ \vec{\gamma}^{!} \qquad \gamma^{!}_{\vec{\varepsilon}}(\epsilon) \overset{\text{def}}{=} \gamma^{!}_{\pi} \circ \epsilon \circ \vec{\alpha}^{!} \tag{65}$$

Proposition 23. *(65), (61) and (64) imply:*

$$\vec{\mathcal{E}}^{\star}(\vec{\subseteq}) \underset{\alpha^{!}_{\vec{\varepsilon}}}{\overset{\gamma^{!}_{\vec{\varepsilon}}}{\leftrightarrows}} \vec{\mathcal{E}}^{!}(\vec{\lesssim}) \tag{66}$$

Approximation and Computational Orderings Observe that $\vec{\lesssim}$ is the abstraction of the approximation ordering $\vec{\subseteq}$:

$$\vec{\varphi} \vec{\lesssim} \vec{\varphi}' \Leftrightarrow \vec{\gamma}^{!}(\vec{\varphi}) \vec{\subseteq} \vec{\gamma}^{!}(\vec{\varphi}') \tag{67}$$

The approximation and computational orderings differ in the concrete semantics but coincide in the abstract semantics[24] :

$$\begin{array}{rclcrcl} \phi \subseteq \gamma^{!}(\varphi) & \Leftrightarrow & \phi \sqsubseteq^{*} \gamma^{!}(\varphi) & \quad & \vec{\phi} \mathrel{\vec{\subseteq}} \vec{\gamma}^{!}(\vec{\varphi}) & \Leftrightarrow & \vec{\phi} \mathrel{\vec{\sqsubseteq}}^{*} \vec{\gamma}^{!}(\vec{\varphi}) \\ \varphi \leq \varphi' & \Leftrightarrow & \gamma^{!}(\varphi) \sqsubseteq^{*} \gamma^{!}(\varphi') & & \vec{\varphi} \mathrel{\vec{\leq}} \vec{\varphi}' & \Leftrightarrow & \vec{\gamma}^{!}(\vec{\varphi}) \mathrel{\vec{\sqsubseteq}}^{*} \vec{\gamma}^{!}(\vec{\varphi}') \end{array} \tag{68}$$

An immediate consequence is that we have the following Galois surjections:

$$\mathcal{F}^{*}(\sqsubseteq^{*}) \underset{\alpha^{!}}{\overset{\gamma^{!}}{\leftrightarrows}} \mathcal{F}^{!}(\leq) \qquad \vec{\mathcal{F}}^{*}(\vec{\sqsubseteq}^{*}) \underset{\vec{\alpha}^{!}}{\overset{\vec{\gamma}^{!}}{\leftrightarrows}} \vec{\mathcal{F}}^{!}(\vec{\leq}) \tag{69}$$

Forward Dependence-Sensitive Strictness Semantics The abstract semantics of a program $\prod_{f\epsilon f} f\vec{v} \equiv \vec{F}[f]$ is specified by $\vec{f}^{!}$ which we would like to define such that $\vec{f}^{*} = \mathrm{lfp}^{\vec{\sqsubseteq}^{*}}_{\vec{\bot}^{*}} \vec{F}^{*} \mathrel{\vec{\subseteq}} \vec{\gamma}^{!}(\vec{f}^{!})$. Corollary 9 provides a method for refining this specification as $\vec{f}^{!} \stackrel{\text{def}}{=} \vec{\sqcup}^{!}_{n\in\mathbb{N}} \vec{F}^{!\,n}(\vec{\bot}^{!})$ by defining the infimum $\vec{\bot}^{!} \stackrel{\text{def}}{=} \vec{\alpha}^{!}(\vec{\bot}^{*})$, the inductive join $\vec{\sqcup}^{!}_{i\in\Delta}\vec{\varphi}_i \stackrel{\text{def}}{=} \vec{\alpha}^{!}(\vec{\sqcup}^{*}_{i\in\Delta}\vec{\gamma}^{!}(\vec{\varphi}_i))$ and the semantics function $\vec{F}^{!}$ such that $\vec{\alpha}^{!} \circ \vec{F}^{*} \circ \vec{\gamma}^{!} \mathrel{\vec{\leq}^{!}} \vec{F}^{!}$. Observe that monotonicity and finiteness of the lattice imply continuity and more precisely convergence below ω in (8). We now simplify these formulæ in order to get the formal definition of $\vec{\bot}^{!}$, $\vec{\sqcup}^{!}$ and $\vec{F}^{!}$. The hand-computation presents no difficulties. It is provided just to show that the method is mathematically constructive. The strictness semantics, that is the formal specification of the abstract interpreter for strictness analysis, is entirely determined by the choice of the collecting semantics in Sec. 2.4 and the choice of the approximations (58) and (69). The derivation of $\vec{\bot}^{!}$, $\vec{\sqcup}^{!}$ and $\vec{F}^{!}$ is nothing else than a refinement of this formal specification. This computation is certainly amenable to automation. Moreover, the replacement/simplification/definition introduction strategies are very similar for different abstract domains so that proof strategies should be definable for a given relational semantics so as to guide the automatic derivation of the abstract semantics.

$$\begin{array}{lll}
- \ \vec{\bot}^{!} & = \vec{\alpha}^{!}(\vec{\bot}^{*}) & \text{by definition of } \vec{\bot}^{!} \\
& = \prod_{f\epsilon f} \alpha^{!}(\vec{\bot}^{*}[f]) & \text{by (57)} \\
& = \prod_{f\epsilon f} \alpha^{!}(\vec{\Upsilon}^{n}_{\bot} \times \{\bot\}) & \text{by (35)} \\
& = \prod_{f\epsilon f} \lambda\vec{\nu}^{!}\cdot\alpha^{!}_{\mathcal{D}}(\{\nu \mid \exists\vec{\nu} \in \gamma^{!}_{\vec{\mathcal{D}}}(\vec{\nu}^{!}) : \langle\vec{\nu},\, \nu\rangle \in (\vec{\Upsilon}^{n}_{\bot} \times \{\bot\})\}) & \text{by (53)} \\
& = \prod_{f\epsilon f} \lambda\vec{\nu}^{!}\cdot\alpha^{!}_{\mathcal{D}}(\{\bot \mid \exists\vec{\nu} \in \gamma^{!}_{\vec{\mathcal{D}}}(\vec{\nu}^{!})\}) & \text{since } \vec{\Upsilon}^{n}_{\bot} \neq \emptyset \\
& = \prod_{f\epsilon f} \lambda\vec{\nu}^{!}\cdot\alpha^{!}_{\mathcal{D}}(\{\bot \mid \exists\vec{\nu} \in \{\vec{\nu}' \in \vec{\mathcal{D}}^{*} \mid \forall v \,\epsilon\, \vec{v} : \vec{\nu}'[v] \in \gamma^{!}_{\mathcal{D}}(\vec{\nu}^{!}[v])\}\}) & \text{by (50)} \\
& = \prod_{f\epsilon f} \lambda\vec{\nu}^{!}\cdot\alpha^{!}_{\mathcal{D}}(\{\bot \mid \exists\vec{\nu} \in \vec{\mathcal{D}}^{*} : \forall v \,\epsilon\, \vec{v} : \vec{\nu}[v] \in \{\nu' \mid \alpha^{!}_{\Upsilon}(\nu') \leq \vec{\nu}^{!}[v]\}\}) & \text{by (48)} \\
& = \prod_{f\epsilon f} \lambda\vec{\nu}^{!}\cdot\alpha^{!}_{\mathcal{D}}(\{\bot\}) & \text{by (47), choosing } \vec{\nu} = \lambda v\cdot\bot \\
& = \prod_{f\epsilon f} \lambda\vec{\nu}^{!}\cdot\alpha^{!}_{\Upsilon}(\bot) & \text{by (48) and definition of lubs} \\
& = \prod_{f\epsilon f} \lambda\vec{\nu}^{!}\cdot 0 & \text{by (47)}
\end{array}$$

[24] whereas in [CC77a] they also coincide in the semantics collecting invariance properties, where non-termination is ignored, and therefore is obtained by further application of the abstraction $\alpha(V) = V \cup \{\bot\}$ to $\mathcal{D}^{*}$.

- $\vec{\sqcup}^{!}{}_{i \in \Delta} \vec{\varphi}_i = \vec{\alpha}^{!}(\vec{\sqcup}^{*}{}_{i \in \Delta} \vec{\gamma}^{!}(\vec{\varphi}_i))$ by definition of $\vec{\sqcup}$
  $= \vec{\nabla}^{!}{}_{i \in \Delta} \vec{\alpha}^{!}(\vec{\gamma}^{!}(\vec{\varphi}_i))$ since $\vec{\alpha}^{!}$ is a complete $\vec{\sqcup}^{*}$-morphism by (69)
  $= \vec{\vee}^{!}{}_{i \in \Delta} \vec{\varphi}_i$ since $\vec{\alpha}^{!} \circ \vec{\gamma}^{!}$ is the identity by (69)
- For a program $\prod_{f \epsilon \vec{f}} f\vec{v} \equiv \vec{F}[f]$, the abstract semantics function $\vec{F}^{!}$ is defined as an upper approximation of $\vec{\alpha}^{!} \circ \vec{F}^{*} \circ \vec{\gamma}^{!}$. We have:

$$\vec{\alpha}^{!} \circ \vec{F}^{*} \circ \vec{\gamma}^{!} = \lambda\vec{\varphi} \cdot \vec{\alpha}^{!}\left(\prod_{f \epsilon \vec{f}} \vec{F}[f]^{*}[\vec{\gamma}^{!}(\vec{\varphi})]\right) \quad \text{by (39)}$$
$$= \lambda\vec{\varphi} \cdot \prod_{f \epsilon \vec{f}} \alpha^{!}(\vec{F}[f]^{*}[\vec{\gamma}^{!}(\vec{\varphi})]) \quad \text{by (57)}$$
$$= \lambda\vec{\varphi} \cdot \prod_{f \epsilon \vec{f}} \alpha^{!}_{\varepsilon}(\vec{F}[f]^{*})[\vec{\varphi}] \quad \text{by (61)}$$

This suggest the definition $\vec{F}^{!} \stackrel{\text{def}}{=} \lambda\vec{\varphi} \cdot \prod_{f \epsilon \vec{f}} \vec{F}[f]^{!}[\vec{\varphi}]$ with the condition that for all $f \epsilon \vec{f}$, we have $\vec{F}[f]^{!} \geq \alpha^{!}_{\varepsilon}(\vec{F}[f]^{*})$. This condition is obviously satisfied by generalization to all expressions as $e^{!} \geq \alpha^{!}_{\varepsilon}(e^{*})$. We proceed by structural induction on the syntax of e. The basis corresponds to the cases when e is reduced to a constant k or variable v:

- $k^{!}[\vec{\varphi}] = \alpha^{!}_{\varepsilon}(k^{*})[\vec{\varphi}] = \alpha^{!} \circ k^{*} \circ \vec{\gamma}^{!}[\vec{\varphi}] = \alpha^{!}(\{\langle\vec{\nu}, \underline{k}\rangle \mid \vec{\nu} \in \vec{\mathcal{D}}^{*}\}) = \lambda\vec{\nu}^{!} \cdot \alpha^{!}_{\mathcal{D}}(\{\underline{k} \mid \vec{\nu} \in \gamma^{!}_{\mathfrak{D}}(\vec{\nu}^{!})\}) = \lambda\vec{\nu}^{!} \cdot \bigvee\{\alpha^{!}_{\Upsilon}(\underline{k}) \mid \vec{\nu} \in \gamma^{!}_{\mathfrak{D}}(\vec{\nu}^{!})\} = \lambda\vec{\nu}^{!} \cdot 1$, by (61), (40), (53), (48) and (47) since $\underline{k} \in \Upsilon$ and $\gamma^{!}_{\mathfrak{D}}(\vec{\nu})$ is not empty.
- $v^{!}[\vec{\varphi}] = \alpha^{!}_{\varepsilon}(v^{*})[\vec{\varphi}] = \alpha^{!} \circ v^{*} \circ \vec{\gamma}^{!}[\vec{\varphi}] = \alpha^{!}(\{\langle\vec{\nu}, \vec{\nu}[v]\rangle \mid \vec{\nu} \in \vec{\mathcal{D}}^{*}\})$ by (61) and (41). By (53) this is equal to $\lambda\vec{\nu}^{!} \cdot \bigvee\{\alpha^{!}_{\Upsilon}(\vec{\nu}[v]) \mid \exists\vec{\nu} \in \gamma^{!}_{\mathfrak{D}}(\vec{\nu}^{!})\}$, hence, by (48) and definition (50) of $\gamma^{!}_{\mathfrak{D}}$, to $\lambda\vec{\nu}^{!} \cdot \bigvee\{\alpha^{!}_{\Upsilon}(\nu) \mid \nu \in \gamma^{!}_{\mathcal{D}}(\vec{\nu}^{!}[v])\}$. By (48), this is equal to $\lambda\vec{\nu}^{!} \cdot \alpha^{!}_{\mathcal{D}}(\gamma^{!}_{\mathcal{D}}(\vec{\nu}^{!}[v])) = \lambda\vec{\nu}^{!} \cdot \vec{\nu}^{!}[v]$ since, by (49), $\alpha^{!}_{\mathcal{D}} \circ \gamma^{!}_{\mathcal{D}}$ is the identity.

For the induction step, we must prove that $e^{!} \geq \alpha^{!}_{\varepsilon}(e^{*})$ where e is $b\vec{e}$, $f\vec{e}$ or $(e_1 \rightarrow e_2, e_3)$. We can assume, by induction hypothesis, that $e^{!}_{i} \geq \alpha^{!}_{\varepsilon}(e^{*}_{i})$, $i = 1, 2, 3$ and that for all $v \epsilon \vec{e}$, $\vec{e}[v]^{!} \geq \alpha^{!}_{\varepsilon}(\vec{e}[v]^{*})$.

- We define $\vec{e}^{!}[\vec{\varphi}^{!}]\vec{\nu}^{!} \stackrel{\text{def}}{=} \prod_{v \epsilon \vec{e}} \vec{e}[v]^{!}[\vec{\varphi}]\vec{\nu}$ and first prove that the induction hypothesis implies that $\vec{e}^{!} \gtrsim \alpha^{!}_{\varepsilon}(\vec{e}^{*})$.

By (65), $\alpha^{!}_{\varepsilon}(\vec{e}^{*}) = \alpha^{!}_{*} \circ \vec{e}^{*} \circ \vec{\gamma}^{!} = \lambda\vec{\varphi}^{!} \cdot \alpha^{!}_{*}(\vec{e}^{*}[\vec{\gamma}^{!}(\vec{\varphi}^{!})])$ which, by (63), equals $\lambda\vec{\varphi}^{!} \cdot \alpha^{!}_{\mathfrak{D}} \circ (\vec{e}^{*}[\vec{\gamma}^{!}(\vec{\varphi}^{!})])^{\star} \circ \gamma^{!}_{\mathfrak{D}} = \lambda\vec{\varphi}^{!} \cdot \lambda\vec{\nu}^{!} \cdot \alpha^{!}_{\mathfrak{D}}\left((\vec{e}^{*}[\vec{\gamma}^{!}(\vec{\varphi}^{!})])^{\star}(\gamma^{!}_{\mathfrak{D}}(\vec{\nu}^{!}))\right)$. By (50), this is equal to $\lambda\vec{\varphi}^{!} \cdot \lambda\vec{\nu}^{!} \cdot \prod_{v \epsilon \vec{v}} \alpha^{!}_{\mathcal{D}}(\{\vec{\nu}'[v] \mid \vec{\nu}' \in (\vec{e}^{*}[\vec{\gamma}^{!}(\vec{\varphi}^{!})])^{\star}(\gamma^{!}_{\mathfrak{D}}(\vec{\nu}^{!}))\}) = \text{LHS}$. By definition of the image $\star$, $\vec{\nu}' \in (\vec{e}^{*}[\vec{\gamma}^{!}(\vec{\varphi}^{!})])^{\star}(\gamma^{!}_{\mathfrak{D}}(\vec{\nu}^{!}))$ is equivalent to $\exists\vec{\nu} \in \gamma^{!}_{\mathfrak{D}}(\vec{\nu}^{!})$: $\langle\vec{\nu}, \vec{\nu}'\rangle \in \vec{e}^{*}[\vec{\gamma}^{!}(\vec{\varphi}^{!})]$ hence, by definition (42) of $\vec{e}^{*}$, to $\exists\vec{\nu} \in \gamma^{!}_{\mathfrak{D}}(\vec{\nu}^{!}) : \forall v \epsilon \vec{e} : \langle\vec{\nu}, \vec{\nu}'[v]\rangle \in \vec{e}[v]^{*}[\vec{\gamma}^{!}(\vec{\varphi}^{!})]$ which, for all $v \epsilon \vec{e}$ implies $\exists\vec{\nu} \in \gamma^{!}_{\mathfrak{D}}(\vec{\nu}^{!}) : \langle\vec{\nu}, \vec{\nu}'[v]\rangle \in \vec{e}[v]^{*}[\vec{\gamma}^{!}(\vec{\varphi}^{!})]$. By (49), $\alpha^{!}_{\mathcal{D}}$ is monotonic, whence LHS $\lesssim \lambda\vec{\varphi}^{!} \cdot \lambda\vec{\nu}^{!} \cdot \prod_{v \epsilon \vec{v}} \alpha^{!}_{\mathcal{D}}(\{\vec{\nu}'[v] \mid \exists\vec{\nu} \in \gamma^{!}_{\mathfrak{D}}(\vec{\nu}^{!}) : \langle\vec{\nu}, \vec{\nu}'[v]\rangle \in \vec{e}[v]^{*}[\vec{\gamma}^{!}(\vec{\varphi}^{!})]\})$. This is $\lambda\vec{\varphi}^{!} \cdot \lambda\vec{\nu}^{!} \cdot \prod_{v \epsilon \vec{v}} \alpha^{!}_{\mathcal{D}}(\{\nu \mid \exists\vec{\nu} \in \gamma^{!}_{\mathfrak{D}}(\vec{\nu}^{!}) : \langle\vec{\nu}, \nu\rangle \in \vec{e}[v]^{*}[\vec{\gamma}^{!}(\vec{\varphi}^{!})]\})$, if we let ν be $\vec{\nu}'[v]$. By definition (53) of $\alpha^{!}$, this is equal to $\lambda\vec{\varphi}^{!} \cdot \lambda\vec{\nu}^{!} \cdot \prod_{v \epsilon \vec{v}} \alpha^{!}(\vec{e}[v]^{*}[\vec{\gamma}^{!}(\vec{\varphi}^{!})])\vec{\nu}$, hence by definition (61) of $\alpha^{!}_{\varepsilon}$ to $\lambda\vec{\varphi}^{!} \cdot \lambda\vec{\nu}^{!} \cdot \prod_{v \epsilon \vec{v}} \alpha^{!}_{\varepsilon}(\vec{e}[v]^{*})[\vec{\varphi}^{!}]\vec{\nu} = \prod_{v \epsilon \vec{v}} \alpha^{!}_{\varepsilon}(\vec{e}[v]^{*}) \lesssim \prod_{v \epsilon \vec{v}} \vec{e}[v]^{!} = \vec{e}^{!}$ by induction hypothesis and definition of $\vec{e}^{!}$.

- If e is $b\vec{e}$ then we must define $b\vec{e}^{!}$ such that $b\vec{e}^{!} \geq \alpha^{!}_{\varepsilon}(b\vec{e}^{*})$. We do this by formal hand-computation. This consists is expanding the term $\alpha^{!}_{\varepsilon}(b\vec{e}^{*})$ in order to let appear the subterm $\alpha^{!}_{\mathcal{E}}(\vec{e}^{\ast})$. By (61), $\alpha^{!}_{\mathcal{E}}(b\vec{e}^{\ast})$ is equal to $\alpha^{!} \circ b\vec{e}^{\ast} \circ \vec{\gamma}^{!} = \lambda\vec{\varphi}^{!}\cdot\alpha^{!}(b\vec{e}^{\ast}[\vec{\gamma}^{!}(\vec{\varphi}^{!})])$. By (53), this is equal to $\lambda\vec{\varphi}^{!}\cdot\alpha^{!}_{\mathcal{D}} \circ (b\vec{e}^{\ast}[\vec{\gamma}^{!}(\vec{\varphi}^{!})])^{\star} \circ \gamma^{!}_{\mathcal{B}} = \lambda\vec{\varphi}^{!}\cdot\lambda\vec{\nu}^{!}\cdot\alpha^{!}_{\mathcal{D}}((b\vec{e}^{\ast}[\vec{\gamma}^{!}(\vec{\varphi}^{!})])^{\star}(\gamma^{!}_{\mathcal{B}}(\vec{\nu}^{!})))$. By definition (43) of $b\vec{e}^{\ast}$, this is equal to $\lambda\vec{\varphi}^{!}\cdot\lambda\vec{\nu}^{!}\cdot\alpha^{!}_{\mathcal{D}}((\vec{e}^{\ast}[\vec{\gamma}^{!}(\vec{\varphi}^{!})] \circ b^{\ast})^{\star}(\gamma^{!}_{\mathcal{B}}(\vec{\nu}^{!})))$ and, since $(\rho \circ \varrho)^{\star}(X) = \varrho^{\star}(\rho^{\star}(X))$, to: $\lambda\vec{\varphi}^{!}\cdot\lambda\vec{\nu}^{!}\cdot\alpha^{!}_{\mathcal{D}}(b^{\ast\star}(\vec{e}^{\ast}[\vec{\gamma}^{!}(\vec{\varphi}^{!})]^{\star}(\gamma^{!}_{\mathcal{B}}(\vec{\nu}^{!})))) = \text{LHS}$. By (51), $\gamma^{!}_{\mathcal{B}} \circ \alpha^{!}_{\mathcal{B}}$ is extensive so that $b^{\ast\star}(\vec{e}^{\ast}[\vec{\gamma}^{!}(\vec{\varphi}^{!})]^{\star}(\gamma^{!}_{\mathcal{B}}(\vec{\nu}^{!}))) \subseteq b^{\ast\star}(\gamma^{!}_{\mathcal{B}}(\alpha^{!}_{\mathcal{B}}(\vec{e}^{\ast}[\vec{\gamma}^{!}(\vec{\varphi}^{!})]^{\star}(\gamma^{!}_{\mathcal{B}}(\vec{\nu}^{!})))))$. Moreover $\alpha^{!}_{\mathcal{D}}$ is monotonic so that $\text{LHS} \leq \lambda\vec{\varphi}^{!}\cdot\lambda\vec{\nu}^{!}\cdot\alpha^{!}_{\mathcal{D}}(b^{\ast\star}(\gamma^{!}_{\mathcal{B}}(\alpha^{!}_{\mathcal{B}} \circ \vec{e}^{\ast}[\vec{\gamma}^{!}(\vec{\varphi}^{!})]^{\star} \circ \gamma^{!}_{\mathcal{B}}(\vec{\nu}^{!})))) = \lambda\vec{\varphi}^{!}\cdot\lambda\vec{\nu}^{!}\cdot\alpha^{!}_{\mathcal{D}} \circ b^{\ast\star} \circ \gamma^{!}_{\mathcal{B}}(\alpha^{!}_{\pi}(\vec{e}^{\ast}[\vec{\gamma}^{!}(\vec{\varphi}^{!})])(\vec{\nu}^{!}))$ by definition (63) of $\alpha^{!}_{\pi}$. By (53), this is $\lambda\vec{\varphi}^{!}\cdot\lambda\vec{\nu}^{!}\cdot\alpha^{!}(b^{\ast})(\alpha^{!}_{\pi} \circ \vec{e}^{\ast} \circ \vec{\gamma}^{!}(\vec{\varphi}^{!})\vec{\nu}^{!})$, which, by (65), is equal to $\lambda\vec{\varphi}^{!}\cdot\lambda\vec{\nu}^{!}\cdot\alpha^{!}(b^{\ast})(\alpha^{!}_{\mathcal{E}}(\vec{e}^{\ast})[\vec{\varphi}^{!}]\vec{\nu}^{!}) = \text{LHS}$. This suggests to define $b^{!}$ as $\alpha^{!}(b^{\ast})$ but in order to allow for further approximations, we assume that $b^{!} \in \vec{\mathcal{D}}^{!} \stackrel{m\leq^{!}}{\longmapsto} \mathcal{D}^{!}$ is such that $b^{!} \geq^{!} \alpha^{!}(b^{\ast})$. By definition of the pointwise ordering $\leq^{!}$, it follows that $\text{LHS} \leq^{!} \lambda\vec{\varphi}^{!}\cdot\lambda\vec{\nu}^{!}\cdot b^{!}(\alpha^{!}_{\mathcal{E}}(\vec{e}^{\ast})[\vec{\varphi}^{!}]\vec{\nu}^{!}) \leq^{!} \lambda\vec{\varphi}^{!}\cdot\lambda\vec{\nu}^{!}\cdot b^{!}(\vec{e}^{!}[\vec{\varphi}^{!}]\vec{\nu}^{!})$ by induction hypothesis $\vec{e}^{!} \vec{\geq} \alpha^{!}_{\mathcal{E}}(\vec{e}^{\ast})$ and monotonicity of $b^{!}$. We define $b\vec{e}^{!} \stackrel{\text{def}}{=} \lambda\vec{\varphi}^{!}\cdot\lambda\vec{\nu}^{!}\cdot b^{!}(\vec{e}^{!}[\vec{\varphi}^{!}]\vec{\nu}^{!})$ so that we have proved that $b\vec{e}^{!} \geq \alpha^{!}_{\mathcal{E}}(b\vec{e}^{\ast})$.

- If e is $f\vec{e}$, $\alpha^{!}_{\mathcal{E}}(f\vec{e}^{\ast}) = \lambda\vec{\varphi}^{!}\cdot\lambda\vec{\nu}^{!}\cdot\alpha^{!}_{\mathcal{D}}((f\vec{e}^{\ast}[\vec{\gamma}^{!}(\vec{\varphi}^{!})])^{\star}(\gamma^{!}_{\mathcal{B}}(\vec{\nu}^{!})))$, as above. By definition (45) of $f\vec{e}^{\ast}$, this is: $\lambda\vec{\varphi}^{!}\cdot\lambda\vec{\nu}^{!}\cdot\alpha^{!}_{\mathcal{D}}((\vec{e}^{\ast}[\vec{\gamma}^{!}(\vec{\varphi}^{!})] \circ \vec{\gamma}^{!}(\vec{\varphi}^{!})[f])^{\star}(\gamma^{!}_{\mathcal{B}}(\vec{\nu}^{!})))$. As above, this is: $\lambda\vec{\varphi}^{!}\cdot\lambda\vec{\nu}^{!}\cdot\alpha^{!}(\vec{\gamma}^{!}(\vec{\varphi}^{!})[f])(\alpha^{!}_{\mathcal{E}}(\vec{e}^{\ast})[\vec{\varphi}^{!}]\vec{\nu}^{!})$. By definition (57) of $\vec{\gamma}^{!}$, this is equal to $\lambda\vec{\varphi}^{!}\cdot\lambda\vec{\nu}^{!}\cdot\alpha^{!}(\gamma^{!}(\vec{\varphi}^{!}[f]))(\alpha^{!}_{\mathcal{E}}(\vec{e}^{\ast})[\vec{\varphi}^{!}]\vec{\nu}^{!})$. But $\alpha^{!} \circ \gamma^{!}$ is the identity, a characteristic property of the Galois surjection (56). We get $\lambda\vec{\varphi}^{!}\cdot\lambda\vec{\nu}^{!}\cdot\vec{\varphi}^{!}[f](\alpha^{!}_{\mathcal{E}}(\vec{e}^{\ast})[\vec{\varphi}^{!}]\vec{\nu}^{!}) \leq^{!} \lambda\vec{\varphi}^{!}\cdot\lambda\vec{\nu}^{!}\cdot\vec{\varphi}^{!}[f](\vec{e}^{!}[\vec{\varphi}^{!}]\vec{\nu}^{!})$ by induction hypothesis $\vec{e}^{!} \vec{\geq} \alpha^{!}_{\mathcal{E}}(\vec{e}^{\ast})$ and monotonicity of $\vec{\varphi}^{!}[f] \in \vec{\mathcal{D}}^{!} \stackrel{m\leq^{!}}{\longmapsto} \mathcal{D}^{!}$. We define $f\vec{e}^{!} \stackrel{\text{def}}{=} \lambda\vec{\varphi}^{!}\cdot\lambda\vec{\nu}^{!}\cdot\vec{\varphi}^{!}[f](\vec{e}^{!}[\vec{\varphi}^{!}]\vec{\nu}^{!})$ so that we have proved that $f\vec{e}^{!} \geq \alpha^{!}_{\mathcal{E}}(f\vec{e}^{\ast})$.

- It e is $(e_1 \rightarrow e_2, e_3)$ then $\alpha^{!}_{\mathcal{E}}((e_1 \rightarrow e_2, e_3)^{\ast})$ can be shown, as above, to be equal to $\lambda\vec{\varphi}^{!}\cdot\lambda\vec{\nu}^{!}\cdot\alpha^{!}_{\mathcal{D}}(((e_1 \rightarrow e_2, e_3)^{\ast}[\vec{\gamma}^{!}(\vec{\varphi}^{!})])^{\star}(\gamma^{!}_{\mathcal{B}}(\vec{\nu}^{!})))$. By definition (44) of $(e_1 \rightarrow e_2, e_3)^{\ast}$, this is:

$$
\begin{aligned}
\lambda\vec{\varphi}^{!}\cdot\lambda\vec{\nu}^{!}\cdot\alpha^{!}_{\mathcal{D}}\Big((&\{\langle\vec{\nu}, \bot\rangle \mid \langle\vec{\nu}, \bot\rangle \in e_1^{\ast}[\vec{\gamma}^{!}(\vec{\varphi}^{!})]\} \\
&\cup \{\langle\vec{\nu}, \Omega\rangle \mid \exists\nu \in \Upsilon^{\Omega} - \{\mathrm{tt}, \mathrm{ff}\} : \langle\vec{\nu}, \nu\rangle \in e_1^{\ast}[\vec{\gamma}^{!}(\vec{\varphi}^{!})]\} \\
&\cup \{\langle\vec{\nu}, \nu\rangle \mid \langle\vec{\nu}, \mathrm{tt}\rangle \in e_1^{\ast}[\vec{\gamma}^{!}(\vec{\varphi}^{!})] \wedge \langle\vec{\nu}, \nu\rangle \in e_2^{\ast}[\vec{\gamma}^{!}(\vec{\varphi}^{!})]\} \\
&\cup \{\langle\vec{\nu}, \nu\rangle \mid \langle\vec{\nu}, \mathrm{ff}\rangle \in e_1^{\ast}[\vec{\gamma}^{!}(\vec{\varphi}^{!})] \wedge \langle\vec{\nu}, \nu\rangle \in e_3^{\ast}[\vec{\gamma}^{!}(\vec{\varphi}^{!})]\})^{\star}(\gamma^{!}_{\mathcal{B}}(\vec{\nu}^{!}))\Big)
\end{aligned}
$$

But $(\rho \cup \varrho)^{\star}(X) = \rho^{\star}(X) \cup \varrho^{\star}(X)$ and by (49), $\alpha^{!}_{\mathcal{D}}$ is the lower adjoint of a Galois connection, hence a complete $\cup$-morphism so that this is:

$$
\begin{aligned}
\lambda\vec{\varphi}^{!}\cdot\lambda\vec{\nu}^{!}\cdot\ &\alpha^{!}_{\mathcal{D}}(\{\langle\vec{\nu}, \bot\rangle \mid \langle\vec{\nu}, \bot\rangle \in e_1^{\ast}[\vec{\gamma}^{!}(\vec{\varphi}^{!})]\}^{\star}(\gamma^{!}_{\mathcal{B}}(\vec{\nu}^{!}))) \\
&\vee \alpha^{!}_{\mathcal{D}}(\{\langle\vec{\nu}, \Omega\rangle \mid \exists\nu \in \Upsilon^{\Omega} - \{\mathrm{tt}, \mathrm{ff}\} : \langle\vec{\nu}, \nu\rangle \in e_1^{\ast}[\vec{\gamma}^{!}(\vec{\varphi}^{!})]\}^{\star}(\gamma^{!}_{\mathcal{B}}(\vec{\nu}^{!}))) \\
&\vee \alpha^{!}_{\mathcal{D}}(\{\langle\vec{\nu}, \nu\rangle \mid \langle\vec{\nu}, \mathrm{tt}\rangle \in e_1^{\ast}[\vec{\gamma}^{!}(\vec{\varphi}^{!})] \wedge \langle\vec{\nu}, \nu\rangle \in e_2^{\ast}[\vec{\gamma}^{!}(\vec{\varphi}^{!})]\}^{\star}(\gamma^{!}_{\mathcal{B}}(\vec{\nu}^{!}))) \\
&\vee \alpha^{!}_{\mathcal{D}}(\{\langle\vec{\nu}, \nu\rangle \mid \langle\vec{\nu}, \mathrm{ff}\rangle \in e_1^{\ast}[\vec{\gamma}^{!}(\vec{\varphi}^{!})] \wedge \langle\vec{\nu}, \nu\rangle \in e_3^{\ast}[\vec{\gamma}^{!}(\vec{\varphi}^{!})]\}^{\star}(\gamma^{!}_{\mathcal{B}}(\vec{\nu}^{!})))
\end{aligned}
$$

By definition $\rho^\star(X) \stackrel{\text{def}}{=} \{\nu \mid \exists\vec{\nu} \in X : \langle\vec{\nu},\, \nu\rangle \in \rho\}$, this is LHS =

$$\begin{aligned}\lambda\vec{\varphi}^{!}\cdot\lambda\vec{\nu}^{!}\cdot\ & \alpha^{!}_{\mathcal{D}}(\{\bot \mid \exists\vec{\nu} \in \gamma^{!}_{\mathcal{B}}(\vec{\nu}^{!}) : \langle\vec{\nu},\, \bot\rangle \in e_1^{*}[\vec{\gamma}^{!}(\vec{\varphi}^{!})]\})\\ & \vee \alpha^{!}_{\mathcal{D}}(\{\Omega \mid \exists\vec{\nu} \in \gamma^{!}_{\mathcal{B}}(\vec{\nu}^{!}) : \exists\nu \in \Upsilon^{\Omega} - \{\text{tt},\, \text{ff}\} : \langle\vec{\nu},\, \nu\rangle \in e_1^{*}[\vec{\gamma}^{!}(\vec{\varphi}^{!})]\})\\ & \vee \alpha^{!}_{\mathcal{D}}(\{\nu \mid \exists\vec{\nu} \in \gamma^{!}_{\mathcal{B}}(\vec{\nu}^{!}) : \langle\vec{\nu},\, \text{tt}\rangle \in e_1^{*}[\vec{\gamma}^{!}(\vec{\varphi}^{!})] \wedge \langle\vec{\nu},\, \nu\rangle \in e_2^{*}[\vec{\gamma}^{!}(\vec{\varphi}^{!})]\})\\ & \vee \alpha^{!}_{\mathcal{D}}(\{\nu \mid \exists\vec{\nu} \in \gamma^{!}_{\mathcal{B}}(\vec{\nu}^{!}) : \langle\vec{\nu},\, \text{ff}\rangle \in e_1^{*}[\vec{\gamma}^{!}(\vec{\varphi}^{!})] \wedge \langle\vec{\nu},\, \nu\rangle \in e_3^{*}[\vec{\gamma}^{!}(\vec{\varphi}^{!})]\})\end{aligned}$$

We now examine each subterm of LHS in turn.

* In the subterm $\alpha^{!}_{\mathcal{D}}(\{\bot \mid C\})$, the condition $C = \exists\vec{\nu} \in \gamma^{!}_{\mathcal{B}}(\vec{\nu}^{!}) : \langle\vec{\nu},\, \bot\rangle \in e_1^{*}[\vec{\gamma}^{!}(\vec{\varphi}^{!})]$ is either true and, by (48) and (47), this is $\alpha^{!}_{\mathcal{D}}(\{\bot\}) = \alpha^{!}_{\Upsilon}(\bot) = 0$ or else C is false and this subterm is $\alpha^{!}_{\mathcal{D}}(\emptyset) = \vee\emptyset = 0$. We have $0 \vee \nu^{!} = \nu^{!}$ since 0 is the infimum of $\mathcal{D}^{!}$ so that this subterm can be eliminated from LHS.

* The situation is the same for term $\alpha^{!}_{\mathcal{D}}(\{\Omega \mid \exists\vec{\nu} \in \gamma^{!}_{\mathcal{B}}(\vec{\nu}^{!}) : \exists\nu \in \Upsilon^{\Omega} - \{\text{tt},\, \text{ff}\} : \langle\vec{\nu},\, \nu\rangle \in e_1^{*}[\vec{\gamma}^{!}(\vec{\varphi}^{!})]\})$ since $\alpha^{!}_{\Upsilon}(\Omega) = 0$.

* For the term $T_2 = \alpha^{!}_{\mathcal{D}}(\{\nu \mid \exists\vec{\nu} \in \gamma^{!}_{\mathcal{B}}(\vec{\nu}^{!}) : \langle\vec{\nu},\, \text{tt}\rangle \in e_1^{*}[\vec{\gamma}^{!}(\vec{\varphi}^{!})] \wedge \langle\vec{\nu},\, \nu\rangle \in e_2^{*}[\vec{\gamma}^{!}(\vec{\varphi}^{!})]\})$, we proceed by cases:
  - If $e_1^{!}[\vec{\varphi}^{!}]\vec{\nu}^{!} = 0$ then $\alpha^{!}_{\mathcal{E}}(e_1^{*})[\vec{\varphi}^{!}]\vec{\nu}^{!} = 0$ since $e_1^{!} \geq \alpha^{!}_{\mathcal{E}}(e_1^{*})$ by induction hypothesis. It follows, by (61), that $\alpha^{!}(e_1^{*}(\vec{\gamma}^{!}[\vec{\varphi}^{!}]))(\vec{\nu}^{!}) = 0$, whence, by (53) that $\alpha^{!}_{\mathcal{D}}(\{\nu \mid \exists\vec{\nu} \in \gamma^{!}_{\mathcal{B}}(\vec{\nu}^{!}) : \langle\vec{\nu},\, \nu\rangle \in e_1^{*}(\vec{\gamma}^{!}[\vec{\varphi}^{!}])\}) = 0$. From (47) and (48) we derive that for all $\vec{\nu} \in \gamma^{!}_{\mathcal{B}}(\vec{\nu}^{!})$ such that $\langle\vec{\nu},\, \nu\rangle \in e_1^{*}(\vec{\gamma}^{!}[\vec{\varphi}^{!}])$, we have $\nu \in \{\bot,\, \Omega\}$ and therefore ν cannot be tt. In this case $T_2 = \alpha^{!}_{\mathcal{D}}(\emptyset) = \vee\emptyset = 0$.
  - If $e_1^{!}[\vec{\varphi}^{!}]\vec{\nu} = 1$ then, by monotonicity of $\alpha^{!}_{\mathcal{D}}$, we have $T_2 \leq \alpha^{!}_{\mathcal{D}}(\{\nu \mid \exists\vec{\nu} \in \gamma^{!}_{\mathcal{B}}(\vec{\nu}^{!}) : \langle\vec{\nu},\, \nu\rangle \in e_2^{*}[\vec{\gamma}^{!}(\vec{\varphi}^{!})]\})$ which, by (53) is equal to $\alpha^{!}(e_2^{*}[\vec{\gamma}^{!}(\vec{\varphi}^{!})])(\vec{\nu}^{!})$ hence, by (61) to $\alpha^{!}_{\mathcal{E}}(e_2^{*})[\vec{\varphi}^{!}])\vec{\nu}^{!}$, $\leq e_2^{!}[\vec{\varphi}^{!}])\vec{\nu}^{!}$ by induction hypothesis.
  - We conclude that $T_2 \leq e_1^{!}[\vec{\varphi}^{!}]\vec{\nu}^{!} \wedge e_2^{!}[\vec{\varphi}^{!}]\vec{\nu}^{!}$ which equals 0 when $e_1^{!}[\vec{\varphi}^{!}]\vec{\nu}^{!} = 0$ and else is $e_2^{!}[\vec{\varphi}^{!}])\vec{\nu}^{!}$.

* The same way, we have $\alpha^{!}_{\mathcal{D}}(\{\nu \mid \exists\vec{\nu} \in \gamma^{!}_{\mathcal{B}}(\vec{\nu}^{!}) : \langle\vec{\nu},\, \text{ff}\rangle \in e_1^{*}[\vec{\gamma}^{!}(\vec{\varphi}^{!})] \wedge \langle\vec{\nu},\, \nu\rangle \in e_3^{*}[\vec{\gamma}^{!}(\vec{\varphi}^{!})]\}) \leq e_1^{!}[\vec{\varphi}^{!}]\vec{\nu}^{!} \wedge e_3^{!}[\vec{\varphi}^{!}]\vec{\nu}^{!}$.

Taking all cases into account, $(e_1 \rightarrow e_2,\, e_3)^{!}$ can be defined as $(e_1^{!} \wedge e_2^{!}) \vee (e_1^{!} \wedge e_3^{!}) = e_1^{!} \wedge (e_2^{!} \vee e_3^{!})$. Observe that this is not the best possible approximation since for example, by taking values into account, we could have: $(\text{true} \rightarrow e_2,\, e_3)^{!} = e_2^{!}$ and $(\text{false} \rightarrow e_2,\, e_3)^{!} = e_3^{!}$.

• In conclusion, the definition of $\vec{f}^{!}$ is given at Fig. 2. According to Cor. 9, our definition of $\vec{f}^{!}$ is safe by construction:

Proposition 24 Connection of Mycroft's forward dependence-sensitive strictness semantics $\vec{f}^{!}$ with the relational semantics $\vec{f}^{*}$ of program $\prod_{f\in\vec{f}} f\vec{v} \equiv \vec{F}[f]$.

$$\vec{\alpha}^{!}(\vec{f}^{*}) = \vec{\alpha}^{!}(\text{lfp}^{\sqsubseteq^{*}}_{\mathcal{I}^{*}} \vec{F}^{*}) \mathrel{\dot{\leq}} \vec{f}^{!} = \text{lfp}^{\dot{\leq}}_{\mathcal{I}^{!}} \vec{F}^{!}$$

Proposition 25 Safeness of Mycroft's strictness analysis method. *If $\vec{\nu}^{!} = \prod_{v\in\vec{v}}(v \in I \rightarrow 0,\, 1)$ and $\vec{f}^{!}[f]\vec{\nu}^{!} = 0$ then $\vec{f}^{*}[f]$ is strict in its parameters I.*

Comments on Mycroft's Strictness Analysis Method One strength of Mycroft's strictness analysis method is that it is dependence-sensitive and therefore can detect dependencies between arguments. For example, if we define $f(x, y, z) \equiv (x \rightarrow y, z)$ then we find that $f^{\natural}(1, 0, 1) = 1$ and $f^{\natural}(1, 1, 0) = 1$ so that f is neither strict in y nor in z but $f^{\natural}(1, 0, 0) = 0$ so that it is jointly strict in y and z. More precisely, Sekar, Mishra and Ramakrishnan have proved [SMR91] that "Mycroft's method will deduce a strictness property for program P if and only the property is independent of any constant appearing in any evaluation of P". Hence the only way to improve its power is to take values of variables into account.

In practice, two methods have been proposed in [CC77c] for solving iteratively the system of equations $\prod_{f \in \vec{f}} \vec{f}[f]^{\natural}\vec{\nu} = \vec{F}^{\natural}(\vec{f}^{\natural})[f]\vec{\nu}$ which are more efficient than the naïve Jacobi iteration $\vec{F}^{\natural n}(\vec{\perp}^{\natural})$, $n \geq 0$:

(a) We can use first-order chaotic iterations [CC77b] for $\vec{f}^{\natural} = \vec{F}^{\natural}(\vec{f}^{\natural})$ (using a tabular representation of functions $\vec{f}^{\natural}$ or a symbolic representation by e.g. BDDs [Bry86]); This method is expensive since $\vec{f}^{\natural}[f](\nu_1, \ldots, \nu_n)$ is computed for all $f \in \vec{f}$, $\nu_i \in \{0, 1\}$ and $i \in \{1, \ldots, n\}$ so that there are $|\text{Dom } \vec{f}|.2^n$ possibilities[25].
(b) It is preferable to use second-order chaotic iterations introduced in [CC77c], page 265[26] to compute the subset of $\vec{f}^{\natural}[f']\vec{\nu}'$ satisfying $\vec{f}^{\natural}[f']\vec{\nu}' = \vec{F}^{\natural}(\vec{f}^{\natural})[f']\vec{\nu}'$ which are needed to answer a given strictness question $\vec{f}^{\natural}[f]\vec{\nu} = 0$. Again sets of arguments can be represented using BDDs [Bry86] and their refinements.

However, in both cases, exponential worst cases cannot be avoided since Hudak and Young have proved a conjecture due to Albert Meyer stating that "the problem of first-order strictness analysis is complete in deterministic exponential time" [HY86].

3.3 Forward Dependence-Free Strictness Analysis

Thomas Johnsson proposed a backward dependence-free strictness analysis method [Joh81], which was recognized by John Hugues to be considerably more efficient than Mycroft's forward analysis [Hug88]: "there are grounds for believing that backward analysis will be considerably more efficient than forward analysis ... Every context function has only one argument". The difference of forward or backward direction of analysis is a specious explanation of efficiency. Efficiency comes from the idea of dependence-free strictness analysis which can be applied independently of the direction of analysis.

À la Johnsson's Abstraction As shown in example 6.2.0.2 of [CC79b], a dependence-sensitive abstract interpretation can always be transformed into an dependence-free one to make the method less expensive by being less precise. For Mycroft's strictness analysis method, the approximation of $\mathcal{F}^{\natural} = \vec{\mathcal{D}}^{\natural} \overset{m \leq}{\longmapsto} \mathcal{D}^{\natural}$ by

[25] $|S|$ is the cardinality of set S

[26] This technique was latter popularized by Jones and Mycroft [JM86] as *minimal function graphs* understood as a denotational semantics collecting the needed subcalls for a main call (although the corresponding naïve Jacobi iteration is inefficient).

$$
\begin{array}{|l|l|}
\hline
\begin{array}{ll}
\nu & : \mathcal{D}^\sharp \stackrel{\mathrm{def}}{=} \{0, 1\} \\
\vec{\nu} & : \vec{\mathcal{D}}^\sharp \stackrel{\mathrm{def}}{=} \prod_{v \in \vec{v}} \mathcal{D}^\sharp \\
\varphi, f^\sharp & : \mathcal{F}^\sharp \stackrel{\mathrm{def}}{=} \vec{\mathcal{D}}^\sharp \overset{m\leq}{\longmapsto} \mathcal{D}^\sharp \\
\vec{\varphi}, \vec{f}^\sharp & : \vec{\mathcal{F}}^\sharp \stackrel{\mathrm{def}}{=} \prod_{f \in \vec{f}} \mathcal{F}^\sharp
\end{array}
&
\begin{array}{l}
e^\sharp : \mathcal{E}^\sharp \stackrel{\mathrm{def}}{=} \vec{\mathcal{F}}^\sharp \overset{m\leq}{\longmapsto} \vec{\mathcal{D}}^\sharp \overset{m\leq}{\longmapsto} \mathcal{D}^\sharp \\
\vec{e}^\sharp : \vec{\mathcal{E}}^\sharp \stackrel{\mathrm{def}}{=} \vec{\mathcal{F}}^\sharp \overset{m\leq}{\longmapsto} \vec{\mathcal{D}}^\sharp \overset{m\leq}{\longmapsto} \vec{\mathcal{D}}^\sharp \\
\vec{F}^\sharp : \vec{\mathcal{F}}^\sharp \overset{m\leq}{\longmapsto} \vec{\mathcal{F}}^\sharp
\end{array}
\\ \hline
\multicolumn{2}{|l|}{
\begin{array}{ll}
k^\sharp[\vec{\varphi}]\vec{\nu} & \stackrel{\mathrm{def}}{=} 1 \\
v^\sharp[\vec{\varphi}]\vec{\nu} & \stackrel{\mathrm{def}}{=} \vec{\nu}[v] \\
b\vec{e}^{\,\sharp}[\vec{\varphi}]\vec{\nu} & \stackrel{\mathrm{def}}{=} b^\sharp(\vec{e}^{\,\sharp}[\vec{\varphi}]\vec{\nu}) \text{ where } b^\sharp \geq^\sharp \alpha^\sharp(b^\natural) \\
(e_1 \to e_2, e_3)^\sharp[\vec{\varphi}]\vec{\nu} & \stackrel{\mathrm{def}}{=} e_1^\sharp[\vec{\varphi}]\vec{\nu} \wedge (e_2^\sharp[\vec{\varphi}]\vec{\nu} \vee e_3^\sharp[\vec{\varphi}]\vec{\nu}) \\
f\vec{e}^{\,\sharp}[\vec{\varphi}]\vec{\nu} & \stackrel{\mathrm{def}}{=} \vec{\varphi}[f](\vec{e}^{\,\sharp}[\vec{\varphi}]\vec{\nu}) \\
\vec{e}^{\,\sharp}[\vec{\varphi}]\vec{\nu} & \stackrel{\mathrm{def}}{=} \prod_{v \in \vec{e}} \vec{e}[v]^\sharp[\vec{\varphi}]\vec{\nu}
\end{array}}
\\ \hline
\multicolumn{2}{|l|}{
\begin{array}{l}
\vec{\bot}^\sharp \stackrel{\mathrm{def}}{=} \prod_{f \in \vec{f}} \lambda\vec{\nu}\cdot 0 \\
\vec{F}^\sharp \stackrel{\mathrm{def}}{=} \lambda\vec{\varphi}\cdot \prod_{f \in \vec{f}} \vec{F}[f]^\sharp[\vec{\varphi}] \\
\vec{f}^\sharp \stackrel{\mathrm{def}}{=} \mathrm{lfp}^{\leq}_{\vec{\bot}^\sharp} \vec{F}^\sharp = \dot{\bigvee}_{n \in \mathbb{N}} \vec{F}^{\sharp n}(\vec{\bot}^\sharp)
\end{array}}
\\ \hline
\end{array}
$$

Fig. 2. Synopsis of the forward dependence-sensitive strictness semantics of $\prod_{f \in \vec{f}} f\vec{v} \equiv \vec{F}[f]$

$$
\begin{array}{|l|l|}
\hline
\begin{array}{ll}
\nu & : \mathcal{D}^\natural \stackrel{\mathrm{def}}{=} \{0, 1\} \\
\varphi, f^\natural & : \mathcal{F}^\natural \stackrel{\mathrm{def}}{=} \prod_{v \in \vec{v}} (\mathcal{D}^\natural \overset{m\leq}{\longmapsto} \mathcal{D}^\natural) \\
\vec{\varphi}, \vec{f}^\natural & : \vec{\mathcal{F}}^\natural \stackrel{\mathrm{def}}{=} \prod_{f \in \vec{f}} \mathcal{F}^\natural
\end{array}
&
\begin{array}{l}
e^\natural : \vec{\mathcal{F}}^\natural \overset{m\leq}{\longmapsto} \prod_{v \in \vec{v}} (\mathcal{D}^\natural \overset{m\leq}{\longmapsto} \mathcal{D}^\natural) \\
\vec{e}^{\,\natural} : \vec{\mathcal{F}}^\natural \overset{m\leq}{\longmapsto} \prod_{v \in \vec{v}} (\mathcal{D}^\natural \overset{m\leq}{\longmapsto} \prod_{v' \in \vec{e}} \mathcal{D}^\natural) \\
\vec{F}^\natural : \vec{\mathcal{F}}^\natural \overset{m\leq}{\longmapsto} \vec{\mathcal{F}}^\natural
\end{array}
\\ \hline
\multicolumn{2}{|l|}{
\begin{array}{lll}
k^\natural[\![\vec{\varphi}]\!][v]\nu & \stackrel{\mathrm{def}}{=} 1 & \\
v^\natural[\![\vec{\varphi}]\!][v]\nu & \stackrel{\mathrm{def}}{=} \nu & \\
v^\natural[\![\vec{\varphi}]\!][v']\nu & \stackrel{\mathrm{def}}{=} 1 & \text{if } v' \neq v \\
b\vec{e}^{\,\natural}[\![\vec{\varphi}]\!][v]\nu & \stackrel{\mathrm{def}}{=} b^\sharp(\vec{e}^{\,\natural}[\![\vec{\varphi}]\!][v]\nu) & \\
(e_1 \to e_2, e_3)^\natural[\![\vec{\varphi}]\!][v]\nu & \stackrel{\mathrm{def}}{=} e_1^\natural[\![\vec{\varphi}]\!][v]\nu \wedge (e_2^\natural[\![\vec{\varphi}]\!][v]\nu \vee e_3^\natural[\![\vec{\varphi}]\!][v]\nu) & \\
f\vec{e}^{\,\natural}[\![\vec{\varphi}]\!][v]\nu & \stackrel{\mathrm{def}}{=} \bigwedge_{v' \in \vec{e}} \vec{\varphi}[f][v'](\vec{e}[v']^\natural[\![\vec{\varphi}]\!][v]\nu) & \\
\vec{e}^{\,\natural}[\![\vec{\varphi}]\!][v]\nu & \stackrel{\mathrm{def}}{=} \prod_{v' \in \vec{e}} \vec{e}[v']^\natural[\![\vec{\varphi}]\!][v]\nu &
\end{array}}
\\ \hline
\multicolumn{2}{|l|}{
\begin{array}{l}
\vec{\bot}^\natural \stackrel{\mathrm{def}}{=} \prod_{f \in \vec{f}} \prod_{v \in \vec{v}} \lambda\nu\cdot 0 \\
\vec{F}^\natural \stackrel{\mathrm{def}}{=} \lambda\vec{\varphi}\cdot \prod_{f \in \vec{f}} \vec{F}[f]^\natural[\![\vec{\varphi}]\!] \\
\vec{f}^\natural \stackrel{\mathrm{def}}{=} \mathrm{lfp}^{\leq}_{\vec{\bot}^\natural} \vec{F}^\natural = \dot{\bigvee}_{n \in \mathbb{N}} \vec{F}^{\natural n}(\vec{\bot}^\natural)
\end{array}}
\\ \hline
\end{array}
$$

Fig. 3. Synopsis of the forward dependence-free strictness semantics of $\prod_{f \in \vec{f}} f\vec{v} \equiv \vec{F}[f]$

$\prod_{v \in \vec{v}} (\mathcal{D}^\natural \overset{m\leq}{\longmapsto} \mathcal{D}^\natural)$ where $\mathcal{D}^\natural \stackrel{\mathrm{def}}{=} \mathcal{D}^\sharp = \{0, 1\}$[27] with $0 \leq 0 < 1 \leq 1$ is the following[28]:

$$\alpha^\natural(\varphi) = \prod_{v \in \vec{v}} \lambda\nu\cdot\varphi(\vec{1}[v \leftarrow \nu]) \qquad \gamma^\natural(\phi) = \lambda\vec{\nu}\cdot \bigwedge_{v \in \vec{v}} \phi[v](\vec{\nu}[v]) \tag{70}$$

Example 4. The dependence-free abstraction of the dependence-sensitive strictness analysis $f^\sharp(x, y, z) = x \wedge (y \vee z)$ of function $f(x, y, z) = (x \to y, z)$ is $f^\natural$ such that

[27] We use the double sharp symbol $^\natural$ to denote the upper approximation of the sharp symbol $^\sharp$.

[28] For all $v' \in \vec{v}$ such that $v' \neq v$, we have $\vec{1}[v \leftarrow \nu][v'] = 1$ and $\vec{1}[v \leftarrow \nu][v] = \nu$.

$f^{\times}[x] = \lambda x \cdot x$, $f^{\times}[y] = \lambda y \cdot 1$ and $f^{\times}[z] = \lambda z \cdot 1$. Its concrete form is $\varphi(x, y, z) = x$ so that one can no longer handle $f(x, y, z)$ nicely. □

Observe that if $\varphi \in \mathcal{F}^{\natural} = \vec{\mathcal{D}}^{\natural} \overset{m\leq}{\longmapsto} \mathcal{D}^{\natural}$ and $\alpha^{\times}(\varphi)[v](1) = 0$ then, by (70), $\varphi(\vec{1}[v \leftarrow 1]) = \varphi(\vec{1}) = 0$ so that, by monotonicity, $\forall \vec{\nu} \in \vec{\mathcal{D}}^{\natural} : \varphi(\vec{\nu}) = 0$ proving that $\forall v' \in \vec{v} : \forall \nu \in \mathcal{D}^{\times} : \alpha^{\times}(\varphi)[v'](\nu) = \varphi(\vec{1}[v' \leftarrow \nu] = 0$. Let us now consider the expression LHS $= \bigwedge_{v' \in \vec{v}} \alpha^{\times}(\varphi)[v'](\vec{1}[v \leftarrow \nu][v']) = \alpha^{\times}(\varphi)[v](\nu) \wedge \bigwedge\{\alpha^{\times}(\varphi)[v'](1) \mid v' \in \vec{v} \wedge v' \neq v\}$. If all $\alpha^{\times}(\varphi)[v'](1)$ are equal to 1 then LHS $= \alpha^{\times}(\varphi)[v](\nu)$. Else, there exists some $v' \in \vec{v}$ such that $\alpha^{\times}(\varphi)[v'](1) = 0$. Then $\alpha^{\times}(\varphi)[v](\nu) = 0$ and once again LHS $= \alpha^{\times}(\varphi)[v](\nu)$. Intuitively, if the dependence-free strictness analysis shows non-termination of a function for all values ν of its parameter v, this conclusion is drawn without knowing the other parameters v' so that in the analysis the proof of non-termination of this function cannot depend on any of its parameters. This leads to the following definition:

$$\mathcal{F}^{\times} \stackrel{\text{def}}{=} \left\{ \phi \in \prod_{v \in \vec{v}} (\mathcal{D}^{\times} \overset{m\leq}{\longmapsto} \mathcal{D}^{\times}) \;\middle|\; \forall v \in \vec{v} : \lambda \nu \cdot \bigwedge_{v' \in \vec{v}} \phi[v'](\vec{1}[v \leftarrow \nu][v']) = \phi[v] \right\} \quad (71)$$

Proposition 26 Connection between the dependence-sensitive and dependence-free strictness semantics.

$$\mathcal{F}^{\natural}(\leq) \underset{\alpha^{\times}}{\overset{\gamma^{\times}}{\leftrightarrows}} \mathcal{F}^{\times}(\dot{\leq})$$

For a program $\prod_{f \in \vec{f}} f\vec{v} \equiv \vec{F}[f]$, $\vec{\mathcal{F}}^{\natural} = \prod_{f \in \vec{f}} \mathcal{F}^{\natural}$ is approximated componentwise by $\vec{\mathcal{F}}^{\times} \stackrel{\text{def}}{=} \prod_{f \in \vec{f}} \mathcal{F}^{\times}$ as follows:

$$\begin{array}{cc} \vec{\mathcal{F}}^{\natural}(\dot{\leq}) \underset{\vec{\alpha}^{\times}}{\overset{\vec{\gamma}^{\times}}{\leftrightarrows}} \vec{\mathcal{F}}^{\times}(\dot{\leq}) & \\ \vec{\alpha}^{\times}(\vec{\phi}) = \prod_{f \in \vec{f}} \alpha^{\times}(\vec{\phi}[f]) \qquad \vec{\gamma}^{\times}(\vec{\varphi}) = \prod_{f \in \vec{f}} \gamma^{\times}(\vec{\varphi}[f]) \end{array} \quad (72)$$

Forward Dependence-Free Strictness Semantics The forward dependence-free strictness semantics $\vec{f}^{\times} \stackrel{\text{def}}{=} \text{lfp}_{\vec{\bot}^{\times}}^{\dot{\leq}^{\times}} \vec{F}^{\times}$ of a program $\prod_{f \in \vec{f}} f\vec{v} \equiv \vec{F}[f]$ is given at Fig. 3.

The equations which have been proposed by guesswork in the literature are not always optimal. For example in [Joh81] `x` is needed in `f(x) ≡ f(x)` but not in `f(x) ≡ f(1)`. In both cases, the algorithm of Fig. 3 leads to the conclusion that `f` is strict in `x` (because it is non-terminating). Analogously, for the program: `f(x, y) ≡ (x → y, f(y, x))` the equations: $f^{\times}[\mathtt{x}]x = x$ and $f^{\times}[\mathtt{y}]y = y \vee f^{\times}[\mathtt{y}]1$ proposed by [Hug88] have the least solution: $f^{\times}[\mathtt{x}]0 = 0$ and $f^{\times}[\mathtt{y}]0 = 1$ whereas the equations of Fig. 3: $f^{\times}[\mathtt{x}]x = x$ and $f^{\times}[\mathtt{y}]y = y \vee (f^{\times}[\mathtt{x}]y \wedge f^{\times}[\mathtt{y}]1)$ are correct and lead to better results: $f^{\times}[\mathtt{x}]0 = 0$ and $f^{\times}[\mathtt{y}]0 = 0$.

The forward dependence-free strictness semantics of Fig. 3 has been obtained constructively by a formal hand-computation, using the abstract interpretation (72) of Mycroft's dependence-sensitive strictness semantics $\vec{f}^{\natural} = \text{lfp}_{\vec{\bot}^{\natural}}^{\dot{\leq}^{\natural}} \vec{F}^{\natural}$ given at Fig. 2 and applying Cor. 9, whence it is correct:

Proposition 27 Connection of the forward dependence-free strictness semantics with the dependence-sensitive semantics.

$$\vec{\alpha}^{\times}(\vec{\alpha}^{\natural}(\vec{f}^{\circledast})) = \vec{\alpha}^{\times}(\vec{\alpha}^{\natural}(\text{lfp}_{\vec{\bot}^{\circledast}}^{\dot{\sqsubseteq}^{\circledast}} \vec{F}^{\circledast})) \dot{\leq} \vec{\alpha}^{\times}(\vec{f}^{\natural}) = \vec{\alpha}^{\times}(\text{lfp}_{\vec{\bot}^{\natural}}^{\dot{\leq}^{\natural}} \vec{F}^{\natural}) \dot{\leq} \vec{f}^{\times} = \text{lfp}_{\vec{\bot}^{\times}}^{\dot{\leq}^{\times}} \vec{F}^{\times}$$

Proposition 28 Safeness of the dependence-free strictness analysis method. *If* $\vec{f}^{*}[f][v]0 = 0$ *then function* f *is strict in its parameter* v.

But for suboptimality and the use of different notations, it is interesting to note that the forward equations of Fig. 3 resemble the backward equations proposed by [Joh81], [Hug88] and [DW90]. T. Johnsson [Joh81] uses the dual notations T, F, $\vee$, & instead of 0, 1, $\wedge$, $\vee$. His equations define $\vec{f}^{*}[f][v]0$ directly. When needed, $\vec{f}^{*}[f][v]1$ is approximated by 1 which is suboptimal. J. Hughes [Hug88] uses the notations S, L, &, $\cup$ instead of 0, 1, $\wedge$, $\vee$. The handling of parameters as $f\vec{e}^{*}[\vec{\varphi}][v]\nu \stackrel{\text{def}}{=} \vec{\varphi}[f][v](\vec{e}[v]^{*}[\vec{\varphi}][v]\nu)$ instead of $f\vec{e}^{*}[\vec{\varphi}][v]\nu \stackrel{\text{def}}{=} \bigwedge_{v' \epsilon \vec{e}} \vec{\varphi}[f][v'](\vec{e}[v']^{*}[\vec{\varphi}][v]\nu)$ is suboptimal. The equations of the "new low-fidelity first-order forward strictness analysis technique" of [DW90] directly provide $f^{\natural}(\nu_1, \ldots, \nu_n) = \bigwedge_{i=1}^{n} f^{*}[v_i]\nu_i$ so that $f^{*}[v_i]\nu_i = f^{\natural}(\vec{1}[v_i \leftarrow \nu_i])$. Forward and backward strictness analysis lead to isomorphic equations hence to isomorphic analysis algorithms. The only difference is between the dependence-sensitive and dependence-free strictness semantics.

3.4 On the Use of Widenings to Mix Dependence-Free and Dependence-Sensitive Forward Strictness Analyses

Since Mycroft's dependence-sensitive strictness analysis is powerful but sometimes expensive and Johnsson's dependence-free strictness analysis is cheaper but less precise, we can attempt a compromise using widenings. One idea is to avoid the combinatorial explosion due to the computation of $f^{\natural}(\nu_1, \ldots, \nu_n)$ with $\nu_i \in \{0, 1\}$ for all $i \in \{1, \ldots, n\}$ by limiting the number of allowed 0 to a bound κ which can be fixed arbitrarily. For $\kappa = \infty$ we would obtain Mycroft's dependence-sensitive strictness analysis whereas with $\kappa = 1$ we would obtain the dependence-free strictness analysis. The value of κ would have to be fixed experimentally or could be left to the user as a way to adjust the compromise between his available time and computing power resources or could vary during the analysis from ∞ down to 1 so as to avoid exponential analysis times. Calls $f^{\natural}\vec{\nu}$ where $\vec{\nu} = (\nu_1, \ldots, \nu_n)$ has less than κ 0-valued ν_i (that is $(\sum_{v \epsilon \nu} \neg\nu[v]) \leq \kappa$ where $\neg 0 = 1$ and $\neg 1 = 0$) would be evaluated by Mycroft's method as given at Fig. 2. Calls $f^{\natural}\vec{\nu}$ with $(\sum_{v \epsilon \nu} \neg\nu[v]) > \kappa$ would be subject to a widening which consists in applying the dependence-free method so that $f^{\natural}\vec{\nu}$ would be over-estimated by $\bigwedge\{f^{\natural}(\vec{1}[v \leftarrow 0]) \mid v \epsilon \vec{\nu} \wedge \vec{\nu}[v] = 0\}$. A less drastic approximation of $f^{\natural}\vec{\nu}$ would be $\bigwedge\{f^{\natural}\vec{\nu}' \mid \vec{\nu}' \geq \vec{\nu} \wedge (\sum_{v \epsilon \nu} \neg\vec{\nu}'[v]) = \kappa\}$ which consists in approximating $f^{\natural}(\nu_1, \ldots, \nu_n)$ with more than κ 0-valued ν_i using the dependence-sensitive method with upper approximations having exactly κ zero-parameters.

4 Conclusion

We have shown that Mycroft's seminal dependence-sensitive [Myc80, Myc81] and Johnsson's dependence-free [Joh81] strictness analyses can be constructed by formal hand-derivation from a relational semantics [CC92] within the Galois connection based abstract interpretation framework [CC77a, CC79b]. This was hardly considered to be possible (e.g. [AH87], page 25) and shows that the difficulties that have been encountered with the formalization of strictness analysis [MN83, Nie88] are not intrinsic but mainly due to denotational semantics. Our methodology for designing an abstract interpreter is constructive and this should be opposed to empirical design methods with a posteriori safeness verification using e.g. logical relations [MJ86].

Abstract interpretation is often opposed to dataflow analysis as being intrinsically costly. This is misunderstanding that widening operators can always be used as a practical compromise between efficiency and precision. For the strictness analysis example, we have suggested a good compromise between efficiency of dependence-free and the precision of the dependence-sensitive strictness analysis algorithms using a user-adjustable threshold for taking partial-dependencies into account.

Acknowledgement We would like to thank Alan Mycroft for numerous judicial comments on the first April 27, 1992 draft of this paper.

References

[AH87] S. Abramsky & C. Hankin, eds. *Abstract Interpretation of Declarative Languages*. Computers and their Applications. Ellis Horwood, 1987.

[Bry86] R. E. Bryant. Graph-based algorithms for boolean function manipulation. *IEEE Trans. Comput.*, C-35(8), 1986.

[CC77a] P. Cousot & R. Cousot. Abstract interpretation: a unified lattice model for static analysis of programs by construction or approximation of fixpoints. In 4th *POPL*, pp. 238–252, Los Angeles, California, 1977. ACM Press.

[CC77b] P. Cousot & R. Cousot. Automatic synthesis of optimal invariant assertions: mathematical foundations. In *ACM Symposium on Artificial Intelligence & Programming Languages*, Rochester, New York, SIGPLAN Notices 12(8):1–12, 1977.

[CC77c] P. Cousot & R. Cousot. Static determination of dynamic properties of recursive procedures. In E.J. Neuhold, ed., *IFIP Conference on Formal Description of Programming Concepts*, St-Andrews, N.B., Canada, pp. 237–277. North-Holland, 1977.

[CC79a] P. Cousot & R. Cousot. Constructive versions of Tarski's fixed point theorems. *Pacific J. Math.*, 82(1):43–57, 1979.

[CC79b] P. Cousot & R. Cousot. Systematic design of program analysis frameworks. In 6th *POPL*, pp. 269–282, San Antonio, Texas, 1979. ACM Press.

[CC92] P. Cousot & R. Cousot. Inductive definitions, semantics and abstract interpretation. In 19th *POPL*, pp. 83–94, Albuquerque, New Mexico, 1992. ACM Press.

[DW90] K. Davis & P. Wadler. Strictness analysis in 4D. In S.L. Peyton Jones, G. Hutton, & C. Kehler Holst, eds., *Functional Programming, Glasgow 1990*, Proc. 1990 Glasgow Workshop on Functional Programming, Ullapool, Scotland, pp. 23–43. Springer-Verlag, 13–15 Aug. 1990.

[Hug88] R. J. M. Hughes. Backwards analysis of functional programs. In A. P. Bjørner D., Ershov & N. D. Jones, eds., *Partial Evaluation and Mixed Computation*, Proceedings IFIP TC2 Workshop, Gammel Avernæs, Denmark, pp. 187–208. Elsevier, Oct. 1988.

[HY86] P. Hudak & J. Young. Higher-order strictness analysis in untyped lambda calculus. In 12th *POPL*, pp. 97–109. ACM Press, Jan. 1986.

[JM81] N. D. Jones & S. S. Muchnick. Complexity of flow analysis, inductive assertion synthesis and a language due to Dijkstra. In S. S. Muchnick & N. D. Jones, eds., *Program Flow Analysis: Theory and Applications*, ch. 12, pp. 380–393. Prentice-Hall, 1981.

[JM86] N. D. Jones & A. Mycroft. Data flow analysis of applicative programs using minimal function graphs: abridged version. In 13th *POPL*, pp. 296–306, St. Petersburg Beach, Florida, 1986. ACM Press.

[Joh81] T. Johnsson. Detecting when call-by-value can be used instead of call-by-need. Research Report LPM MEMO 14, Laboratory for Programming Methodology, Department of Computer Science, Chalmers University of Technology, S-412 96 Göteborg, Sweden, Oct. 1981.

[Mil78] R. Milner. A theory of polymorphism in programming. *J. Comput. Sys. Sci.*, 17(3):348–375, Dec. 1978.

[MJ86] A. Mycroft & N. D. Jones. A relational framework for abstract interpretation. In N. D. Jones & H. Ganzinger, eds., *Programs as Data Objects, Proceedings of a Workshop*, Copenhagen, Denmark, 17-19 Oct. 1985, LNCS 215, pp. 156–171. Springer-Verlag, 1986.

[MN83] A. Mycroft & F. Nielson. Strong abstract interpretation using power domains. In J. Diaz, ed., *Tenth ICALP*, LNCS 154, pp. 536–547. Springer-Verlag, 1983.

[Mos90] P. D. Mosses. Denotational semantics. In J. van Leeuwen, ed., *Formal Models and Semantics*, vol. B of *Handbook of Theoretical Computer Science*, ch. 11, pp. 575–631. Elsevier, 1990.

[Myc80] A. Mycroft. The theory and practice of transforming call-by-need into call-by-value. In B. Robinet, ed., *Proc. Fourth International Symposium on Programming*, Paris, France, 22-24 Apr. 1980, LNCS 83, pp. 270–281. Springer-Verlag, 1980.

[Myc81] A. Mycroft. *Abstract Interpretation and Optimising Transformations for Applicative Programs.* Ph.D. Dissertation, CST-15-81, Department of Computer Science, University of Edinburgh, Edinburgh, Scotland, Dec. 1981.

[Nie88] F. Nielson. Strictness analysis and denotational abstract interpretation. *Inf. & Comp.*, 76(1):29–92, 1988.

[SMR91] R. C. Sekar, P. Mishra, & I. V. Ramakrishnan. On the power and limitation of strictness analysis based on abstract interpretation. In *18th POPL*, pp. 37–48, Orlando, Florida, 1991. ACM Press.

Proofs

▪ **Proof of Prop. 2** By (1), $F^{0}(\bot) = \bot \sqsubseteq F^{1}(\bot)$. Assume, by hypothesis, that $\forall \lambda \leq \lambda' : F^{\lambda}(\bot) \sqsubseteq F^{\lambda'}(\bot)$. It $\lambda' = \beta + 1$ is a successor ordinal then in particular $F^{\beta}(\bot) \sqsubseteq F^{\lambda'}(\bot)$ so that transitivity (1) and monotonicity (2) imply $F^{\lambda}(\bot) \sqsubseteq F^{\lambda'}(\bot)$ $= F^{\beta+1}(\bot) = F(F^{\beta}(\bot)) \sqsubseteq F(F^{\lambda'}(\bot)) = F^{\lambda'+1}(\bot)$. It follows that $\forall \lambda \leq \lambda' + 1 :$ $F^{\lambda}(\bot) \sqsubseteq F^{\lambda'+1}(\bot)$. If λ' is a limit ordinal then $F^{\lambda'}(\bot) = \bigsqcup_{\lambda<\lambda'} F^{\lambda}(\bot)$ so that $\forall \lambda \leq \lambda' : F^{\lambda}(\bot) \sqsubseteq F^{\lambda'}(\bot)$ by (1) and definition of least upper bounds. By transfinite induction on λ', we conclude $\forall \lambda \leq \lambda' : F^{\lambda}(\bot) \sqsubseteq F^{\lambda'}(\bot)$.

We have $F^{0}(\bot) = \bot \leq \gamma(\bot^{\sharp}) = \gamma(F^{\sharp 0}(\bot^{\sharp}))$ by (5). Assume, by hypothesis, that $F^{\lambda}(\bot) \leq \gamma(F^{\sharp\lambda}(\bot^{\sharp}))$. Then (6) implies $F^{\lambda+1}(\bot) = F(F^{\lambda}(\bot)) \leq \gamma(F^{\sharp}(F^{\sharp\lambda}(\bot^{\sharp})))$ $= \gamma(F^{\sharp\lambda+1}(\bot^{\sharp}))$. By (7) this remains true for limit ordinals. Hence, by transfinite induction induction, $\forall \lambda \in \mathbb{O} : F^{\lambda}(\bot) \leq \gamma(F^{\sharp\lambda}(\bot^{\sharp}))$.

Now (1) and (2) imply (3), which, together with (7) for $\lambda = \mathbb{O}$, imply $\mathrm{lfp}_{\bot}^{\sqsubseteq} F =$ $\bigsqcup_{\lambda\in\mathbb{O}} F^{\lambda}(\bot) \leq \gamma(\bigsqcup^{\sharp}_{\lambda\in\mathbb{O}} F^{\sharp\lambda}(\bot^{\sharp}))$.

▪ **Proof of Prop. 3** By (15), $\alpha_1 \circ \Phi \circ \gamma_0 \leq^{\sharp} \Psi$ implies $\forall \varphi \in \mathcal{F}_0^{\sharp} : \alpha_1(\Phi(\gamma_0(\varphi))) \leq^{\sharp}$ $\Psi(\varphi)$ whence $\forall \varphi \in \mathcal{F}_0^{\sharp} : \Phi(\gamma_0(\varphi)) \leq \gamma_1(\Psi(\varphi))$ by (14). In particular for $\varphi = \alpha_0(\phi)$, we have $\forall \phi \in \mathcal{F}_0 : \Phi(\gamma_0(\alpha_0(\phi))) \leq \gamma_1(\Psi(\alpha_0(\phi)))$. But (14) implies $\forall \phi \in \mathcal{F}_0 :$ $\phi \leq \gamma_0(\alpha_0(\phi))$ so that $\forall \phi \in \mathcal{F}_0 : \Phi(\phi) \leq \Phi(\gamma_0(\alpha_0(\phi)))$ since $\Phi \in \mathcal{F}_0 \stackrel{m\leq}{\longmapsto} \mathcal{F}_1$. By transitivity (9), we conclude $\forall\phi \in \mathcal{F}_0 : \Phi(\phi) \leq \gamma_1(\Psi(\alpha_0(\phi)))$ that is $\Phi \leq \gamma_1 \circ \Psi \circ \alpha_0$ by (15).

Reciprocally, if $\Phi \leq \gamma_1 \circ \Psi \circ \alpha_0$ that is $\forall\phi \in \mathcal{F}_0 : \Phi(\phi) \leq \gamma_1(\Psi(\alpha_0(\phi)))$ then $\forall\phi \in \mathcal{F}_0 : \alpha_1(\Phi(\phi)) \leq^! \Psi(\alpha_0(\phi))$ by (14). In particular for $\phi = \gamma_0(\varphi)$, $\forall\varphi \in \mathcal{F}_0^! :$ $\alpha_1(\Phi(\gamma_0(\varphi))) \leq^! \Psi(\alpha_0(\gamma_0(\varphi)))$. But $\alpha_0(\gamma_0(\varphi)) \leq^! \varphi$ by (14) so that $\Psi(\alpha_0(\gamma_0(\varphi))) \leq^!$ $\Psi(\varphi)$ since $\Psi \in \mathcal{F}_0^! \xmapsto{m\leq^!} \mathcal{F}_1^!$. By transitivity (10), we conclude $\forall\varphi \in \mathcal{F}_0^! : \alpha_1(\Phi(\gamma_0(\varphi)))$ $\leq^! \Psi(\varphi)$ that is $\alpha_1 \circ \Phi \circ \gamma_0 \leq^{!\cdot} \Psi$ by (15).

- **Proof of Cor. 4** The proof of (18) is similar to that of Prop. 3, except that in the reciprocal, we have $\Psi(\alpha_0(\gamma_0(\varphi))) = \Psi(\varphi)$ by (17) hence $\Psi(\alpha_0(\gamma_0(\varphi))) \leq^! \Psi(\varphi)$ by (10).

- **Proof of Cor. 5** If we consider a monotone abstract function $\Psi \in \mathcal{F}_0^! \xmapsto{m\leq^!} \mathcal{F}_1^!$ then $\gamma_1 \circ \Psi \circ \alpha_0 \in \mathcal{F}_0 \xmapsto{m\leq} \mathcal{F}_1$ since it is the composition of monotonic functions and its abstraction $\alpha_1 \circ \gamma_1 \circ \Psi \circ \alpha_0 \circ \gamma_0$ is Ψ since $\alpha_0 \circ \gamma_0$ and $\alpha_1 \circ \gamma_1$ are the identity, proving the Galois surjection property (19).

- **Proof of Prop. 6** By (21), $\bot^! \geq^! \alpha(\bot)$ whence $\gamma(\bot^!) \geq \bot$ by (14) proving (5).

If $\varphi \leq \gamma(\phi^!)$ then $\alpha(F(\varphi)) \leq^! \alpha(F(\gamma(\phi^!)))$ since F and α are monotonic by (20) and (14). It follows that $\alpha(F(\varphi)) \leq^! F^!(\phi^!)$ by (22), (15) and transitivity (10) whence $F(\varphi) \leq \gamma(F^!(\phi^!))$ by (14) proving that (6) holds.

If $\forall\beta \leq \beta' < \lambda : \varphi_\beta \sqsubseteq \varphi_{\beta'} \wedge \varphi_\beta \leq \gamma(\phi^!_\beta)$ then $\alpha(\varphi_\beta) \leq^! \phi^!_\beta$ by (14) so that $\alpha\Big(\bigsqcup_{\beta<\lambda} \varphi_\beta\Big) \leq^! \bigsqcup^!_{\beta<\lambda} \phi^!_\beta$ by (23), proving $\bigsqcup_{\beta<\lambda} \varphi_\beta \leq \gamma\Big(\bigsqcup^!_{\beta<\lambda} \phi^!_\beta\Big)$ by (14) whence (7).

By Prop. 2, we conclude that $\mathrm{lfp}^{\sqsubseteq}_{\bot} F \leq^! \gamma\Big(\bigsqcup^!_{\lambda\in\mathbb{O}} F^{!\lambda}(\bot^!)\Big)$.

- **Proof of Lem. 7** By (1), we have $\bot \sqsubseteq \gamma(\bot^!)$ whence $\alpha(\bot) \sqsubseteq^! \bot^!$ by (25) and therefore $\alpha(\bot) = \bot^!$ by (24). By (25), (1) and (24), α is a complete $\bigsqcup$-morphism so that $\alpha(\bigsqcup_{\lambda\in\mathbb{O}} \gamma(\varphi^!_\lambda)) = \bigsqcup^!_{\lambda\in\mathbb{O}} \alpha(\gamma(\varphi^!_\lambda)) = \bigsqcup^!_{\lambda\in\mathbb{O}} \varphi^!_\lambda$ by (25).

- **Proof of Lem. 8** If $\forall\lambda \in \mathbb{O} : \alpha(\varphi_\lambda) \leq^! \phi^!_\lambda$ then $\forall\lambda \in \mathbb{O} : \alpha(\varphi_\lambda) \sqsubseteq^! \phi^!_\lambda$ since $\leq^!$ $= \sqsubseteq^!$ and therefore $\bigsqcup^!_{\lambda\in\mathbb{O}} \alpha(\varphi_\lambda) \sqsubseteq^! \bigsqcup^!_{\lambda\in\mathbb{O}} \phi^!_\lambda$ by (24) and definition of lubs. But (1), (24) and (25) imply that α is a complete $\sqcup$-morphism, a property of Galois connections: $\bigsqcup^!_{\lambda\in\mathbb{O}} \alpha(\varphi_\lambda) = \alpha\Big(\bigsqcup^!_{\lambda\in\mathbb{O}} \varphi_\lambda\Big)$. We get $\alpha\Big(\bigsqcup^!_{\lambda\in\mathbb{O}} \varphi_\lambda\Big) \sqsubseteq^! \bigsqcup^!_{\lambda\in\mathbb{O}} \phi^!_\lambda$ whence $\alpha\Big(\bigsqcup_{\lambda\in\mathbb{O}} \varphi_\lambda\Big) \leq^! \bigsqcup^!_{\lambda\in\mathbb{O}} \phi^!_\lambda$ since $\leq^! = \sqsubseteq^!$.

- **Proof of Cor. 9** By Lem. 7, we have (26) hence $\alpha(\bot) \geq^! \bot^!$ by (24) and $\leq^! =$ $\sqsubseteq^!$. By Lem. 8, we have (23). We conclude by Prop. 6 that (8) holds.

- **Proof of Prop. 13** If $\forall v \in I : \vec{\nu}[v] \in \{\bot, \Omega\}$ and $\varphi(\vec{\nu}) \rightsquigarrow \nu$ then $\langle\vec{\nu}, \nu\rangle \in \varphi$ whence $\langle\vec{\nu}, \nu\rangle \in \varphi'$ since $\varphi \subseteq \varphi'$ that is $\varphi'(\vec{\nu}) \rightsquigarrow \nu$ so that $\nu \in \{\bot, \Omega\}$ since φ' is strict in its parameters $v \in I$ proving that φ is strict in $v \in I$.

- **Proof of Prop. 14** By (48), $\alpha^!_{\mathcal{D}}(V) \leq \nu$ is equivalent to $\bigvee\{\alpha^!_{\mathcal{T}}(\nu') \mid \nu' \in V\} \leq \nu$, that is, by definition of lubs to $\forall\nu' \in V : \alpha^!_{\mathcal{T}}(\nu') \leq \nu$ and to $V \subseteq \{\nu' \mid \alpha^!_{\mathcal{T}}(\nu') \leq \nu\}$, which by (48) is equivalent to $V \subseteq \gamma^!_{\mathcal{D}}(\nu)$.

If $\alpha^!_{\mathcal{T}}$ is surjective then $\forall\nu^! \in \mathcal{D}^! : \exists\nu \in \mathcal{D} : \alpha^!_{\mathcal{T}}(\nu) = \nu^!$ so that $\alpha^!_{\mathcal{D}}(\{\nu\}) = \nu^!$ by (48) proving that $\alpha^!_{\mathcal{D}}$ is surjective.

■ **Proof of Prop. 15** By (50) and definition of the componentwise ordering $\vec{\leq}$, $\alpha^{\natural}_{\vec{\mathcal{D}}}(\vec{V}) \vec{\leq} \vec{\nu}$ is equivalent to $\forall v \in \vec{v} : \alpha^{\natural}_{\mathcal{D}}(\{\vec{\nu}[v] \mid \vec{\nu} \in \vec{V}\}) \leq \vec{\nu}[v]$, hence by (49) to $\forall v \in \vec{v} : \{\vec{\nu}[v] \mid \vec{\nu} \in \vec{V}\} \subseteq \gamma^{\natural}_{\mathcal{D}}(\vec{\nu}[v])$ and to $\forall \vec{\nu} \in \vec{V} : \forall v \in \vec{v} : \vec{\nu}[v] \in \gamma^{\natural}_{\mathcal{D}}(\vec{\nu}[v])$ that is $\vec{V} \subseteq \{\vec{\nu}' \in \vec{\mathcal{D}}^{*} \mid \forall v \in \vec{v} : \vec{\nu}'[v] \in \gamma^{\natural}_{\mathcal{D}}(\vec{\nu}[v])\}$ which, by (50), is equivalent to $\vec{V} \subseteq \gamma^{\natural}_{\vec{\mathcal{D}}}(\vec{\nu})$. By (50), $\alpha^{\natural}_{\vec{\mathcal{D}}}$ is surjective since, by (49), $\alpha^{\natural}_{\mathcal{D}}$ is surjective.

■ **Proof of Prop. 16** If $\alpha^{\natural}_{\Upsilon}$ is surjective then $\forall \nu^{\natural} \in \mathcal{D}^{\natural} : \exists \nu \in \Upsilon^{\Omega}_{\perp} : \alpha^{\natural}_{\Upsilon}(\nu) = \nu^{\natural}$ whence $\alpha^{\natural}_{\Upsilon}(\nu) \leq \nu^{\natural}$ proving by definition (48) that $\forall \nu^{\natural} \in \mathcal{D}^{\natural} : \exists \nu \in \gamma^{\natural}_{\mathcal{D}}(\nu^{\natural}) : \alpha^{\natural}_{\mathcal{D}}(\{\nu\}) = \nu^{\natural}$. It follows that $\forall \vec{\nu}^{\natural} \in \vec{\mathcal{D}}^{\natural} : \exists \vec{\nu} : (\forall v \in \vec{v} : \vec{\nu}[v] \in \gamma^{\natural}_{\vec{\mathcal{D}}}(\vec{\nu}^{\natural}[v])) \wedge (\forall v \in \vec{v} : \alpha^{\natural}_{\vec{\mathcal{D}}}(\{\vec{\nu}[v]\}) = \vec{\nu}^{\natural}[v])$ so that we conclude by (50).

■ **Proof of Prop. 17** By definition of the pointwise ordering $\leq$, $\alpha^{\natural}(\phi) \leq \varphi$ is equivalent to $\forall \vec{\nu} \in \vec{\mathcal{D}}^{\natural} : \alpha^{\natural}(\phi)(\vec{\nu}) \leq \varphi(\vec{\nu})$ hence, by (53) to $\forall \vec{\nu} \in \vec{\mathcal{D}}^{\natural} : \alpha^{\natural}_{\mathcal{D}}\left(\phi^{*}(\gamma^{\natural}_{\vec{\mathcal{D}}}(\vec{\nu}))\right) \leq \varphi(\vec{\nu})$ thus, by definition of * and (48) to $\forall \vec{\nu} \in \vec{\mathcal{D}}^{\natural} : \left(\bigvee\{\alpha^{\natural}_{\mathcal{D}}(\{\nu\}) \mid \exists \vec{\nu}' \in \gamma^{\natural}_{\vec{\mathcal{D}}}(\vec{\nu}) : \langle \vec{\nu}', \nu\rangle \in \phi\}\right) \leq \varphi(\vec{\nu})$ and, by definition of lubs, to $\forall \vec{\nu} \in \vec{\mathcal{D}}^{\natural} : \forall \nu \in \mathcal{D}^{*} : \left(\exists \vec{\nu}' \in \gamma^{\natural}_{\vec{\mathcal{D}}}(\vec{\nu}) : \langle \vec{\nu}', \nu\rangle \in \phi\right) \Rightarrow \left(\alpha^{\natural}_{\mathcal{D}}(\{\nu\}) \leq \varphi(\vec{\nu})\right)$, which can also be written as $\forall \vec{\nu} \in \vec{\mathcal{D}}^{\natural} : \forall \nu \in \mathcal{D}^{*} : \forall \vec{\nu}' \in \vec{\mathcal{D}}^{*} : \left(\langle \vec{\nu}', \nu\rangle \in \phi\right) \Rightarrow \left(\{\vec{\nu}'\} \subseteq \gamma^{\natural}_{\vec{\mathcal{D}}}(\vec{\nu})\right) \Rightarrow \left(\alpha^{\natural}_{\mathcal{D}}(\{\nu\}) \leq \varphi(\vec{\nu})\right)$. By the Galois connection properties (49) and (51), this is equivalent to LHS $\equiv [\forall \vec{\nu} \in \vec{\mathcal{D}}^{\natural} : \forall \nu \in \mathcal{D}^{*} : \forall \vec{\nu}' \in \vec{\mathcal{D}}^{*} : \left(\langle \vec{\nu}', \nu\rangle \in \phi\right) \Rightarrow \left(\left(\alpha^{\natural}_{\vec{\mathcal{D}}}(\{\vec{\nu}'\}) \vec{\leq} \vec{\nu}\right) \Rightarrow \left(\{\nu\} \subseteq \gamma^{\natural}_{\mathcal{D}}(\varphi(\vec{\nu}))\right)\right)]$. Observe that in this formula, we have $\{\nu\} \subseteq \gamma^{\natural}_{\mathcal{D}}(\varphi(\alpha^{\natural}_{\vec{\mathcal{D}}}(\{\vec{\nu}'\})))$ for $\vec{\nu} = \alpha^{\natural}_{\vec{\mathcal{D}}}(\{\vec{\nu}'\})$. reciprocally, if $\{\nu\} \subseteq \gamma^{\natural}_{\mathcal{D}}(\varphi(\alpha^{\natural}_{\vec{\mathcal{D}}}(\{\vec{\nu}'\})))$ then $\alpha^{\natural}_{\vec{\mathcal{D}}}(\{\vec{\nu}'\}) \vec{\leq} \vec{\nu}$, $\varphi \in \mathcal{F}^{\natural}$, (54) and (49) imply $\gamma^{\natural}_{\mathcal{D}}(\varphi(\alpha^{\natural}_{\vec{\mathcal{D}}}(\{\vec{\nu}'\}))) \subseteq \gamma^{\natural}_{\mathcal{D}}(\varphi(\vec{\nu}))$ by monotonicity, whence $\{\nu\} \subseteq \gamma^{\natural}_{\mathcal{D}}(\varphi(\vec{\nu}))$ by transitivity. It follows that LHS is equivalent to $\forall \nu \in \mathcal{D}^{*} : \forall \vec{\nu}' \in \vec{\mathcal{D}}^{*} : \left(\langle \vec{\nu}', \nu\rangle \in \phi\right) \Rightarrow \left(\{\nu\} \subseteq \gamma^{\natural}_{\mathcal{D}} \circ \varphi \circ \alpha^{\natural}_{\vec{\mathcal{D}}}(\{\vec{\nu}'\})\right)$, that is to $\phi \subseteq \{\langle \vec{\nu}', \nu\rangle \mid \nu \in \gamma^{\natural}_{\mathcal{D}}(\varphi(\alpha^{\natural\,\prime}_{\vec{\mathcal{D}}}(\vec{\nu}')))\}$, hence by definition of ϱ and (53), to $\phi \subseteq \gamma^{\natural}(\varphi)$.

■ **Proof of Prop. 18** Let $\varphi \in \mathcal{F}^{\natural} = \vec{\mathcal{D}}^{\natural} \overset{m\vec{\leq}}{\longmapsto} \mathcal{D}^{\natural}$ and $\phi = \{\langle \vec{\nu}, \nu\rangle \mid \vec{\nu} \in \vec{\mathcal{D}}^{*} \wedge \nu \in \gamma^{\natural}_{\mathcal{D}}(\varphi(\alpha^{\natural}_{\vec{\mathcal{D}}}(\{\vec{\nu}\})))\}$. We show that $\alpha^{\natural}(\phi) = \varphi$. By definition, we have $\alpha^{\natural}(\phi) = \lambda \vec{\nu}^{\natural} \cdot \bigvee\{\alpha^{\natural}_{\mathcal{D}}(\{\nu\}) \mid \exists \vec{\nu} \in \gamma^{\natural}_{\vec{\mathcal{D}}}(\vec{\nu}^{\natural}) : \nu \in \gamma^{\natural}_{\mathcal{D}}(\varphi(\alpha^{\natural}_{\vec{\mathcal{D}}}(\{\vec{\nu}\})))\}$ which is equal to $\lambda \vec{\nu}^{\natural} \cdot \bigvee_{\vec{\nu} \in \gamma^{\natural}_{\vec{\mathcal{D}}}(\vec{\nu}^{\natural})} \bigvee\{\alpha^{\natural}_{\mathcal{D}}(\{\nu\}) \mid \nu \in \gamma^{\natural}_{\mathcal{D}}(\varphi(\alpha^{\natural}_{\vec{\mathcal{D}}}(\{\vec{\nu}\})))\}$. By (49), $\alpha^{\natural}_{\mathcal{D}}$ is a complete $\cup$-morphism so that $\alpha^{\natural}(\phi) = \lambda \vec{\nu}^{\natural} \cdot \bigvee_{\vec{\nu} \in \gamma^{\natural}_{\vec{\mathcal{D}}}(\vec{\nu}^{\natural})} \alpha^{\natural}_{\mathcal{D}}(\gamma^{\natural}_{\mathcal{D}}(\varphi(\alpha^{\natural}_{\vec{\mathcal{D}}}(\{\vec{\nu}\}))))$ since $\alpha^{\natural}_{\mathcal{D}}(V) = \bigvee\{\alpha^{\natural}_{\mathcal{D}}(\{\nu\}) \mid \nu \in V\}$. By (49), $\alpha^{\natural}_{\mathcal{D}} \circ \gamma^{\natural}_{\mathcal{D}}$ is the identity so that $\alpha^{\natural}(\phi)$ is equal to $\lambda \vec{\nu}^{\natural} \cdot \bigvee_{\vec{\nu} \in \gamma^{\natural}_{\vec{\mathcal{D}}}(\vec{\nu}^{\natural})} \varphi(\alpha^{\natural}_{\vec{\mathcal{D}}}(\{\vec{\nu}\}))$. $\vec{\nu} \in \gamma^{\natural}_{\vec{\mathcal{D}}}(\vec{\nu}^{\natural})$ implies $\{\vec{\nu}\} \subseteq \gamma^{\natural}_{\vec{\mathcal{D}}}(\vec{\nu}^{\natural})$ hence by (51), $\alpha^{\natural}_{\vec{\mathcal{D}}}(\{\vec{\nu}\}) \vec{\leq} \vec{\nu}^{\natural}$ proving, by monotonicity and definition of lubs, that $\alpha^{\natural}(\phi)(\vec{\nu}^{\natural}) \leq \varphi(\vec{\nu}^{\natural})$. Moreover, by (52) and definition of upper bounds, $\varphi(\vec{\nu}^{\natural}) \leq \alpha^{\natural}(\phi)$ proving, by antisymmetry that we have $\alpha^{\natural}(\phi) = \varphi$ hence $\alpha^{\natural}$ is surjective.

■ **Proof of (67)** The approximation ordering is defined componentwise so that (53) implies that $\vec{\gamma}^{\natural}(\vec{\varphi}) \vec{\sqsubseteq} \vec{\gamma}^{\natural}(\vec{\varphi}')$ is equivalent to $\{\langle \vec{\nu}, \nu\rangle \mid \nu \in \gamma^{\natural}_{\mathcal{D}} \circ \vec{\varphi}[f] \circ \alpha^{\natural}_{\vec{\mathcal{D}}}(\{\vec{\nu}\})\} \subseteq \{\langle \vec{\nu}, \nu\rangle \mid \nu \in \gamma^{\natural}_{\mathcal{D}} \circ \vec{\varphi}'[f] \circ \alpha^{\natural}_{\vec{\mathcal{D}}}(\{\vec{\nu}\})\}$, that is to say $\forall f \in \vec{f} : \forall \vec{\nu} \in \vec{\mathcal{D}}^{*} : \gamma^{\natural}_{\mathcal{D}} \circ \vec{\varphi}[f] \circ \alpha^{\natural}_{\vec{\mathcal{D}}}(\{\vec{\nu}\}) \subseteq \gamma^{\natural}_{\mathcal{D}} \circ \vec{\varphi}'[f] \circ \alpha^{\natural}_{\vec{\mathcal{D}}}(\{\vec{\nu}\})$ hence, by definition of the Galois surjection (49) to $\forall f \in \vec{f} : \forall \vec{\nu} \in \vec{\mathcal{D}}^{*} : \alpha^{\natural}_{\mathcal{D}} \circ \gamma^{\natural}_{\mathcal{D}} \circ \vec{\varphi}[f] \circ \alpha^{\natural}_{\vec{\mathcal{D}}}(\{\vec{\nu}\}) \leq \vec{\varphi}'[f] \circ \alpha^{\natural}_{\vec{\mathcal{D}}}(\{\vec{\nu}\})$.

(49) is a Galois surjection so $\alpha^!_{\mathcal{D}} \circ \gamma^!_{\mathcal{D}}$ is the identity and therefore this is equivalent to $\forall f \in \vec{f} : \forall \vec{\nu} \in \vec{\mathcal{D}}^{\texttt{*}} : \vec{\varphi}[f] \circ \alpha^!_{\vec{\mathcal{D}}}(\{\vec{\nu}\}) \leq \vec{\varphi}'[f] \circ \alpha^!_{\vec{\mathcal{D}}}(\{\vec{\nu}\})$ which implies, by (52), that $\forall f \in \vec{f} : \forall \vec{\nu}^! \in \vec{\mathcal{D}}^! : \vec{\varphi}[f](\vec{\nu}^!) \leq \vec{\varphi}'[f](\vec{\nu}^!)$ that is $\vec{\varphi} \mathrel{\vec{\leq}} \vec{\varphi}'$ componentwise and pointwise. Reciprocally, $\vec{\gamma}^!$ is monotonic by (58).

■ **Proof of (68)** – Assume $\phi \sqsubseteq^{\texttt{*}} \gamma^!(\varphi)$ or equivalently, by (37), $(\phi \cap \top^{\texttt{*}}) \subseteq (\gamma^!(\varphi) \cap \top^{\texttt{*}}) \wedge (\phi \cap \bot^{\texttt{*}}) \supseteq (\gamma^!(\varphi) \cap \bot^{\texttt{*}})$. By (53), we have $\gamma^!(\varphi) = \{\langle \vec{\nu}, \nu \rangle \mid \nu \in \gamma^!_{\mathcal{D}} \circ \varphi \circ \alpha^!_{\vec{\mathcal{D}}}(\{\vec{\nu}\})\}$. By (47) and (48), $\forall \nu^! \in \mathcal{D}^! : \bot \in \gamma^!_{\mathcal{D}}(\nu^!)$ so that $\bot^{\texttt{*}} \subseteq \gamma^!(\varphi)$ whence $\bot^{\texttt{*}} \subseteq (\phi \cap \bot^{\texttt{*}})$ and therefore $\bot^{\texttt{*}} \subseteq \phi$. It follows that $\phi \sqsubseteq^{\texttt{*}} \gamma^!(\varphi)$ is equivalent to $(\phi \cap \top^{\texttt{*}}) \subseteq (\gamma^!(\varphi) \cap \top^{\texttt{*}})$ that is $\phi \subseteq \gamma^!(\varphi)$ since $\phi = \bot^{\texttt{*}} \cup (\phi \cap \top^{\texttt{*}}) \subseteq \bot^{\texttt{*}} \cup (\gamma^!(\varphi) \cap \top^{\texttt{*}})$ $= (\gamma^!(\varphi) \cap \bot^{\texttt{*}}) \cup (\gamma^!(\varphi) \cap \top^{\texttt{*}}) = \gamma^!(\varphi)$. The componentwise extention is obvious.

– For the computational ordering, we have $\vec{\gamma}^!(\vec{\varphi}) \mathrel{\vec{\sqsubseteq}^{\texttt{*}}} \vec{\gamma}^!(\vec{\varphi}') \Leftrightarrow \vec{\gamma}^!(\vec{\varphi}) \mathrel{\vec{\subseteq}} \vec{\gamma}^!(\vec{\varphi}') \Leftrightarrow \vec{\varphi} \mathrel{\vec{\leq}} \vec{\varphi}'$ by (67).

■ **Proof of (69)** By (68) and (56), we have $\phi \sqsubseteq^{\texttt{*}} \gamma^!(\varphi) \Leftrightarrow \phi \subseteq \gamma^!(\varphi) \Leftrightarrow \alpha^!(\phi) \leq (\varphi)$. Likewise by (68) and (58), we have $\vec{\phi} \mathrel{\vec{\sqsubseteq}^{\texttt{*}}} \vec{\gamma}^!(\vec{\varphi}) \Leftrightarrow \vec{\phi} \mathrel{\vec{\subseteq}} \vec{\gamma}^!(\vec{\varphi}) \Leftrightarrow \vec{\alpha}^!(\vec{\phi}) \mathrel{\vec{\leq}} (\vec{\varphi})$.

■ **Proof of Prop. 24** (1) holds since $\vec{\mathcal{F}}^{\texttt{*}}(\vec{\sqsubseteq}^{\texttt{*}})$ is a complete lattice. $\vec{\leq}$ is defined pointwise and $\vec{\mathcal{F}}^!(\vec{\leq})$ is a complete lattice so that (2), (10), (24) and (15) hold. $\vec{\subseteq}$ is a partial order on $\vec{\mathcal{F}}^{\texttt{*}}$ so that (9) holds. (14) follows from (58). (20) that is $\vec{F}^{\texttt{*}} \in \vec{\mathcal{F}}^{\texttt{*}} \xrightarrow{m\subseteq} \vec{F}^{\texttt{*}}$ follows from (46). (22) that is $\vec{F}^! \mathrel{\vec{\geq}} \vec{\alpha}^! \circ \vec{F}^{\texttt{*}} \circ \vec{\gamma}^!$ has been proved above. (25) follows from (69). Finally, $\vec{\leq}^! = \vec{\sqsubseteq}^! = \vec{\leq}$. We conclude by Cor. 9.

■ **Proof of Prop. 25** If $\vec{f}^![f]\vec{\nu}^! = 0$ then by proposition 24, $\vec{\alpha}^!(\vec{f}^{\texttt{*}})[f]\vec{\nu}^! \leq 0$ hence $\alpha^!(\vec{f}^{\texttt{*}}[f])\vec{\nu}^! \leq 0$ by (57) which implies $\alpha^!_{\mathcal{D}}(\{\nu \mid \exists \vec{\nu} \in \gamma^!_{\vec{\mathcal{D}}}(\vec{\nu}^!) : \langle \vec{\nu}, \nu \rangle \in \vec{f}^{\texttt{*}}[f]\}) \leq 0$ by equation (53). By (49), (48) and (47), $\{\nu \mid \exists \vec{\nu} \in \gamma^!_{\vec{\mathcal{D}}}(\vec{\nu}^!) : \langle \vec{\nu}, \nu \rangle \in \vec{f}^{\texttt{*}}[f]\} \subseteq \{\Omega, \bot\}$. By definition of $\vec{\nu}^!$ and (50), $\gamma^!_{\vec{\mathcal{D}}}(\vec{\nu}^!) = \{\vec{\nu} \in \vec{\mathcal{D}}^{\texttt{*}} \mid \forall v \in I : \vec{\nu}[v] \in \{\Omega, \bot\}\}$. We get $\forall \nu \in \mathcal{D}^{\texttt{*}} : \forall \vec{\nu} \in \vec{\mathcal{D}}^{\texttt{*}} : (\forall v \in I : \vec{\nu}[v] \in \{\Omega, \bot\} \wedge \langle \vec{\nu}, \nu \rangle \in \vec{f}^{\texttt{*}}[f]) \Rightarrow \nu \in \{\Omega, \bot\}$ proving, by Def. 12, that $\vec{f}^{\texttt{*}}[f]$ is strict in I.

■ **Proof of Prop. 26** – Assume that $\alpha^{\times}(\varphi) \mathrel{\vec{\leq}} \phi$ then, by (70), for all $v \in \vec{v}$ and $\nu \in \mathcal{D}^!$, we have $\varphi(\vec{1}[v \leftarrow \nu]) \leq \phi[v](\nu)$. For all $\vec{\nu}$ in $\vec{\mathcal{D}}^!$ and $v \in \vec{v}$, we have $\vec{\nu} \mathrel{\vec{\leq}} \vec{1}[v \leftarrow \vec{\nu}[v]]$, hence, by monotonicity and transitivity, $\varphi(\vec{\nu}) \leq \varphi(\vec{1}[v \leftarrow \vec{\nu}[v]]) \leq \phi[v](\vec{\nu}[v])$ whence $\varphi(\vec{\nu}) \leq \bigwedge_{v \in \vec{v}} \phi[v](\vec{\nu}[v]) = \gamma^{\times}(\phi)(\vec{\nu})$.

– Reciprocally, if for all $\vec{\nu} \in \vec{\mathcal{D}}^!$ we have $\varphi(\vec{\nu}) \leq \gamma^{\times}(\phi)(\vec{\nu}) = \bigwedge_{v \in \vec{v}} \phi[v](\vec{\nu}[v])$ then for all $v \in \vec{v}$ and $\nu \in \mathcal{D}^!$, if we let $\vec{\nu} = \vec{1}[v \leftarrow \nu]$ then $\varphi(\vec{1}[v \leftarrow \nu]) \leq \bigwedge_{v' \in \vec{v}} \phi[v'](\vec{1}[v \leftarrow \nu][v']) \leq \phi[v](\nu)$ proving that $\alpha^{\times}(\varphi) \mathrel{\vec{\leq}} \phi$.

– We have $\alpha^{\times}(\gamma^{\times}(\phi)) = \prod_{v \in \vec{v}} \lambda\nu \cdot \gamma^{\times}(\phi)(\vec{1}[v \leftarrow \nu]) = \prod_{v \in \vec{v}} \lambda\nu \cdot \bigwedge_{v' \in \vec{v}} \phi[v'](\vec{1}[v \leftarrow \nu][v'])$ by (70). By (71), this is equal to $\prod_{v \in \vec{v}} \phi[v] = \phi$.

■ **Proof of Prop. 27** – The infimum is $\vec{\bot}^{\times} \stackrel{\text{def}}{=} \vec{\alpha}^{\times}(\vec{\bot}^!)$, which by definition of $\vec{\alpha}^{\times}$ and $\vec{\bot}^!$ can be simplified into $\prod_{f \in \vec{f}} \alpha^{\times}(\vec{\bot}^![f]) = \prod_{f \in \vec{f}} \alpha^{\times}(\lambda\vec{\nu} \cdot 0)$. By definition of $\alpha^{\times}$, this is equal to $\prod_{f \in \vec{f}} \prod_{v \in \vec{v}} \lambda\nu \cdot \lambda\vec{\nu} \cdot 0(\vec{1}[v \leftarrow \nu]) = \prod_{f \in \vec{f}} \prod_{v \in \vec{v}} \lambda\nu \cdot 0$.

– We define the inductive join $\vec{\bigvee}_{i \in \Delta} \vec{\varphi}_i \stackrel{\text{def}}{=} \vec{\alpha}^{\times}(\vec{\bigvee}_{i \in \Delta} \vec{\gamma}^{\times}(\vec{\varphi}_i))$ which by (72) is equal to $\prod_{f\in\vec{f}}\alpha^{\times}((\vec{\bigvee}_{i\in\Delta}\vec{\gamma}^{\times}(\vec{\varphi}_i))[f]) = \prod_{f\in\vec{f}}\alpha^{\times}(\vec{\bigvee}_{i\in\Delta}\gamma^{\times}(\vec{\varphi}_i[f]))$. By (70), this is equal to $\prod_{f\in\vec{f}}\prod_{v\in\vec{v}}\lambda\nu\cdot\vec{\bigvee}_{i\in\Delta}\vec{\bigwedge}_{v'\in\vec{v}}\vec{\varphi}_i[f][v'](\vec{1}[v\leftarrow\nu][v']) = \prod_{f\in\vec{f}}\prod_{v\in\vec{v}}\vec{\bigvee}_{i\in\Delta}\vec{\varphi}_i[f][v]$ by (71).

– We define $\vec{F}^{\times}$ as an upper approximation of $\vec{\alpha}^{\times}\circ\vec{F}^{\natural}\circ\vec{\gamma}^{\times}$ by letting $\vec{F}^{\times}\stackrel{\text{def}}{=}\lambda\vec{\varphi}\cdot\prod_{f\in\vec{f}}\vec{F}[f]^{\times}[\![\vec{\varphi}]\!]$ and by defining $e^{\times}[\![\vec{\varphi}]\!]\geq\alpha^{\times}(e^{\natural}[\![\vec{\gamma}^{\times}(\vec{\varphi})]\!])$ for all $\vec{\varphi}$ in $\vec{\mathcal{F}}^{\times}$. We proceed by case analysis on the expression e:

- $k^{\times}[\![\vec{\varphi}]\!]=\alpha^{\times}(k^{\natural}[\![\vec{\gamma}^{\times}(\vec{\varphi})]\!])=\alpha^{\times}(\lambda\vec{\nu}\cdot 1)=\prod_{v\in\vec{v}}\lambda\nu\cdot 1$.
- $v^{\times}[\![\vec{\varphi}]\!]=\alpha^{\times}(v^{\natural}[\![\vec{\gamma}^{\times}(\vec{\varphi})]\!])=\alpha^{\times}(\lambda\vec{\nu}\cdot\vec{\nu}[v])=\prod_{v'\in\vec{v}}\lambda\nu\cdot\vec{1}[v'\leftarrow\nu][v]$.
- Assume, by induction hypothesis, that we have $\vec{e}[v']^{\times}[\![\vec{\varphi}]\!]\gtrsim\alpha^{\times}(\vec{e}[v']^{\natural}[\![\vec{\gamma}^{\times}(\vec{\varphi})]\!])$ for all $\vec{\varphi}\in\vec{\mathcal{F}}^{\times}$ and $v'\in\vec{e}$. Then $\alpha^{\times}(b\vec{e}^{\natural}[\![\vec{\gamma}^{\times}(\vec{\varphi})]\!])=\alpha^{\times}(\lambda\vec{\nu}\cdot b^{\natural}(\vec{e}^{\natural}[\![\vec{\gamma}^{\times}(\vec{\varphi})]\!]\vec{\nu}))=\alpha^{\times}(\lambda\vec{\nu}\cdot b^{\natural}(\prod_{v'\in\vec{e}}\vec{e}[v']^{\natural}[\![\vec{\gamma}^{\times}(\vec{\varphi})]\!]\vec{\nu}))\lesssim\alpha^{\times}(\lambda\vec{\nu}\cdot b^{\natural}(\prod_{v'\in\vec{e}}\gamma^{\times}(\alpha^{\times}(\vec{e}[v']^{\natural}[\![\vec{\gamma}^{\times}(\vec{\varphi})]\!]))\vec{\nu}))=T$ since $\gamma^{\times}\circ\alpha^{\times}$ is extensive, $b^{\natural}$ and $\alpha^{\times}$ are monotonic. By induction hypothesis and monotonicity, $T\lesssim\alpha^{\times}(\lambda\vec{\nu}\cdot b^{\natural}(\prod_{v'\in\vec{e}}\gamma^{\times}(\vec{e}[v']^{\times}[\![\vec{\varphi}]\!])\vec{\nu}))$ which, by definition (70) of $\alpha^{\times}$ is $\prod_{v\in\vec{v}}\lambda\nu\cdot b^{\natural}(\prod_{v'\in\vec{e}}\gamma^{\times}(\vec{e}[v']^{\times}[\![\vec{\varphi}]\!])\vec{1}[v\leftarrow\nu])$. By definition (70) of $\gamma^{\times}$, this is $\prod_{v\in\vec{v}}\lambda\nu\cdot b^{\natural}(\prod_{v'\in\vec{e}}\bigwedge_{v''\in\vec{v}}\vec{e}[v']^{\times}[\![\vec{\varphi}]\!][v''](\vec{1}[v\leftarrow\nu][v'']))$. According to (71), this can be simplified as $\prod_{v\in\vec{v}}\lambda\nu\cdot b^{\natural}(\prod_{v'\in\vec{e}}\vec{e}[v']^{\times}[\![\vec{\varphi}]\!][v]\nu)$. This is $\prod_{v\in\vec{v}}\lambda\nu\cdot b^{\natural}(\vec{e}^{\times}[\![\vec{\varphi}]\!][v]\nu)$ by defining $\vec{e}^{\times}[\![\vec{\varphi}]\!][v]\nu\stackrel{\text{def}}{=}\prod_{v'\in\vec{e}}\vec{e}[v']^{\times}[\![\vec{\varphi}]\!][v]\nu$ so that $b\vec{e}^{\times}[\![\vec{\varphi}]\!][v]\nu\stackrel{\text{def}}{=}b^{\natural}(\vec{e}^{\times}[\![\vec{\varphi}]\!][v]\nu)$.
- Assume that $e_i^{\times}[\![\vec{\varphi}]\!]\geq\alpha^{\times}(e_i^{\natural}[\![\vec{\gamma}^{\times}(\vec{\varphi})]\!])$ for all $\vec{\varphi}\in\vec{\mathcal{F}}^{\times}$ and $i=1,2,3$. We have $\alpha^{\times}((e_1\rightarrow e_2,e_3)^{\natural}[\![\vec{\gamma}^{\times}(\vec{\varphi})]\!])=\alpha^{\times}(e_1^{\natural}[\![\vec{\gamma}^{\times}(\vec{\varphi})]\!]\wedge(e_2^{\natural}[\![\vec{\gamma}^{\times}(\vec{\varphi})]\!]\vee e_3^{\natural}[\![\vec{\gamma}^{\times}(\vec{\varphi})]\!]))=T$. By Prop. 26, $\alpha^{\times}$ is a complete $\vee$-morphism, hence monotonic whence $T\leq\alpha^{\times}(e_1^{\natural}[\![\vec{\gamma}^{\times}(\vec{\varphi})]\!])\wedge(\alpha^{\times}(e_2^{\natural}[\![\vec{\gamma}^{\times}(\vec{\varphi})]\!])\vee\alpha^{\times}(e_3^{\natural}[\![\vec{\gamma}^{\times}(\vec{\varphi})]\!]))$. By induction hypothesis, this is less than or equal to $(e_1\rightarrow e_2,e_3)^{\times}[\![\vec{\varphi}]\!]\stackrel{\text{def}}{=}e_1^{\times}[\![\vec{\varphi}]\!]\wedge(e_2^{\times}[\![\vec{\varphi}]\!]\vee e_3^{\times}[\![\vec{\varphi}]\!])$.
- Assume, by induction hypothesis, that $\vec{e}[v']^{\times}[\![\vec{\varphi}]\!]\geq\alpha^{\times}(\vec{e}[v']^{\natural}[\![\vec{\gamma}^{\times}(\vec{\varphi})]\!])$ for all $\vec{\varphi}\in\vec{\mathcal{F}}^{\times}$ and $v'\in\vec{e}$. Then $\alpha^{\times}(f\vec{e}^{\natural}[\![\vec{\gamma}^{\times}(\vec{\varphi})]\!])=\alpha^{\times}(\lambda\vec{\nu}\cdot\vec{\gamma}^{\times}(\vec{\varphi})[f](\vec{e}^{\natural}[\![\vec{\gamma}^{\times}(\vec{\varphi})]\!]\vec{\nu}))=\alpha^{\times}(\lambda\vec{\nu}\cdot\gamma^{\times}(\vec{\varphi}[f])(\vec{e}^{\natural}[\![\vec{\gamma}^{\times}(\vec{\varphi})]\!]\vec{\nu}))=\alpha^{\times}(\lambda\vec{\nu}\cdot\bigwedge_{v'\in\vec{v}}\vec{\varphi}[f][v']((\prod_{v''\in\vec{e}}\vec{e}[v'']^{\natural}[\![\vec{\gamma}^{\times}(\vec{\varphi})]\!]\vec{\nu})[v']))$ $=\alpha^{\times}(\lambda\vec{\nu}\cdot\bigwedge_{v'\in\vec{v}}\vec{\varphi}[f][v'](\vec{e}[v']^{\natural}[\![\vec{\gamma}^{\times}(\vec{\varphi})]\!]\vec{\nu}))=T$. By (70), $\alpha^{\times}\circ\gamma^{\times}$ is extensive, so that by monotonicity $T\leq\alpha^{\times}(\lambda\vec{\nu}\cdot\bigwedge_{v'\in\vec{v}}\vec{\varphi}[f][v'](\gamma^{\times}(\alpha^{\times}(\vec{e}[v']^{\natural}[\![\vec{\gamma}^{\times}(\vec{\varphi})]\!]))\vec{\nu}))=T'$. By induction hypothesis $T'\leq\alpha^{\times}(\lambda\vec{\nu}\cdot\bigwedge_{v'\in\vec{v}}\vec{\varphi}[f][v'](\gamma^{\times}(\vec{e}[v']^{\times}[\![\vec{\varphi}]\!])\vec{\nu}))$. By definition of $\gamma^{\times}$, we get $\alpha^{\times}(\lambda\vec{\nu}\cdot\bigwedge_{v'\in\vec{v}}\vec{\varphi}[f][v'](\bigwedge_{v''\in\vec{v}}\vec{e}[v']^{\times}[\![\vec{\varphi}]\!][v''](\vec{\nu}[v''])))$, which, by definition of $\vec{e}^{\times}$, is equal to $\alpha^{\times}(\lambda\vec{\nu}\cdot\bigwedge_{v'\in\vec{v}}\vec{\varphi}[f][v'](\bigwedge_{v''\in\vec{v}}\vec{e}^{\times}[\![\vec{\varphi}]\!][v''](\vec{\nu}[v''])[v']))$, hence, by (70), to $\prod_{v\in\vec{v}}\lambda\nu\cdot\lambda\vec{\nu}\cdot\bigwedge_{v'\in\vec{v}}\vec{\varphi}[f][v'](\bigwedge_{v''\in\vec{v}}\vec{e}^{\times}[\![\vec{\varphi}]\!][v''](\vec{\nu}[v''])[v'])(\vec{1}[v\leftarrow\nu])$. This simplifies into $\prod_{v\in\vec{v}}\lambda\nu\cdot\bigwedge_{v'\in\vec{v}}\vec{\varphi}[f][v'](\bigwedge_{v''\in\vec{v}}\vec{e}^{\times}[\![\vec{\varphi}]\!][v''](\vec{1}[v\leftarrow\nu][v''])[v'])$ by passing the parameters $\vec{\nu}$. By (70), this is $\prod_{v\in\vec{v}}\lambda\nu\cdot\bigwedge_{v'\in\vec{v}}\vec{\varphi}[f][v'](\vec{e}^{\times}[\![\vec{\varphi}]\!][v]\nu[v'])$, which, by definition of $\vec{e}^{\times}$, is $\prod_{v\in\vec{v}}\lambda\nu\cdot\bigwedge_{v'\in\vec{v}}\vec{\varphi}[f][v'](\vec{e}[v']^{\times}[\![\vec{\varphi}]\!][v]\nu)\stackrel{\text{def}}{=}f\vec{e}^{\times}[\![\vec{\varphi}]\!]$. By eliminating the conjunctions $\bigwedge\{\vec{\varphi}[f][v'](\vec{e}[v']^{\times}[\![\vec{\varphi}]\!][v]\nu)\mid v'\in\vec{v}\wedge v'\neq v\}$ we get the suboptimal expression: $\prod_{v\in\vec{v}}\lambda\nu\cdot\vec{\varphi}[f][v](\vec{e}[v]^{\times}[\![\vec{\varphi}]\!][v]\nu)$ which can be used to obtain the equations for the examples given in [Hug88].

▪ **Proof of Prop. 28** If $\vec{f}^{\times}[f][v]0=0$ then $\vec{\alpha}^{\times}(\vec{f}^{\natural})[f][v]0=0$ by Prop. 27 whence $\alpha^{\times}(\vec{f}^{\natural}[f])[v]0=0$ by (72) which implies $\vec{f}^{\natural}[f](\vec{1}[v\leftarrow 0])=0$ by (70) so that, by Prop. 25, function f is strict in its parameter v.

Efficient chaotic iteration strategies with widenings

François Bourdoncle

DIGITAL Paris Research Laboratory
85, avenue Victor Hugo
92500 Rueil-Malmaison — France
Tel: +33 (1) 47 14 28 22

Centre de Mathématiques Appliquées
Ecole des Mines de Paris
Sophia-Antipolis
06560 Valbonne — France

bourdoncle@prl.dec.com

Abstract. Abstract interpretation is a formal method that enables the static and automatic determination of run-time properties of programs. This method uses a characterization of program invariants as least and greatest fixed points of continuous functions over complete lattices of program properties. In this paper, we study precise and efficient *chaotic iteration strategies* for computing such fixed points when lattices are of infinite height and speedup techniques, known as *widening* and *narrowing*, have to be used. These strategies are based on a *weak topological ordering* of the dependency graph of the system of semantic equations associated with the program and minimize the loss in precision due to the use of widening operators. We discuss complexity and implementation issues and give precise upper bounds on the complexity of the intraprocedural and interprocedural abstract interpretation of higher-order programs based on the structure of their control flow graph.

1 Introduction

Abstract interpretation [7, 10, 11] is a formal method that enables the static and automatic determination of run-time properties of programs, such as the range [2, 4, 5] or congruence properties [15] of integer variables, linear inequalities [9] between variables, data aliasing [2, 4, 12, 13], etc. This method is based on a characterization of programs invariants as either least or greatest fixed points of continuous functions over complete lattices, which are classically computed by iterative computations starting from either the smallest element or the largest element of the lattice. Efficient computation of extremal fixed points of functions over lattices of finite height is a classical topic [17, 19]. Unfortunately, abstract interpretation also has to deal with lattices of infinite or very large height. For instance, when the values of the integer variables of a program are coded on n bits, the lattice of intervals, which is used to compute the maximum range of these variables, is of height 2^n and iterative computations of extremal fixed points of functions over this lattice have a worst-case complexity of 2^n, which is unacceptable in practice. Speed-up techniques, known as widening and narrowing [7, 11], have been designed to determine *safe approximations* of extremal fixed points of continuous function over lattices of infinite height, non-complete lattices [9], and even complete partial orders [3, 4]. When the control flow graph of the program being analyzed is known in advance (as is the case for intraprocedural abstract interpretation) the fixed point equation to be solved amounts to a system of equations, each equation being associated with a control point $c \in C$. In this case, widening techniques require that widening (i.e. generalization) operators be applied at each control point of a *set of widening points* W such that every cycle in the dependency graph of the system is cut by at least one widening point. Of course, it is always possible to choose $W = C$, but this leads to very poor results. In this paper, we propose efficient and precise algorithms for computing approximate fixed points of continuous functions over lattices of infinite height by an appropriate use of widening and narrowing operators.

The paper is organized as follows. In section 2, we review the classical notions of widening operators, narrowing operators and chaotic iterations. Then, in section 3, we introduce the notion of *weak topological ordering* of directed graphs, which generalizes the notion of topological ordering of directed acyclic graphs. We show that this notion is very well suited to the design of chaotic iteration strategies with widenings and give the worst-case complexity of the corresponding algorithms. In section 4, we present three algorithms for computing weak topological orderings with different price/performance ratios. In section 5, we apply the previous theoretical framework to the intraprocedural abstract interpretation of programs. Finally, in section 6, we describe a simple algorithm for the interprocedural abstract interpretation of higher-order programs (for which the control flow graph is not known in advance) and deduce its worst-case complexity from a canonical weak topological ordering of the interprocedural call graph.

2 Chaotic iterations

A central problem in the abstract interpretation of a program is to compute the least (or greatest) solution of a system of semantic equations of the form:

$$\begin{cases} x_1 & = & \Phi_1(x_1, \ldots, x_n) \\ & \vdots & \\ x_n & = & \Phi_n(x_1, \ldots, x_n) \end{cases}$$

where each index $i \in C = [1, n]$ represents a *control point* of the program, and each function Φ_i is a continuous function from L^n to L (L being the abstract lattice of program properties) which computes the property holding at point i after one program step executed from any point leading to i. The *dependency graph* of this system is a graph with set of vertices C such that there is an edge $i \rightarrow j$ if Φ_j depends on its i-th parameter, that is, if it is possible to jump from point i to point j by executing a single program step. In most cases, this graph is thus identical to the control flow graph of the program. For the sake of simplicity, we shall suppose that point 1 is the entry point of the program and that every other point is reachable from 1.

A naive algorithm for solving this system consists in applying each equation in parallel until the vector $(x_1, \ldots, x_n)$ stabilizes, starting from the least (or greatest) element of L^n. If the lattice L is of height h, then the lattice L^n is of height $h \cdot n$ and, therefore, at most $h \cdot n^2$ equations can be applied before the solution is reached.

But this algorithm is far from optimal, since it does not follow the control flow of the program and recomputes the program property associated with every control point at each iteration step. However, since each Φ_i is continuous, and hence monotonic, any sequential algorithm *à la* Gauss-Seidel can also be used, provided that every equation is applied infinitely many times. Such algorithms are called *chaotic iteration algorithms* [8, 10], and a particular choice of the order in which the equations are applied is called a *chaotic iteration strategy.*

When the dependency graph is acyclic, an optimal and linear iteration strategy thus consists in applying the equations in any topological ordering of the set of vertices of the dependency graph, but when there are loops in the program, this method is not applicable. Furthermore, even the naive algorithm cannot be effectively applied to compute least fixed points when the height of the abstract lattice is very large or infinite, as for the lattice of intervals.

A speed-up technique, pioneered by Patrick and Radhia Cousot [7, 10, 11] consists in choosing a subset $W \subseteq C$ and replacing each equation $i \in W$ by the equation:

$$x_i \;=\; x_i \nabla \Phi_i(x_1, \ldots, x_n)$$

where "∇" is a widening operator, i.e. a safe approximation of the least upper bound such that for every increasing chain $(l_k)_{k \geq 0}$, the chain $(l'_k)_{k \geq 0}$ defined by $l'_0 = l_0$ and $l'_{k+1} = l'_k \nabla l_{k+1}$ is eventually stable.

When W is such that every cycle in the dependency graph contains at least an element of W, then any chaotic iteration strategy is guaranteed to terminate and stabilize on a safe approximation of the least fixed point (actually a post-fixed point).

Similarly, narrowing operators can be used to improve the post-fixed points determined by widening operators and to compute safe approximations of greatest fixed points.

However, since widening operators generally lead to an important loss in precision, it is essential that W be as small as possible. Unfortunately, the problem of finding a minimal set W, which happens to be a classical problem (*minimal feedback vertex set*), is a NP-complete problem[14], and since the worst-case complexity of the naive algorithm is quadratic, finding this set would be by far too costly. Hence, two distinct problems have to be solved:

- Determine a good iteration strategy, that is, an order in which to apply the equations.
- Determine a good set of widening points W.

The first problem has been addressed by many authors [6, 16, 17, 19] but, to our knowledge, no algorithm exists for finding good sets of widening points, and authors who mention widening operators use them everywhere or improperly [18].

In the next section, we introduce the notion of weak topological ordering of a directed graph and we show that this notion yields an interesting answer to both problems. In particular, and contrary to what is done by many authors, the iteration strategies we propose are guided by the structure of the dependency graph rather than dynamically selected through the use of ad-hoc data structures, such as work lists. We shall see

```
①  x := 0;
②  while x < 10 do begin
      ③  y := 0;
      ④  while y < x do
            ⑤  y := y + 1  ⑥
⑦  end;  ⑧
```

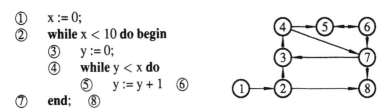

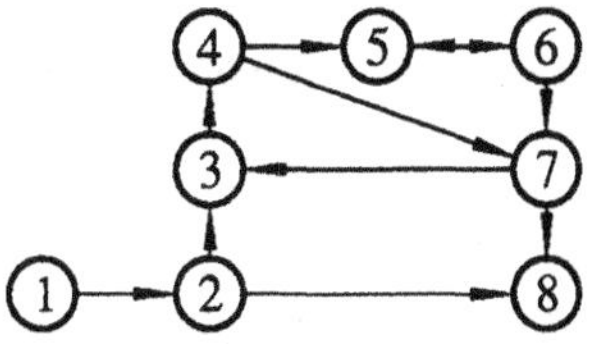

Figure 1: Intraprocedural dependency graph

that this property ensures excellent theoretical upper bounds on the complexities of the resulting algorithms.

3 Weak topological ordering

3.1 Definition

Definition 1 (Hierarchical ordering) *A hierarchical ordering of a set is a well-parenthesized permutation of this set without two consecutive "(".*

A hierarchical ordering of a set C defines a total order $\preceq$ over C. The elements between two matching parentheses are called a *component* and the first element of a component is called the *head*. We call $\omega(c)$, $c \in C$, the set of heads of the nested components containing c, and W the set of components' heads. We define the *depth* of c by $\delta(c) = |\omega(c)|$. An element has depth 0 if it is not contained in a component.

Definition 2 (Weak topological ordering) *A weak topological ordering of a directed graph (w.t.o. for short) is a hierarchical ordering of its vertices such that for every edge $u \rightarrow v$:*

$$(u \prec v \ \wedge \ v \notin \omega(u)) \ \vee \ (v \preceq u \ \wedge \ v \in \omega(u))$$

An edge $u \rightarrow v$ such that $v \preceq u$ is called a *feedback edge*. A w.t.o. of a directed graph is such that the head v of every feedback edge is the head of a component containing its tail u. For instance, a w.t.o. of the dependency graph of figure 1 is:

$$1\ 2\ (\underline{3}\ 4\ (\underline{5}\ 6)\ 7)\ 8$$

This decomposition consists of two nested components with heads 3 and 5 and, for instance, $\omega(1) = \emptyset$, $\omega(6) = \{3, 5\}$, and the feedback edge $7 \rightarrow 3$ is such that $3 \in \omega(7)$. Note that a w.t.o. without parentheses is a topological ordering and that every directed graph over a set of vertices $C = \{1, \ldots, n\}$ always has a trivial w.t.o.:

$$(1\ (2\ (3\ \cdots\ (n))))$$

with n nested components. The following theorem shows that every w.t.o. naturally defines an admissible set of widening points.

Theorem 3 (Widening points) *The set W of components' heads of a w.t.o. of the dependency graph of a system of semantic equations is an admissible set of widening points.*

Proof. Let $c_1 \to \cdots \to c_k \to c_1$, be a cycle of k distinct elements. If $k = 1$, then $c_1 \preceq c_1$ and $c_1 \to c_1$, and thus $c_1 \in \omega(c_1) \subseteq W$. If $k > 1$, then either there exist $i < j$ such that $c_j \prec c_i$ and thus $c_j \in \omega(c_i) \subseteq W$, or $c_1 \prec \cdots \prec c_k$ and thus $c_1 \prec c_k$ and $c_k \to c_1$, which shows that $c_1 \in \omega(c_k) \subseteq W$.

In either case, the cycle is thus cut by at least one widening point, which proves the theorem. ■

Of course, since widenings are costly in terms of precision, one should attempt to minimize the cardinal of W, and the trivial w.t.o. is not very interesting in this respect.

3.2 Iteration strategies

We have seen that a w.t.o. of a dependency graph is useful for determining sets of widening points, but every w.t.o. also defines at least two chaotic iteration strategies. The first strategy, called the *iterative strategy*, simply applies the equations in sequence and "stabilizes" outermost components whereas the second strategy, called the *recursive strategy*, recursively stabilizes the subcomponents of every component every time the component is stabilized.

For instance, the w.t.o. "1 2 ($\underline{3}$ 4 ($\underline{5}$ 6) 7) 8" of the graph of figure 1 yields the iterative strategy:

$$1\ 2\ [\underline{3}\ 4\ \underline{5}\ 6\ 7]^*\ 8$$

where $[\,]^*$ is the "iterate until stabilization" operator, and the recursive strategy:

$$1\ 2\ [\underline{3}\ 4\ [\underline{5}\ 6]^*\ 7]^*\ 8$$

It is easy to see that these strategies are correct, since for every vertex v of depth 0 and every edge $u \to v$, u is necessarily listed before v, i.e. $u \prec v$, and the value of u used in the computation of v already has its final value, which implies that it is not necessary to iterate over the vertices of depth 0. Note that this idea forms the heart of the method described in Jones [16], where the strongly connected components are listed in topological order (c.f. section 4.3). However, our approach is superior in that we also give algorithms for computing fixed points of strongly connected systems of semantic equations instead of using the brute-force, $O(n^2)$ algorithm.

Theorem 4 (Iterative strategy) *For the iterative strategy, the stabilization of an outermost component of a w.t.o. can be detected by the stabilization of its widening points.*

Proof. Let us suppose that the w.t.o. consists of a single outermost component. We are going to show that after applying the equations in sequence, either one (at least) of the the semantic values associated to a widening point has increased, or none has changed.

Suppose that the contrary holds, and that the value associated to a vertex $v \notin W$ has changed. Then by definition of a w.t.o., every edge $u \to v$ is such that $u \prec v$. Thus, there exists at least one vertex $u \prec v$ whose value has changed since the last iteration.

But since $u \notin W$ by hypothesis, the inductive application of the same argument to u shows that the head of the component has changed, which is absurd. ■

Theorem 5 (Recursive strategy) *For the recursive strategy, the stabilization of a component of a w.t.o. can be detected by the stabilization of its head.*

Proof. Let us suppose that the w.t.o. consists of a single outermost component and that, after applying the equations and stabilizing the sub-components, the value associated to the head of the component remains unchanged after recomputation. Then the argument used to prove the fact that no iteration is necessary over the vertices of depth 0 shows that the values associated to the vertices within the component won't change when the equations are applied once more. Therefore, the stabilization of the component's head imply the stabilization of the entire component. ■

These two theorems show that the iterative and recursive strategies minimize the number of comparisons between elements of the lattice L of program properties, which can be very useful when this test is very costly, as with the abstract interpretation of functional or logic programs for instance. Also, note that even though the ordering $\sqsubseteq$ over L is not explicitly used by the algorithm, it can be used to improve precision and force the convergence of the computation to the first post-fixed point when the widening operator is not *stable* [3, 4], i.e. does not satisfy:

$$\forall x, y \in L \,:\, y \sqsubseteq x \implies x \nabla y = x$$

The following theorem gives an upper bound on the complexity of each strategy.

Theorem 6 (Complexity) *When the lattice L is of finite height h, or when the increasing chains built by the widening operator are at most of length h, the maximum complexity of the iterative iteration strategy for a strongly connected graph is:*

$$h \cdot |C| \cdot |W|$$

and the maximum complexity of the recursive iteration strategy is:

$$h \cdot \sum_{c \in C} \delta(c)$$

Proof. Since theorem 4 shows that each iteration yields a strictly greater element over the lattice $L^{|W|}$ of height $h \cdot |W|$, the first result is trivial. The second result is easily proved inductively by showing that each equation (i.e. vertex) of depth k is applied at most $h \cdot k$ and each sub-component of depth $k+1$ is stabilized at most $h \cdot k$ times. A detailed proof can be found in Bourdoncle [4]. ■

The complexity of the recursive iteration strategy is thus a linear function of the sum of the individual depths of the vertices of the graph. It is interesting to remark that since $\delta(c) \leq |W|$ for every vertex c, the upper bound of the recursive iteration strategy is always better than that of the iterative strategy. Practice shows that the recursive strategy is indeed almost always better than the iterative strategy. Furthermore, it is

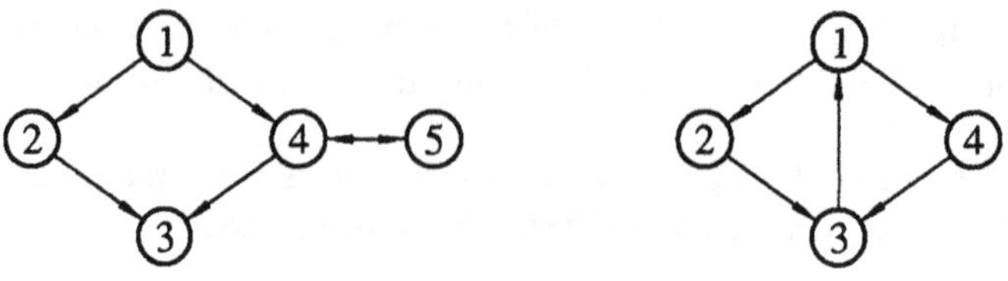

Figure 2: Flow graphs

clear that the worst-case of each strategy for a program of size n, which is obtained with the trivial w.t.o., is $h \cdot n^2$ for the iterative strategy and:

$$h \cdot (1+2+\cdots+n) \;=\; \frac{h \cdot n \cdot (n+1)}{2}$$

for the recursive strategy. These results show that not only does the number of widening points impact on the precision of the fixed point computation, but it also impacts on the cost of the analysis. An essential goal is thus to minimize the number of widening points as well as the sum of the individual depths of the graph's vertices.

The following section presents three algorithms with different price/performance ratios that can be used to compute weak topological orderings of directed graphs and relate them to previous works.

4 Algorithms

4.1 Depth-first numbering

A first idea for building a non-trivial w.t.o. of a graph is to use a depth-first numbering of this graph, which can be obtained in linear time, and 1) open a parenthesis before every head b of edges $a \rightarrow b$ whose tail a has a greater number than b, 2) close all the parentheses after the last vertex. For instance, this algorithm would yield the following result on the graph of figure 1:

$$1\ 2\ (\underline{3}\ 4\ (\underline{5}\ 6\ 7\ 8))$$

which is better than the trivial w.t.o. but not as good as the "optimal" ordering:

$$1\ 2\ (\underline{3}\ 4\ (\underline{5}\ 6)\ 7)\ 8$$

Also, note that this algorithm has a tendency to overestimate the number of widening points. For instance, the graph on the left of figure 2 would yield the following w.t.o.:

$$1\ 2\ (\underline{3}\ (\underline{4}\ 5))$$

with two widening points, although the graph has a single cycle. However, it is shown in Bourdoncle [4] that the set of heads of the *retreating edges* (c.f. [1], p. 661) of the depth-first spanning tree, i.e. edges $a \rightarrow b$ such that b is an ancestor of a in the tree, is a smaller admissible set of widening points. For the graph on the left of figure 2, the only retreating edge is $5 \rightarrow 4$ and $\{4\}$ is thus a set of widening points.

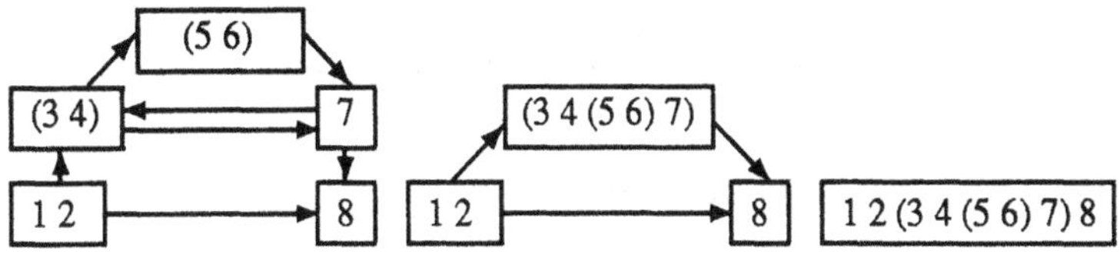

Figure 3: Limit flow graph

Consequently, the w.t.o. decomposition can be used as the basis of the iteration strategy whereas widening points can be detected through the retreating edges of the depth-first spanning tree (which can be easily found by using a stack of "currently visited vertices" during the depth-first visit of the graph).

In spite of its drawbacks, the advantage of this algorithm is that it is very simple and incremental, and can be applied even when the graph is not known in advance as for the abstract interpretation of higher-order programs (c.f. section 6).

4.2 Reducible graphs

When the dependency graph is *reducible* [1], as is always the case for structured programming languages without goto statement, it has been suggested [11] to choose as widening points the heads of the *intervals* of the graph which are also the head of a back-edge.

This idea can be pushed a step further to build a w.t.o. of the graph and, hence, an iteration strategy. The idea consists in computing the *limit flow graph* obtained by iteratively collapsing the graph into its *intervals* (c.f. [1], p. 666). When the graph is reducible, this process is guaranteed to converge to a limit graph reduced to a single vertex containing all the vertices of the original graph. This process is illustrated figure 3 for the reducible graph of figure 1 and gives the following result:

$$1\ 2\ (\underline{3}\ 4\ (\underline{5}\ 6)\ 7)\ 8$$

Note that an interval I is parenthesized only when there exists a feedback edge $u \rightarrow v$ from a vertex $u \in I$ to the *header* v of I.

To prove that the resulting decomposition of the graph is a w.t.o., the only thing to show is that the head v of every feedback edge $u \rightarrow v$ is the header of an interval containing u.

First, remark that the property trivially holds when u and v belong to the same interval of level 1. So let K denote the interval who first "merged" the two distinct intervals I containing u and J containing v during the computation of the limit flow graph. It is known (c.f. [1], prop. 2, p. 669) that v is necessarily the header of J. Now, if J is the first "vertex" of K (as for the interval (3 4) of figure 3) then v is the head of K and the property holds. Otherwise, I and J are proper "vertices" of K, J is listed before I (since $u \rightarrow v$ is a feedback edge) and there is an edge from I to J, which is incompatible with the fact that J has been added to K before I.

This algorithm has a worst-case complexity equal to $O(\delta \cdot \varepsilon \cdot \alpha(\varepsilon))$ where δ is the *depth* of the graph, ε is the number of edges, and α is the inverse of Ackerman's function (which is nearly constant).

Finally, note that the iteration strategies built using this algorithm are similar to the ones described in Burke [6] in a data-flow analysis framework.

4.3 Strongly connected components

Reducible graphs have been extensively studied in the literature but, unfortunately, interprocedural dependency graphs are not reducible in general. For instance, the graph of figure 5, which is an unfolded version of the graph of function "Fact", is not reducible since the head $6'$ of the back-edge $5' \to 6'$ does not dominate its tail, i.e. there is a path from 1 to $5'$ that does not go through $6'$.

The base of the algorithm we propose in this case is an algorithm due to R.E. Tarjan [20] to compute in linear time the strongly connected components of a directed graph. Since Tarjan's algorithm computes a list of (possibly trivial) strongly connected components in topological order, the basic idea of the algorithm, given figure 4, is to recursively apply Tarjan's algorithm to each non-trivial component after having removed its head b and all the back-edges of the form $a \to b$. Note that "::" denotes the list constructor operator.

Theorem 7 *The algorithm of figure 4 computes a w.t.o. of any directed graph.*

The proof is omitted here for the sake of brevity and can be found in Bourdoncle [4]. Note that when the graph is reducible, the interval-based algorithm can give better results. For instance, depending on the order in which the graph's vertices are visited, the algorithm of figure 4 gives one of the following results for the graph on the right of figure 2:

$$(\underline{1}\ 2\ 3\ 4)$$
$$(\underline{1}\ 4\ 3\ 2)$$

whereas the interval-based algorithm gives the optimum result:

$$(\underline{1}\ 2\ 4\ 3)$$

However, excellent decompositions are obtained for non-reducible graphs, such as the graph of figure 5:

$$(\underline{1}\ 4\ 1'\ 4')\ 2'\ 3'\ 2\ 3\ (\underline{6}\ 5'\ 6'\ 5)$$
$$(\underline{1}\ 4\ 1'\ 4')\ 2\ 3\ 2'\ 3'\ (\underline{6'}\ 5\ 6\ 5')$$

The worst-case complexity of this algorithm is $\delta \cdot \varepsilon$, where δ is the maximum depth of the graph's vertices and ε is the number of edges. Hence, this algorithm does not cost more than the fixed point computation itself, and its complexity is a linear function of the intrinsic complexity of the graph.

Finally, note that to our knowledge, other hierarchical decompositions of directed graphs into strongly connected components [21] are not weak topological orderings and, hence, cannot be used to perform chaotic iteration strategies with widenings.

```
function Partition
    var vertex, partition
begin
    foreach vertex ∈ vertices_G do
        DFN[vertex] ← 0
    NUM ← 0
    partition ← nil
    Visit(root_G, partition)
    return partition
end
function Component(in vertex)
    var succ, partition
begin
    partition ← nil
    foreach succ ∈ succ_G[vertex] do
        if DFN[succ] = 0 then
            Visit(succ, partition)
    return (vertex :: partition)
end
function Visit(in vertex, inout partition)
    var head, min, succ, element, loop
begin
    push(vertex)
    head ← DFN[vertex] ← NUM ← NUM + 1
    loop ← false
    foreach succ ∈ succ_G[vertex] do
        if DFN[succ] = 0 then
            min ← Visit(succ, partition)
        else min ← DFN[succ]
        if min ≤ head then
            head ← min
            loop ← true
    if head = DFN[vertex] then
        DFN[vertex] ← +∞
        pop(element)
        if loop then
            while element ≠ vertex do
                DFN[element] ← 0
                pop(element)
            partition ← Component(vertex) :: partition
        else partition ← vertex :: partition
    return head
end
```

Figure 4: Hierarchical decomposition of a directed graph into strongly connected components and subcomponents.

```
function Fact(n : integer) : integer;
begin
① if n = 0 then
     ② Fact := 1 ③
   else
     ④ Fact := n * Fact(n−1) ⑤
⑥
end;
```

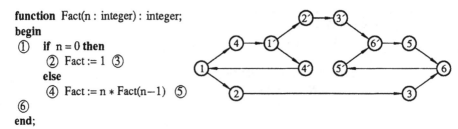

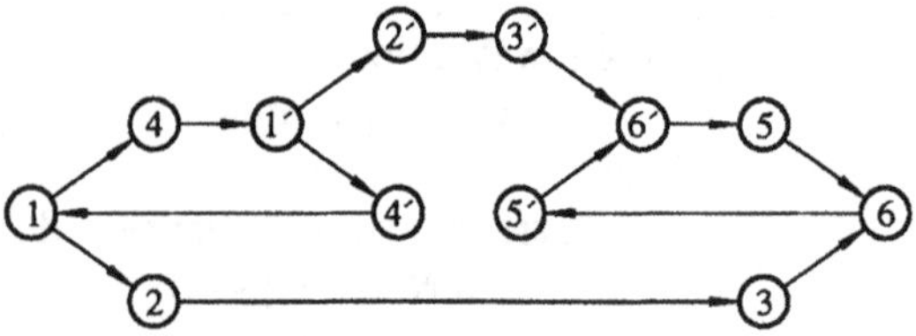

Figure 5: Interprocedural dependency graph

5 Intraprocedural abstract interpretation

The algorithm of figure 4 is thus directly applicable to intraprocedural abstract interpretation, and has been implemented in the abstract debugger *Syntox* [2, 4, 5]. The advantage of using a w.t.o. is that it is computed once at the beginning of the analysis and that the resulting chaotic iteration strategy minimizes the use of widening operators as well as the number of tests needed to detect the stabilization of iterative computations. Furthermore, the algorithm has a predictable worst-case complexity, which is n for a program without loops and:

$$h \cdot p \cdot (n - \frac{p-1}{2})$$

for a program with p nested loops.

6 Interprocedural abstract interpretation

The method of the previous section is applicable to the interprocedural abstract interpretation of first-order program but cannot be used for higher-order programs for which the dependency graph is not known in advance. However, the incremental algorithm of section 4.1 can be used, and since the w.t.o. determined by this algorithm has the generic form:

$$a_1 \cdots (a_{k_1} \cdots (a_{k_2} \cdots (a_{k_3} \cdots)))$$

the iterative strategy seems easier to implement and corresponds to a straightforward depth-first execution of the program. Furthermore, if we note that each point a_{k_i} is necessarily either an entry point of a procedure, the head of an intraprocedural loop or the return point of a procedure call, we have the following property on the worst-case complexity of the interprocedural abstract interpretation of a program.

Theorem 8 *If a program has p procedures, n control points, c procedure calls and l intraprocedural loops, then the abstract interpretation over a lattice of height h using the iterative iteration strategy has a worst-case complexity of:*

$$h \cdot (p + c + l) \cdot n = \rho \cdot h \cdot n^2$$

where $\rho \leq 1$ is sum $c/n + l/n$ of the densities of procedure calls and intraprocedural loops and of the inverse of the average size n/p of procedures. Furthermore, if each *procedure of a higher-order program has at most m procedural formal parameters, then the computation of the interprocedural call graph of the program has a worst-case complexity of* $\rho \cdot m \cdot p \cdot n^2$.

Note that, as hinted in section 4.1, the depth-first visit of the interprocedural dependency graph allows for the on-line determination of a better set of widening points than $\{a_{k_1}, a_{k_2}, \ldots\}$ and, in practice, it is sufficient to use widening operators at the head of intraprocedural loops and at the entry and exit points of formally recursive procedures, as opposed to the return points of procedure calls ([4], p. 52).

Note that the algorithm proposed by Le Charlier et al. [18], which is in fact a particular implementation of *basic functional partitioning* [3], uses widening operators at the entry point of every logic predicate but not at the exit point. This is justified in their context, since they use *noetherian* domains with no infinite strictly increasing chain, but their algorithm can loop when the abstract domain is of infinite height.

7 Conclusion

In this paper, we have addressed the problem of the efficient computation of least and greatest fixed points of continuous functions over lattices of infinite height using widening and narrowing operators.

We have introduced the notion of weak topological ordering of directed graphs and shown how this notion can be used to determine admissible sets of widening points and design efficient chaotic iteration strategies.

We have given several algorithms with different price/performance ratios to compute weak topological orderings of directed graphs and shown how to apply them to the intraprocedural and interprocedural abstract interpretation of higher-order languages.

Further work will be to design an incremental version of the algorithm of figure 4 to handle the interprocedural abstract interpretation of higher-order programs more efficiently.

References

[1] Alfred V. Aho, Ravi Sethi and Jeffrey D. Ullman: "Compilers — Principles, Techniques and Tools", Addison-Wesley Publishing Company (1986)

[2] François Bourdoncle: "Interprocedural Abstract Interpretation of Block Structured Languages with Nested Procedures, Aliasing and Recursivity", *Proc. of the International Workshop PLILP'90*, Lecture Notes in Computer Science 456, Springer-Verlag (1990)

[3] François Bourdoncle: "Abstract Interpretation By Dynamic Partitioning", *Journal of Functional Programming*, Vol. 2, No. 4 (1992)

[4] François Bourdoncle: "Sémantiques des langages impératifs d'ordre supérieur et interprétation abstraite", *Ph.D. dissertation*, Ecole Polytechnique (1992)

[5] François Bourdoncle: "Abstract Debugging of Higher-Order Imperative Languages", *Proc. of SIGPLAN '93 Conference on Programming Language Design and Implementation* (1993)

[6] Michael Burke: "An Interval-Based Approach to Exhaustive and Incremental Interprocedural Data-Flow Analysis", *ACM Transactions on Programming Languages and Systems*, Vol. 12, Num. 3 (1990) 341–395

[7] Patrick and Radhia Cousot: "Abstract Interpretation: a unified lattice model for static analysis of programs by construction or approximation of fixpoints", *Proc. of the 4th ACM Symp. on POPL* (1977) 238–252

[8] Patrick Cousot: "Asynchronous iterative methods for solving a fixpoint system of monotone equations", Research Report IMAG-RR-88, Université Scientifique et Médicale de Grenoble (1977)

[9] Patrick Cousot and Nicolas Halbwachs: "Automatic discovery of linear constraints among variables of a program", *Proc. of the 5th ACM Symp. on POPL* (1978) 84–97

[10] Patrick Cousot: "Méthodes itératives de construction et d'approximation de points fixes d'opérateurs monotones sur un treillis. Analyse sémantique de programmes", *Ph.D. dissertation*, Université Scientifique et Médicale de Grenoble (1978)

[11] Patrick Cousot: "Semantic foundations of program analysis", in Muchnick and Jones Eds., *Program Flow Analysis, Theory and Applications*, Prentice-Hall (1981) 303–343

[12] Alain Deutsch: "On determining lifetime and aliasing of dynamically allocated data in higher-order functional specifications", *Proc. of the 17th ACM Symp. on POPL* (1990)

[13] Alain Deutsch: "A Storeless Model of Aliasing and its Abstractions using Finite Representations of Right-Regular Equivalence Relations", *Proc. of the IEEE'92 International Conference on Computer Languages*, IEEE Press (1992)

[14] Michael. R. Garey and David S. Johnson: "Computers and Intractability: A Guide to the Theory of NP-completeness", W.H. Freeman and Company (1979)

[15] Philippe Granger: "Static analysis of arithmetical congruences", *International Journal of Computer Mathematics* (1989) 165–190

[16] Larry G. Jones: "Efficient Evaluation of Circular Attribute Grammars", *ACM Transactions on Programming Languages and Systems*, Vol. 12, Num. 3 (1990) 429–462

[17] B. Le Charlier, K. Musumbu and P. Van Hentenryck: "A generic abstract interpretation algorithm and its complexity analysis", in K. Furukawa editors, *Proc. of the Eight International Conference on Logic Programming*, MIT Press (1991) 64–78

[18] B. Le Charlier and P. Van Hentenryck: "A universal top-down fixpoint algorithm", Technical Report 92-1, Institute of Computer Science, University of Namur, Belgium (1992)

[19] R.A. O'Keefe: "Finite fixed-point problems", in J.-L. Lassez editor, *Proc. of the Fourth International Conference of Logic Programming*, MIT Press (1987) 729–743

[20] R.E. Tarjan: "Depth-first search and linear graph algorithms", *SIAM J. Comput.*, 1 (1972) 146–160

[21] R.E. Tarjan: "An Improved Algorithm for Hierarchical Clustering Using Strong Components", *Information Processing Letters 17*, Elsevier Science Publishers B.V. (1983) 37–41

SEMANTIC ANALYSIS OF INTERVAL CONGRUENCES

François Masdupuy
Centre de Recherche en Informatique
École des Mines de Paris
35, rue Saint-Honoré
77305 Fontainebleau cedex, France
masdupuy@ensmp.fr

Abstract

This paper describes a new non relational semantic analysis of program integer variables. Interval congruence analysis is designed using Cousot's abstract interpretation framework, its model generalizes integer intervals and integer cosets by the definition of coset congruences. The use of a widening operator defined on rational approximations of the integer model ensures fast convergences of the iteration process whereas the diversity of patterns of the modeled integer sets increases the accuracy of the analysis.

1 The Global Design of the Analysis

The motivation for designing a new non relational integer semantic analysis is first to be able, using only one analysis, to discover program invariant approximations which would have been determined either by the interval or by the congruence analysis. This corresponds to automatically deciding, during the static analysis, which one of these analyses is convenient for every program point and every integer variable. The other goal is to determine invariant approximations where both interval and congruence analyses would have failed, an example of such situation is given at the end of this paper. The rest of this section presents Cousot's abstract interpretation framework and its particular use in interval congruence analysis.

Following [CC79] and the notational conventions of [CC92b] let's take as standard semantics an operational one consisting of a transition system $(S, \tau, \iota, \varsigma)$ where S is a set of states, τ a transition relation binding a state to its possible successors, $\iota \subseteq S$ a set of initial states and $\varsigma \subseteq S$ a set of final states. For example for a program with m control points operating on n distinct integer variable S is potentially $[1, m] \times \mathbb{Z}^n$. Then the forward collecting semantics consists in considering sequences of finite partial execution traces starting with an initial state and where two consecutive states verify the transition relation. In order to discuss of invariance properties of programs we can approximate the forward collecting semantics by the descendant states of the initial states, considering sets of states occurring in the original sequences of finite partial execution traces.

The concrete semantic domain is the powerset of the set S. Only the integer variables are of interest for our analysis, hence in first approximation the set of considered states will be $\mathbb{Z}^n$ where n is the number of variables in the program and the concrete semantic domain the powerset $\mathbb{P}(\mathbb{Z}^n)$ (standard semantics). For the present analysis, the characterized states are non relationally approximated and hence $\mathbb{P}(\mathbb{Z}^n)$ is replaced by $\mathbb{P}(\mathbb{Z})^n$. The next approximation consists in considering only few integer subsets that are coset congruences of CC, hence $\mathbb{P}(\mathbb{Z})$ is now approximated by CC. Then for machine representation requirements, the coset congruences will be denoted as rational subsets (using the set IC of interval congruences) the intersection of which with $\mathbb{Z}$ will consist in the corresponding coset congruences. Two abstractions have to be considered that are the one between $\mathbb{P}(\mathbb{Z})$ and CC and the other between CC and IC. CC is the abstract domain of the first although it is the concrete domain of the second. The first abstraction is modeled with a concretization function $\gamma_0 : CC \longrightarrow \mathbb{P}(\mathbb{Z})$ giving the meaning of a coset congruence in terms of integer subsets, the approximation ordering is therefore induced by the set inclusion relation on the powerset of $\mathbb{Z}$. The latter connection between concrete and abstract domain is established via a pair of abstraction and concretization functions (α, γ).

The exact invariant of the program are shown to be solution of a fixpoint equation, unfortunately, it is very often uncomputable. We must use an approximation of the exact properties, here the set of coset congruences CC ($CC \subset \mathbb{P}(\mathbb{Z})$). The meaning $\gamma_0(C)$ of an element of $C \in CC$ is the integer subsets containing its elements (up to the machine-representation isomorphism) since C approximates all its integer subsets. Hence one integer subset has several possible approximations and unfortunately no least upper bound operator exists in CC so following [CC92c] the relation between concrete (integer subsets) and abstract (coset congruences) properties cannot be modelized by a Galois connection and a widening operator on IC, a rational approximation of CC, has to be introduced[1].

All the proofs of theorems cited in the paper and of the correctness of the widening operator or abstract primitives are given in [Mas92b]. The paper is organized as follow, first some notations are recalled, then the concrete model is designed, the next section provides the abstract one and finally an abstract interpretation using both of them is built, we conclude with an example.

2 Notations

$\mathbb{Q}_{-\infty} \stackrel{def}{=} \mathbb{Q} \cup \{-\infty\}$, $\mathbb{Q}_{+\infty} \stackrel{def}{=} \mathbb{Q} \cup \{+\infty\}$, and $\mathbb{Z}_{-\infty} \stackrel{def}{=} \mathbb{Z} \cup \{-\infty\}$, $\mathbb{Z}_{+\infty} \stackrel{def}{=} \mathbb{Z} \cup \{+\infty\}$ where $-\infty$ and $+\infty$ are considered as limits respectively on $\mathbb{Q}$ and $\mathbb{Z}$. The usual operators (sum, product,...) on $\mathbb{Q}$ and $\mathbb{Z}$ are canonically extended to $\mathbb{Q}_{-\infty}$, $\mathbb{Q}_{+\infty}$, $\mathbb{Z}_{-\infty}$, $\mathbb{Z}_{+\infty}$ and $\lceil -\infty \rceil = \lfloor -\infty \rfloor = -\infty$ and $\lceil +\infty \rceil = \lfloor +\infty \rfloor = +\infty$. Following the context $[-\infty, +\infty]$ is $\mathbb{Q}_{-\infty} \cup \mathbb{Q}_{+\infty}$ or $\mathbb{Z}_{-\infty} \cup \mathbb{Z}_{+\infty}$. The greatest common divisor is always non negative. The integer coset $a + q\mathbb{Z}$ with integer representative a and

[1] Notice that (α, γ) is not a Galois connection since α cannot choose a best approximation and provides only a minimal element of the abstract set soundly approximating a concrete element. It is a good analysis in order to illustrate the general framework of [CC92b]

modulo q is the set $\{a + kq,\ k \in \mathbb{Z}\}$. The rational coset $a + q\mathbb{Z}$ corresponds to the set $\{a + kq,\ k \in \mathbb{Z}\}$ where $(a, q) \in \mathbb{Q}^2$. The relation $l \equiv u \mod m$, which is equivalent to $u - l \in m\mathbb{Z}$, is shortened to $l \overset{m}{=} u$. The least upper bound (in the lattice of cosets, see [Gra89]) of $a + q\mathbb{Z}$ and $a' + q'\mathbb{Z}$ noted $a + q\mathbb{Z} \sqcup a' + q'\mathbb{Z}$, is given by $a + \gcd(a' - a, q, q')\mathbb{Z}$. The notions of divisor, multiple and operators such as gcd are canonically extended to rational numbers (a divides b if there exists an integer k such that $b = ka$). An inverse of the integer θ with respect to the integer m, when it exists, is noted $\theta^{-1}_{(m)}$ and satisfies $\theta\theta^{-1} \in 1 + m\mathbb{Z}$. $\theta^{-1}_{(m)}$ is noted θ^{-1} when there is no ambiguity on the modulo. An inverse of θ with respect to m exists when $\gcd(\theta, m) = 1$; it is a direct consequence of Bezout's theorem[2]. For the rest of this paper, the convention is that an inverse $\theta^{-1}_{(m)}$ of an offset θ is always taken with respect to the modulo m of their coset congruence, if there is no possible ambiguity (see definition 1 for the definitions of coset congruence and offset).

3 The set CC of Coset Congruences on $\mathbb{Z}$

Interval analysis of [CC76] and congruence analysis of [Gra89, Gra91b] are quite orthogonal concepts. This leads to the definition of a third analysis with the basic idea of generalizing the two first to the notion of coset congruence. The basic components of coset congruences (and two degenerate cases of the general definition) are integer intervals and integer cosets. To fill the gap between these two kinds of elements, general coset congruences are introduced. A coset congruence is a set of arithmetical cosets with the same modulo and whose representatives are separated by an offset such that the common modulo and the offset are prime numbers.

Definition 1 (Coset congruence $\theta.[l, u](m)$) *Let $l \in \mathbb{Z}_{-\infty}$, $u \in \mathbb{Z}_{+\infty}$ and $m, \theta \in \mathbb{Z}$ be integers such that $\gcd(\theta, m) = 1$ and $m = 0$ implies $\theta = 1$. The coset congruence $\theta.[l, u](m)$ of offset θ, lower bound l, upper bound u and modulo m is defined by*

$$\theta.[l, u](m) \overset{def}{=} \begin{cases} [-\infty, u] \cup [l, +\infty] & \text{if } l > u \text{ and } m = 0, \quad (1) \\ \bigcup_{l \le \kappa \le u} \kappa\theta + m\mathbb{Z} & \text{otherwise.} \quad (2) \end{cases}$$

CC is the set of coset congruences.

A very important remark for the rest of the discussion about coset congruences is that, when the modulo is non zero, since the offset and modulo are prime numbers by definition, the single cosets $\kappa\theta + m\mathbb{Z}$ used in definition case (2) are all distincts for m consecutive values of κ. Hence taking a sufficiently wide interval $[l, u]$ provides a way to represent $\mathbb{Z}$.

One motivation to define such a surprising integer model is that a coset congruence is exactly the intersection with $\mathbb{Z}$ of a much more intuitive model defined on the set of

[2] Let a and b be two integers; there exist integers u and v such that

$$u.a + v.b = \gcd(a, b)$$

rational numbers: the interval congruences that are defined in section 4. In particular, we will see that the primality between the common modulo and the offset separating the different representatives comes from that intersection. Another, more practical, reason that leads us to approximate CC with rational interval congruences is that comparison (set inclusion) test on CC is for the moment particularly inefficient[3].

The different kinds of integer sets considered in the preceding definition are illustrated by the following figure:

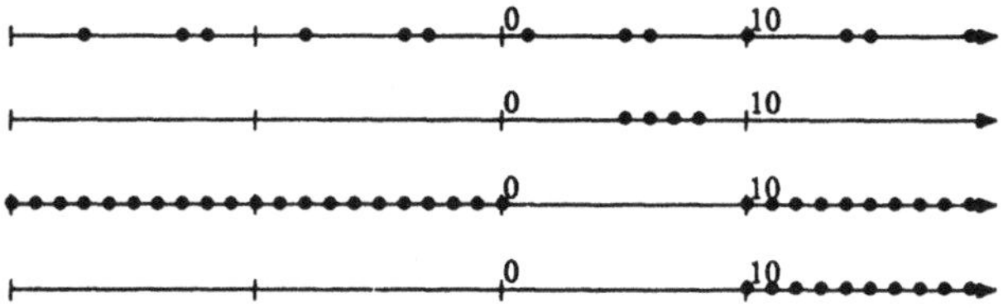

(3)

(4)

(5)

(6)

The general case (3), corresponding to $5.\,[1,3]\,(9)$, is the set of the three integer cosets $5+9\mathbb{Z}$, $10+9\mathbb{Z}=1+9\mathbb{Z}$ and $15+9\mathbb{Z}=6+9\mathbb{Z}$. The case (4), where the modulo is zero and the representative bounds well ordered, is noted $1.\,[5,8]\,(0)$ and corresponds to the integer interval $[5,8]$. The case (5), where the modulo is zero and the representative bounds inverse ordered, corresponds to definition (1) and is noted $1.\,[10,0]\,(0)$. Finally, the case (6) represents the set of integers greater than 10 and is the coset congruence $1.\,[10,+\infty]\,(0)$.

For the relation $\subseteq$ induced by the set inclusion relation, CC is a preorder. An equivalence relation $\approx\ =\ \subseteq \wedge \supseteq$ is defined to build a partial order on the quotient set $CC/_{\approx}$ (for example $2.\,[7,9]\,(-11) \approx 9.\,[2,4]\,(11)$).

Theorem 2 (Coset congruence equivalence relation $\approx$) *Let C_1 and C_2 be two coset congruences, their equivalence $C_1 \approx C_2$ is computable in constant time.*

If we arbitrary choose a representation of the empty set and of the set of integers by a coset congruence, if we remark that apart from them, coset congruences of zero modulo form equivalence classes with only one element, and finally if we consider the coset congruences with positive modulo, offset and lower bound positive and smaller than the modulo, we are able to exhibit a canonical normalization algorithm, the normalization operator is noted $\|\ \|$.

A complement operator, noted $\overline{}$, which provides the negation of a property is provided. For example $\overline{5.\,[1,3]\,(9)} = 5.\,[4,9]\,(9)$

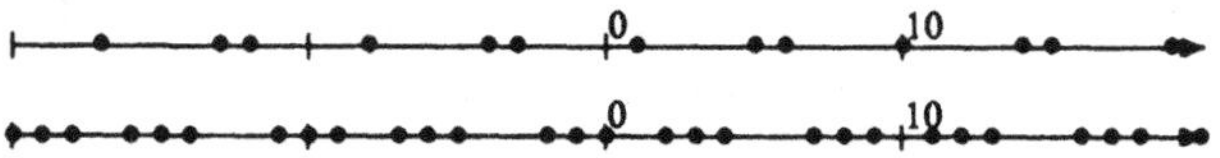

Only few analyses like parity, sign or logic program groundness analysis [CC92a] provide such a complementation characteristic and it will be shown to be very useful in the section 5.3 on abstract primitives. Although such a property necessitates the consideration of complement of finite integer interval and hence complicates the expressions

[3] No efficient (constant time) algorithm has been found by me

concerning coset congruences, such a characteristic feature is kept for analysis accuracy motives. The complement of a coset congruence corresponds to the set of integer cosets not contained in the original one, the property $\forall C \in CC/_{\approx}\ \overline{\overline{C}} \approx C$ holds.

Because we are not able to compare efficiently coset congruences[4] and, moreover, fur the purpose of approximate join operator (see bellow), we need to choose between non comparable ones, a measure of accuracy ι is defined. It partially orders CC using an approximation of the "size" of the integer set where this size is close to the probability that an integer is in the coset congruence. It is an arbitrary order defined on $CC/_{\approx}$ which is used by the approximate join operator (defined bellow) to make an arbitrary choice between two rational interval congruences.

The set $CC/_{\approx}$ Of coset congruences described above has only few interesting algebraic properties; it is a complete partial order with an infimum and a supremum. Its major drawback is a lack of least upper bound and of efficient comparison algorithm between its elements. In addition CC is not closed under intersection and cannot be completed by intersecting its elements (because of the size of the resulting set). These are good motivations to introduce a new approximation, the rational sets of IC which provide efficient algorithms.

4 The set IC of Interval Congruences on $\mathbb{Q}$

The goal of this section is to define a rational model based on the use of a set of rational arithmetical cosets with consecutive representatives.

Definition 3 (Interval congruences $[a,b]\,(q)$) *Let $a \in \mathbb{Q}_{-\infty}$, $b \in \mathbb{Q}_{+\infty}$ and $q \in \mathbb{Q}$ be rational numbers. The interval congruence $[a,b]\,(q)$ of lower bound a, upper bound b, and modulo q is defined by*

$$\begin{cases} [a,+\infty] \cup [-\infty,b] & \text{if } a > b \text{ and } q = 0 \\ \{x, x = x_0 + kq, \exists x_0 \in \mathbb{Q} \cap [a,b],\ k \in \mathbb{Z}\} & \text{otherwise} \end{cases}$$

IC is the set of interval congruences.

In the following, when we need to consider an interval congruence $[a,b]\,(\frac{\nu}{\delta})$, we implicitly take non negative integers ν and δ such that $\gcd(\nu,\delta) = 1$.

An interval congruence is both an infinitely and regularly dispersed set of rational intervals, or equivalently a set of rational cosets with "consecutive" representatives; therefore $[a,b]$ is called the representative of the interval congruence. For all non negative rational q, $IC(q)$ is the set of interval congruences of modulo q. Two interval congruences with different representatives may denote the same rational set. Notice that the set of interval congruences contains the set of rational cosets (where the lower and upper bounds are equal) and the set of rational intervals (where the modulo is zero and the upper bound greater than the lower bound). An example of such a rational set is given below:

[4]Of course the naive algorithm consisting in partitioning $\theta.\,[l,u]\,(m)$ into arithmetical cosets to test the inclusion of $\theta.\,[l,u]\,(m)$ in an other coset congruence is possible but very expensive (except if in practice the lower and upper bounds are very closed).

$$\left[\frac{1}{2},\frac{3}{4}\right]\left(\frac{5}{2}\right)=\bigcup_{k\in\mathbb{Z}}\left[\frac{1+5k}{2},\frac{3+10k}{4}\right]$$

and illustrated by

In the following, we implicitly consider the usual operators on interval congruences of zero modulo (usual rational intervals) that are the sum, the difference of two intervals and the product of an interval with a scalar.

Definition 4 (Interval congruences comparison $\subseteq_\sharp$) *The comparison relation $\subseteq_\sharp$ on IC is the extension to IC of the partial order relation on $\mathbb{P}(\mathbb{Q})$ induced by set inclusion.*

$\subseteq_\sharp$ is a preorder relation, it is computable in constant time.

Theorem & Definition 5 (Equivalence relation $\approx_\sharp$) *Let $I_1 = [a_1, b_1](q_1)$ and $I_2 = [a_2, b_2](q_2)$ be two interval congruences. They represent the same rational set ($I_1 \subseteq_\sharp I_2 \wedge I_2 \subseteq_\sharp I_1$), noted $I_1 \approx_\sharp I_2$ if and only if they are either both empty ($q_i \neq 0, b_i < a_i$ for $i \in \{1,2\}$) or the set of rational numbers (($q_i = 0 \wedge b_i = -a_i = +\infty) \vee (q_i \neq 0 \wedge b_i - a_i \geq |q_i|$) for $i \in \{1,2\}$) or have modulos with the same absolute value $|q_1|$ and satisfy $b_2 - b_1 = a_2 - a_1 \in |q_1|\mathbb{Z}$. $\approx_\sharp$ is an equivalence relation on IC.*

The preceding theorem states that $(IC, \subseteq_\sharp)$ is a preorder. Hence from now and to avoid notational complications, we will note an equivalence class of $IC/_{\approx_\sharp}$ by one of its representative and the partial order on $IC/_{\approx_\sharp}$ by $\subseteq_\sharp$. There is no need here for a normalization operator as in CC since operators on IC are compatible with the equivalence relation.

5 Abstract Interpretation of Interval Congruences

5.1 Semantic Operators

Now that a concrete domain CC and an abstract one IC have been designed, we have to bind them using a pair of abstraction and concretization functions in order to give the meaning of abstract elements and to prove that their respective orders are coherent.

The choice of an interval congruence representing a given coset congruence is formalized by the abstraction function: the chosen abstract element is one of the minimal approximations of the concrete one. Given one coset congruence, there are many interval congruences containing it (provided by the soundness relation); there are still many containing exactly the integers corresponding to the original coset congruence; finally there are still many of these of minimum representative width (informally the difference between the upper and the lower bounds).

Definition 6 (Abstraction α) *The abstraction function is the following:*

$$\begin{array}{rccc} \alpha & : & CC/_{\approx} & \rightarrow & IC/_{\approx_I} \\ & & \theta.[l,u](m) & \mapsto & \left[\frac{l}{\theta^{-1}}, \frac{u}{\theta^{-1}}\right]\left(\frac{m}{\theta^{-1}}\right) \end{array}$$

where $0 < \theta^{-1} < |m|$ is an inverse of θ with respect to m and with the convention that the inverse of 0 with respect to 1 is 1.

Following Bezout's theorem[5], the abstraction function is always defined (θ^{-1} always exists). For example $\alpha(5.[1,3](9)) = \left[\frac{1}{2}, \frac{3}{2}\right]\left(\frac{9}{2}\right)$ that is represented by

$\left[\frac{6}{7}, \frac{8}{7}\right]\left(\frac{9}{7}\right)$ is an other minimal interval congruence containing $5.[1,3](9)$ and no more integers. It is of course non comparable with $\left[\frac{1}{2}, \frac{3}{2}\right]\left(\frac{9}{2}\right)$. This illustrates the lack of a best approximation of an element of CC with an interval congruence.

In the other way, the concretization function associates a concrete element with an abstract one giving its meaning. It is optimal if $\gamma \circ \alpha$ is the identity (which is in fact provided by theorem 8).

Definition 7 (The concretization γ) *The concretization function γ from $IC/_{\approx_I}$ into $CC/_{\approx}$ maps every interval congruence $[a,b]\left(\frac{\nu}{\delta}\right)$ on the coset congruence*

$$\begin{cases} 1.[1,0](1) & \textit{if } \nu = 0 \textit{ and } b \geq a \textit{ and } \lceil a \rceil > \lfloor b \rfloor \\ \left\|\delta^{-1}.[\lceil a\delta \rceil, \lfloor b\delta \rfloor](\nu)\right\| & \textit{otherwise} \end{cases}$$

where δ^{-1} is an inverse of δ with respect to ν.

We see that considering rational interval congruences provides a much more powerful description of concrete properties than only considering integer interval congruences the definition of which would have been quite similar to definition 3 replacing $\mathbb{Q}$ by $\mathbb{Z}$. This is a direct consequence of the strict inclusion of the presumed integer interval congruences in IC. An example of concretization is possibly:

$$\gamma\left(\left[\frac{3}{4}, \frac{3}{2}\right]\left(\frac{9}{4}\right)\right) = 7.[3,6](9) = (1 + 9\mathbb{Z}) \cup (3 + 9\mathbb{Z}) \cup (6 + 9\mathbb{Z}) \cup (8 + 9\mathbb{Z})$$

The next step is to state that the concretization function corresponds to our initial wish that is to express the integer subset of an interval congruence.

Theorem 8 (Correctness of γ) *The meaning $\gamma(I)$ of an interval congruence I is its intersection with $\mathbb{Z}$.*

$$\forall I \in IC \ \gamma(I) = I \cap \mathbb{Z}$$

[5]See footnote 2

Unfortunately, the pair of maps (α, γ) is not a Galois connection hence the usual framework of [CC77] cannot be used and [CC92b] shall be used instead.

In order to provide a unique representation of semantically equivalent abstract properties, a normalization is introduced.

Definition 9 (Normalization η) *The normalization operator η on $IC/_{\approx_1}$ is defined by $\eta = \alpha \circ \gamma$.*

The normalization operator replaces an abstract property by a more precise one or by a non comparable one, but without increasing the accuracy of the corresponding concrete elements. If the result is smaller than the original interval congruence, then the analysis will be more precise and if it is non comparable, the experimentation will justify the use of such a normalization in practice. For example

$$\eta\left(\left[\frac{3}{4}, \frac{5}{4}\right]\left(\frac{9}{7}\right)\right) = \alpha\,(5.\,[1,3]\,(9)) = \left[\frac{1}{2}, \frac{3}{2}\right]\left(\frac{9}{2}\right)$$

The rational intervals not containing any integers have been removed by the normalization and the modulo has increased.

5.2 Abstract Operators

As it will be illustrated in the definition of the widening operator, the only really needed conversion consists in finding the smallest interval congruence of $IC(q)$ containing a given interval congruence when the new modulo divides the one of the original congruence. For reasons that will appear in the widening definition, the result of a conversion operation must absolutely have the new modulo (even in the degenerate cases).

Definition 10 (Conversion to a divisor of the modulo Conv) *Let $I = [a, b]\,(q)$ be an interval congruence and q' be a rational number such that q' divides q, the conversion of I to modulo q' is defined by*

$$Conv_{q'}(I) \stackrel{def}{=} \begin{cases} [a, a+q']\,(q') & \textit{if } b < a \textit{ and } q = 0 \textit{ and } q' \neq 0 \\ [a, b]\,(q') & \textit{otherwise} \end{cases}$$

This conversion algorithm is optimal in the sense that it gives the smallest interval congruence containing the original one and of given modulo.

The next step is to find an algorithm that determines, given two interval congruences, a minimal element containing both of them. If they are comparable, the problem has an optimal solution and will not be considered. Otherwise the interval congruences are converted to a common modulo and two different possible upper bounds are compared using the accuracy function ι on their meaning. Hence the main question is to find a minimal upper bound for two interval congruences with same modulo.

Definition 11 (Interval-like join $\sqcup_{[\,]}$) *Let I_1 and I_2 be two interval congruences of $IC(q)$ and $[\gamma, \gamma']\,(q)$ the interval congruence of modulo q containing I_1 and I_2 and of minimal value of the difference between its upper and lower bounds, if $[\gamma, \gamma']\,(q)$ exists, then it is the interval-like join $I \sqcup_{[\,]} J$.*

$\sqcup_{[\,]} : IC(q) \times IC(q) \rightarrow IC(q)$ is a partial map, it is not defined for $[3,4](7)$ and $[6,8](7)$ for example.

An alternative to the interval join $\sqcup_{[\,]}$ naturally defined for two interval congruences of same modulo is the congruence join $\sqcup_{\dots}$ that first converts them to a modulo's divisor following definition 10 and then makes an interval join. The new modulo is chosen such that the converted representatives overlap. Let us consider that the distance between the representative centers of two interval congruences I_1 and I_2 is the minimum of the distances separating the center of one representative of I_1 from the center of one representative of I_2.

Definition 12 (Congruence-like join $\sqcup_{\dots}$) *Let $I_1 = [a_1, b_1](q_1)$ and $I_2 = [a_2, b_2](q_2)$ be two non comparable interval congruences of same modulo $q = q_1 = q_2$. Let r be the divisor of q that is the smallest rational closest to the distance d between I_1 and I_2 representative centers. The congruence-like join $I_1 \sqcup_{\dots} I_2$ is $[-\infty, +\infty](0)$ if d is zero, it is defined by*

$$I_1 \sqcup_{\dots} I_2 \stackrel{def}{=} Conv_r(I_1) \sqcup_{[\,]} Conv_r(I_2)$$

if $(q \neq 0 \ \vee\ a_2 \leq b_1 \ \vee\ b_1 < a_1 \ \vee\ a_1 \leq b_2 \ \vee\ a_1 + b_1 \neq a_2 + b_2)$ and otherwise $[a_1, b_2](0)$ or $[a_2, b_1](0)$.

For both join algorithms, the following examples can be considered:

$$[3,4](7) \sqcup_{\dots} [5,6](7) \quad = \quad \left[\frac{5}{4}, \frac{5}{2}\right]\left(\frac{7}{4}\right) \tag{7}$$

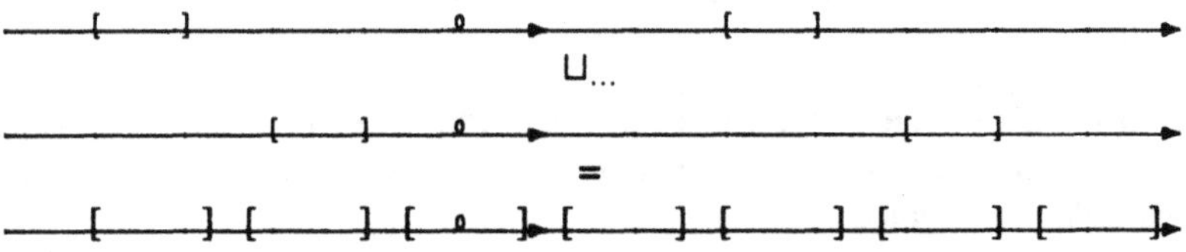

$$[3,4](7) \sqcup_{\dots} [6,8](7) \quad = \quad \left[\frac{5}{2}, \frac{9}{2}\right]\left(\frac{7}{2}\right) \tag{8}$$

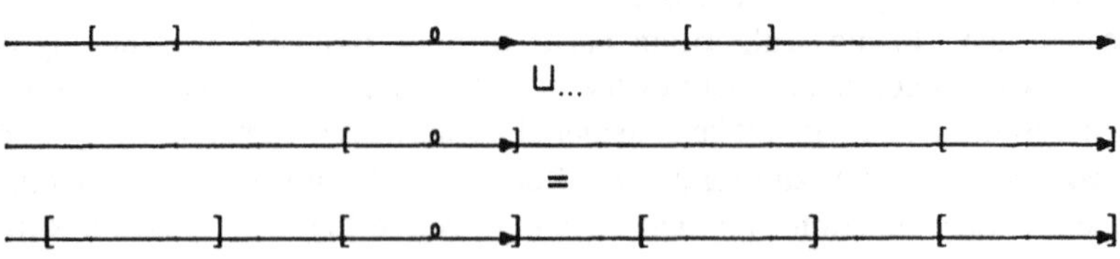

when

$$[3,4](7) \sqcup_{[\,]} [6,7](7) \quad = \quad [3,7](7) \tag{9}$$

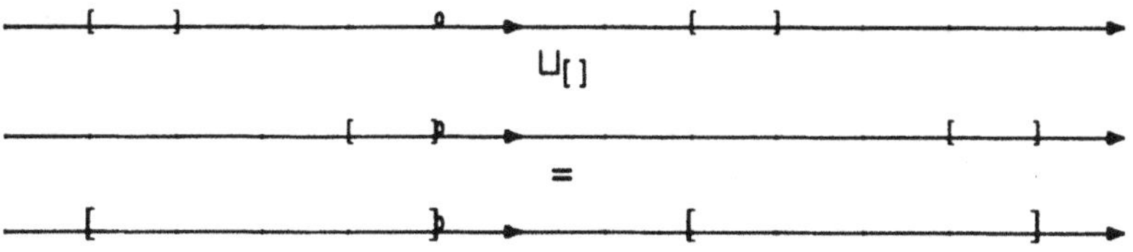

$$[3,4]\,(7) \sqcup_{[\,]} [5,6]\,(7) \quad = \quad [3,6]\,(7) \tag{10}$$

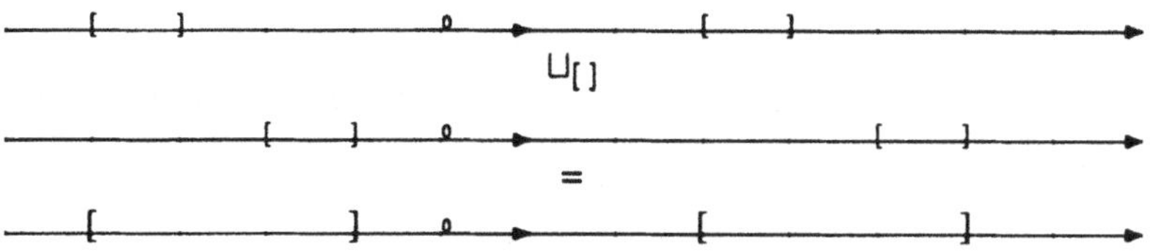

Intuitively comparing the examples (7) and (10), the interval join seems to be more adapted to this case, when comparing the examples (8) and (9) the congruence join seems to be closer to the exact union on rational sets.

An operator $\downarrow$ is introduced that estimates, given two interval congruences, which one is the most informative of the two, in other words which one contains the smallest density of integers. It is naturally defined using $\iota \circ \gamma$.

Finally, the approximation $\sqcup$ of the least upper bound operator on IC is simply the greatest element when they are comparable, the congruence-like join when the interval-like join is not defined, and otherwise is chosen among the interval- and congruence-like joins using the operator $\downarrow$.

Similarly, an approximate meet operator $\sqcap$ is defined using interval-like $\sqcap_{[\,]}$ and congruence-like $\sqcap_{\dots}$ meet operators.

Now a widening operator has to be defined in order to ensure convergence and accelerate and guarantee rapid termination of the iteration process for fixpoint computation.

These two features are part of the definition of the following operator derived from the widening operators on intervals and rational arithmetical cosets from [CC76] and [Gra91a].

Definition 13 (Widening on IC) *Let $I_1 = [a_1, b_1]\,(q_1)$ and $I_2 = [a_2, b_2]\,(q_2)$ be two interval congruences. Their widening $I_1 \nabla I_2$ is defined by*

$$\begin{cases} \left[\frac{\lceil a_2\delta\rceil - 1}{\delta}, \frac{\lfloor b_2\delta\rfloor + 1}{\delta}\right]\left(\frac{\nu}{\delta}\right) & \text{if} \quad q_1 = q_2 = \frac{\nu}{\delta} \neq 0 \quad (11) \\ [a_2, a_2 + q_2]\,(q_2) & \text{if} \quad 0 \neq q_1 \neq q_2 \neq 0 \quad (12) \\ [a, b]\,(0) & \text{if} \quad \begin{cases} q_1 = q_2 = 0 \\ b_1 \geq a_1 \ \wedge \ b_2 \geq a_2 \end{cases} \quad (13) \\ I_1 \sqcup I_2 & \text{otherwise} \quad (14) \end{cases}$$

where, if $a_2 < a_1$ then $a = -\infty$ else $a = a_1$ and if $b_1 < b_2$ then $b = +\infty$ else $b = b_1$.

Notice that in order to be more precise than a sign analysis, the widening on two finite rational intervals only has to jump to zero before extrapolating the infinite values if the infinite extrapolation value is not of the same sign as the original one. This additional feature does not figure in the widening definition for a matter of clarity.

First recall the classical widening operator used on interval with the examples:

$$\begin{array}{rcl} [2,3]\,(0)\,\nabla\,[2,7]\,(0) & = & [2,+\infty]\,(0) \\ [3,10]\,(0)\,\nabla\,[1,10]\,(0) & = & [0,10]\,(0) \\ [-10,-3]\,(0)\,\nabla\,[-10,3]\,(0) & = & [-10,+\infty]\,(0) \end{array}$$

Then the congruence-like behavior of our widening operator is illustrated by:

$$\left[\frac{2}{3},6\right]\left(\frac{190}{77}\right)\nabla\left[\frac{1}{3},1\right](5) = \left[\frac{1}{3},\frac{16}{3}\right](5) \approx_I [-\infty,+\infty]\,(0)$$

and finally the last kind of widening process (apart from the approximate join operator) is examplified in:

$$\left[\frac{1}{5},\frac{3}{5}\right]\left(\frac{8}{5}\right)\nabla\left[\frac{1}{7},\frac{5}{7}\right]\left(\frac{8}{5}\right) = \left[0,\frac{4}{5}\right]\left(\frac{8}{5}\right)$$

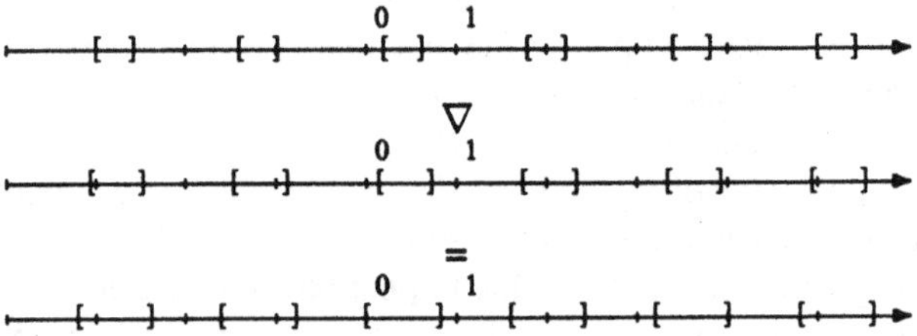

where at most three more applications of ∇ leads to an interval congruence containing $\mathbb{Z}$ (look at the respective meaning of the originals and resulting interval congruences).

5.3 Abstract Primitives

Defining first abstractions of integer sum and product by a constant allows us to deal with assignments of affine expressions to integer variables. Then abstracting a given class of tests gives the possibility to take into account control flow information in the analysis.

The following abstract primitives have been chosen to be sound, i.e. if F is the concrete primitive and ϕ the abstract one, we must have $F \leq \gamma \circ \phi \circ \alpha$.

Definition 14 (Abstract sum $\oplus$) *Let $[a_1,b_1]\,(q_1)$ and $[a_2,b_2]\,(q_2)$ be two interval congruences, their abstract sum, noted $[a_1,b_1]\,(q_1)\oplus[a_2,b_2]\,(q_2)$, is $[1,0]\,(1)$ if the concretization of one operand is the empty set and otherwise is defined by*

$$\begin{cases} \eta\left([a_1+a_2,b_1+b_2]\,(\gcd(q_1,q_2))\right) & \text{if } q_i \neq 0 \vee a_i \leq b_i,\ i\in\{1,2\} \\ [0,0]\,(1) & \text{otherwise} \end{cases}$$

Notice that the definition of abstract sum is commutative which seems natural, unfortunately the abstract sum is not exact, i.e. generally $I_1 + I_2 \subset I_1 \oplus I_2$. An example of non zero modulo sum is

$$\left[\frac{3}{2},\frac{9}{2}\right](14)\oplus\left[\frac{11}{2},7\right](21) = \eta\left(\left[0,\frac{9}{2}\right](7)\right) = [0,4]\,(7)$$

where it is visible that normalizing the operands before doing the abstract sum would have led to a more precise result ($[1, 4]\,(7)$) by not accumulating "errors" on the bounds of the interval congruences. That's why the result is normalized.

Definition 15 (Abstract product $\odot$) *Let λ be an integer and $I = [a, b]\,(q)$ an interval congruence, their abstract product is $[1, 0]\,(1)$ if the meaning of I is empty and otherwise*

$$\lambda \odot [a, b]\,(q) \stackrel{def}{=} \begin{cases} \eta\,([\lambda a, \lambda b]\,(\lambda q)) & \textit{if} \quad \lambda > 0 \\ [0, 0]\,(0) & \textit{if} \quad \lambda = 0 \\ \eta\,([\lambda b, \lambda a]\,(\lambda q)) & \textit{if} \quad \lambda < 0 \end{cases}$$

Examples are: $-2 \odot [-\infty, 5]\,(0) = [-10, +\infty]\,(0)$ when $2 \odot [2, 4]\,(6) = [4, 8]\,(12)$. Other arithmetical abstract primitives could be defined such as product by a rational, modulo and euclidean division.

Definition 16 (Abstract test with an IC condition) *Let I_1 be an abstract context preceding a test with the condition $x \equiv [a_2, b_2] \mod q_2$. The abstract entry context in the true branch of the conditional is*

$$I_1 \sqcap [a_2, b_2]\,(q_2)$$

while the abstract entry context in the false branch of the conditional is

$$I_1 \sqcap \alpha(\overline{\gamma([a_2, b_2]\,(q_2))})$$

Notice that the test condition can easily be extended to an equivalent linear equation by first approximating it. A major improvement with respect to the existing analyses using congruence properties on integers is that the negation of the natural condition is also quite natural. Recall that the meaning of the rational interval congruences is its integer points.

The semantic analysis of coset congruences is non comparable with interval analysis and conjunction of interval and congruence analysis.

6 Example

Let us consider the following program

```
            i := 0;
{1:}        while test do
{2:}          x := 3*i;
{3:}          if (i mod 2) = 0 then
{4:}            y := 3*i + 1
{5:}          else
{6:}            y := 3*i + 3
{7:}          endif;
{8:}          A[x,y] := A[x+1,y+1] + A[x+2,y+2];
              i := i + 1
{9:}        endwhile;
{10:}
```

point	first iteration	second iteration	third iteration
{1:}	$(\{0\},\emptyset,\emptyset)$	$(\{0\},\emptyset,\emptyset)$	$(\{0\},\emptyset,\emptyset)$
{2:}	$(\{0\},\emptyset,\emptyset)$	$(\mathbb{Z}^+,\{0\},\mathbb{Z}^+)$	$(\mathbb{Z},0+3\mathbb{Z},\{0,1\}+6\mathbb{Z})$
{3:}	$(\{0\},\{0\},\emptyset)$	$(\mathbb{Z}^+,0+3\mathbb{Z},\mathbb{Z}^+)$	$(\mathbb{Z},0+3\mathbb{Z},\{0,1\}+6\mathbb{Z})$
{4:}	$(\{0\},\{0\},\emptyset)$	$(0+2\mathbb{Z},0+3\mathbb{Z},\mathbb{Z}^+)$	$(0+2\mathbb{Z},0+3\mathbb{Z},\{0,1\}+6\mathbb{Z})$
{5:}	$(\{0\},\{0\},\{1\})$	$(0+2\mathbb{Z},0+3\mathbb{Z},1+6\mathbb{Z})$	$(0+2\mathbb{Z},0+3\mathbb{Z},1+6\mathbb{Z})$
{6:}	$(\emptyset,\emptyset,\emptyset)$	$(1+2\mathbb{Z},0+3\mathbb{Z},\mathbb{Z}^+)$	$(1+2\mathbb{Z},0+3\mathbb{Z},\{0,1\}+6\mathbb{Z})$
{7:}	$(\emptyset,\emptyset,\emptyset)$	$(1+2\mathbb{Z},0+3\mathbb{Z},0+6\mathbb{Z})$	$(1+2\mathbb{Z},0+3\mathbb{Z},0+6\mathbb{Z})$
{8:}	$(\{0\},\{0\},\{1\})$	$(\mathbb{Z},0+3\mathbb{Z},\{0,1\}+6\mathbb{Z})$	$(\mathbb{Z},0+3\mathbb{Z},\{0,1\}+6\mathbb{Z})$
{9:}	$(\{1\},\{0\},\{1\})$	$(\mathbb{Z},0+3\mathbb{Z},\{0,1\}+6\mathbb{Z})$	$(\mathbb{Z},0+3\mathbb{Z},\{0,1\}+6\mathbb{Z})$
{10:}	$(\{0,1\},\{0\},\{0,1\})$	$(\mathbb{Z},0+3\mathbb{Z},\{0,1\}+6\mathbb{Z})$	$(\mathbb{Z},0+3\mathbb{Z},\{0,1\}+6\mathbb{Z})$

Table 1: Detailed iteration process : (E, F, G) at line `n:` and in column "i th iteration" stands for: during iteration i at program point `n:` the values of `i`, `x` and `y` are approximated respectively by the integer sets E, F and G. The safe static approximation is given in the last column where the fixed point is reached.

---

where `i`, `x` and `y` are integer variables, `A` an array of dimension 2 and `test` a boolean expression that is not taken into account by the analysis.

The analyzed program instead of being very complex or requiring all the subtilities of the interval congruence analysis illustrates the basic idea of our analysis. The exact information to approximate in this program is congruence like, but not quite, since the test inserted in the loop makes it fail; only interval congruences can take this kind of information into account.

The iteration process is summarized in table 1. Each element of the represented tuples stands for the meaning uniquely associated with the corresponding abstract interval congruence in the iteration process. The singletons stand for null modulo integer cosets and the sets $E + q\mathbb{Z}$ for the integer cosets of modulo q and representative in E.

The iteration starts without knowing anything about the variables. Then the abstract primitives and the widening are used to determine the other columns values. Notice that the congruence behavior of the join is preferred at point `2:` (it detects that `i` is in fact the loop index) when the interval behavior is preferably chosen at points `8:` and `10:`. The fourth iteration giving the same results than the third one (telling the analyzer that the fixpoint is reached) has to be computed by the analyzer but has not been represented here.

The important result of analyzing this program with interval congruences is that the three references to the array `A` are shown to be independent. It is easy to see automatically that

$$\begin{aligned}
0+3\mathbb{Z}\times(0+6\mathbb{Z}\cup 1+6\mathbb{Z}) \cap 1+3\mathbb{Z}\times(1+6\mathbb{Z}\cup 2+6\mathbb{Z}) &= \emptyset \\
0+3\mathbb{Z}\times(0+6\mathbb{Z}\cup 1+6\mathbb{Z}) \cap 2+3\mathbb{Z}\times(2+6\mathbb{Z}\cup 3+6\mathbb{Z}) &= \emptyset \\
1+3\mathbb{Z}\times(1+6\mathbb{Z}\cup 2+6\mathbb{Z}) \cap 2+3\mathbb{Z}\times(2+6\mathbb{Z}\cup 3+6\mathbb{Z}) &= \emptyset
\end{aligned}$$

Such results are useful for example for automatic vectorization.

7 Conclusion

The analysis presented here can be used both in the area of data dependence search as well as for local memory optimization. A relational equivalent analysis has been designed in [Mas92a] for which no intermediate model such as CC has been designed resulting in a loss of precision. Semantic analysis of interval congruences is not comparable with interval analysis and conjunction of interval and congruence analysis.

References

[CC76] P. Cousot and R. Cousot. Static determination of dynamic properties of programs. In Paris Dunod, editor, *Proc. of the second International Symposium on programming*, pages 106–130, 1976.

[CC77] P. Cousot and R. Cousot. Abstract interpretation : a unified lattice model for static analysis of programs by construction of approximation of fixpoints. In *4th Annual ACM Symposium on Principles of Programming Languages*, pages 238–252, Los Angeles, January 1977.

[CC79] P. Cousot and R. Cousot. Systematic design of program analysis frameworks. In *6th Annual ACM Symposium on Principles of Programming Languages*, pages 269–282, 1979.

[CC92a] P. Cousot and R. Cousot. Abstract interpretation and application to logic programs. *Journal of Logic Programming*, 13(2–3), 1992.

[CC92b] P. Cousot and R. Cousot. Abstract interpretation frameworks. *Journal of Logic and Computation*, 1992.

[CC92c] P. Cousot and R. Cousot. Comparing the galois connection and widening/narrowing approaches to abstract interpretation. Technical Report LIX/RR/92/09, Laboratoire d'Informatique de l'X, Ecole Polytechnique, 91128 Palaiseau cedex, France, 1992.

[Gra89] P. Granger. Static analysis of arithmetical congruences. *Intern. J. Computer Math.*, 30:165–190, 1989.

[Gra91a] P. Granger. *Analyses sémantiques de congruence*. PhD thesis, Ecole Polytechnique, Palaiseau, July 1991.

[Gra91b] P. Granger. Static analysis of linear congruence equalities among variables of a program. In *International Joint Conference on Theory and Practice of Software Development*, volume 493 of *Lecture Notes on Computer Science*, pages 169–192. Springer Verlag, 1991.

[Mas92a] F. Masdupuy. Array operations abstraction using semantic analysis of trapezoid congruences. In *International Conference on Supercomputing*, July 1992.

[Mas92b] F. Masdupuy. Semantic analysis of rational interval congruences. Research Report LIX/RR/92/05, Ecole Polytechnique, 91128 Palaiseau, 1992.

Polymorphic Typing for Call-By-Name Semantics

B. Monsuez

Laboratoire d'Informatique de l'École Normale Supérieure. (LIENS)
45, rue d'Ulm 75005 Paris, FRANCE
monsuez@dmi.ens.fr

(Extended Abstract)

1 Introduction

The ML-polymorphic type system is sound, efficient, and supports a quite rich and flexible type system. Since this is one of the most efficient algorithms that infer types in the absence of any type declaration, the ML-type system is suitable for interactive languages. Consequently, it has been used in other languages like Hope [BuMcQ80], Miranda [Tur85] or Prolog [MyOK84]. But we should not conclude that the ML-type system is the definitive answer: for lazy-languages, ML-types are too restrictive since they do not preserve β-reduction. If we compute the ML-type of the expression $S\ K$ and its β-reduced form "$\lambda xy.y$", we find out that:

$$\begin{array}{ll} S & : \lambda xyz.(xz)(yz)\ : \\ & \qquad (\alpha \rightarrow \beta \rightarrow \gamma) \rightarrow (\alpha \rightarrow \beta) \rightarrow \alpha \rightarrow \gamma. \\ K & : \lambda xy.x\ :\ \alpha \rightarrow \beta \rightarrow \alpha. \\ S\ K & : (\alpha \rightarrow \beta) \rightarrow \alpha \rightarrow \alpha, \\ & \quad \text{but } \lambda xy.y\ :\ \alpha \rightarrow \beta \rightarrow \beta. \end{array}$$

Much theoretical work has been done to define more generous type systems that avoid this kind of restrictions. Coppo [Cop80] showed how to define a type system which is much more flexible than ML's, and preserves β-reduction, but this type system is undecidable. As a consequence of the lack of appropriate type systems, lazy languages like LML [Joh87], or GAML [Ma91] still use Milner's algorithm which is better suited for a call-by-value semantic.

The purpose of this work is to show that, despite the appearances the evaluation mechanism (call-by-name or call-by-value) has a great influence on the type inference system you can construct for a particular language.

For traditionally statically typed languages that include type constraints as part of their definition, the language evaluation mechanism is immaterial for the purpose of type-checking. The language designer distinguishes between two distinct semantic levels: the dynamic semantics describing the runtime evaluation process, and the static semantics describing types with respect to the dynamic semantics.

It is not the only way to interpret type inference algorithms. Another way is to consider type inference as an ***abstract interpretation*** of the dynamic semantics of the language. Some work has already been done by Jones and Mycroft. In [JoMy85], they showed that the Hindley/Milner type system is an ***abstract interpretation*** of the untyped λ-calculus. More recently, A. Aiken and B. Murphy used the formalism of abstract interpretation to define a type inference algorithm for FL [AiMu91]. In [Mon92a], we show how to use ***abstract interpretation*** in a generic way to define type systems, and how to use widening operators [Co78] to ensure algorithm termination. In the same paper, we show how to retrieve the ML-type algorithm using this generic method.

In this paper, we construct a more precise type system for call-by-name languages. We use the same abstraction scheme as the one described in [Mon92a], but instead of starting from a call-by-value semantics we start from a call-by-name semantics. Moreover we show that the Hindley/Milner type inference algorithm is an abstract interpretation of this larger type system. As a consequence, this type inference algorithm is more generous[1] and is therefore better-suited for lazy languages.

2 Type inference by Abstract Interpretation

With respect to the formalism of *abstract interpretation*, a type is an abstraction of a set of values ensuring an error-free execution.

On the one hand, a type can be interpreted as an upper approximation. In ML for instance, the expression $\lambda x.x + 1$ has the type "num→num". But the most precise description of the behavior of a function is given by its functional graph: $\{(x, x+1) : x \in \text{num}\}$. Now we observe that the functional graph is strictly included in $\{\text{num}\times\text{num}\}$, and therefore we can conclude that "num→num" is an upper approximation of the behavior of the function.

On the other hand, a type cannot always be interpreted as an upper approximation. The ML-expression $(\lambda fx.f(fx))(\lambda x.0)$ has the type "num→num". Since this function returns the numerical value 0 for any kind of arguments, this type is certainly not the most general. The previous expression has the far more general type: "α →num". This time, we observe that the ML principal type is a subset of a more general type.

The framework of *abstract interpretation* requires an abstract value to be either an upper or lower approximation. In [Mon92a], we have shown that characterizing a type as a well-behaved abstract value is much more subtle. We split the abstraction into two steps (i.e. an upper approximation followed by a lower approximation). First we define a huge type system T_g whose elements are the closest upper approximation of the set of values insuring an error-free execution. These general types are not effective in general, so that we can neither infer types nor perform type-checking. In order to define a useful type system, we have to restrict ourselves to less general types. Thus, we have to abstract the general types to smaller type—like ML-types.

To summarize, a type is an abstract value obtained by combining two pairs of abstraction functions. The first one is an upper approximation, the second one is a dual lower approximation.

$$\text{Values} \underset{\gamma}{\overset{\alpha}{\rightleftarrows}} T_g \underset{\gamma^*}{\overset{\alpha^*}{\rightleftarrows}} T_{ML}$$

In general, the lattice of abstract types does not verify the decreasing chain condition (any strictly decreasing chain is finite). The main consequence is that in the

[1] The types are still elements of the ML-type algebra, but their meaning is quite different.

most general case there is no guaranty of algorithm termination. Aiken and Murphy in [AiMu91] or Karnomori and Horiuchi in [KaHo88] make use of a heuristics to guaranty algorithm termination. In [Mon92a] we show that the use of widening operators ensures convergence of the abstract interpretation for infinite type systems. Various choices of widening operators lead to different type systems: for example we can obtain Hindley/Milner computable types, or we can accommodate recursive polymorphic types.

3 A simple applicative language

We do not deal with ML directly, which is a full size programming language. We consider instead a small untyped λ-calculus with constants, which is more or less the kernel of ML-like languages and is referred to as mini-ML in [Ka88]. The main difference with mini-ML is that the evaluation strategy uses call-by-name instead of call-by-value. In addition to the numerical and boolean sets of values we need to introduce values to handle functions and expressions. We represent a function by a closure: that is a triple whose components are an environment, a formal parameter, and a body. We represent a parameter whose evaluation is delayed by a frozen expression, that is a pair composed by an environment and a syntactical expression. The semantics of our language is given in the formalism of big-step-structural-operational semantics [Plo81][Ka88]. Figure 1 lists the rules describing the semantics of our small language. In contrast with the rules given in [Mon92a], the semantic values of an expression are only evaluated when this evaluation is required by the computation.

$$\rho \vdash \text{number } N \Rightarrow N \qquad \rho \vdash \text{true} \Rightarrow \mathit{true}$$

$$\rho \vdash \text{false} \Rightarrow \mathit{false} \qquad \rho \vdash \lambda P.E \Rightarrow \underline{\mathit{closure}} < \lambda P.E, \rho >$$

$$\frac{\rho \vdash E_1 \Rightarrow \mathit{true} \quad \rho \vdash E_2 \Rightarrow \alpha}{\rho \vdash \text{ if } E_1 \text{ then } E_2 \text{ else } E_3 \Rightarrow \alpha} \qquad \frac{\rho \vdash E_1 \Rightarrow \mathit{false} \quad \rho \vdash E_3 \Rightarrow \alpha}{\rho \vdash \text{ if } E_1 \text{ then } E_2 \text{ else } E_3 \Rightarrow \alpha}$$

$$\frac{\rho \vdash_{val\_of} \text{ ident } I \Rightarrow \alpha}{\rho \vdash \text{ ident } I \Rightarrow \alpha} \qquad \frac{\begin{array}{c}\rho \vdash E_1 \Rightarrow \underline{\mathit{closure}} < \lambda P.E, \rho' > \\ \rho'(P \leftarrow \underline{\mathit{frozen}} < E_2, \rho >) \vdash E \Rightarrow \beta\end{array}}{\rho \vdash E_1\ E_2 \Rightarrow \beta}$$

$$\frac{\rho(P \leftarrow \underline{\mathit{frozen}} < E_1, \rho >) \vdash E_2 \Rightarrow \beta}{\rho \vdash \text{ let } P = E_1 \text{ in } E_2 \Rightarrow \beta} \qquad \frac{\begin{array}{c}\rho(P \leftarrow \underline{\mathit{frozen}} < E_1, \rho >) \vdash E_1 \Rightarrow \alpha \\ \rho(P \leftarrow \alpha) \vdash E_2 \Rightarrow \beta\end{array}}{\rho \vdash \text{ let rec } P = E_1 \text{ in } E_2 \Rightarrow \beta}$$

$$\frac{\rho' \vdash E' \Rightarrow \alpha}{\rho \vdash \underline{\mathit{frozen}} < E', \rho' > \Rightarrow \alpha}$$

Fig. 1. A call-by-name semantics of mini-ML

4 Definition of a General Type System

In this section, we briefly summarize the general type construction presented in [Mon92a].

The untyped universe of computational objects can be structured into subsets of uniform behavior. These subsets can be referred to as types. We can at least distinguish between two kinds of types:

- the first ones that we call "basic types" are defined totally independently from the other types. For example, the set of boolean values "bool".
- the second ones that we call "computational types" depend upon the definition of the other types. For instance, functional type denotes a similar behavior of some functional expressions with respect to types.

Because of this distinction we decompose the definition of a general type system in two steps. The first one consists of abstracting the values to their abstract types. The second one consists of computing types that describe the behavior of expressions over types instead of over values.

4.1 Basic Types

We distinguish between four collections of values: closures, frozen expressions, numerical/boolean values and errors. Again, errors should be sorted in two classes: on the one hand **Illegal Domain** occurs when a predefined operator cannot evaluate a value and cannot evaluate any value of the same collection (e.g. 1/true); on the other hand **Illegal Value**$_D$ occurs when a predefined operator cannot evaluate a value of the collection D but can evaluate a value of the same collection D (e.g. 1/0). To perform typing of expressions like $\lambda x.1/x$, the second class of errors should be ignored. But since types are meant to increase program safety, the first class should be abstracted into a demoniac state **wrong**.

The abstraction process consists of abstracting numerical and boolean values to the abstract types "num" and "bool". Closures still get abstracted to closures so that all the values present in their environment get replaced by their basic types. In the same way, frozen expressions get abstracted to frozen expressions so that all the values present in their environment get replaced by their basic types. Additionally, conditional expressions can return values of different types $\tau_1, \ldots, \tau_n$. In the following, we denote the meet of the types $\tau_1, \ldots, \tau_n$ by $\tau_1 \oplus \ldots \oplus \tau_n$.

Definition 1. Let C be the set of all abstract closures, and let F be the set of all abstract frozen expressions. The lattice of error-free basic types is the powerset $\wp(\{\text{num}, \text{bool}\} \cup C \cup F)$ ordered by set-inclusion. The finite or infinite set $\{\tau_1, \ldots, \tau_n \ldots\}$ is denoted by the basic type $\oplus_i \tau_i$.

The lattice of basic types is the lattice of error-free basic types plus the new top element: **wrong**.

The abstraction of pairs of values and environments that denote a state is defined as follows:

$$\begin{array}{lll}
\alpha_v & \text{numerical value} & \rightarrow \text{num} \\
& \text{boolean value} & \rightarrow \text{bool} \\
& \underline{\text{closure}} < \lambda P.E, \rho > & \rightarrow \underline{\text{closure}} < \lambda P.E, \alpha_e(\rho) > \\
& \underline{\text{frozen}} < E, \rho > & \rightarrow \underline{\text{frozen}} < E, \alpha_e(\rho) > \\
& \textbf{Illegal Domain} & \rightarrow \textbf{wrong} \\
& \textbf{Illegal Value } t & \rightarrow \alpha_v(t) \\
\alpha_e & \rho = \ldots (x_i \leftarrow k_i) \ldots & \rightarrow \rho' = \ldots (x_i \leftarrow \alpha_v(k_i)) \ldots
\end{array}$$

For example, $\rho \vdash$ if b then $\lambda x.x$ else 2 has the basic type: $\underline{\text{closure}}(\alpha_e(\rho), \lambda x.x) \oplus$ num.

4.2 General Types

The behavior of a function in a given environment can be described either by its closure or by its functional graph:

$$\underline{\text{closure}}(\rho, \lambda P.E) \text{ or } \{(v, v') \mid \rho(P \leftarrow v) \vdash E \Rightarrow v'\}.$$

Obviously the functional graph is an abstraction of the $\underline{\text{closure}}(\rho, \lambda P.E)$. By analogy, we define the *type* of a function to be its functional graph over *types*.

Definition 2. The general type of closures is defined as the biggest "set" of types that verifies the following properties:

$$\begin{array}{l}
\text{type_of}(\underline{\textit{closure}}(\rho, \lambda P.E)) = \\
\quad \{(\tau, \tau') \ : \text{ such that either} \\
\quad \left\{\begin{array}{l}
\text{basic_type_of}(\tau) \neq \emptyset \\
\rho(P \leftarrow \text{basic_type_of}(\tau)) \vdash E \not\Rightarrow \textbf{wrong} \\
\\
\tau' = \max\limits_{v \in \text{basic_type_of}(\tau)} (\text{type_of}(v' \text{ such that } \rho(P \leftarrow v) \vdash E \Rightarrow v'))
\end{array}\right. \\
\quad \text{or} \\
\quad \left\{\begin{array}{l}
\text{basic_type_of}(\tau) = \emptyset \\
\rho(P \leftarrow \bigcup_{\tau'' \sqsubseteq \tau} \text{basic_type_of}(\tau'')) \vdash E \not\Rightarrow \textbf{wrong} \\
\\
\tau' = \max\limits_{v \in \bigcup_{\tau'' \sqsubseteq \tau} \text{basic_type_of}(\tau'')} (\text{type_of}(v' \text{ such that } \rho(P \leftarrow v) \vdash E \Rightarrow v'))
\end{array}\right. \\
\quad \}.
\end{array}$$

The function "basic_type_of(τ)" returns the set of the basic types with general type τ, in other words, "basic_type_of(τ) = ${\alpha'_v}^{-1}(\tau)$". But there exist some general types τ such that no basic type is abstracted to this general type τ. For instance, there is no basic type which has the general type {num, bool}. Since the conditional (denoted by $\oplus$) is handled as an internal choice, no expression in our language can be handled only as a numerical and also a boolean value with respect to our semantics. For completeness sake, when there is no basic type with general type τ, we use the pair (τ, τ') to express that for all inputs with general types lower than τ, the evaluation of the function returns a basic type with general type lower or equal to τ'. Otherwise, we use the pair (τ, τ') to denote that for all inputs with general type equal to τ, the evaluation of the function returns a basic type with general type lower or equal to τ'.

The function **max** returns the greatest upper bound of a set of types. If the computation does not terminate for any value of type τ then the function **max** returns the set of all types[2].

Following the same intuition, we define the type of a frozen expression to be the type of the semantic value associated to this frozen expression.

Definition 3. The general types of frozen expressions are computed using the following injection:

$$\text{type\_of}(\underline{frozen}(\rho, E)) = \alpha'_v(\{v \text{ such that } \rho \vdash E \Rightarrow v\}).$$

As a consequence of the previous definitions, general types are elements of the domain defined by the following recursive *domain equation*:

$$D \cong ((D \otimes D))^{\natural} \oplus D,$$

where the symbol $\natural$ denotes the constructor of Plotkin's powerdomain [GuSc90]. The initial approximation is the flat domain D_0 containing the elements {num, bool $\oplus$ num, bool} and the emptyset as bottom. We upward close this domain D by adding the new top element **wrong**: $\forall \tau \in D, \tau \sqsubseteq$ **wrong**.

Finally the abstraction of pair of values and environment is defined as follows:

$$\begin{array}{ll} \alpha'_v : \text{num} & \rightarrow \text{num} \\ \quad\;\; \text{bool} & \rightarrow \text{bool} \\ \quad\;\; \underline{\text{closure}}(\rho, \lambda P.E) & \rightarrow \text{type\_of}(\underline{\text{closure}}(\rho, \lambda P.E)) \\ \quad\;\; \underline{\text{frozen}}(\rho, E) & \rightarrow \text{type\_of}(\underline{\text{frozen}}(\rho, E)) \\ \quad\;\; \tau_1 \oplus \tau_2 & \rightarrow \alpha'_v(\tau_1) \oplus \alpha'_v(\tau_2) \\ \alpha'_e : \rho = \ldots (x_i \leftarrow k_i) & \rightarrow \rho' = \ldots (x_i \leftarrow \alpha'_v(k_i)) \end{array}$$

For example, $\rho \vdash$ if b then $\lambda x.x$ else 2 has the general type: $\{(\tau, \tau) : \tau \in T_g\} \oplus$num.

4.3 Connections among evaluation strategies and general types

As previously mentioned the type of a function denotes the lowest upper approximation of the domain and the image of a function. Since the domain—and consequently the image—of a function directly depends upon the evaluation mechanism, the type is determined by the evaluation mechanism. Moreover, since general types are the most precise upper approximation with respect to the introduced basic types, the abstraction induces few information losses about the evaluation strategy. With respect to call-by-name evaluation strategy, it preserves β-reduction for all strong normalizing expressions.

Theorem 4. *For all terminating expressions P, Q, R such that R is the β-reduced form of P Q, the general type of R is equal to the result of the abstract application of the general type of the expression P to the general type of the expression Q.*

Example

[2] Observe that in ML we return the type α for a non terminating expression since we cannot infer any constraint on the output.

The type of the expression K is $\{(\tau_1,(\tau_2,\tau_1)) \ : \ \tau_1,\tau_2 \in T_g\}$. Every expression of type $\{(\tau_1,(\tau_2,\tau_1)) \ : \tau_1,\tau_2 \in T_g\}$ returns a value that has the same type as the first argument, whatever the second argument is. This means that if the evaluation of the second argument loops, then the function still terminates. Subsequently, the second argument is not evaluated.

If for every expression E of type $\{(\tau_1,(\tau_2,\tau_1)) \ : \ \tau_1,\tau_2 \in T_g\}$ we evaluate the expression $S\ E$ then we obtain an expression S_E such that for all syntactically valid expressions E', the evaluation of $S_E\ E'$ returns the same value V_E whatever the expression E' is. The type of this value V is independent from the expression E and equal to $\{(\tau,\tau) \ : \ \tau \in T_g\}$.

Finally, the evaluation of S applied to the expression K of type $\{(\tau_1, (\tau_2,\tau_1)) : \tau_1,\tau_2 \in T_g\}$ returns an expression S' which has type $\{(\tau_1, (\tau_2,\tau_2)) : \tau_1,\tau_2 \in T_g\}$ — i.e. the type of the expression $\lambda xy.y$, the β-reduced form of $S\ K$.

5 Back to the ML-type algebra

The information about the evaluation strategy does not get lost in the definition of General Types. But, since the computation of the General Types is not effective, the previously type system is of no use for the purpose of type checking or type inference. Building such algorithms requires much simpler types. Therefore, we had to approximate the previous defined types by more restrictive but computable types.

Since our goal is not to define a new type system, but to investigate the dependence between the interpretation of a type — and consequently the type system — and the call-by-name language semantics, we decided to restrict our General Types to the ML-polymorphic type algebra.

5.1 Definition of ML-types

Let M be the set of monomorphic types τ:

$$\tau : \mathrm{num}|\mathrm{bool}|\tau \rightarrow \tau'|\tau \oplus \tau'$$

Let S be the set of type schemes σ:

$$\sigma : \tau|\alpha \ldots\ldots|\sigma \rightarrow \sigma'|\sigma \oplus \sigma'$$

5.2 Interpretation of ML-types

In previous work [JoMy85][Mon92a], a type scheme was identified with a set of monomorphic types:

$$\mathrm{inj}_{Mono} \ : \ \sigma \in S \ \rightarrow \ \{\tau \ : \ \tau \in M \text{ and } \tau \text{ instances of } \sigma\}.$$

When we compute the type of the expression $E_1\ E_2$, the expression E_1 has the type $\sigma_1 \rightarrow \sigma_1'$ and the expression E_2 has the type σ_2. Moreover the type of the expression $E_1\ E_2$ is the set σ_3 of all the monomorphic instances τ_2 that verify the properties:

$\tau_1 \in \sigma_2$ and $\tau_1 \rightarrow \tau_2 \in \sigma_1 \rightarrow \sigma_1'$ For example, the type of the expression $S\ K$ with respect to this model is the set: $\{(\sigma_1 \rightarrow \sigma_2) \rightarrow \sigma_1 \rightarrow \sigma_1 \ :\ \sigma_1, \sigma_2 \in M\}$. Therefore, the monomorphic interpretation of type schemes is too restrictive to modelize a call-by-name evaluation strategy.

The only way to overcome this restriction is to identify polymorphic types with sets of polymorphic types too. In the previous example, the expression $S\ K$ could not be correctly typed since there was no type instance $(\tau \rightarrow \tau')$ of S where the type τ matches the type of the expression K. To avoid this, we should define an injection from the set of type schemes to the powerset $\wp(S)$ of type schemes.

Definition 5. We define the injection inj from the set of type schemes S to the powerset $\wp(S)$, defined by a relation R over the set S:

$$\mathrm{inj}\ :\ \sigma \in S\ \rightarrow \{s\ :\ s \in S, sR\sigma\}.$$

If we analyze the set of type schemes yielded in the general type, we find two distinct subsets: on the one hand, the set of all the instances of the canonical ML-polymorphic types; on the other hand, the subset of the pairs of polymorphic type schemes describing the call-by-name—or lazy—interpretation of the function.

For example, the general type of S yields the two subsets:

$$\underbrace{\{(\sigma_1 \rightarrow \sigma_2 \rightarrow \sigma_3) \rightarrow (\sigma_1 \rightarrow \sigma_2) \rightarrow \sigma_1 \rightarrow \sigma_3\ :\ \sigma_1, \sigma_2, \sigma_3 \in S\}.}_{\text{non "lazy" interpretation}}$$
$$\cup$$
$$\underbrace{\{(\sigma_1 \rightarrow \sigma_2 \rightarrow \sigma_3) \rightarrow \sigma_4 \rightarrow \sigma_1 \rightarrow \sigma_3\ :\ \sigma_2 \text{ and } \sigma_3 \text{ are independent}\}}_{\text{"lazy" interpretation}}$$

Definitely, the type "$\alpha \rightarrow \beta \rightarrow \beta$" is a valid element of the set describing the "lazy" evaluation of SK.

The main work is to find a relation Ω that characterizes nearly all the elements of the previous sets as Ω-instances of the canonical ML-polymorphic type. To achieve this goal, we suppose that we have defined an ω-reduction. The ω-reduction computes the non-evaluated subterms of a type and replaces it with fresh free variables. Subsequently, we characterize a polymorphic type scheme σ as:

- the set of all its instances (non "lazy" evaluation).
- the set of all the types $\sigma_1 \rightarrow \sigma_2$ where there exist the types σ_1', σ_2' such that the type $\sigma_1 \rightarrow \sigma_2'$ is an instance of the type σ and the type $\sigma_1' \rightarrow \sigma_2$ is the ω-reduced form of the type $\sigma_1 \rightarrow \sigma_2'$ ("lazy" evaluation).

All the elements of the previously defined set are called Ω-instances of the type scheme σ.

5.3 Introducing the ω-reduction

We call ω-variable a type variable such that the value whose type is denoted by this type variable is not used to compute the result of the function. In other words: whatever the value of this variable is, the output will be the same. The variable β of the type scheme $\alpha \rightarrow \beta \rightarrow \alpha$ is an ω-variable.

When applying a function of type $(\alpha \rightarrow \beta \rightarrow \alpha) \rightarrow (\alpha \rightarrow \beta) \rightarrow \alpha \rightarrow \alpha$ to a function of type $\alpha \rightarrow \beta \rightarrow \alpha$, the result R of this application has the type $(\alpha \rightarrow \beta) \rightarrow \alpha \rightarrow \alpha$. We have the additional information that the variable β is the same as the one of the type $\alpha \rightarrow \beta \rightarrow \alpha$. Moreover, the result of the function $\alpha \rightarrow \beta$ can only be used as an input[3] of the function $\alpha \rightarrow \beta \rightarrow \alpha$. Since the variable β is an ω-variable, the output of the function $\alpha \rightarrow \beta$ will not be evaluated. Therefor, the evaluation strategy does not require that the first argument of the function R is a function. The term $(\alpha \rightarrow \beta)$ can be replaced by an ω-variable, and the type of the result of the application is rewritten in $\omega \rightarrow \alpha \rightarrow \alpha$. Whatever the input of the argument is, the function still returns a function of type $\alpha \rightarrow \alpha$.

5.4 Computing ω-types

On figure 2, we first list the different rules required to determine if a type variable of a given type expression is an ω-variable or not. This set of rules computes a tuple of four boolean functions for each type scheme:

- VarV: v type variable $\rightarrow$ true iff the term t is equal to v.
- ArgV: v type variable $\rightarrow$ true iff the term t is equal to $v \rightarrow t'$.
- ArgV: $v|t_i \rightarrow$ true iff t_i is a subterm of t and $t_i = v \rightarrow t'_i$.
- RetV: v type variable $\rightarrow$ true iff the term t is equal to $t' \rightarrow v$.
- RetV: $v|t_i \rightarrow$ true iff t_i is a subterm of t and $t_i = t'_i \rightarrow v$.
- ΩV: v type variable $\rightarrow$ true iff the variable is useless for the computational process at this state of the analysis.

What we call a useless variable or ω-variable of an expression is a variable (rule 1) such that:

- variable is the input of a function but is not involved in the evaluation of this function (rule 2). (i.e. this variable does not appear in the type of the result of this function).
- this variable is the output of a function but is not present as one of the inputs of this same function (rule 3).
- this variable is neither the output of two functions that are arguments of the main function (rule 4), nor an output of the argument of the main function (rule 5).

On figure 3, we list the two rules that replace the whole functional term returning an ω-variable in a type scheme by an ω-variable.

We define the ω-reduced form as the injection from S to S which is obtained by replacing all the ω-variables by fresh free type variables. For example, in the type $(\alpha \rightarrow \beta \rightarrow \alpha) \rightarrow (\alpha \rightarrow \beta) \rightarrow \alpha \rightarrow \alpha$ the variable β is an ω-variable and this type can be rewritten as follows: $(\alpha \rightarrow \omega \rightarrow \alpha) \rightarrow \omega \rightarrow \alpha \rightarrow \alpha$. Hence, this type has the ω-reduced type: $(\alpha \rightarrow \gamma \rightarrow \alpha) \rightarrow \delta \rightarrow \alpha \rightarrow \alpha$.

[3] Note that the language does not include operations returning "unit", or operation sequences.

1) $$\frac{v \text{ type variable}}{v \Rightarrow < \mathrm{VarV}(v) = \mathrm{true}, \mathrm{ArgV}, \mathrm{RetV}, \Omega\mathrm{V} >}$$

2) $$\frac{\begin{array}{c} t_1 \Rightarrow < \mathrm{VarV}_1(v) = \mathrm{true}, \mathrm{ArgV}_1, \mathrm{RetV}_1, \Omega\mathrm{V}_1 > \\ t_2 \Rightarrow < \mathrm{VarV}_2(v) = \mathrm{false}, \mathrm{ArgV}_2, \mathrm{RetV}_2, \Omega\mathrm{V}_2 > \end{array}}{t_1 \to t_2 \Rightarrow < \mathrm{VarV}(v) = \mathrm{false}, \mathrm{ArgV}, \mathrm{RetV}, \Omega\mathrm{V}(v) >} \quad \begin{array}{l} \mathrm{ArgV}(v|t_2) = \mathrm{ArgV}_2(v) \\ \mathrm{ArgV}(v) = \mathrm{true} \\ \mathrm{RetV}(v|t_2) = \mathrm{RetV}_2(v) \\ \Omega\mathrm{V}(v) = \Omega\mathrm{V}_1(v) \wedge (\neg\mathrm{ArgV}_2(v) \\ \wedge \neg\mathrm{RetV}_2(v) \wedge \Omega\mathrm{V}_2(v)) \end{array}$$

3) $$\frac{\begin{array}{c} t_1 \Rightarrow < \mathrm{VarV}_1(v) = \mathrm{false}, \mathrm{ArgV}_1, \mathrm{RetV}_1, \Omega\mathrm{V}_1 > \\ t_2 \Rightarrow < \mathrm{VarV}_2(v) = \mathrm{true}, \mathrm{ArgV}_2, \mathrm{RetV}_2, \Omega\mathrm{V}_2 > \end{array}}{t_1 \to t_2 \Rightarrow < \mathrm{VarV}(v) = \mathrm{false}, \mathrm{ArgV}, \mathrm{RetV}, \Omega\mathrm{V}(v) >} \quad \begin{array}{l} \mathrm{ArgV}(v|t_1) = \mathrm{ArgV}_1(v) \\ \mathrm{RetV}(v|t_1) = \mathrm{RetV}_1(v) \\ \mathrm{RetV}(v) = \mathrm{true} \\ \Omega\mathrm{V}(v) = \Omega\mathrm{V}_2(v) \wedge (\neg\mathrm{ArgV}_1(v) \\ \wedge \neg\mathrm{RetV}_1(v) \wedge \Omega\mathrm{V}_1(v)) \end{array}$$

4) $$\frac{\begin{array}{c} t_1 \Rightarrow < \mathrm{VarV}_1(v) = \mathrm{true}, \mathrm{ArgV}_1, \mathrm{RetV}_1, \Omega\mathrm{V}_1 > \\ t_2 \Rightarrow < \mathrm{VarV}_2(v) = \mathrm{true}, \mathrm{ArgV}_2, \mathrm{RetV}_2, \Omega\mathrm{V}_2 > \end{array}}{t_1 \to t_2 \Rightarrow \begin{array}{l} < \mathrm{VarV}(v) = \mathrm{false}, \mathrm{ArgV}(v) = \mathrm{true}, \\ \mathrm{RetV}(v) = \mathrm{true}, \Omega\mathrm{V} = \mathrm{false} > \end{array}}$$

5) $$\frac{\begin{array}{c} t_1 \Rightarrow < \mathrm{VarV}_1(v) = \mathrm{false}, \mathrm{ArgV}_1, \mathrm{RetV}_1, \Omega\mathrm{V}_1 > \\ t_2 \Rightarrow < \mathrm{VarV}_2(v) = \mathrm{false}, \mathrm{ArgV}_2, \mathrm{RetV}_2, \Omega\mathrm{V}_2 > \end{array}}{t_1 \to t_2 \Rightarrow < \mathrm{VarV}(v) = \mathrm{false}, \mathrm{ArgV}, \mathrm{RetV}, \Omega\mathrm{V} >} \quad \begin{array}{l} \mathrm{ArgV}(v|t_1|s_i^1) = \mathrm{ArgV}_1(v|s_i^1), \\ \mathrm{ArgV}(v|t_2|s_i^2) = \mathrm{ArgV}_2(v|s_i^2), \\ \mathrm{RetV}(v|t_1|s_i^1) = \mathrm{RetV}_1(v|s_i^3), \\ \mathrm{RetV}(v|t_2|s_i^2) = \mathrm{RetV}_2(v|s_i^4), \\ \text{where } s_i^1, s_i^2, s_i^3, s_i^4 \text{ subterm of} \\ t_1, t_2, t_3, t_4, \\ \Omega\mathrm{V}(v) = \mathrm{Cond}\Omega \end{array}$$

$$\begin{aligned} \mathrm{Cond}\Omega = {} & (\Omega\mathrm{V}(v)_1 \vee (\forall s_i^1 \text{ subterm of } t_1, \neg\mathrm{ArgV}_1(v|s_i) \wedge \neg\mathrm{RetV}_1(v|s_i))) \\ & \wedge (\Omega\mathrm{V}(v)_2 \vee (\forall s_i^2 \text{ subterm of } t_2, \neg\mathrm{ArgV}_2(v|s_i) \wedge \neg\mathrm{RetV}_2(v|s_i))) \\ & \wedge (\forall s_i^1 \text{ subterm of } t_1, \neg\mathrm{RetV}_1(v|s_i)) \end{aligned}$$

Fig. 2. Omega rules

5.5 Abstraction to ML-types

As in [Mon92a], we abstract the types τ of T_g to the intersection of all the greatest elements of S that are lower or equal to τ:

$$\alpha'' : \tau \in T_g \longrightarrow \bigcap_{\substack{\{\sigma' | \sigma' \in S \text{ such that} \\ \sigma' \subset \tau \wedge (\sigma'' \supset \sigma' \Rightarrow \sigma'' \not\subset \tau)\}}} \sigma'$$

The concretization γ is the identity.

Figure 4 lists the abstract rules describing the behavior of a program for the ML-polymorphic types.

In order to compute the type $\sigma \to \sigma'$ of a function $\lambda P.E$, we had to verify that for all types $\sigma_0 \to \sigma_0'$ Ω-instances of the type $\sigma \to \sigma'$ the evaluation of $\rho(P \leftarrow \sigma_0) \vdash E$ returns the type σ_0'. But this can be avoided. Evaluation of the expression $\rho(x \leftarrow \alpha) \vdash$

$$(t_1 \to \omega) \to t_2 \Rightarrow \omega \to t_2$$

$$\frac{t_1 \Rightarrow t'_1 \qquad t_2 \to t'_2}{t_1 \to t_2 \Rightarrow t'_1 \to t'_2}$$

Fig. 3. Type variable elimination rules

$$\rho \vdash \text{number } N \Rightarrow \text{num} \qquad \rho \vdash \text{true} \Rightarrow \text{bool}$$

$$\rho \vdash \text{false} \Rightarrow \text{bool}$$

$$\frac{\forall \sigma_1^i \to \sigma_2'^i \sqsubseteq \sigma_1 \to \sigma_2 \quad (P \leftarrow \sigma_1^i) \vdash E \Rightarrow \sigma_2'^i \text{ and } \sigma'_2 \sqsubseteq \sigma_2 \quad \sigma'_2 = \text{ nearest lower approximation of } \bigcup_i \sigma_2'^i}{\lambda P.E \Rightarrow \sigma_1 \to \sigma_2}$$

$$\frac{\rho \vdash_{val\_of} \text{ ident } I \Rightarrow \alpha}{\rho \vdash \text{ ident } I \Rightarrow \alpha}$$

$$\frac{\rho \vdash E_1 \Rightarrow \text{ bool} \quad \rho \vdash E_2 \Rightarrow \alpha \quad \rho \vdash E_3 \Rightarrow \beta}{\rho \vdash \text{ if } E_1 \text{ then } E_2 \text{ else } E_3 \Rightarrow \alpha \oplus \beta}$$

$$\frac{\rho \vdash E_2 \Rightarrow \sigma_2 \quad \rho \vdash E_1 \Rightarrow \oplus_{i \in I} \sigma_1^i \quad \sigma_2^i \in \sigma_2 \quad (\sigma_2^i \to \sigma'^{i}_2)\ \Omega - \text{instances } \sigma_1^i}{\rho \vdash E_1 E_2 \Rightarrow \oplus_{i \in I} \sigma'^{i}_2}$$

$$\frac{\rho \vdash E_1 \Rightarrow \alpha \quad \rho(P \leftarrow \alpha) \vdash E_2 \Rightarrow \beta}{\rho \vdash \text{ let } P = E_1 \text{ in } E_2 \Rightarrow \beta}$$

$$\frac{\rho(P \leftarrow \alpha) \vdash E_1 \Rightarrow \alpha \quad \rho(P \leftarrow \alpha) \vdash E_2 \Rightarrow \beta}{\rho \vdash \text{ let rec } P = E_1 \text{ in } E_2 \Rightarrow \beta}$$

Fig. 4. The dynamic semantics over ML polymorphic types

α with the additional information that for each t in T_{ML} the function $\lambda x.x$ returns t, allows us to conclude that the function $\lambda x.x$ has the type $\alpha \to \alpha$.

To simplify the computation, we collect all the equality relations between the subterms of the type schemes. This is sufficient to ensure the error-free computation of the value whose type is an instance of the principal type of the function. A priori it does not seem to insure that the evaluation of the Ω-instances will be error-free. But such an evaluation error can only happen when we evaluate an expression whose type has been reduced to such an Ω-type. In this case the whole expression cannot be the argument of a function whose argument type is not an ω-variable. Either this is a predefined operator like the equality test, or the type of the function argument and the type of the computed argument are not the same. Since there is an equality relation between the type of the function argument and the type of the computed argument, this second hypothesis cannot occur.

Lemma 6. If the evaluation of a function with the pair of type and equality relation between types $(\sigma, \Re)$ as argument returns the pair $(\sigma', \Re')$; if all ω-variables are arguments of functions which accept ω-variables as valid types, then the function has type $(\sigma_0 \to \sigma'_0)$. The type σ_0 (resp. σ'_0) is an instance of σ (resp. σ'), and each instance of

$\sigma_0 \rightarrow \sigma'_0$ *verifies the relation* $\Re$ *and* $\Re'$ *between the term* σ *and* σ'.

Therefore, we can use with no modification the abstraction scheme described in [Mon92a] that characterizes the principal type as the greatest fixpoint. Hence, we can use the algorithm described in the same paper [Mon92a] to infer the principal type of ML-expressions.

6 Marked and Unmarked types

In the previous section we have enhanced the meaning of a type scheme. However, if we get for some ML-expressions "greater types" than Hindley/Milner's , some other valid ML-expressions have a less general type than Hindley/Milner's one. In the worst case, they have no type at all. For example, we can define two functions that have the same Hindley/Milner's type in ML, but the first one is strict and the other is not strict in its arguments. A good example of such a function is the expression S_1 : $\lambda xyz.$ if $yz = yz$ then $xz(yz)$ else $xz(yz)$. The subset of $P(S)$ included in the general type of S_1 is the set $s_1 = \{\sigma\ :\ \sigma$ instance of $(\alpha \rightarrow \beta \rightarrow \gamma) \rightarrow (\alpha \rightarrow \beta) \rightarrow \alpha \rightarrow \gamma\}$. Since there is no type scheme which denotes a non trivial subset with respect to the previous abstraction, the expression S_1 does not have a principal type.

This is so because some predefined overloaded operator require to evaluate the expression to compute the result. Henceforth, the connection between the output type and the input type is not preserved, since we do not have any way to infer the information concerning this evaluation. For soundness reasons, we cannot relax the analysis so that in this particular case the evaluation operates in a call-by-name fashion.

Since we cannot get rid of this kind of operators—for example the equality test—we need to assume that the evaluation happened in a call-by-value fashion. Therefore, we slightly modify the definition of the general type of a function by the introduction of marked type variables. A type variable that is not marked denotes the type of a value which is argument of a predefined operator operating in a call-by-name fashion. In the same way, we introduce marked subterms of type schemes s^* that denote the type of values which are argument of a predefined operator that is supposed to operate in a call-by-value fashion.

All what we have to do now is to modify the definition of type schemes:

$$\sigma\ :\ \tau|\alpha\ldots|\sigma \rightarrow \sigma'|\sigma \oplus \sigma'|\sigma^*.$$

Since a marked term is argument of an operator that requires the evaluation of its argument, this term cannot contain any ω-variable. By extension, all the type variables present in the marked term should be considered as marked too, and cannot be ω-variables. The first rule describing the computation of ω-variables is replaced by the two following ones:

$$\frac{v \text{ non-marked type variable}}{v \Rightarrow <\mathrm{VarV}(v)\ =\ \mathrm{true}, \mathrm{ArgV}, \mathrm{RetV}, \Omega\mathrm{V}>} \quad (1)$$

$$\frac{v \text{ marked type variable}}{v \Rightarrow <\mathrm{VarV}(v)\ =\ \mathrm{true}, \mathrm{ArgV}, \mathrm{RetV}, \Omega\mathrm{V}\ =\ \mathrm{false}>} \quad (2)$$

From now on, the set of all the instances of a type scheme σ denotes the same set as the set of all the Ω-instances of the type scheme σ^*.

Property 7. *If σ is the Hindley/Milner type of a ML-expression E, then the call-by-name type of the expression E denotes an upper set of the set denoted by σ.*

For example, the expression $S_1 : \lambda xyz.$ if $yz = yz$ then $xz(yz)$ else $xz(yz)$ has type $(\alpha \to \beta \to \gamma) \to (\alpha \to \beta^*) \to \alpha \to \gamma$ which matches exactly the subset of $P(S)$ included in the general type of $S_1 : \{\sigma \ : \ \sigma$ instance of $(\alpha \to \beta \to \gamma) \to (\alpha \to \beta) \to \alpha \to \gamma\}$.

Remark. Like every approximation, some information gets lost, and we do not have every time the type we would expect.

First example is the function $S_2 : \lambda xyz.$ if $y = y$ then $xz(yz)$ else $xz(yz)$. y is argument of the equality operation. Because of the previous restriction, the type of y is marked, and cannot contain any ω-variables. Then the inferred type is $(\alpha \to \beta \to \gamma) \to (\alpha \to \beta)^* \to \alpha \to \gamma$ denoting the set $\{\sigma \ : \ \sigma$ instance of $(\alpha \to \beta \to \gamma) \to (\alpha \to \beta) \to \alpha \to \gamma\}$. If we compute the General Type and then abstract it to the corresponding ML-type, we find out the more general type: $\{\sigma \ : \ \sigma \ \Omega -$ instance of $(\alpha \to \beta \to \gamma) \to (\alpha \to \beta) \to \alpha \to \gamma\}$.

A second example, which has nothing to do with marked or unmarked variables, illustrates the limit of the concept of ω-variables. The expression $S_f : \lambda xyzf.\ x(fz)(fyz)$ has type $(\alpha \to \alpha \to \beta) \to (\gamma \to \gamma) \to \gamma \to (\gamma \to \alpha) \to \beta$. If we apply this function S_f to K then the useful instance is $(\gamma \to \gamma) \to \gamma \to (\gamma \to \alpha) \to \alpha$. This type does not include any ω-variable, so that the expression $S_f\ K$ requires the first argument to be a function of type $\gamma \to \gamma$ although this function is never evaluated.

7 Conclusion

We have presented herein an "interpretation" of the ML-type algebra leading to more general types for call-by-name functional languages.

In contrast to the standard approach for designing type systems, the approach based on *Abstract Interpretation* mimics in a more precise way the evaluation strategy described by the dynamic semantics. For this reason, it enables us to understand what is going on. In the logical framework, the ML type algebra can only modelize call-by-value function application, because of a certain rigidity of this framework. As we have seen, Abstract Interpretation is more flexible and allows us to slightly modify the meaning of ML-types, and hence preserves valuable information about the evaluation strategy.

More generally, while this approach for defining a call-by-name, a call-by-need, and a call-by-value version of the same type system can naturally be used with the ML type algebra, it cam also be used with other type algebras like the Fractional Types[Mon92b]—a subset of Coppo's intersection types. This same approach can be extended to mimics other semantically relevant properties in a better way. And before all, it pleads for a better interaction of the dynamic semantics and its associated type system.

References

[AiMu91] A. Aiken, B. Murphy, *Static Type Inference in a Dynamically Typed Language*, Conf. rec of the 18th ACM Symposium on POPL, Orlando Florida 279-290, Jan 1991.

[BuMcQ80] R. Burstall, D. MacQueen and D. Sanella, *Hope: An experimental applicative language*, Conf. rec. of the 1980 Lisp Conference, Standford 1980

[Cop80] M. Coppo, M. Dezani-Ciancaglini, *An extension of the Basic Functionality Theory for the l-Calculus* Notre Dame, Journal of Formal Logic 21 (4), 1980.

[Co78] P. Cousot, *Méthodes itératives de construction et d'approximation de points fixes d'opérateurs monotones sur un treillis, analyse sémantique des programmes.* Thèse d'Etat, Grenoble 1978.

[CoCo77] P. & R. Cousot , *Abstract Interpretation: a unified lattice model for static analysis of program by construction of approximate fixpoints*, Conf. rec of the 4th ACM Symposium on POPL, Los Angeles Calif, 239-252, Jan 1977.

[CoCo79] P. & R. Cousot , *Systematic design of program analysis framework*, Conf. rec of the 6th ACM Symposium on POPL, Jan 1979.

[GuSc90] C.A. Gunther D.S. Scott *Semantic domains* in J. van Leeuwen (ed.) Formal Models and Semantics. vol. B of *Handbook of Theoretical Computer Science*, Elsevier 1990.

[HaMcQMi86] R. Harper ,D. MacQueen, R. Milner *Standard ML* Report ECS-LFCS-86-2, Edinburgh University.

[Joh87] Th. Johnsson, *Compiling Lazy Functional Language*, PhD Thesis Chalmers University of Technology, 1987.

[JoMy85] N.D. Jones, A. Mycroft, *A relational framework for abstract interpretation* in "Program as Data Objects" Proc. of a workshop in Copenhagen Oct. 1985. Ed. Ganziger and Jones. LNCS 215 Springer Verlag.

[Ka88] G. Kahn, *Natural Semantics*, "Programming of Future Generation Computers" edited by K. Fuchi, M. Nivat, Elsevier Science Publishers B .V. (North-Holland) 1988.

[KaHo88] T. Karnomori, K. Horiuchi *Polymorphic Type Inference in Prolog by Abstract Interpretation*, in Logic Programming '87, Tokyo, LNCS 315 Springer Verlag.

[Ma91] L. Maranget *GAML: a Parallel Implementation of Lazy ML*, Conf. rec of the 5th ACM Conference on FPCA, August 1991, LNCS 523 Springer Verlag.

[McQ84] D. MacQueen, G. Plotkin, R. Sethi *An ideal model for recursive polymorphic types*, Conf. rec of the 11th Symposium on POPL, Jan 1984.

[Mon92a] B. Monsuez *Polymorphic Typing by Abstract Interpretation* Conf. rec. of the 12th Conference on FST&TCS, December 1992, New Dehli, India, LNCS 652 Springer Verlag.

[Mon92b] B. Monsuez *Fractional Types* Prooceding of the Workshop on Static Analysis, Septembre 1992, Bordeaux, France, BIGRE 82.

[MyOK84] A. Mycroft, R. O'Keefe *A polymorphic Type System for Prolog*, Artificial Intelligence 23, 1984, 195-307.

[Plo81] G. D. Plotkin, *A Structural Approach to Operational Semantics*, DAIMI FN-19, Computer Science Department, Aarhus, Denmark, September 1981.

[Tur85] D. Turner, *Miranda, a non-strict functional language with polymorphic types* in Functional Programming Languages and Computer Architecture, Nancy, LNCS 201 Springer Verlag.

Logic Program Testing Based on Abstract Interpretation

Lunjin Lu and Peter Greenfield

The University of Birmingham, Birmingham B15 2TT, U.K.

Abstract. In this paper we present a logic program testing algorithm. This algorithm tests a program against a particular program property each time it is applied. A program property is derived from the program and is compared with the desired program property and the difference between them is used to guide test data generation.

1 Introduction

A logic program is intended to compute a set of ground atoms each of which falls into a relation. When a logic program has been written, it is necessary to make sure that the program computes the right set of ground atoms. The set of ground atoms that the program is intended to compute is called the intended semantics of the program whilst the set of ground atoms that the program actually computes is called the actual semantics of the program. When the actual semantics of the program is not equal to its intended semantics, the program contains at least one bug symptom. A ground atom in the actual semantics but not in the intended semantics is called an inconsistency symptom and a ground atom in the intended semantics but not in the actual semantics is called an insufficiency symptom.

While program testing has been an intensive research topic in the imperative programming community, it does not appear to have received much attention in the logic programming community; we have found little literature on logic program testing. We propose a logic program testing algorithm that tests the program against program properties. Each time the program is tested, a program property such as the type of the program is derived from the program and this derived program property is then compared with the desired program property. The difference between the derived program property and the desired program property is then used to limit the range from which a set of test data is selected. The motivation is twofold. Firstly, testing the program with all possible input values is infeasible because there is usually infinite number of input values. Secondly, the bugs in the program have manifestations in the properties of the program. We derive program properties from the program by means of abstract interpretation and hence we refer to program properties as abstract semantics. The derived program property will be called the actual abstract semantics of the program and the desired program property will be called the intended abstract semantics of the program.

2 Semantics and Abstract Interpretation

A language L is the set of programs that can be constructed from the grammar rules of L. The semantics of a program in L is a point in a semantic domain $\mathcal{D}$ and is given by a semantic function $\mu : L \mapsto \mathcal{D}$. Taking the definite logic program language as an example, L is the set of all definite logic programs, the semantic domain $\mathcal{D}$ is the powerset of the Herbrand base of L and the semantic function μ is the least fixpoint semantic function $\lambda P.lfp \circ T_P$ given by van Emden and Kowalski [21] where lfp is the least fixpoint operator.

A semantic domain $\mathcal{D}$ is a complete lattice $\mathcal{D}(\sqsubseteq, \bot, \top, \sqcup, \sqcap)$ where $\sqsubseteq$ is a partial order on $\mathcal{D}$ such that any subset X of $\mathcal{D}$ has a least upper bound $\sqcup X$ and a greatest lower bound $\sqcap X$ and $\bot = \sqcup\emptyset = \sqcap\mathcal{D}$ and $\top = \sqcap\emptyset = \sqcup\mathcal{D}$ [19]. A semantic domain $\mathcal{D}(\sqsubseteq, \bot, \top, \sqcup, \sqcap)$ is usually a powerset domain. $\mathcal{D} = \wp(D)$ for some base domain D, say the Herbrand base of the definite logic program language, $\sqsubseteq$ is set inclusion $\subseteq$, $\bot$ is the empty set $\emptyset$, $\top$ is the base domain D, $\sqcup$ is set union $\cup$ and $\sqcap$ is set intersection $\cap$.

The semantic domains that we consider in this paper are powerset domains. We now define the notion of the correctness of a program with respect to a semantic function and the intended semantics of the program on the corresponding semantic domain.

Definition 1. Let P be a program in P, $\mu : L \mapsto \mathcal{D}$ a semantic function and I the intended semantics of P on $\mathcal{D}$. When $\mu(P) = I$, P is said to be correct wrt μ and I. When $\mu(P) \neq I$, P is incorrect wrt μ and I. P is said to be inconsistent wrt μ and I if $\mu(P) - I \neq \emptyset$. Any $A \in \mu(P) - I$ is said to be an inconsistency symptom of P wrt μ and I. P is said to be insufficient wrt μ and I if $I - \mu(P) \neq \emptyset$. Any $A \in I - \mu(P)$ is said to be an insufficiency symptom of P wrt μ and I.

Abstract interpretation is a general scheme for static program analysis. The basic idea is that program analysis consists in approximating the standard semantic function $\mu_c : L \mapsto \mathcal{D}_c$ by a non-standard semantic function $\mu_a : L \mapsto \mathcal{D}_a$. The standard domain $\mathcal{D}_c$ is called the concrete domain, and the standard semantic function $\mu_c : L \mapsto \mathcal{D}_c$ is called the concrete semantic function. The non-standard domain $\mathcal{D}_a$ is called the abstract domain and the non-standard semantic function $\mu_a : L \mapsto \mathcal{D}_a$ is called the abstract semantic function. The foundation of abstract interpretation has been laid by Cousot and Cousot [5, 6]. In the logic programming community, Mellish [16], Jones et al. [10], Bruynooghe [4] and Marriott et al. [14] have developed several frameworks for abstract interpretation of logic programs and found many applications of these frameworks such as in occur check [20], mode inference [15, 16, 3] and type inference [4, 3].

Let the concrete semantic function be $\mu_c : L \mapsto \mathcal{D}_c(\sqsubseteq_c, \bot_c, \top_c, \sqcup_c, \sqcap_c)$ and the abstract semantic function be $\mu_a : L \mapsto \mathcal{D}_a(\sqsubseteq_a, \bot_a, \top_a, \sqcup_a, \sqcap_a)$. If there is an abstraction function $\alpha : \mathcal{D}_c \mapsto \mathcal{D}_a$ and a concretization function $\gamma : \mathcal{D}_a \mapsto \mathcal{D}_c$ such that $\alpha \circ \mu_c \sqsubseteq_a \mu_a$ and $\mu_c \sqsubseteq_c \gamma \circ \mu_a$, then μ_a is called an abstract interpretation of L relative to μ_c and (α, γ). We assume that $\mathcal{D}_c(\sqsubseteq_c, \bot_c, \top_c, \sqcup_c, \sqcup_c) = \wp(D_c)(\subseteq, \emptyset, D_c, \cup, \cap)$ and $\mathcal{D}_a(\sqsubseteq_a, \bot_a, \top_a, \sqcup_a, \sqcup_a) = \wp(D_a)(\subseteq, \emptyset, D_a, \cup, \cap)$ since both $\mathcal{D}_c$ and $\mathcal{D}_a$ are powerset domains. We also assume that α is a setwise function from D_c to D_a and γ is a setwise function from D_a to $\wp(D_c)$ such that $\gamma(x) = \{y | y \in D_c \textit{ and } \alpha(y) = x\}$. Definition 1 applies to both the concrete semantics by letting $\mu = \mu_c$ and $I = I_c$ and the abstract semantics by letting $\mu = \mu_a$ and $I = I_a$.

Each of $\mu_c(P)$ and I_c is usually a set of infinite size. This renders it infeasible to compare $\mu_c(P)$ with I_c. It is possible to compare $\alpha \circ \mu_c(P)$ with $\alpha \circ I_c$ if D_a is finite.

3 Algorithm

In this section, we discuss the relationship between bug symptoms of P wrt μ_c and I_c and bug symptoms of P wrt μ_a and I_a and present a testing algorithm for logic programs.

Lemma 2. *Let μ_a be an abstract interpretation relative to μ_c and (α, γ).*

(1) *If $\alpha \circ \mu_c(P) - \alpha \circ I_c \neq \emptyset$, then P is inconsistent* wrt *μ_c and I_c. For each inconsistency symptom $A_a \in \alpha \circ \mu_c(P) - \alpha \circ I_c$ of P* wrt *μ_a and I_a, there is an inconsistency symptom $A_c \in \mu_c(P) - I_c$ of P* wrt *μ_c and I_c such that $\alpha(A_c) = A_a$.*
(2) *If $\alpha \circ I_c - \alpha \circ \mu_c(P) \neq \emptyset$, then P is insufficient* wrt *μ_c and I_c. For each insufficiency symptom $A_a \in \alpha \circ I_c - \alpha \circ \mu_c(P)$ of P* wrt *μ_a and I_a, there is an insufficiency symptom $A_c \in I_c - \mu_c(P)$ of P* wrt *μ_c and I_c such that $\alpha(A_c) = A_a$.*
(3) *If $\alpha \circ \mu_c(P) \neq \alpha \circ I_c$, then P is incorrect* wrt *μ_c and I_c.*

Proof. We prove (1) by contradiction. Assume P is consistent wrt μ_c and I_c. By definition 1, $\mu_c(P) - I_c = \emptyset$. $\alpha \circ \mu_c(P) - \alpha \circ I_c = \emptyset$. This is in contradiction with the premise of (1). Therefore, P is inconsistent wrt μ_c and I_c. This completes the proof of (1). (2) can be proved in a similar way and (3) results from (1) and (2). □

By lemma 2, $\alpha \circ \mu_c(P) \neq \alpha \circ I_c$ implies that P is incorrect wrt μ_c and I_c. The comparison between $\alpha \circ \mu_c(P)$ and $\alpha \circ I_c$ is feasible because D_a is of finite size. However, there is no direct way of calculating $\alpha \circ \mu_c(P)$ or $\alpha \circ I_c$ in finite time unless both of $u_c(P)$ and $u_c(P_{spec})$ are finite. So, instead of comparing $\alpha \circ \mu_c(P)$ and $\alpha \circ I_c$, algorithm 1 compares $\mu_a(P)$ and I_a and uses the difference between them as a guide to the generation of test data.

A program is tested with respect to a given program property. To do this, algorithm 1 takes as an input the intended abstract semantics I_a corresponding to the program property. I_a is finite and can either be specified or be delivered by an abstract oracle. The abstract oracle is capable of deciding whether an element in $\mu_a(P)$ is in I_a and whether any element in I_a is missing from $\mu_a(P)$.

Algorithm 1. *Let μ_a be an abstract interpretation relative to μ_c and (α, γ).*

- *Inputs: P the program, I_a the intended abstract semantics and I_c the intended concrete semantics.*
- *Outputs: W_c a set of inconsistency symptoms of P* wrt *μ_c and I_c, and M_c a set of insufficiency symptoms of P* wrt *μ_c and I_c.*
- *Specifications for auxillary functions:*
  - *$\gamma'(A_a)$ terminates and $\gamma'(A_a)$ is a finite subset of $\gamma(A_a)$;*
  - *solve/2 terminates and solve(A_c, P) implies $A_c \in \mu_c(P)$;*
  - *verify/2 terminates and verify(A_c, I_c) implies $A_c \in I_c$.*
- *Algorithm:*

  $W_a := \mu_a(P) - I_a$
  $W_c := \emptyset$
  for each $A_a \in W_a$ *do*
      $T_{A_a} := \gamma'(A_a))$
      for each $A_c \in T_{A_a}$ *do*
          if solve(A_c, P) *then* $W_c := W_c \cup \{A_c\}$

  $M_a := I_a - \mu_a(P)$
  $M_c := \emptyset$
  for each $A_a \in M_a$ *do*
      $T_{A_a} := \gamma'(A_a)$
      for each $A_c \in T_{A_a}$ *do*
          if verify(A_c, I_c) *then* $M_c := M_c \cup \{A_c\}$

  return W_c, M_c

Another input to algorithm 1 is the intended concrete semantics I_c. I_c is delivered by a concrete oracle. The concrete oracle is queried by auxiliary function *verify/2* in order to decide whether an element in $\mu_c(P)$ is in I_c. The concrete oracle must declare $A_c \notin I_c$ for any A_c not in I_c whilst it may either declare $A_c \in I_c$ or declare $A_c \notin I_c$ for any A_c in I_c. If the concrete oracle makes a mistake by declaring an A_c in I_c to be not in I_c then this mistake will only reduce the number of the insufficiency symptoms of P *wrt* μ_c and I_c that algorithm 1 returns but will not compromise the soundness of algorithm 1. This tolerance is meant to ensure the termination of a mechanised concrete oracle.

Auxiliary function *solve/2* emulates the concrete semantic function μ_c. The specification for *solve/2* allows *solve*(A_c, P) to evaluate into *false* when $A_c \in \mu_c(P)$. This is designed to ensure the termination of the emulation of the concrete semantic function μ_c.

Auxiliary function $\gamma'(A_a)$ generates test data by returning a finite subset of $\gamma(A_a)$. An implementation of γ' would be dependent on the program property against which the program is tested and need to adopt certain testing hypotheses such as the uniformity hypothesis and the regularity hypothesis [2, 1, 13].

Lemma 3. *Let μ_a be an abstract interpretation relative to μ_c and (α, γ), and W_c and M_c be results from algorithm 1.*

(1) If $A_c \in W_c$ then A_c is an inconsistency symptom of P wrt *μ_c and I_c.*
(2) If $A_c \in M_c$ then A_c is an insufficiency symptom of P wrt *μ_c and I_c.*

Proof. Consider (1). $A_c \in W_c$ implies $\alpha(A_c) \in \mu_a(P) - I_a$ by algorithm 1. So, $A_c \notin I_c$. We also know from algorithm 1 that $A_c \in \mu_c(P)$. So, A_c is an inconsistency symptom of P wrt μ_c and I_c by definition 1. This completes the proof of (1). (2) can be proved in a similar way. □

solve/2 is not used to detect insufficiency symptoms of P wrt μ_c and I_c because $A_c \in \gamma(I_a - \mu_a(P))$ implies $A_c \notin I_a$ and *verify/2* is not used to detect inconsistency symptoms of P wrt μ_c and I_c because $A_c \in \gamma(\mu_a(P) - I_a)$ implies $A_c \notin I_c$. This typically results in a significant gain in efficiency and reduction in user interaction. The price for the gain in efficiency and the reduction in user interaction is that P is not tested on any input from $\gamma(\mu_a(P) \cap I_a)$. Because $\mu_a(P)$ and I_a are supersets of $\alpha \circ \mu_c(P)$ and $\alpha \circ I_c$ respectively, a discrepancy between $\mu_a(P)$ and I_a does not necessarily imply any discrepancy between $\mu_c(P)$ and I_c. In other words, that A_a is a bug symptom of P wrt μ_a and I_a does not imply that there is some $A_c \in \gamma(A_a)$ such that A_c is a bug symptom of P wrt μ_c and I_c. Algorithm 1 deals with this uncertainty by using *solve/2* and *verify/2*.

Algorithm 1 uses the difference between $\mu_a(P)$ and I_a as a guide to test data generation. Algorithm 1 does not test P on any input from outside $\gamma(\mu_a(P) \cup I_a)$ because such an input cannot be a bug symptom of P wrt μ_c and I_c. Algorithm 1 does not test P on any input from $\gamma(\mu_a(P) \cap I_a)$ in order to reduce user interaction. Because there is a loss of information during abstract interpretation, there is probably some bug symptom $A_c \in \gamma(\mu_a(P) \cap I_a)$ of P wrt μ_c and I_c. So, algorithm 1 is incomplete. Incompleteness is not a weakness peculiar to algorithm 1. It has been proved that complete testing cannot in general be achieved [9, 17]. In our case, this incompleteness may be alleviated by testing P against multiple program properties.

4 Examples

Algorithm 1 applies to any logic programming language L, any concrete semantic function μ_c, any abstract semantic function μ_a, any abstraction function α and any concretization function γ as long as μ_a is an abstract interpretation of L relative to μ_c and (α, γ). We now illustrate algorithm 1 by taking as L the definite logic programming language and concrete semantic function μ_c of L as the least fixpoint semantic function in [21] which is restated here. Let P be a definite logic program, B_P the Herbrand base of P. Then $D_c = B_P$ and $\mu_c = \lambda P.lfp \circ T_P$ where $T_P : \wp(B_P) \mapsto \wp(B_P)$ is the following function.

$$T_P(I) = \left\{ A_0\sigma \middle| \begin{array}{l} \exists\sigma.\exists A_0 \leftarrow A_1, \ldots, A_m \in P. \\ (A_1\sigma \in I, \ldots, A_m\sigma \in I, A_0\sigma \in B_P) \end{array} \right\} \quad (1)$$

Let D_a be the abstract domain of interest and α be an abstraction function $\alpha : B_P \mapsto D_a$. Lu and Greenfield [12] formulate an abstract semantic function $\mu_a = \lambda\alpha.\lambda P.lfp \circ \Phi_{P,\alpha}$, where $\Phi_{P,\alpha}$ is defined below. $\mu_a = \lambda\alpha.\lambda P.lfp \circ \Phi_{P,\alpha}$ is an abstract interpretation relative to $\mu_c = \lambda P.lfp \circ T_P$ and $(\alpha, \gamma = \lambda A_a.\{A_c|\alpha(A_c) = A_a\})$.

$$\Phi_{P,\alpha}(I_a) = \left\{ \alpha(A_0\sigma) \middle| \begin{array}{l} \exists\sigma.\exists C = A_0 \leftarrow A_1, \ldots, A_m \in P. \\ (\alpha(A_1\sigma) \in I_a, \ldots, \alpha(A_m\sigma) \in I_a) \end{array} \right\} \tag{2}$$

We illustrate algorithm 1 through depth abstraction [18] and type abstraction [11, 8]. Examples 1 and 2 illustrate algorithm 1 through depth abstraction and use an executable specification P_{spec} for P. $solve/2$ is implemented as a depth-bound meta-interpreter for Prolog [8] and we let $I_a = \mu_a(P_{spec})$ and $verify(A_c, I_c) = solve(A_c, P_{spec})$ because P_{spec} is executable. γ' is implemented such that $\gamma'(A_a)$ replaces each occurrence of _ by a new variable.

A depth abstraction function partitions B_P into a finite number of equivalent classes. Two ground terms belong to the same class if and only if they are identical to a certain depth k. For example, $p(f(a), g(b))$ is equivalent to $p(f(b), g(a))$ to depth 2. Let d_k represent the depth k abstraction function. d_k replaces each sub-term of depth k with an _ where _ denotes any term. We have $d_2(p(f(a), g(b)) = d_2(p(f(b), g(a)) = p(f(\_), g(\_))$.

Example 1. Suppose we have the following specification and its implementation.

$$P_{spec} = \left\{ \begin{array}{l} p(a,b),\ p(X,Y) \leftarrow q(Y,X), \\ q(a,b),\ q(r(r(X)), s(s(Y))) \leftarrow q(X,Y) \end{array} \right\}$$

$$P1 = \{p(a,b),\ p(X,Y) \leftarrow q(X,Y),\ q(a,b),\ q(r(X), s(Y)) \leftarrow q(X,Y)\}$$

Let the abstract semantic function be $\mu_a = \lambda P.lfp \circ \Phi_{P,d_2}$ by setting $\alpha = d_2$. We have

$$I_a = \{p(a,b), q(a,b),\ p(b,a),\ q(r(\_), s(\_)),\ p(s(\_), r(\_))\}$$

$$\mu_a(P1) = \{p(a,b),\ q(a,b),\ q(r(\_), s(\_)),\ p(r(\_), s(\_))\}$$

and $I_a \neq \mu_a(P1)$. By applying algorithm 1, we have

$$W_a = \mu_a(P1) - I_a = \{p(r(\_), s(\_))\}$$

$$M_a = I_a - \mu_a(P1) = \{p(b,a),\ p(s(\_), r(\_))\}$$

and

$$W_c = \{p(r(a), s(b))\}$$

$$M_c = \{p(s(s(b)), r(r(a))),\ p(b,a)\}$$

The following example shows how the use of multiple program properties is useful in testing.

Example 2. Suppose we have the same specification as example 1 and we have the implementation that results from changing the positions of the arguments in the body goal $q(X,Y)$ of the second clause of $P1$ in example 1.

$$P2 = \{p(a,b),\ p(X,Y) \leftarrow q(Y,X),\ q(a,b),\ q(r(X),s(Y)) \leftarrow q(X,Y)\}$$

It can be shown that algorithm 1 does not find any inconsistency or insufficiency symptom when $\alpha = d_2$ because $I_a = \mu_a(P2)$. We now use another abstract semantic function $\mu_a = \lambda P.lfp \circ \Phi_{P,d_3}$ by setting $\alpha = d_3$. Then

$$I_a = \{p(a,b),\ q(a,b),\ p(b,a),\ q(r(r(\_)),s(s(\_))),\ p(s(s(\_)),r(r(\_)))\}$$

$$\mu_a(P2) = \left\{\begin{array}{l} p(a,b),\ q(a,b),\ p(b,a),\ q(r(a),s(b)),\ p(s(b),r(a)), \\ q(r(r(\_)),s(s(\_))),\ p(s(s(\_)),r(r(\_))) \end{array}\right\}$$

$I_a \neq \mu_a(P2)$ and hence algorithm 1 is applicable. We have

$$W_a = \{q(r(a),s(b)),\ p(s(b),r(a))\}$$

$$M_a = \{\}$$

and

$$W_c = \{p(s(b),r(a)),\ q(r(a),s(b))\}$$

$$M_c = \{\}$$

The new program property, corresponding to $\alpha = d_3$, has helped to find two inconsistency symptoms of P wrt μ_c and I_c that are not found when the program is tested with respect to the program property corresponding to $\alpha = d_2$.

The above two examples have used an executable specification. The convenience of using an executable specification is that both the intended concrete semantics I_c and the intended abstract semantics I_a can be derived from the specification. However, the specification for a program is rarely given as an executable program. Instead, a number of program properties may be specified. In example 3, I_a is a type specification for P, γ' enumerates ground terms of a type using a depth-bound constraint, *solve*/2 is implemented as a depth-bound meta-interpreter and *verify*/2 is implemented by means of imposing queries on the programmer.

Example 3. Let $P = \{double(0,0), double(s(X),s(Y)) \leftarrow double(X,Y)\}$. P is a buggy implementation for the specification that $double(X,Y)$ is true if and only if $Y = 2 * X$. We follow Frühwirth in using a particular predicate ':/2' for type declarations [8]. The type declarations for the functors in P are

$$\{0 : even,\ s(X) : even \leftarrow X : odd,\ s(X) : odd \leftarrow X : even\}$$

$0 : even$ means that 0 is of type *even*. $s(X) : even \leftarrow X : odd$ means that if X is of type *odd* then $s(X)$ is of type *even*. A possible type specification for P is

$$I_a = \{double(odd, even),\ double(even, even)\}$$

Let $\mu_a = \lambda P.lfp \circ \Phi_{P,type}$ by setting $\alpha = type$ where $type(A_c) = A_a$ if and only if $A_c : A_a$ is true. The derived type for P is

$$\mu_a(P) = \{double(odd, odd), double(even, even)\}$$

$I_a \neq \mu_a(P)$ and hence algorithm 1 is applicable. We have

$$W_a = \{double(odd, odd)\}$$
$$M_a = \{double(odd, even)\}$$

and

$$W_c = \{double(s(0), s(0))\}$$
$$M_c = \{double(s(0), s(s(0)))\}$$

The following two queries are asked by *verify*/2:

```
Is double(s(0),0) valid? n.
Is double(s(0),s(s(0))) valid? y.
```

5 Adaptations

Examples 1 and 2 use an executable specification and example 3 uses a type specification and requires the programmer to verify if a ground atom A_c is valid in I_c or not. There are other cases that are of concern. In this section, we show how to adapt algorithm 1 to cater for these different cases.

5.1 Incomplete specifications

We first consider the case where we have an executable specification P_{spec} but P_{spec} does not cover all the procedures of P.

Examples 1 and 2 assume that each procedure of P is specified by P_{spec}. This is restrictive because this assumption requires that a decomposition of P_{spec} is available and is followed by the programmer, whilst P_{spec} should only be concerned with the functionality of P rather than the way that the functionality is to be attained. When the decomposition of P differs from that of P_{spec}, it is sensible to test only those procedures of P that are specified by P_{spec}. Algorithm 1 applies with minor modification: assign to W_a and M_a only those elements of D_a that correspond to the procedures of P specified by P_{spec}.

5.2 Correctness assertions

There are cases where some procedures of the program are known to be correct. For example, the program may contain some procedures from validated libraries or the programmer may have obtained sufficient confidence in some procedures during debugging. This kind of knowledge can be conveyed incremently to algorithm 1 and be used to reduce the size of W_a and M_a. Algorithm 1 can be improved by deleting from W_a and M_a those elements of D_a that correspond to the procedures that are known to be correct.

6 Conclusion and further work

We have presented an algorithm (algorithm 1) that tests a program with respect to a given program property. The program properties appropriate to algorithm 1 are those derivable by means of abstract interpretation. Instead of comparing the actual concrete semantics and the intended concrete semantics of the program, algorithm 1 first compares the actual abstract semantics and the intended abstract semantics of the program and then uses the difference between them to guide test data generation. We have also shown how algorithm 1 can be adapted to cater for different cases.

Yan [22] studies declarative testing using a reliable test data set. What is not covered by Yan is how a reliable test data set is selected. Informally, a test set T is reliable for program P and specification P_{spec} if that P computes the same value as P_{spec} upon each point over T implies that P computes the same value as P_{spec} upon each point at which P is defined. A finite reliable test data set T for program P and specification P_{spec} exists, but there is no effective procedure to either generate or recognise a reliable test set [9, 17].

Dershowitz et. al [7] study logic program testing when an executable specification for the program is available. The program is exhaustively compared with the specification when a goal to be tested has been given. This is impractical because both the intended concrete semantics of the program specified by the specification and the actual concrete semantics of the program are usually infinite.

Logic program testing is an area which has rarely been looked at. There are only a few publications in this area. It does not appear that there has been any other work on using abstract interpretation in in logic program testing. Therefore, there are many topics to be addressed. Firstly, we need to find out what program properties are effective. We can test the program against as many program properties as possible. The more program properties against which the program is tested, the more confidence we have in the correctness of the program. However, there must be a trade-off between the number of the program properties against which the program is tested and the cost at which this testing is conducted. Therefore, we need to find out a set of program properties that are effective in exposing bugs. Secondly, we also need to know what program property is more useful than other program properties in exposing a particular kind of bug.

Acknowledgements

The first author is supported by the Sino-British Friendship Scholarship Scheme. Thanks are extended to Dr M. Ducassé for useful comments on an earlier version of this paper.

References

1. G. Bernot, M.C. Gaudel, and B. Marre. Software testing based on formal specifications: a theory and a tool. LRI research report no. 581, April 1991.
2. L. Bouge, N. Choquet, L. Fribourg, and M.-C. Gaudel. Test sets generating from algebraic specifications using Logic Programming. *Journal of System and Software*, 6(4):343–360, 1986.
3. M. Bruynooghe and G. Janssens. An instance of abstract interpretation integrating type and mode inferencing. In R.A. Kowalski and K.A. Bowen, editors, *Proceedings of the fifth International Conference and Symposium on Logic Programming*, pages 669–683. The MIT Press, 1988.
4. M. Bruynooghe, G. Janssens, A. Callebaut, and B. Demoen. Abstract interpretation: towards the global optimisation of Prolog programs. In *Proceedings of the 1987 Symposium on Logic Programming*, pages 192–204. The IEEE Society Press, 1987.
5. P. Cousot and R. Cousot. Abstract interpretation: a unified framework for static analysis of programs by construction or approximation of fixpoints. In *Proceedings of the fourth annual ACM symposium on Principles of programming languages*, pages 238–252, Los Angeles, California, 1977.
6. P. Cousot and R. Cousot. Systematic design of program analysis frameworks. In *Proceedings of the sixth annual ACM symposium on Principles of programming languages*, pages 269–282, San Antonio, Texas, 1979.
7. N. Dershowitz and Y.-J. Lee. Deductive debugging. In *Proceedings of 1987 Symposium of Logic Programming*, pages 298–306. The IEEE Computer Society Press, 1987.
8. T.W. Früehwirth. Using meta-interpreters for polymorphic type checking. In M. Bruynooghe, editor, *Proceedings of the Second Workshop on Meta-programming in Logic*, pages 339–351, Leuven, Belgium, April 4-6 1990.
9. W.E. Howden. Reliability of the path analysis testing strategy. *IEEE Transactions on Software Engineering*, 2(3):208–215, 1976.
10. N.D. Jones and H. Søndergarrd. A semantics-based framework for abstract interpretation of Prolog. In S. Abramsky and C. Hankin, editors, *Abstract interpretation of declarative languages*, pages 123–142. Ellis Horwood Limited, 1987.
11. T. Kanamori and K. Horiuchi. Type inference in Prolog and its application. In *Proceedings of the ninth International Joint Conference on Artificial Intelligence*, pages 704–707, 1985.
12. L. Lu and P. Greenfield. Abstract fixpoint semantics and abstract procedural semantics of definite logic programs. In *Proceedings of IEEE Computer Society 1992 International Conference on Computer Languages*, pages 147–154, Oakland, California, USA, April 20-23 1992. IEEE Computer Society Press.
13. B. Marre. Toward automatic test data set selection using algebraic specifications and Logic Programming. In K. Furukawa, editor, *Proc. of the eighth International Conference on Logic Programming*, pages 202–219, Paris, France, 1991. ALP, MIT Press.
14. K. Marriott and H. Søndergaard. Bottom-up abstract interpretation of logic programs. In R.A. Kowalski and K.A. Bowen, editors, *Proceedings of the fifth International Conference and Symposium on Logic Programming*, pages 733–748. The MIT Press, 1988.
15. C. Mellish. Some global optimisations for a Prolog compiler. *The Journal of Logic Programming*, 2(1):43–66, 1985.

16. C. Mellish. Abstract interpretation of Prolog programs. In S. Abramsky and C. Hankin, editors, *Abstract interpretation of declarative languages*, pages 181–198. Ellis Horwood Limited, 1987.
17. L.J. Morell. A theory of fault-based testing. *IEEE Transactions on Software Engineering*, 16(8):844–857, 1990.
18. T. Sato and H. Tamaki. Enumeration of success patterns in logic programs. *Theoretical Computer Science*, 34(1):227–240, 1984.
19. D.A. Schmidt. *Denotational Semantics: a methodology for language development.* Allyn and Bacon, Inc., 1986.
20. H. Søndergaard. An application of abstract interpretation of logic programs. In *Proceedings of the European symposium on programming.* Springer-Verlag, 1986.
21. M.H. van Emden and R.A. Kowalski. The semantics of predicate logic as a programming language. *Artificial Intelligence*, 23(10):733–742, 1976.
22. S.Y. Yan. Declarative testing in logic programming. In *Proceeding of the third Australian software engineering conference*, pages 423–435, 1987.

Analysis of Some Semantic Properties for Programs of the Applicative Language AL

V. Sabelfeld*

Institute of Informatics Systems
Siberian Division of Russian Academy of Sciences
Av. Acad. Lavrentyev, 6, Novosibirsk 630090, Russia
E-mail: vks@isi.itfs.nsk.su

Abstract. This paper describes an approximation semantics $[\![p]\!]$ for programs p of the applicative language AL [Br 92, De 92]. We show that the approximation semantics of each program term can be effectively computed. The approximation semantics is safe in the sense that if $[\![p]\!]X = Y$ for classes X, Y of values, then $\forall x \in X\ p(x) \in Y$. These features allow us to deduce some assertions about the behaviour of p from computed approximation semantics $[\![p]\!]$ and also to detect certain semantic errors in p. For example, if $[\![p]\!] = \bot$ holds for an applicative program p, then we can conclude that the program p never terminates. In the last section a method for finding the so called inessential program constructions is described. Deletion of inessential constructions does not change the meaning of a specification. The presence of such inessential constructions in a program can be regarded as an anomaly, indicating a possible program error.

1 Introduction

Many programs contain semantic errors. That is why program tools detecting semantic errors are needed more often than a verifier program. For applicative specifications, the possibility of earlier detection of semantic errors seems to be even more significant than for imperative programs. In this paper, an approach to the problem of detection of semantic errors in applicative specifications of the language AL [Br 92, De 92] is described. This approach is based on ideas of global flow analysis algorithm proposed first by G. Kildall [Ki 73] and can be regarded as a simplified version of the abstract interpretation notion [Co 77] applied to property analysis of applicative programs with partial operations and nondeterminism.

At the first step, a notion of an approximation of a function defined by a specification is introduced. To this end, the object set D of each sort is partitioned into a finite number of disjoint subsets (classes) : $D = D_1 \cup \ldots \cup D_n$. We choose the unique class containing x as an approximation $||x||$ of an element $x \in D$. Some classes of error values like *division by zero*, *pred(0)*, *negative result of subtraction*, *first(ε)*, *rest(ε)* etc. can also be taken into the consideration.

* This research has been carried out in 1992 when the author worked at the Institut für Informatik, Technische Universität München, Germany

Unfortunately, one cannot effectively calculate the approximations of functions described by applicative programs, although these approximations are less detailed than the original semantic functions. Therefore, at the second step we replace all primitive operations in a given specification p by corresponding approximations. What we get is an approximation program $\tilde{p}$ and the fixpoint semantics of $\tilde{p}$ is called the approximation semantics $[\![p]\!]$ of the given program p. This approximation semantics has the following useful properties :

- The approximation semantics of each program term can be effectively computed.
- The approximation semantics is safe in the sense that $\|p\| \subseteq [\![p]\!]$ holds for all program terms p.

These features allow us to deduce some assertions about the original semantics $[p]$ from computed approximation semantics $[\![p]\!]$ of a program p. For example, if $[\![p]\!] = \perp\!\!\!\perp$ holds for an applicative program p, then we can conclude that the program p never terminates.

Since the problem of finding semantic errors in applicative programs is unsolvable, it is desirable to detect certain "suspicious" program points which may contain an error with a large probability. In the last section, a method for finding the so called inessential program constructions is described. Inessential constructions can be deleted without any effect on the semantics of a specification. The presence of such inessential constructions in a program can be regarded as an anomaly and as a consequence of an possible semantic error.

2 The language AL

The applicative language AL described in [De 92] leans on the language AMPL designed by Broy in [Br 86]. The syntactical rules for AL below are taken from [De 92].

⟨program⟩	::=	**program**	⟨prg_id⟩ ≡ ⟨stream⟩* → ⟨stream⟩+ : {⟨agent⟩|⟨function⟩}* ⟨eq_sys⟩
		end	
⟨agent⟩	::=	**agent**	⟨agt_id⟩ ≡ ⟨object⟩*, ⟨stream⟩* → ⟨stream⟩+ : ⟨eq_sys⟩
		end	
⟨function⟩	::=	**funct**	⟨fct_id⟩ ≡ ⟨object⟩* → ⟨sort⟩ : ⟨exp⟩
		end	
⟨stream⟩	::=	**chan** ⟨sort⟩ ⟨str_id⟩+	
⟨object⟩	::=	⟨sort⟩⟨obj_id⟩+	
⟨eq_sys⟩	::=	⟨equation⟩+	
⟨equation⟩	::=	⟨str_id⟩+ ≡ ⟨exp ⟩	
⟨exp⟩	::=	ε | ⊥ | ⟨obj_id⟩|⟨str_id⟩|⟨primitive object⟩ |	

$\langle exp\rangle \& \langle exp\rangle \mid \{\mathbf{ft} \mid \mathbf{rt} \mid \mathbf{isempty}\}.\langle exp\rangle \mid$
$\langle exp\rangle [\![\langle exp\rangle \mid \mathbf{if}\ \langle exp\rangle\ \mathbf{then}\ \langle exp\rangle\ \mathbf{else}\ \langle exp\rangle\ \mathbf{fi} \mid$
$\{\langle agt\_id\rangle \mid \langle fct\_id\rangle \mid \langle primitive\ function\rangle\}\ (\langle exp\rangle^*)$

Streams of messages are used to represent communication histories of channels. Input and output channels are distinguished. These channels are the communication links between the system components. Stream processing functions are called agents.

AL is a typed language. Every object or stream belongs to some type, which semantically means that it is an element of a particular domain. For example, **chan Nat** stands for N^ω, the prefix ordered domain of streams over N.

$D^\perp = D \cup \{\perp\}, D^\omega = D^* \cup (D^* \times \{\perp\}) \cup D^\infty$.

The partial order $\sqsubseteq$ on domain of streams is defined by

$$s_1 \sqsubseteq s_2 \stackrel{\text{def}}{=} s_1 = s_2 \vee \exists s_3 \in D^* \exists s_4 \in D^\omega : s_1 = s_3 \frown \langle\perp\rangle \wedge s_2 = s_3 \frown s_4.$$

On the domain of streams the following primitive operations are defined :
$.\&.: D^\perp \times D^\omega \to D^\omega$, $\mathbf{ft} : D^\omega \to D^\perp$, $\mathbf{rt} : D^\omega \to D^\omega$, $\mathbf{isempty} : D^\omega \to Bool^\perp$.
From the axiomatic definition of the semantics of these operations follows, that the operation & is not strict in the second argument.
$\forall d \in D\ \forall s \in D^\omega$
$d\&s = d \frown s$, $\perp\&s = s$,
$\mathbf{ft}.d\&s = d$, $\mathbf{rt}.d\&s = s$, $\mathbf{isempty}.d\&s = \mathbf{false}$,
$\mathbf{ft}.\varepsilon = \perp$, $\mathbf{rt}.\varepsilon = \perp$, $\mathbf{isempty}.\varepsilon = \mathbf{true}$,
$\mathbf{ft}.\perp = \perp$, $\mathbf{rt}.\perp = \perp$, $\mathbf{isempty}.\perp = \perp$.
The semantics of a program (agent, function) p whose body is an equation system

$$ES = \{s_1 \equiv S_1[x, s_1, \ldots, s_n], \ldots, s_n \equiv S_n[x, s_1, \ldots, s_n]\}$$

is defined by the least fixpoint solution of this system

$$F_\delta(ES[x]) = \{\lambda x. fix \begin{pmatrix} s_1 \equiv f_1[x, s_1, \ldots, s_n] \\ \ldots \\ s_n \equiv f_n[x, s_1, \ldots, s_n] \end{pmatrix} : f_i \in F_\delta(S_i)\}.$$

3 Approximations of functions

Let D be the object set of a sort **S**. We partition the set D into a finite number of disjoint subsets (classes) $D_1, \ldots, D_n$,

$$D = \bigcup_{i=1}^{n} D_i,\ i \neq j \Rightarrow D_i \cap D_j = \emptyset$$

and consider the boolean algebra

$$\langle K_{\mathbf{S}}, \cup, \cap, \perp\!\!\!\perp, \square\rangle,$$

where $K_{\mathbf{S}}$ is the minimal set closed under the following conditions:

- $\forall i(1 \leq i \leq n) D_i \in K_{\mathbf{S}}$
- $k_1, k_2 \in K_{\mathbf{S}} \Rightarrow k_1 \cup k_2, k_1 \cap k_2 \in K_{\mathbf{S}}$.

Here $\perp\!\!\!\perp$ and $\square$ are the notations for polymorphic constants of the empty set and of the whole of the object set, respectively. For elements of the set D and functions over D, the notion of approximation will be introduced. The *approximation* $\|x\|$ *of an element* $x \in D$ is the minimal class from $K_{\mathbf{S}}$ containing x :

$$\|x\| \stackrel{\text{def}}{=} \bigcap\{k : x \in k\}.$$

The subset relation $\subseteq$ is used for the partial order on domain $K_{\mathbf{S}}$: $k_1 \leq k_2 \stackrel{\text{def}}{=} k_1 \subseteq k_2$. Under this order relation, all completely ordered chains in $K_{\mathbf{S}}$ will be finite since the number n of subsets D_i in D is finite.

The *approximation* $\|f\| : K_{\mathbf{S}} \to K_{\mathbf{T}}$ *of a function* $f : D_1^{\mathbf{S}} \to D_2^{\mathbf{T}}$ is defined by

$$k \neq \perp\!\!\!\perp \to \|f\|k \equiv \bigcup_{x \in k} \|fx\|, \text{ and } \|f\|\perp\!\!\!\perp \equiv \|f\perp\|.$$

The approximation of each function is monotonic: if $k_1 \leq k_2$, then in the case $k_1 = \perp\!\!\!\perp$ we obtain

$$\|f\|\perp\!\!\!\perp = \|f\perp\| \leq \bigcup_{x \in k_2} \|fx\| = \|f\|k_2,$$

and in the case $k_1 \neq \perp\!\!\!\perp$ we have

$$\|f\|k_1 = \bigcup_{x \in k_1} \|fx\| \leq \bigcup_{x \in k_2} \|fx\| = \|f\|k_2.$$

Since all completely ordered chains in $K_{\mathbf{S}}$ are finite, all approximations are continuous. Approximation of the identity function $id : D \to D$, $id(x) = x$, is obviously the identity $ID_{K_{\mathbf{S}}} : K_{\mathbf{S}} \to K_{\mathbf{S}}$, $ID_{K_{\mathbf{S}}}(k) = k$, and $\|\perp\| = \perp\!\!\!\perp$. Now the partial order $\leq$ on object set approximations can be extended on approximation functions $f, g : K_{\mathbf{S}} \to K_{\mathbf{T}}$ by $f \leq g \stackrel{\text{def}}{=} \forall k : K_{\mathbf{S}} \, . \, f(k) \leq g(k)$.

Let L be an applicative language, containing at least the primitive sorts **Nat**, **Bool** and the corresponding primitive operations $\{+, -, =, \mathbf{if}\}$. A program of the language L is a system of recursive equations. The semantics $[p]$ of such a program p is defined by the least fixpoint solution of the corresponding equation system.

Theorem 1 *There is no algorithm to evaluate approximations for programs of the language L.*

Proof. Let $[p]$ be the semantics of an applicative program $p \in L$. Then $\|p\| = \perp\!\!\!\perp$ holds iff $[p] = \perp$ which means that the program p never terminates. This halting problem is known to be unsolvable for programs of the language L. $\square$

But if a program p uses only sorts with finite object sets, then the approximation $\|p\| = [p]$ of p obviously can be effectively computed.

4 Approximation semantics

Assume the approximations $||f||$ of all primitive operations f of the language L to be fixed. We define the *approximation semantics* $[\![p]\!]$ of a term p in the following way :

- $[\![x]\!] \stackrel{\text{def}}{=} ID_{K_{\mathbf{S}}}$ for a variable x of a sort $\mathbf{S}$.
- $[\![\perp]\!] \stackrel{\text{def}}{=} \perp$.
- $[\![f]\!] \stackrel{\text{def}}{=} ||f||$ for primitive functions f of the language L.
- $[\![f \circ g]\!] \stackrel{\text{def}}{=} [\![f]\!] \circ [\![g]\!]$.
- $[\![f [\!] g]\!] \stackrel{\text{def}}{=} [\![f]\!] \cup [\![g]\!]$.
- If $E = \{x_i \equiv t_i\}_{i=1}^m$ is an equation system, then $[\![E]\!]$ is the minimal (relative to $\subseteq$) solution of the equation system $\tilde{E} = \{[\![x_i]\!] \equiv [\![t_i]\!]\}_{i=1}^m$.

Theorem 2 $||p|| \leq [\![p]\!]$ *holds for all terms p of the language L.*

Proof. By induction on the structure of p. For variables p, term $p = \perp$ and primitive functions p, the inequality $||p|| \leq [\![p]\!]$ holds since $||p|| = [\![p]\!]$ by definition. Let $\sqsubseteq$ be the partial order on object domains, and let f, g be arbitrary terms of the language L. Then we can complete the induction using the following relations between approximation of a term and approximation semantics of this term.

1. If $||f|| \leq [\![f]\!]$ and $||g|| \leq [\![g]\!]$, then

$$||f \circ g|| \leq ||f|| \circ ||g|| \leq [\![f]\!] \circ [\![g]\!] = [\![f \circ g]\!].$$

2. If $[f] \sqsubseteq [g]$, $||f|| \leq [\![f]\!]$ and $||g|| \leq [\![g]\!]$, then

$$||[f] \sqcup [g]|| = ||g|| \leq [\![g]\!] = [\![f]\!] \cup [\![g]\!].$$

3. For the nondeterministic choice construction $f [\!] g$, if $||f|| \leq [\![f]\!]$ and $||g|| \leq [\![g]\!]$, then

$$||f [\!] g|| = ||f|| \cup ||g|| \leq [\![f]\!] \cup [\![g]\!] = [\![f [\!] g]\!]$$

holds.

4. If $E = \{\overline{x} \equiv F(\overline{x})\}$ is an equation system, where $\overline{x} = \langle x_1, \ldots, x_m \rangle$ and $||F|| \leq [\![F]\!]$, then

$$||\mathbf{fix}\ F|| = ||\bigsqcup_{i=0}^{\infty} F^i|| = \bigcup_{i=0}^{\infty} ||F^i|| \leq \bigcup_{i=0}^{\infty} ||F||^i \leq \bigcup_{i=0}^{\infty} [\![F]\!]^i = \mathbf{fix}\ [\![F]\!]\ . \ \square$$

5 Approximation semantics for AL-terms

In this section, we describe the approximation semantics of the language AL. To this end, we have to define the approximations of the primitive functions of the language AL [Br 92, De 92]. Partitioning an object set of a sort into disjoint subsets, we introduce some classes, which correspond to error values. The notation of the polymorphic constant $\diamond$ is used for the union of all classes except for error classes. The approximation semantics of a program (an agent, a function) p defined by an equation system E (the body of p) is obtained from $[\![E]\!]$ by means of restriction to variables declared as formal parameters and result variables in the declaration of p.

5.1 The sort $K_{\mathbf{Bool}}$.

$K_{\mathbf{Bool}}$= { **T**, **F**, **E**, $\perp$}, where **T**={**true**}, **F**={**false**}, and **E** is the class of error values.

$$[\![if]\!](x,y,z) = \begin{cases} y, & \text{if } x = \mathbf{T}, \\ z, & \text{if } x = \mathbf{F}, \\ \mathbf{E}, & \text{if } x = \mathbf{E}, \\ \perp, & \text{if } x = \perp. \end{cases}$$

∧	T	F	E	⊥
T	T	F	E	⊥
F	F	F	E	⊥
E	E	E	E	⊥
⊥	⊥	⊥	⊥	⊥

∨	T	F	E	⊥
T	T	T	E	⊥
F	T	F	E	⊥
E	E	E	E	⊥
⊥	⊥	⊥	⊥	⊥

	¬
T	F
F	T
E	E
⊥	⊥

→	T	F	E	⊥
T	T	F	E	⊥
F	T	T	E	⊥
E	E	E	E	⊥
⊥	⊥	⊥	⊥	⊥

≡	T	F	E	⊥
T	T	F	E	⊥
F	F	T	E	⊥
E	E	E	E	⊥
⊥	⊥	⊥	⊥	⊥

5.2 The sort $K_{\mathbf{Nat}}$.

$K_{\mathbf{Nat}}$={ **0**, **1**, **>1**, **E** ,$\perp$}, where **0** = { 0 }, **1** = { 1 }, **>1**= $\{i : i > 1\}$, **E** is the class of error values, and **>0** = **>1**∪ **1**. For the sort $K_{\mathbf{Nat}}$, it would be useful to consider at least three different classes of error values: *division by zero*, *pred(0)*, *negative result of subtraction* . But, with a view to retain the visibility, we have not done this here.

	succ	*pred*
0	1	E
1	>1	0
>1	>1	>0
E	E	E
⊥	⊥	⊥

+	0	1	>1	E	⊥
0	0	1	>1	E	⊥
1	1	>1	>1	E	⊥
>1	>1	>1	>1	E	⊥
E	E	E	E	E	⊥
⊥	⊥	⊥	⊥	⊥	⊥

−	0	1	>1	E	⊥
0	0	E	E	E	⊥
1	1	0	E	E	⊥
>1	>1	>0	□	E	⊥
E	E	E	E	E	⊥
⊥	⊥	⊥	⊥	⊥	⊥

×	0	1	>1	E	⊥
0	0	0	0	0	⊥
1	0	1	>1	E	⊥
>1	0	>1	>1	E	⊥
E	0	E	E	E	⊥
⊥	⊥	⊥	⊥	⊥	⊥

÷	0	1	>1	E	⊥
0	E	0	0	E	⊥
1	E	1	0	E	⊥
>1	E	>1	◇	E	⊥
E	E	E	E	E	⊥
⊥	⊥	⊥	⊥	⊥	⊥

mod	0	1	>1	E	⊥
0	E	0	0	E	⊥
1	E	0	1	E	⊥
>1	E	0	◇	E	⊥
E	E	E	E	E	⊥
⊥	⊥	⊥	⊥	⊥	⊥

$=$	0	1	>1	E	$\perp$
0	T	F	F	E	$\perp$
1	F	T	F	E	$\perp$
>1	F	F	$\diamond$	E	$\perp$
E	E	E	E	E	$\perp$
$\perp$	$\perp$	$\perp$	$\perp$	$\perp$	$\perp$

$\leq$	0	1	>1	E	$\perp$
0	T	T	T	E	$\perp$
1	F	T	T	E	$\perp$
>1	F	F	$\diamond$	E	$\perp$
E	E	E	E	E	$\perp$
$\perp$	$\perp$	$\perp$	$\perp$	$\perp$	$\perp$

$<$	0	1	>1	E	$\perp$
0	F	T	T	E	$\perp$
1	F	F	T	E	$\perp$
>1	F	F	$\diamond$	E	$\perp$
E	E	E	E	E	$\perp$
$\perp$	$\perp$	$\perp$	$\perp$	$\perp$	$\perp$

Note for example that $pred(succ(x)) = x$, but $[\![pred(succ(x))]\!] \neq [\![x]\!]$.

5.3 The sort $K_{\mathbf{Stream(S)}}$.

$K_{\mathbf{Stream(S)}} = \{\ \varepsilon, \mathbf{E}, \perp\} \cup \{pr(k) : k \in K_{\mathbf{S}} \setminus \{\perp, \mathbf{E}\}\}$, where $\varepsilon = \{\varepsilon\}$ is the class of the empty stream, $pr(k)$ is a class of nonempty streams, beginning with a value from (nonempty) class k, **E** is the class of error values. Again, it would be useful to consider at least two different error classes **ft.**ε and **rt.**ε.

	ft	rt	isempty
ε	E	E	T
$pr(k)$	k	$\square$	F
E	E	E	E
$\perp$	$\perp$	$\perp$	$\perp$

$$x\&y = \begin{cases} pr(x), & \text{if } x \neq \mathbf{E} \text{ and } x \neq \perp, \\ \mathbf{E}, & \text{if } x = \mathbf{E}, \\ \perp, & \text{if } x = \perp. \end{cases}$$

For example, for streams of elements of sorts $K_{\mathbf{Bool}}$ and $K_{\mathbf{Nat}}$ the operation "&" is defined in the following way :

&	ε	pr(T)	pr(F)	E	$\perp$
T	pr(T)	pr(T)	pr(T)	pr(T)	pr(T)
F	pr(F)	pr(F)	pr(F)	pr(F)	pr(F)
E	E	E	E	E	E
$\perp$	$\perp$	$\perp$	$\perp$	$\perp$	$\perp$

&	ε	pr(0)	pr(1)	pr(>1)	E	$\perp$
0	pr(0)	pr(0)	pr(0)	pr(0)	pr(0)	pr(0)
1	pr(1)	pr(1)	pr(1)	pr(1)	pr(1)	pr(1)
>1	pr(>1)	pr(>1)	pr(>1)	pr(>1)	pr(>1)	pr(>1)
E	E	E	E	E	E	E
$\perp$	$\perp$	$\perp$	$\perp$	$\perp$	$\perp$	$\perp$

5.4 Example

Two agents A and B have the following definitions :
agent $A \equiv$ **chan Nat** $x \rightarrow$ **chan Nat** $y : y \equiv B(x)$ **end**,
agent $B \equiv$ **chan Nat** $x \rightarrow$ **chan Nat** $y : y \equiv A(x)$ **end**.
The calculation of the approximation semantics of A and B gives immediately $[\![A]\!] \equiv [\![B]\!] \equiv \perp$, that means A and B to be erroneous (a deadlock).

5.5 Example

For a definition of a function *fac*
funct *fac* $\equiv$ **Nat** $x \to$ **Nat :**
if $x = 0$ **then** 1 **else** $x \times fac(x-1)$ **fi**
end,
the calculation of the approximation semantics gives the following intermediate results $[\![fac]\!]^i$:

	$[\![fac]\!]_0$	$[\![fac]\!]_1$	$[\![fac]\!]_2$	$[\![fac]\!]_3$
0	$\perp$	1	1	1
1	$\perp$	$\perp$	1	1
>1	$\perp$	$\perp$	>1	>1
E	$\perp$	**E**	**E**	**E**
$\perp$	$\perp$	$\perp$	$\perp$	$\perp$

6 Essential terms, definitions and variables

The *essential occurrence set* of a program p is the minimal set of occurrences of terms, definitions and equations, closed under the following conditions:

- All program result variables are essential.
- Each equation occurrence containing an essential variable in the left-hand side is essential.
- The right-hand side of an essential equation is an essential term occurrence.
- If a call to a function or to an agent is essential, then the definition of this function resp. agent is essential.
- The body of an essential function definition is essential.
- Each result variable of an essential agent definition is essential.
- Each subterm occurrence in an essential occurrence of a term which is not a call to an agent or to a function, is an essential term occurrence.
- Each definition of an essential variable is essential.
- If an agent or a function has an essential formal parameter, then the corresponding actual parameter in each essential call to this agent or to this function is essential.
- All other occurrences of terms and definitions of the program p are inessential.

6.1 Example

program *Example* $\equiv$ **chan Nat** $i \to$ **chan Nat** j **:**
funct *fact* $\equiv$ **Nat** $m, n \to$ **Nat :**
if $m{=}0$ **then** 1 **else** $m \times fact(m-1, dummy(m,n)+n)$ **fi**
end,
funct *dummy* $\equiv$ **Nat** $m, n \to$ **Nat :**
$dummy(n, m)$
end,

agent A $\equiv$**chan Nat** $x \rightarrow$ **chan Nat** y :
$y \equiv \mathbf{rt}.x, u \equiv \mathbf{ft}.x\&v, v \equiv 0\&u$
end,
$j \equiv fact(\mathbf{ft}.i), \bot)\&A(i)$
end.
In this example, the definition of the function *dummy*, the second formal parameter of the function *fact* and the second actual parameter in all calls to *fact* are inessential. In the definition of the agent A only the first equation is essential.

References

[Br 86] Broy M. A Theory for Nondeterminism, Parallelism, Communication, and Concurrency. - Theoretical Computer Science 45, 1986, p. 1–61.

[Br 92] Broy M., Dederichs F., Dendorfer C., Fuchs M., Gritzner T.F., Weber R. The Design of Distributed Systems - An Introduction to FOCUS. Bericht TUM-I9202, Technische Universität München, Januar 1992, 65 S.

[Co 77] Cousot P. and Cousot R. Abstract interpretation: a unified lattice model for static analysis of programs by construction or approximation of fixpoints. - Conference Record of the 4-th Annual ACM SIGPLAN-SIGACT Simposium on Principles of Programming Languages, Los Angeles, California, 1977, p. 238–252.

[De 92] Dederichs F. Transformation verteilter Systeme : Von applikativen zu prozeduralen Darstellungen. PhD thesis, Institut für Informatik, Technische Universität München, Bericht TUM-I9227, August 1992, 147 S.

[Ki 73] Kildall G. A unified approach to to global program optimization.- Conference Record of the Annual ACM SIGPLAN-SIGACT Simposium on Principles of Programming Languages, Boston, Massachusetts, October 1973, p. 194–206.

Communication as Unification in Process Algebras : Operational Semantics

Philippe Jorrand
CNRS

Institut IMAG - LIFIA
46, avenue Félix Viallet - 38000 Grenoble - France
e-mail : jorrand@lifia.imag.fr

Abstract. Process algebras are formal systems aimed at the abstract description of computing devices organized as collections of components which can operate in parallel and cooperate by communicating values among them. In classical process algebras, communication is by rendez-vous, where symmetric proposals made by two processes meet synchronously : a local variable proposed by a receiving process is bound to a value proposed by a sending process. In this paper, this binary rendez-vous with only one variable on one side and one value on the other side is viewed as a mere special case of a more general situation for communication by synchronous rendez-vous. An arbitrary number of processes may offer terms to each others : if these terms have common instances, communication can indeed take place, and amounts to applying the unifying substitutions to the processes involved. The syntax and operational semantics of process algebras with this general view of communication are formally defined. The operational semantics show how this generalization leads to a clean formalization of the notion of global variables in process algebras. Applications are presented, which show that these algebras implement an original computing paradigm, where computation is achieved solely by means of communication.

1 INTRODUCTION

Process algebras are formal systems aimed at the abstract description of computing devices organized as collections of components (*processes*) which can operate in parallel and cooperate by communicating values among them. Given a set of primitive processes, more complex processes are built hierarchically by means of *operators* for combining processes.

This is not new : CSP [Hoare 78, Hoare 85], CCS [Milner 80, Milner 89], ACP [Bergstra & Klop 86], etc., are examples of process algebras. Some of these algebras have given rise to the design of practical languages for parallel programming : Occam [May 83] has its origins with CSP, and Lotos [Brinksma 88] with CCS.

CCS can be taken as a typical example of classical process algebras. CCS processes perform sequences of *actions*, where an action α is a *communication* of one value through one *port*. Ports are labelled according to a convention where $a, b, c, \ldots, l, \ldots$ are *input* port labels, whereas their complements $\sim a, \sim b, \sim c \ldots, \sim l, \ldots$ are *output* port labels. If $u, v, w, \ldots$ are values and x is a variable, $l(x)$ denotes any one of the input actions $l(u), l(v), l(w), \ldots$, i.e. the input of any value into the variable x through port l, whereas $\sim l(v)$ denotes the output action $\sim l(v)$, i.e. the output of v through port $\sim l$. In addition to these input and output, observable actions, there is an internal, *silent action* denoted by τ. The algebra of CCS has six operators for combining processes according to the following syntax, where p and q are processes, e denotes an action (i.e. e has one of the forms $l(x)$, $\sim l(v)$ or τ), L is a set of port labels and ϕ a *relabelling* function (i.e. a function from port labels to port labels which is consistent with complementary labels : $\phi(\sim l) = \sim\phi(l)$) :

$$p, q ::= Nil \mid e \bullet p \mid p+q \mid p|q \mid p\backslash L \mid p[\phi]$$

The interpretation of processes is given by the *operational semantics* of CCS. The basic idea is that, when a process p performs an action α, it becomes a new process q, in turn able to perform new actions. The possibility for p to perform such an execution step, called a *transition*, is denoted by $p \overset{\alpha}{\longrightarrow} q$ in the definition of the semantics. The operational semantics therefore associates *transition systems* with processes. Following the technique proposed by [Plotkin 81], this transition system is defined inductively, by means of inference rules where p, q, r and s are processes, α and β are actions, l is a port label, λ stands for l or $\sim l$ (with $\sim\sim\lambda = \lambda$), x is a variable, v a value, L a set of port labels and where the relabelling function ϕ is extended to actions (i.e. $\phi(\lambda(\ldots)) = \phi(\lambda)(\ldots)$ and $\phi(\tau) = \tau$) :

Inactive process : There is no α nor p such that $Nil \overset{\alpha}{\longrightarrow} p$

Prefixed processes :
$$\frac{}{l(x) \bullet p \overset{l(v)}{\longrightarrow} p_{[x \leftarrow v]}} \qquad \frac{}{\sim l(v) \bullet p \overset{\sim l(v)}{\longrightarrow} p} \qquad \frac{}{\tau \bullet p \overset{\tau}{\longrightarrow} p}$$

Non deterministic processes :
$$\frac{p \overset{\alpha}{\longrightarrow} r}{p+q \overset{\alpha}{\longrightarrow} r} \qquad \frac{q \overset{\beta}{\longrightarrow} s}{p+q \overset{\beta}{\longrightarrow} s}$$

Compound processes :
$$\frac{p \text{ —}\alpha\text{→ } r}{p|q \text{ —}\alpha\text{→ } r|q} \qquad \frac{q \text{ —}\beta\text{→ } s}{p|q \text{ —}\beta\text{→ } p|s} \qquad \frac{p \text{ —}\lambda(v)\text{→ } r \quad q \text{ —}{\sim}\lambda(v)\text{→ } s}{p|q \text{ —}\tau\text{→ } r|s}$$

Restricted processes :
$$\frac{p \text{ —}\alpha\text{→ } q}{p\backslash L \text{ —}\alpha\text{→ } q\backslash L}$$
(where $\alpha=\tau$, or $\alpha=\lambda(...)$ and $\lambda,{\sim}\lambda \notin L$)

Relabelled processes :
$$\frac{p \text{ —}\alpha\text{→ } q}{p[\phi] \text{ —}\phi(\alpha)\text{→ } q[\phi]}$$

Nil and prefixed processes are the primitive processes in CCS. A prefixed process of the form $l(x)\bullet p$ can input any value v in x, and v is substituted for x in the rest p of its behavior. Two rules define the transitions of non deterministic processes : if the first action of $p+q$ is an action of p (resp. of q), then $p+q$ follows the behavior of p (resp. of q) and forgets about q (resp. p). The transitions of a compound processes $p|q$ are defined by three rules. The first two rules tell that its components p and q can behave independently in an asynchronous fashion. The third rule tells that p and q can communicate synchronously whenever they are both ready to perform actions on ports with complementary labels : the composition of the input action $l(v)$ with the output action ${\sim}l(v)$ yields the internal silent action τ. Given the semantics of prefixed process with input actions denoted by $l(x)$, this does indeed implement communication. A restricted process $p\backslash L$ can perform all actions of p which do not use a label in L. Finally, a relabelled process $p[\phi]$ behaves like p where port labels are modified by the relabelling function ϕ.

In addition to the above operators of its process algebra, CCS also allows definitions of *process identifiers* : if *Id* is an identifier and p a process, $Id \equiv p$ defines the process identifier *Id*. This identifier may be used with the obvious interpretation :

Process identifiers :
$$\frac{p \text{ —}\alpha\text{→ } q}{Id \text{ —}\alpha\text{→ } q}$$
(where the definition $Id \equiv p$ has been made)

Communication between processes is a composition of actions that these processes are ready to perform. From the point of view of expressiveness, what actions are and how they can be composed is crucial in the design of a process algebra. In CCS-like algebras, an action transports one value, in one predefined direction (in or out), through one port. Exactly two actions on ports with complementary labels can be composed, yielding the silent action τ. Communication is thus restricted to its simplest and most naive form : exactly two partners, one sender and one receiver, and exactly one value flowing unidirectionally from the former to the latter.

Communication is synchronous. This idea, which was already present in the original "rendez-vous" communication of CSP [Hoare 78], is conceptually very useful since situations where a value has been sent but "not yet" received never have to be considered. This also implies that communication becomes the means of synchronization among processes. As a consequence, in CCS-like algebras, not more than two processes can act synchronously.

There exist process algebras which provide a more general view of synchronization, like SCCS [Milner 83] and Meije [Boudol 85]. In these algebras, actions are still *atomic*, which means that they still occur within one transition. But the set of actions form a group, with a commutative and associative product $\alpha\beta$ which is an action consisting in the simultaneous occurrences of the *particle* actions α and β. In these algebras, an action can thus have a complex particle structure, e.g. $\alpha\beta\gamma\delta$, while still remaining an atomic action. Whenever $\alpha=\lambda(v)$ and $\beta=\sim\lambda(v)$, their product $\alpha\beta$ is the unit τ of the group. Such compound actions are performed by a *synchronous product* of processes, built by a product operator $p\times q$ in the process algebra :

$$\textit{Synchronous processes :}\quad \frac{p \xrightarrow{\alpha\ldots\beta} r \qquad q \xrightarrow{\gamma\ldots\delta} s}{p\times q \xrightarrow{\alpha\ldots\beta\gamma\ldots\delta} r\times s}$$

where, in $\alpha\ldots\beta\gamma\ldots\delta$, each pair of complementary particles $\lambda(v)$ and $\sim\lambda(v)$ yields a τ. However, each pairwise communication is still directed : one value from one sender to one variable in one receiver.

This can be generalized : pairwise communications need not be restricted to unidirectional transmissions of single values and more than one process may act as receivers for values sent by more than one sender. This requires a generalized notion of communication. Section 2 defines such a notion of communication in a CCS-like algebra where more than one ground term may flow within a single communication between two processes. When the communicated terms are not restricted to be ground, this gives rise to a notion of global variables : this is defined in section 3 and the ground case is shown to be, as should be expected, a special interpretation of the non-ground case. Then, this generalized form of communication is introduced into an SCCS-like algebra in section 4, where several communications of the previous CCS-like algebras are joined into coarser grains of synchronous communications among at least two processes. Two applications are presented in section 5 : parallel functional machines and parallel logical machines.

2 COMMUNICATION AS UNIFICATION

A CCS process $P \equiv l(x) \bullet r$ offers on its port l to receive a value into its variable x. This value may come from a process $Q \equiv \sim l(v) \bullet s$, which offers on its port $\sim l$ to send a value v. In the compound process $P|Q$, they become able to share these offers : v can be communicated from Q to P by having P perform $l(v)$ and become $r_{[x \leftarrow v]}$, and Q perform $\sim l(v)$ and become s. In the compound process, this yields the silent action τ. The same notion of communication is used in SCCS, with the additional possiblity that more than one pair of complementary offers can be shared within a single action. But there is no reason to stay with pairs of mutual offers containing one variable on one side and one value on the other side. This can be easily generalized, and made more symmetric, by having offers from both sides contain *terms* of a free term algebra $T_{F,X}$ where F is a set of *function* names with an arity function $r : F \rightarrow N$, $F_n = \{f \in F \mid r(f) = n\}$, and X is a set of *variable* names :

(i) $\forall x \in X, x \in T_{F,X}$

(ii) $\forall c \in F_0, c \in T_{F,X}$

(iii) $\forall f \in F_n$, n≥1, and $\forall t_1, \ldots, t_n \in T_{F,X}, f(t_1, \ldots, t_n) \in T_{F,X}$

$T_F = T_{F,\emptyset}$ is the set of *ground terms* on F and F_0, the set of *constants*, is assumed to be non empty. Now, the sharing of two offers $l(t)$ and $\sim l(u)$, with $t,u \in T_{F,X}$, by processes $P \equiv l(t) \bullet r$ and $Q \equiv \sim l(u) \bullet s$ in the compound process $P|Q$ no longer results in the mere substitution of one value for one variable in the rest of the behavior of one partner of the communication. If t and u have at least one common instance $v = t_\sigma = u_\theta$, where σ and θ are substitutions, P can perform $l(v)$ and become r_σ, while Q can perform $\sim l(v)$ and becomes s_θ (with some precautions about variable names). In $P|Q$, actions $l(v)$ and $\sim l(v)$ can be composed and yield the silent action τ. If t and u have no common instance (i.e. v does not exist), the offers cannot be shared and there is no corresponding τ-transition in $P|Q$.

This generalization has a number of nice consequences. First, communication is bidirectional : several values (sub-terms) may be communicated at once and they flow in directions defined by σ and θ. This implies that input and output are no longer syntactically prescribed : the notion of complementary port labels is no longer needed. Second, CCS-like communications are a special case with offers of the form $l(x)$ and $l(v)$, where x is a variable and v a constant. Third, synchronization without actual communication of values is given for free by the composition of actions $l(t)$ and $l(t)$, where t is an arbitrary ground term. Last but not least, the unifying substitutions are not necessarily ground : this is a source of added expressiveness.

An algebra described in [Pletat 86] already makes an operational junction between algebraic data types and CCS. It generalizes the notion of communication by having it done through unification, but it restricts itself to communications of ground terms. The algebra FP2 [Schnoebelen & Jorrand 89] also contains this idea, with an SCCS-like approach to action composition, allowing n-party communication and synchronization. However, FP2 is also restricted to ground communications and has a static view of process composition, with no recursive process definitions, a rather severe limitation to expressiveness.

In the process algebra introduced next, CCS_U (CCS with unification) communication and synchronization still involve two processes, like in CCS, but they are achieved through unification of terms. In this section, the unifying substitutions are first restricted to be ground, leading to the *ground semantics* of CCS_U. Then, in section 3, this restriction is removed, giving rise to the *non-ground semantics* of CCS_U, where non-ground terms can be communicated.

2.1 Terms and substitutions

Elementary syntactic properties and notations for terms and substitutions are used in the definitions of semantics of process algebras and in the applications. The *positions* of a term t of $T_{F,X}$, written $\mathrm{pos}\ t$, are a subset of $\mathbb{N}_+^*$ inductively defined by :

(i) $\forall x \in X$, $\mathrm{pos}\ x = \{\varepsilon\}$
(ii) $\forall c \in F_0$, $\mathrm{pos}\ c = \{\varepsilon\}$
(iii) $\forall f \in F_n$, $n \geq 1$, $\mathrm{pos}\ f(t_1,\ldots,t_n) = \{\varepsilon\} \cup \{i.p \mid i=1,\ldots,n \wedge p \in \mathrm{pos}\ t_i\}$

Thus $\mathrm{pos}\ t$ is a non-empty prefix-closed set of strings of positive integers, where ε denotes the empty string and "." denotes concatenation. Positions are used to denote paths to subterms : if $p \in \mathrm{pos}\ t$, t/p is the subterm of t at position p, defined by $t/\varepsilon = t$ and $f(t_1,\ldots,t_n)/(i.p) = t_i/p$. Subterm replacement is also relative to a position, where, if t and u are terms, $t[p \leftarrow u]$, the replacement in t of t/p by u, is defined by :

(i) $t[\varepsilon \leftarrow u] = u$
(ii) $f(t_1,\ldots,t_i,\ldots,t_n)[i.p \leftarrow u] = f(t_1,\ldots,t_i[p \leftarrow u],\ldots,t_n)$

The set of variables occuring in a term t is $\mathrm{var}\ t$ and the *head* $\mathrm{hd}\ t$ of t is the element of $F \cup X$ at position ε of t :

(i) $\forall x \in X$, $\mathrm{hd}\ x = x$
(ii) $\forall c \in F_0$, $\mathrm{hd}\ c = c$
(iii) $\forall f \in F_n$, $n \geq 1$, $\mathrm{hd}\ f(t_1,\ldots,t_n) = f$

A *substitution* is a function $\sigma : X \to T_{F,X}$ which is the identity almost everywhere. For notational convenience, x_σ will be written instead of $\sigma(x)$. The *domain* of σ is the set of variables where σ is not the identity : $\mathrm{dom}\ \sigma = \{x \in X \mid x_\sigma \neq x\}$. The substitution σ/U, where U is a set of variables, is σ restricted to U : $x_{\sigma/U} = x_\sigma$ if $x \in U$, identity otherwise. Substitutions are extended in the obvious way to morphisms over terms :

(i) $\forall c \in F_0, c_\sigma = c$

(ii) $\forall f \in F_n, n \geq 1, f(t_1,\ldots,t_n)_\sigma = f(t_{1\sigma},\ldots,t_{n\sigma})$

If $\mathrm{dom}\ \sigma = \{x_1,\ldots,x_n\}$ and $x_{i\sigma} = t_i$, $i=1,\ldots,n$, σ can be denoted by $[x_1 \leftarrow t_1,\ldots,x_n \leftarrow t_n]$. The *range* of σ is $\mathrm{rng}\ \sigma = \bigcup_{i=1..n} \mathrm{var}\ t_i$. A substitution $\sigma : X \to T_F$ is a *ground substitution.* A substitution σ is *idempotent* iff $\mathrm{dom}\ \sigma \cap \mathrm{rng}\ \sigma = \emptyset$. Two idempotent substitutions σ and θ are *compatible* over a set V of variables iff $\forall x \in V \cap \mathrm{dom}\ \sigma \cap \mathrm{dom}\ \theta$, $x_\sigma = x_\theta$. The *sum* $\sigma+\theta$ of two idempotent substitutions compatible over V with $\mathrm{dom}\ \sigma \cup \mathrm{dom}\ \theta \subseteq V$ and such that $\mathrm{dom}\ \sigma \cap \mathrm{rng}\ \theta = \emptyset$ and $\mathrm{dom}\ \theta \cap \mathrm{rng}\ \sigma = \emptyset$ is defined by :

$$\forall x \in \mathrm{dom}\ \sigma,\ x_{(\sigma+\theta)} = x_\sigma \text{ and } \forall x \in \mathrm{dom}\ \theta,\ x_{(\sigma+\theta)} = x_\theta$$

Clearly, $\mathrm{dom}\ (\sigma+\theta) = \mathrm{dom}\ \sigma \cup \mathrm{dom}\ \theta$, and $\sigma+\theta = \theta+\sigma$ is idempotent. A *permutation* is a bijective substitution $\pi = [x_1 \leftarrow y_1,\ldots,x_n \leftarrow y_n]$ where the y_i's are variable names. In the definition of the non-ground semantics of CCS_U, permutations will used to implement consistent renamings of variables, similar to α-conversion in λ-calculus. The notation $\pi[U,V]$, where U and V are sets of variable names, stands for a family of idempotent permutations. In $t_{\pi[U,V]}$, a permutation π from this family is applied to t which renames only a subset of $\mathrm{var}\ t$ and which is such that :

(i) $\mathrm{dom}\ \pi = \mathrm{var}\ t \cap (U-V)$

(ii) $\mathrm{rng}\ \pi \cap (\mathrm{var}\ t \cup V) = \emptyset$

That is, $t_{\pi[U,V]}$ leaves unchanged the variable names in t which belong to V, while renaming all those which belong only to U into new variable names which belong neither to V nor to $\mathrm{var}\ t$.

2.2 Syntax of CCS_U

The operators of CCS_U have the same syntax as those of CCS. The only differences are : (i) $l(t)$, where l is a port label and t is a term of $T_{F,X}$, denotes actions of the form $l(t_\sigma)$, and (ii) a simplification due to the disappearance of complementary port labels. As in CCS, τ denotes the silent action.

Let p and q be processes, e denote an action, L be a set of port labels, ϕ be a function from port labels to port labels and Id be an identifier :

$$p, q \ ::= \ Nil \ \mid \ e\bullet p \ \mid \ p+q \ \mid \ p|q \ \mid \ p\backslash L \ \mid \ p[\phi] \ \mid \ Id$$

From now on, terms of process algebras will be called *P-terms*, for "process-terms", while terms in $T_{F,X}$ will simply be called *terms*. If p is a P-term, var p, p_σ and $p_{\pi[U,V]}$ are obvious extensions to P-terms of the same operations on terms.

2.3 Ground semantics of CCS_U

The ground operational semantics of CCS_U define an interpretation of P-terms. In the ground semantics, actions either are τ, or have the form $l(v)$, where l is a port label anf v a ground term. As in CCS, *Nil* is the inactive process and the inference rules for non deterministic processes, relabelled processes and process identifiers have the same form as in the semantics of CCS. Let p, q, r and s be P-terms, t be a term, σ be a ground substitution, α and β be actions :

Prefixed processes :

$$\frac{}{l(t)\bullet p \xrightarrow{l(t_\sigma)} p_\sigma} \ \text{(where dom } \sigma = \text{var } t) \qquad \frac{}{\tau\bullet p \xrightarrow{\tau} p}$$

(pfp1)

Compound processes :

$$\frac{p \xrightarrow{\alpha} r}{p|q \xrightarrow{\alpha} r|q} \qquad \frac{q \xrightarrow{\beta} s}{p|q \xrightarrow{\beta} p|s} \qquad \frac{p \xrightarrow{l(v)} r \quad q \xrightarrow{l(v)} s}{p|q \xrightarrow{\tau} r|s}$$

(cmp3)

Restricted processes :

$$\frac{p \xrightarrow{\alpha} q}{p\backslash L \xrightarrow{\alpha} q\backslash L} \quad \text{(where } \alpha=\tau, \text{ or } \alpha=l(...) \text{ and } l \notin L)$$

The condition dom σ=var t on the rule (pfp1) for prefixed processes guarantees at the same time that t_σ is indeed a ground term and that no variables in var p other than those of var t are modified by σ.

Together with the rule (cmp3) for compound processes, the rule (pfp1) for prefixed processes provides some insight into a notion of *scope* for variables in CCS_U. For example, according to (pfp1), $k(f(a,x))\bullet l(x)\bullet Nil \xrightarrow{k(f(a,b))} l(b)\bullet Nil$ holds, with the substitution $[x\leftarrow b]$. Similarly for $k(f(x,b))\bullet m(x)\bullet Nil \xrightarrow{k(f(a,b))} m(a)\bullet Nil$, with $[x\leftarrow a]$. Then, $(k(f(a,x))\bullet l(x)\bullet Nil) \mid (k(f(x,b))\bullet m(x)\bullet Nil) \xrightarrow{\tau} (l(b)\bullet Nil) \mid (m(a)\bullet Nil)$ also holds according to (cmp3) (this is also an example of bidirectional communication).

Thus, the variable x in the left operand of "|" is not the same as the variable x in the right operand. This is reflected in the following definitions :

(i) Within $e \bullet p$, an occurrence of x in e is a *defining occurrence* of x with scope p,
(ii) If $x \in \mathrm{var}\, e$, an occurrence of x in p is a *bound occurrence* of x for $e \bullet p$,
(iii) If $x \notin \mathrm{var}\, e$, an occurrence of x in p is a *free occurrence* of x for $e \bullet p$.

However, in $Id \equiv n(x) \bullet ((k(f(a,x)) \bullet l(x) \bullet Nil) \mid (k(f(x,b)) \bullet m(x) \bullet Nil))$, the occurrence of x in $n(x)$ is a defining occurrence of x with a scope syntactically identical to the above compound process. All occurrences of x in this compound process become bound occurrences of x for Id. Now, x is the same variable in both operands of the compound process. As a consequence, the situation is not the same as above : $f(a,x)$ and $f(x,b)$ should no longer be considered as having common instances. This is reflected in the semantics : $Id \xrightarrow{n(b)} (k(f(a,b)) \bullet l(b) \bullet Nil) \mid (k(f(b,b)) \bullet m(b) \bullet Nil)$ holds according to (pfp1), yielding a compound process with a different semantics, since (cmp3) may no longer be applied.

3 NON-GROUND COMMUNICATION

The *non-ground semantics* of CCS_U allows communications of non-ground terms. This implies that the condition on rule (pfp1) for prefixed processes in the ground semantics, telling that σ be a ground substitution, be removed. However, simply removing this condition without further precautions leads to some pitfalls, with respect to variable names, which can be shown by means of simple examples :

(i) With the possibility of non-ground communications, the process $k(x) \bullet l(f(x,y)) \bullet p$, where p is an arbitrary process, can perform the communication $k(y)$, with the non-ground substitution $\sigma = [x \leftarrow y]$, and become $(l(f(x,y)) \bullet p)_\sigma$. This is a variable name conflict : since y in $k(y)$ corresponds in fact to a variable communicated by some other process, y in $\mathrm{rng}\, \sigma$ is not the same as y in $l(f(x,y)) \bullet p$. It is necessary to rename y in $l(f(x,y)) \bullet p$ before applying σ to this P-term.

(ii) In the process $(l(x) \bullet p) \mid (m(x) \bullet q)$, where p and q are arbitrary processes, the two occurrences of x are independent defining occurrences : they can be bound to different terms by means two different substitutions due to two independent communications. When this process is embedded in a context like $k(x) \bullet ((l(x) \bullet p) \mid (m(x) \bullet q))$, the same two occurrences of x become bound occurrences : whatever substitution σ is applied to x in $k(x)$ is also applied to these two occurrences. In other words, when $(l(x)\bullet p)\,|\,(m(x)\bullet q)$ is placed in this context, both occurrences of x refer to the same variable shared by $l(x)\bullet p$ and $m(x)\bullet q$. As long as σ is ground, this is not an issue, as could be seen in the ground semantics. But if σ is not ground, e.g. $\sigma=[x\leftarrow y]$, occurrences of y in the resulting process $((l(x)\bullet p)\,|\,(m(x)\bullet q))_\sigma = (l(y)\bullet p_\sigma)\,|\,(m(y)\bullet q_\sigma)$ can no longer be viewed as independent defining occurrences : in spite of the fact that $(l(y)\bullet p_\sigma)\,|\,(m(y)\bullet q_\sigma)$ is no longer embedded in a context providing a defining occurrence of y, both occurrences refer to the same variable y shared by $l(y)\bullet p_\sigma$ and $m(y)\bullet q_\sigma$. As a consequence, for example, if $l(y)\bullet p_\sigma$ performs $l(y)_\theta$ independently of $m(y)\bullet q_\sigma$ and becomes $p_{\sigma\circ\theta}$, the substitution θ must be propagated and applied also to the occurrences of y in $m(y)\bullet q_\sigma$: y is a *global variable* for $l(y)\bullet p_\sigma$ and $m(y)\bullet q_\sigma$. To keep track of this new situation, it is necessary to maintain a record of global variables created by non-ground communications.

3.1 Non-ground transitions

In the ground semantics, transitions had the form $p\text{—}\alpha\rightarrow q$, where p and q are processes and α is an action. Here, two addional kinds of information are required within transitions in order to handle global variables. First, it is necessary to keep track of the creation and deletion of global variables : this will be achieved by explicitly telling, for each transition, what is the set of currently relevant global variables and how this transition modifies that set. Second, when a component of a compound process performs a transition which applies a substitution to global variables, it is necessary to propagate this substitution to other components of the same compound process : this will be achieved by recording in each transition the substitution that it applies to global variables. This leads to a new form of transitions for the non-ground semantics :

$$X : p \text{—}\alpha,\sigma\rightarrow Y : q$$

This should be read : "In the context of the set X of global variables, the process p can perform the action α, while applying the substitution σ to global variables in X, and becomes the process q in the context of the set Y of global variables".

3.2 Non-ground semantics of CCS_U

The syntax of CCS_U is unchanged. Let p, q, r and s be P-terms, l be a port label, t a term, σ and θ idempotent substitutions, α and β actions, L a set of port labels, ϕ a function from port labels to port labels, *Id* an identifier and X, Y, U and V sets of variable names.

Inference rules for the non-ground semantics :

Prefixed processes :

(pfp1) $$\frac{}{X:\ l(t)\bullet p\ \ \text{—}l(t_\sigma),\sigma/X\!\rightarrow\ (X-\mathrm{dom}\ \sigma)\cup \mathrm{rng}\ \sigma:\ p'_\sigma}$$ (where $\mathrm{dom}\ \sigma = \mathrm{var}\ t$, $\mathrm{rng}\ \sigma \cap X = \varnothing$, and $p' = p_{\pi[\mathrm{rng}\ \sigma,\ X]}$)

(pfp2) $$\frac{}{X:\tau\bullet p\ \text{—}\tau,[]\!\rightarrow\ X:p}$$

Non deterministic processes :

(ndp1) $$\frac{X:p\text{—}\alpha,\sigma\!\rightarrow Y:r}{X:p{+}q\ \text{—}\alpha,\sigma\!\rightarrow\ Y:r}$$

(ndp2) $$\frac{U:q\text{—}\beta,\theta\!\rightarrow V:s}{U:p{+}q\ \text{—}\beta,\theta\!\rightarrow\ V:s}$$

Compound processes :

(cmp1) $$\frac{X:p\text{—}\alpha,\sigma\!\rightarrow Y:r}{X:p\,|\,q\ \text{—}\alpha,\sigma\!\rightarrow\ Y:r\,|\,q'_\sigma}$$ (where $q' = q_{\pi[Y,\ X]}$)

(cmp2) $$\frac{U:q\text{—}\beta,\theta\!\rightarrow V:s}{U:p\,|\,q\ \text{—}\beta,\theta\!\rightarrow\ V:p'_\theta\,|\,s}$$ (where $p' = p_{\pi[V,\ U]}$)

(cmp3) $$\frac{X:p\text{—}l(t),\sigma\!\rightarrow Y:r,\qquad X:q\text{—}\bar{l}(t),\theta\!\rightarrow V:s}{X:\ p\,|\,q\ \ \text{—}\tau,\sigma{+}\theta\!\rightarrow\ (Y-\mathrm{dom}\ \theta)\cup(V-\ \mathrm{dom}\ \sigma):\ r_\theta\,|\,s_\sigma}$$

Restricted processes :

(rsp) $$\frac{X:p\text{—}\alpha,\sigma\!\rightarrow Y:r}{X:p\backslash L\text{—}\alpha,\sigma\!\rightarrow Y:r\backslash L}$$ (where $\alpha=\tau$, or $\alpha=l(\ldots)$ and $l\notin L$)

Relabelled processes :

(rlp) $$\frac{X:p\text{—}\alpha,\sigma\!\rightarrow Y:r}{X:p[\phi]\text{—}\phi(\alpha),\sigma\!\rightarrow Y:r[\phi]}$$

Process identifiers :

(pid) $$\frac{\varnothing:p\text{—}\alpha,[]\!\rightarrow Y:r}{X:Id\text{—}\alpha,[]\!\rightarrow Y:r}$$ (where the definition $Id \equiv p$ has been made)

These rules deserve some comments, which are made below by distinguishing two sorts of related issues : treatment of global variables and avoiding conflicts of variable names.

3.3 Treatment of global variables

Rule (pfp1). This is the base case. The substitution σ is non longer required to be ground, but its domain is still required to be var t, as in the ground semantics. This means that all variables in t are either replaced by a more instanciated term or renamed : the technical reason for this will become clear in rule (cmp3). In order to reflect correctly the notion of communication (i.e. binding of the variables in t to information provided by some other process), this substitution must be idempotent : no variable from var t may appear in rng σ. Similarly, no variable from X, the set of global variables in the context of which the transition is performed, may appear in rng σ. Some of the variables in dom σ belong to X : σ/X appearing in the transition will be used to propagate σ restricted to X to processes sharing these global variables with $l(t)\bullet p$ (see rules (cmp1), (cmp2) and (cmp3)). The resulting process after the transition is p_σ, modulo some renamings of variables to avoid conflicting variable names (see 3.4 below). For this new process, the variables which are in dom σ are no longer global variables, since σ is idempotent, but those in rng σ must be considered as new global variables, because the new process is no longer placed in the context of a P-term providing defining occurrences for them.

Rule (pfp2). No substitution is applied when a P-term of the form $\tau\bullet p$ performs its silent transition : no substitution (the empty substitution) has to be propagated to processes sharing global variables with $\tau\bullet p$ and the set of global variables is not modified.

Rules (ndp1) *and* (ndp2). Trivial.

Rule (cmp1). When the component process p of a compound process $p \mid q$ performs a transition independently of the other component q, some of the global variables in X shared by p and q may be in the domain of the substitution applied within this transition : this is represented by σ in the transition performed by p. In order to reflect these substitutions for global variables consistently in r and q, σ must be propagated to q and applied in q_σ, modulo some renamings of variables. Symmetric comment for rule (cmp2).

Rule (cmp3). The component processes p and q share global variables in X. There are two problems : define the substitution attached to the transition from $p \mid q$ and build the set of new global variables after this transition.

The transitions performed by p and q, respectively, as given in the premises of this rule, are necessarily of the following forms, respectively :

$$X : p \ \dashv(u_{\sigma'}),\sigma'/X\rightarrow \ (X - \text{dom}\,\sigma') \cup \text{rng}\,\sigma' : r$$

$$\text{and} \quad X : q \ \dashv(v_{\theta'}),\theta'/X\rightarrow \ (X - \text{dom}\,\theta') \cup \text{rng}\,\theta' : s$$

where u and v are terms, with $u_{\sigma'}=v_{\theta'}=t$, $\sigma'/X=\sigma$ and $\theta'/X=\theta$, $(X - \text{dom}\,\sigma') \cup \text{rng}\,\sigma' = Y$ and $(X - \text{dom}\,\theta') \cup \text{rng}\,\theta' = V$. The substitution for global variables attached to the τ-transition performed by $p \mid q$ is clearly $\sigma+\theta=\sigma'/X+\theta'/X$. For such a sum to be defined, σ' and θ' must compatible over X, which is a most desired property : σ' and θ' must indeed apply consistent bindings to global variables. But $\sigma+\theta$ should also be idempotent : this comes as a consequence of the condition on rule (pfp1) saying that $\text{dom}\,\sigma'=\text{var}\,u$ and $\text{dom}\,\theta'=\text{var}\,v$, thus $\text{rng}\,\sigma'=\text{var}\,u_{\sigma'}=\text{var}\,v_{\theta'}=\text{rng}\,\theta'$ and, since σ' and θ' are idempotent, $\text{dom}\,\sigma' \cap \text{rng}\,\theta'=\emptyset$ and $\text{dom}\,\theta' \cap \text{rng}\,\sigma'=\emptyset$. Since $\sigma=\sigma'/X$ and $\theta=\theta'/X$, σ and θ are idempotent and their domains and ranges are contained in the respective domains and ranges of σ' and θ'. Hence, $\text{dom}\,\sigma \cap \text{rng}\,\theta=\emptyset$, $\text{dom}\,\theta \cap \text{rng}\,\sigma=\emptyset$, and $\sigma+\theta$ is idempotent.

Similarly to rules (cmp1) and (cmp2), var p may contain global variables in $X \cap \text{dom}\,\theta$ and var q global variables in $X \cap \text{dom}\,\sigma$. The substitutions applied to these variables by θ and by σ respectively must be propagated to r and to s respectively.

Since var u and var v may contain different subsets of X, the two premises may delete different sets of global variables from X. Because of the propagations of θ to r and of σ to s, the variables in dom θ are longer global for r_θ and those in dom σ are no longer global for s_σ. Thus, the global variables which remain relevant for $r_\theta \mid s_\sigma$ are those of the set $(Y - \text{dom}\,\theta) \cup (V - \text{dom}\,\sigma)$.

Rules (rsp) *and* (rlp). Trivial.

Rule (pid). In the definition $Id \equiv p$ of a process identifier *Id*, var p is a set of variables local to this definition. Thus, when an arbitrary sequence of transitions has eventually placed *Id* in the context of a set X of global variables, the variables in X are not relevant for *Id*, even if some of them happen to have the same names as variables of var p : transitions from *Id* must be considered as transitions of p started in an empty context of global variables, leading to a process r placed in a new set Y of "fresh" global variables.

3.4 Avoiding conflicts of variable names

Rule (pfp1). As in the ground case, the substitution σ applied to t must also be applied to p. Variable names in rng σ are chosen so that σ is idempotent : they are all distinct from variables in var t, but some of them may happen to be identical to variable names in var p, although they bear no relation with them. Thus, there may be conflicts of variable names. These conflicts are avoided by renaming in p all variables which belong to rng σ (none of them belong to X) into new variable names. These new names must of course not belong already to var p, since this would create further conflicts, but they must also not belong X, since they would unduly be considered as global variables : this is guaranteed by $p_{\pi[\text{rng}\,\sigma,X]}$.

Rule (cmp1). The transition performed by p independently of q may create new global variables. All these variables are relevant for r, but only some of them for q : after the compound process $p\,|\,q$ has performed the transition due to p, the set Y may contain new global variables which are relevant for r but not for q. Thus, the variables in q which happen to belong to Y must be renamed, except those which already belonged to X, because they are still relevant global variables for q and dom σ is among them. Since Y necessarily contains rng σ, no new conflict can be created by applying σ to q after this renaming. Symmetric comment for rule (cmp2).

Rule (cmp3). As opposed to rules (cmp1) and (cmp2) no renaming is necessary here before applying θ and σ respectively to r and s. Refering to the previous comment on rule (cmp3), the condition dom σ=var t on rule (pfp1) implies that the new global variables created by the two premises are the same (rng σ' = rng θ'). As a consequence, all required renamings have already been performed within r and s, so that no conflicts may be created by applying θ and σ.

3.5 Interpretations of the non-ground semantics

When starting the interpretation of a P-term in the non-ground semantics, the initial set of global variables is $\varnothing$. When a prefixed process performs a transition, the set of global variables is enriched by rng σ, where σ is the substitution applied in the transition. If σ is always ground, sets of global variables will always be empty, transitions will always be $\varnothing : p \xrightarrow{\alpha,[]} \varnothing : q$ and no renaming will occur. Thus, if σ is always ground, the applied inference rules of the non-ground semantics are isomorphic to those of the ground semantics. Therefore, the ground semantics is a special case of the non-ground semantics : it is the *ground interpretation* of the non-ground semantics, which is obtained from the non-ground semantics by putting a restriction on rule (pfp1), saying that σ must be ground.

Other interpretations are possible, depending on conditions restricting the choice of substitutions. One of them is the *mgu interpretation* (most general unifier interpretation), which puts its restriction on rule (cmp3). There, since the two premises $X : p \xrightarrow{l(t),\sigma} Y : r$ and $X : q \xrightarrow{l(t),\theta} V : s$ of rule (cmp3) are in fact of the respective forms :

$$X : p \xrightarrow{l(u_{\sigma'}),\sigma} Y : r$$

$$\text{and} \quad X : q \xrightarrow{l(v_{\theta'}),\theta} V : s$$

where u and v are terms, with $u_{\sigma'}=v_{\theta'}=t$, the mgu interpretation restricts t to be the *most general instance* of u and v.

4 MULTI-DIRECTIONAL COMMUNICATION

In the compound processes of CCS_U, communications are between two processes. Since communication is by unification, this allows bi-directional exchanges of terms within a single communication. This can be generalized and lead to multi-directional communication, using SCCS-like compositions of actions and processes, in a process algebra $SCCS_U$.

4.1 Composition of actions in $SCCS_U$

In $SCCS_U$, action α in transitions of the form $p \xrightarrow{\alpha} q$ (or $X : p \xrightarrow{\alpha,\sigma} Y : q$, in the non-ground case) is now made of *particles* which occur synchronously, each particle having the form of a CCS_U action :

$$\alpha = \alpha_1 \ldots \alpha_m, \quad \alpha_i = l_i(t_i), \; i=1\ldots m$$

where l_i is a port label and t_i a term. Within an action, no two particles may have the same port label. Therefore, an action in $SCCS_U$ is a function from port labels to terms : dom α is the set of port labels $\{l_1,\ldots,l_m\}$ and $\alpha(l_i)=t_i$, $i=1\ldots m$.

Two actions $\alpha=\alpha_1\ldots\alpha_m$ and $\beta=\beta_1\ldots\beta_n$ can be composed in the context of a set V of variables. Their composition $V{:}\alpha{\times}\beta$ is defined iff : (i) if they contain particles with the same port labels, the same term is offered by each pair of such particles, and (ii) they share no other variables than those of V or of their common particles :

(i) $\forall l \in \text{dom}\,\alpha \cap \text{dom}\,\beta, \; \alpha(l) = \beta(l)$

(ii) $\text{var}\,\alpha \cap \text{var}\,\beta \subseteq V \cup \bigcup_{l \in \text{dom}\,\alpha \cap \text{dom}\,\beta} \text{var}\,\alpha(l)$

The resulting action collapses each pair of common particles into τ, which can be viewed now as an action with empty domain, while joining the remaining particles of α and β into a new action. That is, $\gamma = V{:}\alpha{\times}\beta$ is such that :

$$\text{dom } \gamma = \text{dom } \alpha \cup \text{dom } \beta - \text{dom } \alpha \cap \text{dom } \beta$$
$$\forall l \in \text{dom } \gamma \cap \text{dom } \alpha,\ \gamma(l) = \alpha(l)$$
$$\forall l \in \text{dom } \gamma \cap \text{dom } \beta,\ \gamma(l) = \beta(l)$$

4.2 Syntax of $SCCS_U$

Let p and q be P-terms of $SCCS_U$, e denote and action (i.e. e is of the form $l_1(t_1)...l_n(t_n)$, or e is τ), L be a set of port labels, ϕ be a function from port labels to port labels and Id be an identifier. P-terms of $SCCS_U$ are defined by :

$$p, q \ ::= \ Nil \ \mid \ e{\bullet}p \ \mid \ p{+}q \ \mid \ p{\times}q \ \mid \ p\backslash L \ \mid \ p[\phi] \ \mid \ Id$$

4.3 Semantics of $SCCS_U$

The non-ground semantics is given directly. The inference rules (ndp1), (ndp2), (rsp), (rlp) and (pid) for $SCCS_U$ have exactly the same form as for CCS_U. The other inference rules for $SCCS_U$ are :

Prefixed processes :

(pfp) $$\frac{}{X : e{\bullet}p \ \text{—}e_\sigma,\sigma/X{\rightarrow} \ (X - \text{dom } \sigma) \cup \text{rng } \sigma : p'_\sigma} \quad (\text{where dom } \sigma = \text{var } e,\ \text{rng } \sigma \cap X = \varnothing, \text{ and } p' = p_{\pi[\text{rng } \sigma,\, X]})$$

Compound processes :

(cmp1) $$\frac{X : p \text{—}\alpha,\sigma{\rightarrow}\ Y : r}{X : p{\times}q \ \text{—}\alpha,\sigma{\rightarrow}\ Y : r{\times}q'_\sigma} \quad (\text{where } q' = q_{\pi[Y,\, X]})$$

(cmp2) $$\frac{U : q \text{—}\beta,\theta{\rightarrow}\ V : s}{U : p{\times}q \ \text{—}\beta,\theta{\rightarrow}\ V : p'_\theta{\times}s} \quad (\text{where } p' = p_{\pi[V,\, U]})$$

(cmp3) $$\frac{X : p \text{—}\alpha,\sigma{\rightarrow}\ Y : r, \qquad X : q \text{—}\beta,\theta{\rightarrow}\ V : s}{X : p{\times}q \ \text{—rng}(\sigma{+}\theta){:}\alpha{\times}\beta,\sigma{+}\theta{\rightarrow} \ (Y - \text{dom } \theta) \cup (V - \text{dom } \sigma) : r'_\theta{\times}s'_\sigma} \quad (\text{where } r' = r_{\pi[V,\, X]}, \text{ and } s' = s_{\pi[Y,\, X]})$$

Since τ is now an action with empty domain, there is no special inference rule for $\tau{\bullet}p$. Except for rule (cmp3), where the condition $\text{rng } \sigma' = \text{rng } \theta'$ is no longer satisfied, global variables and renamings are handled the same way as in CCS_U. Similarly, this semantics also has a ground interpretation and a mgu interpretation. CCS_U is a special case of $SCCS_U$, where actions are restricted to contain at most one particle and where "|" is "×" restricted to processes using the same port label in rule (cmp3).

4.4 Multi-directional communication in $SCCS_U$

An example is used to show that more than two processes may participate synchronously in one generalized communication in $SCCS_U$, and that terms flow multi-directionally among the processes involved. The following definitions are given, where *j, k, l, m, n* are port labels, *f, g, h* are function names, *x, y, z, u, v, w, t* are variables, *a, b, c* are constants, *p', q', r'* and *s'* are arbitrary P-terms :

$$
\begin{aligned}
p &\equiv j(f(x,y))\ k(f(x,b))\ l(g(x,y)) \bullet p' \\
q &\equiv l(g(z,u))\ m(u) \bullet q' \\
r &\equiv k(f(v,w))\ m(c)\ n(h(v,w)) \bullet r' \\
s &\equiv n(h(a,t)) \bullet s'
\end{aligned}
$$

Using the ground interpretation of the semantics of $SCCS_U$ to keep it simple, the inference rules yield the following derivation for the P-term $(p{\times}q{\times}r{\times}s)\backslash\{k,l,m,n\}$:

$$\text{(pfp/pid)}\ \frac{}{p \xrightarrow{j(f(a,c))\ k(f(a,b))\ l(g(a,c))} p'_{[x\leftarrow a,y\leftarrow c]}} \qquad \text{(pfp/pid)}\ \frac{}{q \xrightarrow{l(g(a,c))\ m(c)} q'_{[z\leftarrow a,u\leftarrow c]}}$$

$$\text{(cmp3)}\ \frac{}{p{\times}q \xrightarrow{j(f(a,c))\ k(f(a,b))\ m(c)} p'_{[x\leftarrow a,y\leftarrow c]} \times q'_{[z\leftarrow a,u\leftarrow c]}}$$

$$\text{(pfp/pid)}\ \frac{}{r \xrightarrow{k(f(a,b))\ m(c)\ n(h(a,b))} r'_{[v\leftarrow a,w\leftarrow b]}}$$

$$\text{(cmp3)}\ \frac{}{p{\times}q{\times}r \xrightarrow{j(f(a,c))\ n(h(a,b))} p'_{[x\leftarrow a,y\leftarrow c]} \times q'_{[z\leftarrow a,u\leftarrow c]} \times r'_{[v\leftarrow a,w\leftarrow b]}}$$

$$\text{(pfp/pid)}\ \frac{}{s \xrightarrow{n(h(a,b))} s'_{[t\leftarrow b]}}$$

$$\text{(cmp3)}\ \frac{}{p{\times}q{\times}r{\times}s \xrightarrow{j(f(a,c))} p'_{[x\leftarrow a,y\leftarrow c]} \times q'_{[z\leftarrow a,u\leftarrow c]} \times r'_{[v\leftarrow a,w\leftarrow b]} \times s'_{[t\leftarrow b]}}$$

$$\text{(rsp)}\ \frac{}{(p{\times}q{\times}r{\times}s)\backslash\{k,l,m,n\} \xrightarrow{j(f(a,c))} (p'_{[x\leftarrow a,y\leftarrow c]} \times q'_{[z\leftarrow a,u\leftarrow c]} \times r'_{[v\leftarrow a,w\leftarrow b]} \times s'_{[t\leftarrow b]})\backslash\{k,l,m,n\}}$$

Considering now, from a more concrete point of view, where the constants come from and go to, what happens synchronously among the four processes is :

- *a* comes from *s* and, through *r* and *p*, goes to *q* ;
- *b* come from *p* and, through *r*, goes to *s* ;
- *c* comes from *r* and, through *q*, goes to *p*, which "outputs" *f(a,c)* through port *j*.

5 APPLICATIONS

Two applications are shown. The first one, which deals with abstract parallel functional machines for interpreting functions defined by rewrite rules, relies on the ground interpretation of CCS_U. The second one, which describes abstract parallel logical machines for interpreting Horn clauses with full "AND" parallelism, relies on the mgu interpretation of $SCCS_U$. The functional machine is built by translating rewrite rules into CCS_U and the logical machine by translating Horn clauses into $SCCS_U$.

5.1 Parallel functional machines

Ground CCS_U is applied to the description of abstract parallel functional machines for a class of functions defined by rewrite rules, namely functions defined by means of *constructor-based equational presentations*. These machines rely entirely on communications to "mimic" computations by rewriting. Furthermore, they model the natural parallelism which is inherent in functional programming : arguments to functions are evaluated independently.

5.1.1 Constructor-based presentations

Given a set of function names F, an *equation* over F is a couple (u,v), written $u=v$, of two terms of $T_{F,X}$, and an *equational presentation* is a set E of equations over F. Equations may be oriented as *rewrite rules*, written $u \rightarrow v$, where $\mathrm{var}\ v \subseteq \mathrm{var}\ u$. A *rewrite system* is a set R of rewrite rules over F. Classical properties of rewrite systems can be found in [Huet & Oppen 80].

Following [Guttag & Horning 78, Huet & Hulot 82], a constructor-based presentation is an equational presentation E where F is partitioned into $C+D$, where the elements of C are called *constructors* and those of D, *defined functions*. In a constructor-based presentation, the equations of E used as rewrite rules yield a rewrite system R such that :

(1) R is terminating,
(2) R is left linear,
(3) Left-hand sides have the form $f(t_1,\ldots,t_n)$ where $f \in D$ and $t_i \in T_{C,X}$, for $i=1,\ldots,n$,
(4) No two left-hand sides are unifiable,
(5) Every term of $T_F - T_C$ is reducible by R.

A consequence of this definition is that R is convergent (i.e. R is terminating and every term has a unique normal form). Another consequence is that no term of T_C is reducible, which implies that T_C is the set of normal forms. This property is known as "sufficient completeness" of the presentation. Asserting that a presentation is a constructor-based presentation is in general undecidable because of the termination condition, though there exist standard tools [Dershowitz 87] that may be used. In practice, however, the termination requirement is often dropped : some computations may simply fail to terminate, but confluence is not affected, so that the behavior of the system remains consistent.

Constructor-based presentations constitute a rather natural technique for defining and applying functions. Evaluation amounts to reduce a term to its normal form, that is to a term of T_C : in that sense, every term containing at least one occurrence of an element of D is an *expression* to be evaluated, while every term of T_C is a *value*. The "constructor discipline" enjoys many interesting properties, making it well suited for programming languages [O'Donnel 86]. It is usually not felt as too strong a restriction, and there exist automatic methods to translate more general presentations into "equivalent" ones respecting this discipline [Thatte 85].

5.1.2 *Example of constructor-based presentation*

Let $F=C+D$, with $C=\{0,s\}$ and $D=\{add,fib\}$: T_C is the set of natural numbers 0, $s(0)$, $s(s(0))$, ..., while $add(i,j)$ is intended to compute the sum of i and j and $fib(i)$ the ith Fibonacci number. It is well known that these functions can be defined by a constructor-based presentation :

$$add(0,j) \rightarrow j$$
$$add(s(i),j) \rightarrow s(add(i,j))$$
$$fib(0) \rightarrow s(0)$$
$$fib(s(0)) \rightarrow s(0)$$
$$fib(s(s(i))) \rightarrow add(fib(s(i)),fib(i))$$

5.1.3 *From presentations to processes*

Translating a constructor-based presentation R into a CCS_U process is a syntactic manipulation, where the following notations are used in translation rules :

- "$w \twoheadrightarrow p$" stands for "w can be translated into p", where w is a syntactic construction from a presentation and p a syntactic construction in CCS_U ;
- id f is a unique process identifier associated with every symbol $f \in D$;

- def f is the pair (f,R) for every function f defined by R ;
- rules $f = \{u \to v \in R \mid \text{hd } u = f\}$;
- For every $f \in F$, $f(t_I)$ stands for $f(t_1, ..., t_n)$ if $r(f)=n$, for f if $r(f)=0$;
- $\forall t \in T_{F,X}$, fct $t = \{j \in \text{pos } t - \{\varepsilon\} \mid \text{hd } (t/j) \in D\}$;
- Let $J=\{j,...\}$ be a set of positions :
  - $\forall j \in J$, p, p_j are processes, k, l, l_j are distinct port labels, x_j are distinct variables and ϕ_j are relabelling functions such that $\phi_j(l)=l_j$;
  - $\forall t \in T_{F,X}$, "$t/J \twoheadrightarrow p_J$" stands for "$t/j \twoheadrightarrow p_j$,...", $j \in J$, and "$t[J \leftarrow x_J]$" stands for "$t[j \leftarrow x_j]$...", $j \in J$;
  - "$l_J(x_J) \circ p$" stands for the prefixed process $l_j(x_j) \bullet ... \bullet p$", $j \in J$;
  - "$p_J[\phi_J] \mid p$" stands for the compound process "$p_j[\phi_j] \mid ... \mid p$", $j \in J$;
  - "$p \backslash \{l_J\}$" stands for the restricted process "$p \backslash \{l_j, ...\}$", $j \in J$.

Translation rules :

$$\frac{\text{rules } f \twoheadrightarrow p}{\text{def } f \twoheadrightarrow \text{id } f \equiv p} \qquad \frac{u_1 \to v_1 \twoheadrightarrow p_1, ..., u_m \to v_m \twoheadrightarrow p_m}{\{u_1 \to v_1, ..., u_m \to v_m\} \twoheadrightarrow p_1 + ... + p_m} \qquad \frac{\to v \twoheadrightarrow p}{u \to v \twoheadrightarrow k(u) \bullet p}$$

$$\frac{x \in X}{\to x \twoheadrightarrow l(x) \bullet Nil} \qquad \frac{c \in C,\ c(t_I)/J \twoheadrightarrow p_J,\ J = \text{fct } c(t_I)}{\to c(t_I) \twoheadrightarrow (p_J[\phi_J] \mid l_J(x_J) \circ l(c(t_I)[J \leftarrow x_J]) \bullet Nil) \backslash \{l_J\}}$$

$$\frac{f \in F,\ f(t_I) \twoheadrightarrow p}{\to f(t_I) \twoheadrightarrow p} \qquad \frac{f \in F,\ f(t_I)/J \twoheadrightarrow p_J,\ J = \text{fct } f(t_I)}{f(t_I) \twoheadrightarrow (\text{id } f \mid (p_J[\phi_J] \mid l_J(x_J) \circ k(f(t_I)[J \leftarrow x_J]) \bullet Nil) \backslash \{l_J\}) \backslash \{k\}}$$

For example, according to the first three translation rules, the translation of the definition of *add* has the form :

$$\begin{aligned} \text{id } add \equiv\ & k(add(0,j)) \bullet p_1 \\ +\ & k(add(s(i),j)) \bullet p_2 \end{aligned}$$

where p_1 and p_2 are appropriate translations of the right-hand sides of rules. An application of *add*, e.g. $add(s(s(0)),s(0))$, is translated into :

$$(\text{id } add \mid k(add(s(s(0)),s(0))) \bullet Nil) \backslash \{k\}$$

Since no two left-hand sides are unifiable, this selects only one branch in the non deterministic process id *add*, in this case the second branch, and evaluates $p_{2[i \leftarrow s(0), j \leftarrow s(0)]}$: p_2 is defined so that the result comes out of a port labelled "l".

This is a general pattern in the translation. Given a term $f(t_1, ..., t_n)$ to be evaluated, where $f \in D$, the process $\mathrm{id}\, f$ receives its arguments, once they have been evaluated (i.e. reduced, by other processes, to terms in T_C), through a port labelled "k" and returns the result through a port labelled "l". According to the last translation rule, this term is translated into a P-term of the form :

$$(\mathrm{id}\, f \mid p \mid (e \bullet k(f(t_1, ..., t_n)[j \leftarrow x_j]...[\,]) \bullet Nil)\backslash\{l_j,...\})\backslash\{k\}$$

where p is a compound process which evaluates independently all subterms at positions $j \neq \varepsilon$ in $f(t_1, ..., t_n)$ which are applications of defined functions, and where the results of these independent evaluations are collected by the sequence of prefixes e into the variables x_j, ... : each component process $p_j[\phi_j]$ in p returns a ground term through its port l_j and the communications with all prefixes $l_j(x_j)$ in e define a ground substitution σ with domain $\{x_j,...\}$. After that, the appropriate branch in the definition of $\mathrm{id}\, f$ is selected by $(\mathrm{id}\, f \mid k(f(t_1, ..., t_n)[j \leftarrow x_j]...[\,])_\sigma \bullet Nil)\backslash\{k\}$ and the final result returned through the port l of $\mathrm{id}\, f$.

5.1.4 *Example of translation*

The constructor-based presentation from the previous example can be translated into the following definitions of process identifiers :

$$
\begin{aligned}
\mathrm{id}\, add \equiv\ & k(add(0,j)) \bullet l(j) \bullet Nil \\
+\ & k(add(s(i),j)) \bullet ((\mathrm{id}\, add \mid k(add(i,j)) \bullet Nil)\backslash\{k\}[\phi_1] \mid l_1(x_1) \bullet l(s(x_1)) \bullet Nil)\backslash\{l_1\}
\end{aligned}
$$

$$
\begin{aligned}
\mathrm{id}\, fib \equiv\ & k(fib(0)) \bullet l(s(0)) \bullet Nil \\
+\ & k(fib(s(0))) \bullet l(s(0)) \bullet Nil \\
+\ & k(fib(s(s(i)))) \bullet (\mathrm{id}\, add \mid ((\mathrm{id}\, fib \mid k(fib(s(i))) \bullet Nil)\backslash\{k\}[\phi_1] \\
& \qquad \mid (\mathrm{id}\, fib \mid k(fib(i)) \bullet Nil)\backslash\{k\})[\phi_2] \\
& \qquad \mid l_1(x_1) \bullet l_2(x_2) \bullet k(add(x_1,x_2)) \bullet Nil)\backslash\{l_1,l_2\})\backslash\{k\}
\end{aligned}
$$

5.2 Parallel logical machines

The mgu interpretation of $SCCS_U$ is applied to the description of abstract parallel machines for interpreting Horn clause programs. These machines must be viewed as very rough and naive sketches. They implement full "AND-parallelism", but no "OR-parallelism", nor any search strategy. However, they are able to produce exactly the least Herbrand model of the Horn clauses they represent : they are correct, but "weakly" complete because of uncontrolled non determinism. The essential purpose of this example is to show the added expressiveness brought by global variables and multi-directional communication.

5.2.1 *The language of Horn clauses*

Let $\mathcal{P}$ be a set of *predicate* names, with an arity function $r : \mathcal{P} \to \mathbb{N}$, and $\mathcal{P}_n = \{P \in \mathcal{P} \mid r(P) = n\}$. The set of A of *atoms* is defined as follows :

(i) $\forall P \in \mathcal{P}_0, P \in A$
(ii) $\forall P \in \mathcal{P}_n$, $n \geq 1$, and $\forall t_1, \ldots, t_n \in T_{F,X}$, $P(t_1, \ldots, t_n) \in A$

Horn clauses may take four different forms :

Assertions :	$H \Leftarrow$
Conditional clauses :	$H \Leftarrow G_1 \ldots G_n$
Goals :	$\Leftarrow G_1 \ldots G_n$
Empty clause :	$\Leftarrow$

where $H, G_1, \ldots, G_n$ are atoms. H is called the *head* of the clause where it appears and $G_1, \ldots, G_n$ the *subgoals*. A Horn clause *program* S is a set of assertions and conditional clauses (program clauses). Executing a program amounts to solving a goal with this program : a presentation of Horn clause semantics can be found in [Lloyd 84].

5.2.2 *Example of Horn clause program*

Let $\mathcal{P}=\{Add, Fib\}$ and $F=\{s,0\}$: $Add(x,y,z)$ holds iff z is the sum of x and y, and $Fib(x,y)$ holds iff y is the xth Fibonacci number. This is defined by the following Horn clause program :

$Add(0,y,y) \Leftarrow$

$Add(s(x),y,s(z)) \Leftarrow Add(x,y,z)$

$Fib(0,s(0)) \Leftarrow$

$Fib(s(0),s(0)) \Leftarrow$

$Fib(s(s(x)),y) \Leftarrow Fib(s(x),u)\ Fib(x,v)\ Add(u,v,y)$

5.2.3 *From Horn clauses to processes*

For translating a Horn clause program S and a goal G into $SCCS_U$ processes, notations similar to the previous application are used, where operations on terms are extended in the obvious way to atoms :

- "$w \rightarrowtail p$" stands for "w can be translated into p", where w is a syntactic construction from Horn clauses and p a syntactic construction in $SCCS_U$;

- id P is a unique process identifier associated with every symbol $P \in \mathcal{P}$;
- def P is the pair (P,S) for every predicate P defined by S;
- clauses $P = \{H \Leftarrow G_1 ... G_n \in S, n \geq 0 \mid \text{hd } H = P\}$;
- $C_1, ..., C_m$ are program clauses and $p_1, ..., p_m$ are processes;
- Let $I=\{1,...,n\}$ be a set of indices, $n \geq 0$, and p be a process:
  - "$H \Leftarrow G_I$" stands for "$H \Leftarrow G_1 ... G_n$" and "$\Leftarrow G_I$" for "$\Leftarrow G_1 ... G_n$";
  - $\forall i \in I$, k, k_i, l_i are distinct port labels and ϕ_i relabelling functions : $\phi_i(k)=k_i$;
  - "id (hd $G_I)[\phi_I] \times p$" stands for "id (hd $G_1)[\phi_1] \times ... \times$ id (hd $G_n)[\phi_n] \times p$";
  - "$\{k_I(G_I)\} \bullet p$" stands for "$k_1(G_1)...k_n(G_n) \bullet p$", similarly for "$\{l_I(G_I)\} \bullet p$";
  - "$p \backslash \{k_I\}$" stands for "$p \backslash \{k_1,...,k_n\}$".

Translation rules :

$$\frac{\text{clauses } P \rightarrow p}{\text{def } P \rightarrow \text{id } P \equiv p} \qquad \frac{C_1 \rightarrow p_1, \; ..., \; C_m \rightarrow p_m}{\{C_1, ..., C_m\} \rightarrow p_1 + ... + p_m}$$

$$\frac{}{H \Leftarrow G_I \rightarrow k(H) \bullet (\text{id (hd } G_I)[\phi_I] \times \{k_I(G_I)\} \bullet Nil) \backslash \{k_I\}}$$

$$\frac{}{\Leftarrow G_I \rightarrow (\text{id (hd } G_I)[\phi_I] \times \{k_I(G_I)\} \bullet \{l_I(G_I)\} \bullet Nil) \backslash \{k_I\}}$$

5.2.4 Example of translation

The Horn clause program from the previous example would be translated into the following definitions of process identifiers :

$$\begin{aligned} \text{id } Add \equiv \; & k(Add(0,y,y)) \bullet Nil \\ & + k(Add(s(x),y,s(z))) \bullet (\text{id } Add[\phi_1] \times k_1(Add(x,y,z)) \bullet Nil) \backslash \{k_1\} \end{aligned}$$

$$\begin{aligned} \text{id } Fib \equiv \; & k(Fib(0,s(0))) \bullet Nil \\ & + k(Fib(s(0),s(0))) \bullet Nil \\ & + k(Fib(s(s(x)),y)) \bullet (\text{id } Fib[\phi_1] \times \text{id } Fib[\phi_2] \times \text{id } Add[\phi_3] \\ & \qquad \times k_1(Fib(s(x),u)) \; k_2(Fib(x,v)) \; k_3(Add(u,v,y)) \bullet Nil) \backslash \{k_1,k_2,k_3\} \end{aligned}$$

The goal $\Leftarrow Fib(s(s(0)),w)$ would be translated into :

$$(\text{id } Fib[\phi_1] \times k_1(Fib(s(s(0)),w)) \bullet l_1(Fib(s(s(0)),w)) \bullet Nil) \backslash \{k_1\}$$

In order to show how the inference rules of the non-ground semantics of $SCCS_U$ yield a derivation which solves this goal, with the mgu interpretation, the following abbreviations are introduced :

- A stands for *Add*, F for *Fib* and parentheses are omitted in $s(s(...))$;
- A_1, A_3, F_1, F_2 stand respectively for id $Add[\phi_1]$, id $Add[\phi_3]$, id $Fib[\phi_1]$, id $Fib[\phi_2]$;
- $\backslash\{k_1\}$ and $\backslash\{k_1,k_2,k_3\}$ and rules (ndp1), (ndp2), (rsp), (rlp) and (pid) are not shown.

(pfp) $\vdash$ $\varnothing : F_1 \xrightarrow{k_1(F(ss0,w')),\ []} \{w'\} : F_1 \times F_2 \times A_3 \times k_1(F(s0,u))\ k_2(F(0,v))\ k_3(A(u,v,w')) \bullet Nil$

(pfp) $\vdash$ $\varnothing : k_1(F(ss0,w)) \bullet l_1(F(ss0,w)) \bullet Nil \xrightarrow{k_1(F(ss0,w')),\ []} \{w'\} : l_1(F(ss0,w')) \bullet Nil$

(cmp3) ―――――――――――――――――――――――――――――――――――――――

$\varnothing : F_1 \times k_1(F(ss0,w)) \bullet l_1(F(ss0,w)) \bullet Nil \xrightarrow{\tau,\ []}$
$\{w'\} : F_1 \times F_2 \times A_3 \times k_1(F(s0,u))\ k_2(F(0,v))\ k_3(A(u,v,w')) \bullet Nil \times l_1(F(ss0,w')) \bullet Nil$

(pfp) $\vdash$ $\{w'\} : F_1 \xrightarrow{k_1(F(s0,s0)),\ []} \varnothing : Nil$

(pfp) $\vdash$ $\{w'\} : F_2 \xrightarrow{k_2(F(0,s0)),\ []} \varnothing : Nil$

(cmp3) ――――――――――――――――――――――――

$\{w'\} : F_1 \times F_2 \xrightarrow{k_1(F(s0,s0))\ k_2(F(0,s0)),\ []} \varnothing : Nil$

(pfp) $\vdash$ $\{w'\} : A_3 \xrightarrow{k_3(A(s0,s0,sz')),\ []} \{z'\} : A_1 \times k_1(A(0,s0,z')) \bullet Nil$

(cmp3) ――――――――――――――――――――――――――――――――――

$\{w'\} : F_1 \times F_2 \times A_3 \xrightarrow{k_1(F(s0,s0))\ k_2(F(0,s0))\ k_3(A(s0,s0,sz')),\ []}$
$\{z'\} : A_1 \times k_1(A(0,s0,z')) \bullet Nil$

(pfp) $\vdash$ $\{w'\} : k_1(F(s0,u))\ k_2(F(0,v))\ k_3(A(u,v,w')) \bullet Nil$
$\xrightarrow{k_1(F(s0,s0))\ k_2(F(0,s0))\ k_3(A(s0,s0,sz')),\ [w' \leftarrow sz']} \{z'\} : Nil$

(cmp3) ―――――――――――――――――――――――――――――――――――――――

$\{w'\} : F_1 \times F_2 \times A_3 \times k_1(F(s0,u))\ k_2(F(0,v))\ k_3(A(u,v,w')) \bullet Nil$
$\xrightarrow{\tau,\ [w' \leftarrow sz']} \{z'\} : A_1 \times k_1(A(0,s0,z')) \bullet Nil$

(cmp1) ―――――――――――――――――――――――――――――――――――――――

$\{w'\} : F_1 \times F_2 \times A_3 \times k_1(F(s0,u))\ k_2(F(0,v))\ k_3(A(u,v,w')) \bullet Nil \times l_1(F(ss0,w')) \bullet Nil$
$\xrightarrow{\tau,\ [w' \leftarrow sz']} \{z'\} : A_1 \times k_1(A(0,s0,z')) \bullet Nil \times l_1(F(ss0,sz')) \bullet Nil$

(pfp) $\vdash$ $\{z'\} : A_1 \xrightarrow{k_1(A(0,s0,s0))} \varnothing : Nil$

(pfp) $\vdash$ $\{z'\} : k_1(A(0,s0,z')) \bullet Nil \xrightarrow{k_1(A(0,s0,s0)),\ [z' \leftarrow s0]} \varnothing : Nil$

(cmp3) ――――――――――――――――――――――――――

$\{z'\} : A_1 \times k_1(A(0,s0,z')) \bullet Nil \xrightarrow{\tau,\ [z' \leftarrow s0]} \varnothing : Nil$

(cmp1) ―――――――――――――――――――――――――――――――――――――――

$\{z'\} : A_1 \times k_1(A(0,s0,z')) \bullet Nil \times l_1(F(ss0,sz')) \bullet Nil \xrightarrow{\tau,\ [z' \leftarrow s0]} \varnothing : l_1(F(ss0,ss0)) \bullet Nil$

(pfp) $\vdash$ $\varnothing : l_1(F(ss0,ss0)) \bullet Nil \xrightarrow{l_1(F(ss0,ss0)),\ []} \varnothing : Nil$

According to the semantics of Horn clause programming, the resolution of the goal $\Leftarrow Fib(s(s(0)),w)$ starts with the third clause defining Fib. The first application of rule (cmp3) does precisely that : selecting the third branch in the non deterministic process F_1 used in the goal process, yielding a compound process composed of three new subgoals to be solved and of $l_1(F(ss0,w')) \bullet Nil$. The three subgoals are represented by a prefixed process with a three particle action, composed with three processes corresponding to the relevant clause definitions, and the process $l_1(F(ss0,w')) \bullet Nil$ will eventually produce the solved goal. This compound process is in the context of a "fresh" global variable $\{w'\}$.

The fourth application of rule (cmp3) solves the first two subgoals and starts solving the third subgoal. For the first two subgoals, this is done with the second and third branches of F_1 and F_2 respectively, selected by the second application of rule (cmp3), and for the last subgoal with the second branch of A_3, selected by the third application of rule (cmp3). In the resulting transition, the substitution $[w' \leftarrow sz']$ is applied to the global variable w'. This occurs independently of $l_1(F(ss0,w')) \bullet Nil$ which is in the same compound process, and which shares w' with the process having done this transition : the first application of rule (cmp1) propagates the substitution, yielding the compound process $A_1 \times k_1(A(0,s0,z')) \bullet Nil \times l_1(F(ss0,sz')) \bullet Nil$ in the context of $\{z'\}$.

The new subgoal $A(0,s0,z')$ is solved by the last application of rule (cmp3) independently of $l_1(F(ss0,sz')) \bullet Nil$ and the last application of rule (cmp1) propagates $[z' \leftarrow s0]$, yielding $l_1(F(ss0,ss0)) \bullet Nil$ which eventually outputs the solved goal.

This example shows the general pattern of the way these $SCCS_U$ abstract machines solve goals with Horn clauses. These machines are correct and weakly complete in the following sense. Let S be a program and $G = \Leftarrow G_1 ... G_n$ be a goal. When the process obtained by translation from G arrives at Nil, using the clause processes from the translation of S, the last action it performs is of the form $(l_1(G_1)...l_n(G_n))_\sigma$. Then, if σ is a ground substitution, the atoms $G_{1\sigma}, ..., G_{n\sigma}$ are elements of the least Herbrand model of S. Conversely, if $G_{1\sigma}, ..., G_{n\sigma}$ are elements of the least Herbrand model of S, then there exist a derivation in the semantics of $SCCS_U$ which terminates at Nil with the last action being $(l_1(G_1)...l_n(G_n))_\sigma$. However, because these machines provide no strategy for guiding the choice among the non deterministic branches of clause processes, there may exist derivations of the goal process which fail to arrive at Nil, even if G is solvable.

Acknowledgements. I am indebted to Bent Thomsen, from ECRC in Munich, who has pointed out a number of flaws in the treatment of variables in a very early version of this work and to Zineb Habbas, who has proposed another notion of global variables in process algebras [Habbas 92] and with whom I had the opportunity to discuss this work.

REFERENCES

[Bergstra & Klop 86] J.A. Bergstra and J.W. Klop. Algebra of Communicating Processes. In J.W. De Bakker et al., editors, *Proc. CWI Symp. Math. and Comp. Sci.*, North Holland.

[Boudol 85] G. Boudol. Notes on Algebraic Calculi of Processes. In K. Apt, editor, *Logics and Models of Concurrent Systems*, NATO ASI Series f13.

[Brinksma 88] E. Brinksma. *Information Processing Systems - Open Systems Interconnection - LOTOS - A Formal Description Technique Based upon the Temporal Ordering of Observational Behavior.* Draft International Standard ISO 8807.

[Dershowitz 87] N. Dershowitz, Termination of Rewriting. *Journal of Symbolic Computation*, **3** (1).

[Guttag & Horning 78] J.V. Guttag and J.J. Horning. The Algebraic Specification of Abstract Data Types. *Acta Informatica*, **10**.

[Habbas 92] Z. Habbas. *Une Algèbre de Processus pour un Calcul Basé sur la Déduction*. PhD Thesis, Grenoble University.

[Hoare 78] C.A.R. Hoare. Communicating Sequential Processes. *Communications of the ACM*, **21** (8).

[Hoare 85] C.A.R. Hoare. *Communicating Sequential Processes*. Prentice Hall, International Series in Computer Science.

[Huet & Hullot 82] G. Huet and J.M. Hullot. Proofs by Induction in Equational Theories with Constructors. *Journal of Computer and System Sciences*, **25** (2).

[Huet & Oppen 80] G. Huet and D. Oppen. Equations and Rewrite Rules : a Survey. In R. Book, editor, *Formal Language Theory : Perspectives and Open Problems*, Academic Press.

[Lloyd 84] J.W. Lloyd. *Foundations of Logic Programming*. Springer-Verlag.

[May 83] D. May. OCCAM. *SIGPLAN Notices*, **13** (4).

[Milner 80] R. Milner. *A Calculus of Communicating Systems*. LNCS 92, Springer-Verlag.

Milner 83] R. Milner. Calculi for Synchrony and Asynchrony. *Theoretical Computer Science*, **25**.

[Milner 89] R. Milner. *Communication and Concurrency*. Prentice Hall, International Series in Computer Science.

[O'Donnel 86] M.J. O'Donnel. *Equational Logic as a Programming Language*. MIT Press.

[Pletat 86] U. Pletat. Algebraic Specification of Abstract Data Types and CCS : an Operational Junction. In. *Proc. Sixth IFIP Workshop on Protocol Specification, Testing and Verification*.

[Plotkin 81] G.D. Plotkin. *A Structural Approach to Operational Semantics*. Aarhus University, Dept. of Computer Science, Research Report No. DAIMI-FN-19.

[Schnoebelen & Jorrand 89] Ph. Schnoebelen and Ph. Jorrand. Principles of FP2. Term Algebras for Specification of Parallel Machines. In J.W. de Bakker, editor, *Languages for Parallel Architectures : Design, Semantics, Implementation Models*, Wiley.

[Thatte 85] S.R. Thatte. On the Correspondence between two Classes of Reduction Systems. *Information Processing Letters*, **20**.

Functional Development of Massively Parallel Programs*

Peter Pepper, Jürgen Exner, Mario Südholt

Technische Universität Berlin
Fachbereich Informatik
e-mail: pepper@cs.tu-berlin.de

Abstract

The programming of parallel systems requires a much higher level of abstraction than is provided by traditional programming languages and programming techniques. Therefore we aim at an approach that utilizes the clarity and power of mathematical concepts to the highest possible extent. In this paper we investigate, how a combination of higher-order functions with particular data structure specifications can be used for the programming of parallel algorithms.

1 Introduction

Parallel computers have matured into very nice and useful tools – if you only look at the hardware. *The big problem is the software.* There is no decent methodology that would allow us to tackle the programming of parallel algorithms in a satisfactory fashion. (And parallelizing FORTRAN compilers obviously are no solution either.)

Our claim is that *the use of formal development methods is mandatory* in this area. The reason is that the programming task is much more complex than in the sequential case, and that testing is even more hopeless. We also claim that *functional languages are an ideal means for representing parallel algorithms.* (Whether this is also true for distributed systems with client-server relationships, is an open question.)

Based on the preeminent trends in the technological development we make the following assumptions about our target architecture:

- We presuppose a MIMD architecture with distributed memory.
- The number of processors is by orders of magnitude smaller than the size of the data. (We feel that the search for an optimal number p of processors for a given problem of size N is only of academic nature, since for real problems such a p by far exceeds any feasible machine size.)

* This work was partially sponsored by the "Bundesministerium für Forschung und Technologie (BMFT)" as part of the project "KORSO – Korrekte Software" and by the CEC as part of the ESPRIT Basic Research Working Group 6112 "COMPASS".

– A SPMD paradigm (or rather "SFMD" paradigm) is the only way to achieve massive parallelism, since we can easily program a *s*ingle *f*unction operating on *m*ultiple *d*ata. But it is virtually impossible to program millions of different functions.

These basic premises have, however, unpleasant effects. With a shared--memory system, life is (relatively) easy: Whenever a processor needs an element, it "simply" fetches it from the global memory. Therefore the programmer "only" has to worry about the proper synchronization of data accesses. With a distributed-memory system life is much harder. Now the programmer also has to worry about the location of the data elements: In spite of the ever improving hardware tricks – such as virtual shared memories – there still remain considerable latencies, if the processors have to fetch data objects from "remote" memories. Therefore the distribution of the data elements over the various local memories is of preeminent concern here.

To some degree, this allocation problem can be alleviated by combining advanced techniques for hardware, operating systems, and compilers. But in the end it is the programmer who "understands the application domain and can make better data and computation partitioning decisions" [Gri93]. However, if these decisions have to be expressed on a too low level, the programming task becomes unmanageable.

Therefore we aim at an approach, where these problems can be addressed on as high a level as possible. Our ideas can be outlined as follows:

Data distribution is the key problem to be solved. In many cases, this means a partitioning of the given data structure, but we feel that a generalization to "covers" – allowing overlapping subobjects – is often helpful. For the formalization of these concepts we employ ideas that have been brought into the realm of program development by the work of Y.V. Srinivas [Sri91].
We study various forms of concrete covers, and we investigate the possibility of describing these covers by the traditional techniques of algebraic specifications.

Higher-order functions provide a compact way of describing algorithms that work in parallel on distributed data. We consider a number of such functions – known under buzzwords such as "map", "filter", "reduce" – that are classically used for this purpose.

Communication is described explicitly – when needed – by means of "stream processing".

The methodology of our approach is based on the following development style:

- We start from a functional specification of the programming task.
- Then we choose a suitable cover (data distribution) and appropriate higher-order functions that solve this problem on the cover ("divide and conquer").
- Then we extract the communication pattern. To this end we study two main approaches:

    - In some cases the form of the higher-order functions used in the solution determines the communication requirements.
    - In other cases the overlapping parts in the cover entail communication patterns.
  - Finally, we transform the resulting functional expressions into explicit stream processing.

Even though its characterization still has to be somewhat vague, we will introduce a short-hand terminology for the kind of constructions we are seeking here:

Definition 1. A **skeleton**[2] is a combination of covers and higher-order functions that meets the following requirements:

- It fosters the elegant formulation of algorithms.
- It allows efficient implementations on various (parallel) architectures.

For the time being, we content ourselves with this vague characterization, because our foremost concern is not to define terminology, but to identify those skeletons that are practically relevant, and to devise corresponding programming methods. In other words, there is no abstract definition of the notion "skeleton". It rather is characterized implicitly as the set of useful schemas that we have found so far.

2 Data Distribution

Our first goal is to come up with a useful theory and methodology for data distribution. What we discuss in the sequel are first steps into this direction. As of today, we can present the fundamental ideas of our approach as well as an outline of the underlying theory, but the full details still remain to be worked out.

We think that our ideas can be made sufficiently clear by referring to an intuitive understanding of concepts such as "object", "subobject", "gluing together", and so forth. But we also want to point out that these concepts have a clear formalization within the realm of category theory. However, since we do not want to burden the reader with the quite heavy-weight mathematical apparatus, we merely sketch the connection here. (For an extensive treatment of the category-theoretic foundations we refer to the work of Y.V. Srinivas [Sri91], on which we base the following treatment.)

2.1 Objects, Subobjects, and Covers

We consider here the general concept of "data structures", which comprises all the well-known examples from programming, such as sets, sequences, graphs,

[2] We dislike the name, but it has been established in the literature e.g. in [DFH+92] or [Ski92], adapting a notion introduced in [Col89].

trees, matrices, and so forth. For reasons of brevity we refer to such data structures here as "**objects**".[3]

Definition 2. Each object has three aspects that are of concern to us: Its **shape** S, its **content** C, and the **mapping** $\gamma : S \rightarrow C$ from shape to content.

Examples: The concept of "shape" covers our intuitive understanding of the layout of a data structure. For instance, with a "matrix" we associate in general a rectangular layout, and similar associations exist for triangular matrices, diagonal matrices, band matrices, and so forth. The same holds for trees, balanced trees, graphs, sequences, and the like. (See Fig. 1.)

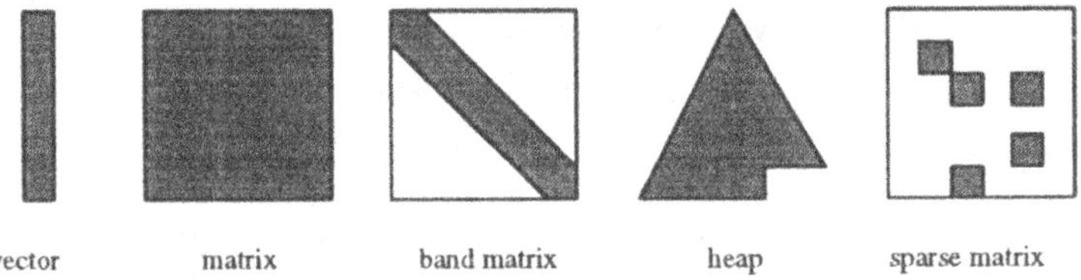

Fig. 1. Various shapes

There are various means for the *formal description of shapes*. For instance, matrices are characterized by an index domain that is the direct product of two intervals from $\mathbb{Z}$. Restricted matrices such as diagonal matrices or triangular matrices can be characterized by suitably chosen subdomains. (The "data fields" found in the work of Y. Choo [YiC92] are an excellent example of an elaborate utilization of index domains. Similarly, [PM91] discuss the use of high-level descriptions for the development of matrix algorithms.) Analogously, the shape of a tree can be determined in various ways, e.g. by a suitable encoding of the "child" or "parent" relation. And the shape of special trees such as balanced trees or heaps can be described by suitably chosen restrictions.

The important point here is that we need not go into the details of how shapes may be encoded technically. All we need is the fact that such a concept exists and that it can be made precise for each concrete object under consideration.

Definition 3. A **subobject** B of a given object A, denoted as $B \hookrightarrow A$, is a *subshape* S_B of S_A together with the corresponding restriction of the contents mapping. The **subshape** S_B is an exact "localization" of B within the overall object A. Evidently, subobject formation has to be compositional; that is, the composition of $C \hookrightarrow B \hookrightarrow A$ yields the embedding $C \hookrightarrow A$.

Examples: The intuitive notions of "subsequences", "submatrices", "subtrees" etc. are covered by this concept of subobject (see Fig. 2). Note that our definition also solves the problem of multiple occurrences of, say, a given tree t' within another tree t: Each instance of t' is a *different* subobject.

[3] They must not be confused with the notion of "objects" as used in object-oriented programming. We rather think in terms of objects in the sense of category theory.

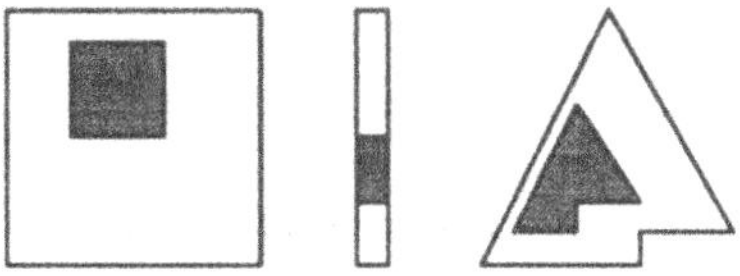

Fig. 2. Subobjects

We call those subobjects of a given object A that cannot be decomposed further **(atomic) components**. For instance, in a matrix these components are the individual elements (which we identify with the (1,1)-submatrices), in a graph these components are the individual nodes *and* edges.

Note that there are two different ways of characterizing subobjects: One works directly by restricting the shape (e.g. "lower triangular submatrix"), and the other works indirectly by considering the contents (e.g. "the submatrix of all positive elements").

Definition 4. The combination of two (sub)objects such that the common parts of their shapes are identified is called **gluing** here. We denote it by $A \circledcirc B$.

Note that gluing need not produce an object of the same kind. For instance, gluing of two trees need not yield a tree, or gluing of two matrices need not yield a matrix (see Fig. 3). So we may have to pass on to more general types. In some situations it is also possible to circumvent the problem by appropriate embeddings; a typical example are matrices which are filled with zeroes.

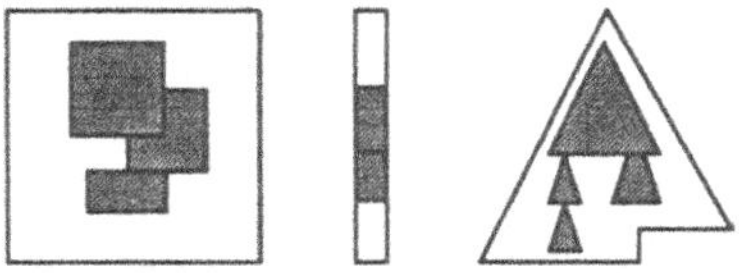

Fig. 3. Gluing of subobjects

Definition 5. A **cover** of a given object A is a set $\mathcal{C}_A = \{ B \mid B \hookrightarrow A \}$ of subobjects of A such that the gluing of these subobjects reconstructs A.

In the short-hand notation of section 2.2 below this reconstruction property can be denoted as the reduction $A = \circledcirc / \mathcal{C}_A$. Obviously, such covers are a *generalization of the concept of partitionings*. In most applications, the special case of disjoint subobjects will do; but in some situations it helps to allow an overlapping of subobjects.

Srinivas [Sri91] points out that one obtains a so-called Grothendieck topology by assigning to each object A a collection $\mathcal{GT}_A$ of covers, which fulfil the following properties: (1) The maximal cover is in the collection. (2) The restriction of a cover C_A to a subobject $B \hookrightarrow A$ is a cover of B. (3) Covers of covers are again (finer) covers. Note that the collection of the atomic components of A constitutes the finest cover of A.

Covers can also be used to provide "multiple views" of objects. We consider situations such as

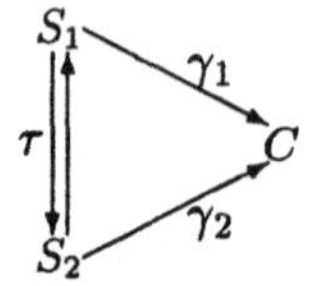

Here, τ is the *translation* between different shapes. The generalization to more than 2 shapes is obvious.

Examples: A well-known situation, where multiple shapes play an essential role for the understanding of an algorithm, is the *heap sort*. Here we have an array that shall be sorted. However, we superimpose on this array a special tree structure, namely that of a heap. And the translation between the two shapes is given by the correspondences *left child* $\leftrightarrow 2 \cdot i$ and *right child* $\leftrightarrow 2 \cdot i + 1$. We will encounter further situations of this kind in the sequel, for instance the row- and column-oriented views of matrices.

2.2 Generalizing Divide-and-Conquer: "Skeletons"

What is the purpose of the above constructions? The answer simply is: In a parallel environment we want to distribute the work over many (relatively) independent processes. That is, we want to apply a so-called divide-and-conquer strategy. Traditionally, this strategy is only employed to reduce a given problem to two or three "smaller" instances of itself, but now we aim at a multitude of such instances. And the following notions shall help us to treat this multitude in a manageable fashion.

First of all we employ an idea the formalization of which goes back to work of F. von Henke [vH75], and which has been elaborated and made popular in particular by R. Bird [Bir87], [Bir88].[4] There are certain *higher-order functions* ("functionals") that allow the formulation of algorithms on a very abstract level, thus leading to a concise programming style with a distinct algebraic flavour. Typical examples of such functionals are:

Map ("apply-to-all"): This functional applies its argument function to all components of its argument object. This argument object may be a set, a sequence, a matrix, etc. We write this in the form $f \star S$. For example:

$$f \star \{a, b, c\} \quad = \quad \{f(a), f(b), f(c)\}$$

We take the liberty here to apply this functional also to n-ary functions. For instance, given two sequences A, B the application $+ \star (A, B)$ yields a sequence which is the pointwise sum of the two given sequences.

[4] It may also be argued that the basics of this idea are already implemented in APL.

Zip For reasons of readability we sometimes also use an alternative infix notation (that is known as the "zip operator") for the "apply-to-all" operator; for example:

$$A \overset{+}{\curlyvee} B \stackrel{\text{def}}{=} + \star(A, B)$$

Reduce ("accumulate-over-all"): This functional uses its (binary) argument function to accumulate all components of its argument object. We write this in the form $\oplus/\, S$, where $\oplus$ is the given argument function. For example:

$$\oplus/\,\{a, b, c\} = a \oplus b \oplus c$$

Note that the operation $\oplus$ has to conform to the given structure. For instance, if we work on sets, then $\oplus$ has to be associative and commutative, and it has to have a neutral element '0' such that $\oplus/\,\emptyset = 0$ is well-defined. (Otherwise, the application of / has to be restricted to nonempty structures.)

Note that the above characterizations leave out a great deal of technical issues and distinctions that need to be dealt with for every concrete kind of structure under consideration.

Now we need to relate these ideas to the above concepts of subobjects and covers. Srinivas [Sri91] employs the category-theoretic concept of *sheaves* (see [BW90]), based on the observations that "sheaf theory studies the global consequences of locally defined properties" and that "the value of a map on an object can be uniquely obtained from its values on any cover of that object". The formal definition of sheaves works with natural transformations on contravariant functors (see [BW90] and [Sri91]), but for our purposes an informal characterization will do.

Definition 6. A **sheaf** assigns to every object A a set S_A such that for every cover C_A every subobject $A_i \hookrightarrow A$ is associated *in a "compatible" way* to some function[5] $S_A \rightarrow S_{A_i}$ between the corresponding sets.

We do not want to build up the whole category-theoretic apparatus here, since the details have to be individually clarified for each application anyhow. Instead, we only illustrate the principle by a nice example taken from [Sri91]. Consider a graph G and the embedding $G_i \hookrightarrow G$ of subgraphs. The coloring of graphs is a sheaf (see Fig. 4): It associates with each graph the set of all its k-colorings, and it associates to each subgraph embedding the restriction of the k-colorings to the subgraph. Here two colorings $c \in S_i$ and $c' \in S_j$ of subgraphs G_i and G_j are *compatible* if they agree on the common nodes of G_i and G_j.

Evidently, the compatibility requirement of sheaves is crucial for their usability: Compatible solutions for subproblems can be composed into solutions of the full problem. In the simplest case, we then have the following kind of situation:

$$f(A) = \oplus/\,(g \star C_A)$$

[5] Note the opposite direction of the arrow.

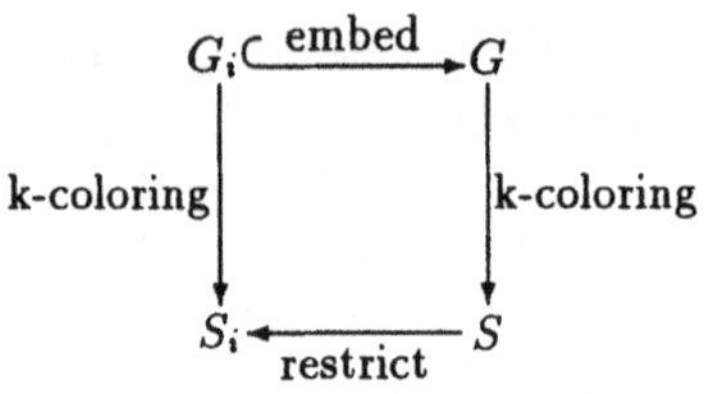

Fig. 4. Graph coloring as a sheaf

Here, $f(A)$ is the problem to be solved, C_A is some cover of the object A, g is a function to be applied to the subobjects, and $\oplus$ composes the subsolutions into a full solution. Note that g has to produce compatible subsolutions.

The main idea now can be described as follows:

- Evidently, the computation of g on each subobject in the cover C_A can be distributed over the parallel processors.
- The overlapping parts of the subobjects entail certain compatibility requirements that usually ask for some communication between the processors.
- The composition of the resulting subsolutions often can be performed in parallel as well, under the proviso of a suitably chosen communication policy.

But we will see in our later examples that this basic scheme comes in numerous variations and with intricate complications.

2.3 Towards an Algebra of Data Distribution

In the previous section we have given an outline of the mathematical principles lying behind our approach. Now we want to sketch, how this method can be assisted by formal specification and development techniques.

Mathematically, covers are sets of subobjects, where subobjects are exactly localized within the encompassing objects. Technically, we have to represent this concept by using the tools that are provided by programming languages. This means that covers are represented as structured objects themselves. That is, instead of e.g. a set of submatrices (= cover) we may actually work with a matrix of matrices (= representation of cover); this matrix representation takes care of the "localization" of the submatrices.

Since there are all kinds of combinations possible, we have to set up a framework that is highly polymorphic. This framework should enable us to represent e.g. a matrix as a sequence of sequences ("row-major view" or "column-major view"), as a sequence of matrices ("block row view" or "block column view"), as a matrix of matrices ("tiling"), and so forth. This genericity is covered by the following algebraic specification:

```
SPEC Cover = {SORT S[α], C[β], U[γ]
              FUN split : S[α]→C[U[α]]
              FUN glue : C[U[α]]→S[α]
```

$$\text{AXM } \mathtt{glue} \circ \mathtt{split} = \mathtt{id} \quad \}$$

Note that the operation **glue** usually can be represented as a reduction using the binary gluing operator, that is, **glue** = $\oplus/$. This often leads to standard skeletons, thus enabling further parallelization.

Classical instances of this generic specification are e.g. (where $+\!\!+$ stands for the concatenation of sequences)

```
SPEC Cover-Seq-by-Blocks =
      { Cover WITH (S = seq, C = seq, U = seq)
        AXM glue = ++/   }
```

Here we partition the sequence into q consecutive subsequences: $\mathtt{split(S)} = \langle S_1, \ldots, S_q \rangle \Rightarrow S = S_1 +\!\!+ \ldots +\!\!+ S_n$. We could constrain this specification further by e.g. requiring that the subsequences are of approximately equal sizes.

```
SPEC Cover-Seq-by-Interleaving =
      { Cover WITH (S = seq, C = seq, U = seq)
        AXM split(S) = ⟨S1, ..., Sq⟩
              ⇒ split(x++S) = ⟨x++Sq, S1, ..., Sq-1⟩ }
```

Here we split the sequence in such a way that the subsequences are formed by taking every q-th element of S. Consequently, gluing is interleaving.

```
SPEC Cover-Matrix-by-Rows =
      {Cover WITH (S = matrix, C = seq, U = seq)
       AXM split(S) =≪ matrix as sequence of rows ≫ }
```

```
SPEC Cover-Matrix-by-Block-Rows =
      {Cover WITH (S = matrix, C = seq, U = matrix)
       AXM split(S) =≪ matrix as sequence of submatrices ≫ }
```

```
SPEC Cover-Matrix-by-Tiles =
      {Cover WITH (S = matrix, C = matrix, U = matrix)
       AXM split(S) =≪ tiling of matrix ≫   }
```

These last examples merely hint at some of the possibilites that we have for the splitting of matrices. Such situations will be discussed in greater detail later on. Note that these splitting specifications do not yet determine, whether the splitting is a partitioning or not.

But the mere specification of covers is not sufficient to obtain skeletons. We also need to consider suitable higher-order functions on them. So we expect to have properties such as

```
SPEC Skeleton1 = {ENRICH Cover BY
                      AXM ∀A : S[α]. f ⋆ A = (f ⋆) ⋆ split(A) }
```

SPEC **Skeleton**$_2$ = {ENRICH **Cover** BY

$$\text{AXM } \forall \mathtt{A}, \mathtt{B} : \mathtt{S}[\alpha].\ \mathtt{A} \overset{\oplus}{\curlyvee} \mathtt{B} \;=\; \mathtt{split(A)} \overset{\oplus}{\curlyvee} \mathtt{split(B)}\ \}$$

We should point out that the logical – and also the physical – connectivity of the processors can be described by the same structured specification techniques as introduced for data structures above. But we do not detailize this aspect here further.

3 Examples

Any presentation of a programming methodology should be accompanied by illustrative examples. Therefore we present in the sequel a few selected programming problems, thus extending our collection of case studies performed in [Pep93]. There is not enough room for treating all the examples in full detail; hence, we will only sketch their essential aspects.

3.1 Prefix Sums

In the first one of our examples we reconsider our treatment in [Pep93], now emphasizing the sheaf-oriented approach more strongly. The computational structure of this problem can be found in so many algorithms that it is sometimes claimed to be one of the most often invoked subroutines in parallel programs.

Problem: We are given a sequence S of numbers and we shall compute the sequence of the sums of the initial subsequences of S; that is, the result sequence $\mathtt{Z} = \mathtt{psums(S)}$ shall fulfil the property (where n is the length of S):

$$Z_i = \sum_{j=1}^{i} S_i \qquad \text{for } i = 1, \ldots, n$$

(1) Following [Bir87], [Bir88] we denote by (**inits** S) the sequence of the initial subsequences of S. (The fact that this happens to be a cover of S is of no relevance here.) Then our task is specified as follows:

$$\mathtt{psums(S)} \quad = \quad (+/) \star (\mathtt{inits\ S})$$

By a simple induction it is immediately seen that the solution $\mathtt{R} = \mathtt{psums(S)}$ has the following property (where $\mathtt{x}{+\!\!+}\mathtt{S}$ stands for the concatenation of the element **x** to the sequence S):

$$\mathtt{R} \;=\; (+/) \star (\mathtt{inits\ S}) \quad \Longleftrightarrow \quad \mathtt{R} \;=\; \mathtt{i}(0{+\!\!+}\mathtt{R}) \overset{+}{\curlyvee} \mathtt{S}$$

Here we employ the convention that $\mathtt{A} \overset{+}{\curlyvee} \mathtt{B}$ ignores the excessive part of the longer one of the input sequences.

(2) Now suppose that we have p processors at our disposal. Therefore we chose a *covering* C_S that consists of p subsequences $\mathtt{S}_1, \ldots, \mathtt{S}_p$. For this splitting we choose **Cover-Seq-by-Interleaving**. This entails the property:

$$\mathtt{split}(S) = (S_1, \ldots, S_p) \iff \mathtt{split}(x +\!\!+ S) = (x +\!\!+ S_p, S_1, \ldots, S_{p-1})$$

By using the skeleton with $\mathtt{split}(A \overset{\oplus}{\curlyvee} B) = \mathtt{split}(A) \overset{\oplus}{\curlyvee} \mathtt{split}(B)$ we can compute the following relationship for R:

$$\begin{aligned}
&\text{Let } \mathtt{split}(R) = (R_1, \ldots, R_p) \text{ and } \mathtt{split}(S) = (S_1, \ldots, S_p):\\
&\mathtt{split}(R) = \mathtt{split}(0 +\!\!+ R \overset{+}{\curlyvee} S)\\
&\qquad = \mathtt{split}(0 +\!\!+ R) \overset{+}{\curlyvee} \mathtt{split}(S)\\
&\qquad = (0 +\!\!+ R_p, R_1, \ldots, R_{p-1}) \overset{+}{\curlyvee} (S_1, S_2, \ldots, S_p)\\
&\qquad = (0 +\!\!+ R_p \overset{+}{\curlyvee} S_1,\ R_1 \overset{+}{\curlyvee} S_2,\ \ldots,\ R_{p-1} \overset{+}{\curlyvee} S_p)
\end{aligned}$$

This yields the system of equations

$$\begin{aligned}
R_1 &= 0 +\!\!+ R_p \overset{+}{\curlyvee} S_1,\\
R_2 &= R_1 \overset{+}{\curlyvee} S_2,\\
&\ldots\\
R_p &= R_{p-1} \overset{+}{\curlyvee} S_p
\end{aligned}$$

(3) Unfortunately, there still is a causal dependency through $R_1, \ldots, R_p$, which prohibits a true parallelization. But by introducing some auxiliary sequences this deficiency can be resolved: Therefore we introduce (where $E = (0, 0, \ldots, 0)$ is a sequence consisting of zeroes only)

$$\begin{aligned}
Q_1 &= E \overset{+}{\curlyvee} S_1,\\
Q_2 &= Q_1 \overset{+}{\curlyvee} S_2,\\
&\ldots\\
Q_p &= Q_{p-1} \overset{+}{\curlyvee} S_p
\end{aligned}$$

Note that $Q_1 = S_1$. Using these sequences, we can now easily deduce the following equations (based on the associativity of '+'):

$$\begin{aligned}
R_1 &= 0 +\!\!+ R_p \overset{+}{\curlyvee} Q_1,\\
R_2 &= R_1 \overset{+}{\curlyvee} S_2 = (0 +\!\!+ R_p \overset{+}{\curlyvee} Q_1) \overset{+}{\curlyvee} S_2 = 0 +\!\!+ R_p \overset{+}{\curlyvee} (Q_1 \overset{+}{\curlyvee} S_2) = 0 +\!\!+ R_p \overset{+}{\curlyvee} Q_2\\
&\ldots\\
R_p &= R_{p-1} \overset{+}{\curlyvee} S_p = \ldots = 0 +\!\!+ R_p \overset{+}{\curlyvee} Q_p
\end{aligned}$$

This system represents true parallelism (see [Pep93]). Initially, each processor P_i has to receive its subsequence S_i. It is then responsible for computing Q_i and R_i. The sequence R_p has to be broadcast to all processors.

Evidently, this parallelization does not bring about a reasonable gain, if the splitting and gluing takes as much time as the addition. However, the above derivation can be applied to more complex functions as well (as long as they constitute a monoid). And in many cases the algorithm has to be applied in a situation, where the data are already distributed among the processors.

3.2 Iterative Solution of Systems of Linear Equations

A system of linear equations is usually given in the form (where '×' is used for both matrix-vector multiplication and matrix-matrix multiplication, and also for scalar-matrix multiplication)

$$A \times x = b$$

Here A is a matrix, and b and x are vectors. The solution vector x can either be found by direct methods such as "Gaussian elimination" or by iterative methods. In [Pep93] we have presented a formal derivation for the former approach, now we consider the latter.

Jacobi and Gauss-Seidel Iteration We start by some mathematical considerations. To begin with, we represent the matrix A as the sum of three matrices, a lower triangular matrix L, a diagonal matrix D, and an upper triangular matrix U (see Figure 5).

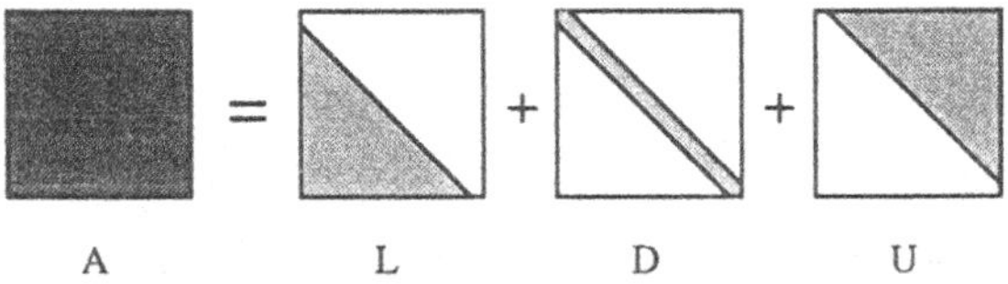

Fig. 5. Sum representation of matrices

The so-called **Jacobi iteration** results from the following transformation:

$$\begin{aligned} b = A \times x &= (L + D + U) \times x \\ \vdash x &= D^{-1} \times (b - (L + U) \times x) \end{aligned}$$

The last one of these equations is used for the iteration process: Starting from an initial vector $x^{(0)}$ we compute

$$x^{(t+1)} = D^{-1} \times (b - (L + U) \times x^{(t)})$$

The **Gauss-Seidel iteration** results from the following transformation:

$$\begin{aligned} b = A \times x &= (L + D + U) \times x \\ \vdash (L + D) \times x &= b - U \times x \end{aligned}$$

Again, the last one of these equations is used for the iteration process: Starting from an initial vector $x^{(0)}$ we compute

$$x^{(t+1)} = D^{-1} \times (b - (L \times x^{(t+1)} + U \times x^{(t)}))$$

It may help to visualize the situations in these algorithms graphically (see Figures 6 and 7). Evidently, the Gauss-Seidel iteration only works, if we fill the vector $x^{(t+1)}$ from top to bottom; and this adds strong sequential dependencies to the algorithm. *Therefore the Gauss-Seidel iteration is not suited for parallelization – unless the matrix is sparse.* (This is unfortunate, since the Gauss-Seidel iteration usually converges faster than the Jacobi iteration.)

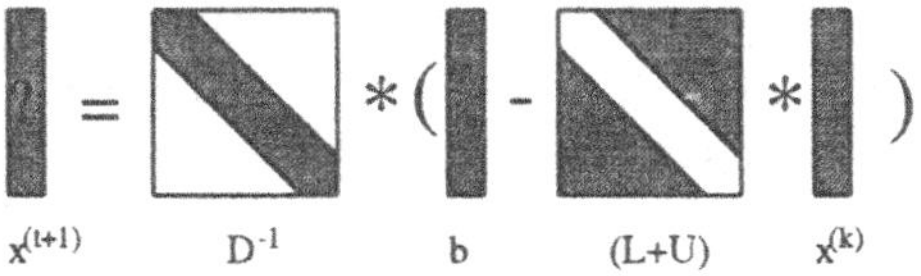

Fig. 6. Jacobi iteration

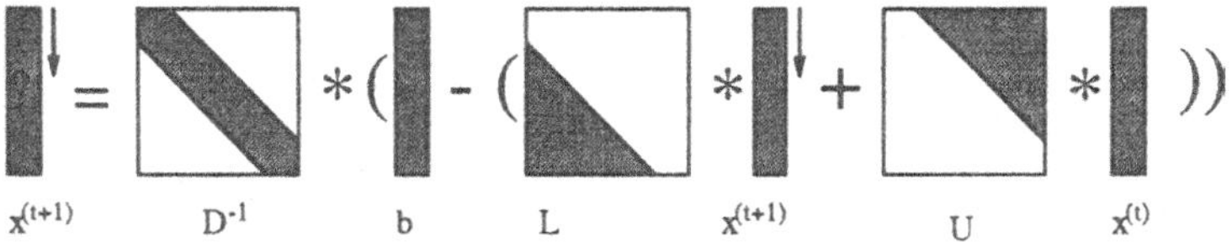

Fig. 7. Gauss-Seidel iteration

Parallelization of the Jacobi Iteration The core of the equation

$$x^{(t+1)} = D^{-1} \times (b - (L + U) \times x^{(t)})$$

is the matrix-vector product. And this is easily amenable to parallelization. In the remainder of this section we therefore concentrate on variations of the matrix-vector product.

Parallel Matrix-Vector Product We want to parallelize the computation of

$$y = B \times x$$

We must take into consideration that this algorithm is usually embedded into larger tasks. And these encompassing tasks may necessitate different forms of data distribution for the matrix B. Therefore we study in the sequel three variants of this multiplication, one for a row-oriented distribution, one for a column-oriented distribution, and one for a tiling of the matrix.

As usual, we assume that we have p processors at our disposal, for $p \ll N$. Therefore we have to find a suitable cover of size p for the matrix B. We prepare the various algorithms by some simple calculations of linear algebra. Suppose that we can represent B as the sum of p matrices: $B = (B_1 + \ldots + B_p)$. Then we have

$$y = B \times x = (B_1 + \ldots + B_p) \times x = (B_1 \times x + \ldots + B_p \times x)$$

Row-oriented view When we choose *split* in such a way that a block row cover is generated, then we obtain the situation of Fig. 8 (a). That is, we may compose y from the pieces $y_i = B_i \times x$. (Note that we also construct a matching cover for the result vector y.) Using our functionals, this reads

$$\begin{aligned} & \texttt{split(y)} = (\times x) \star \texttt{split(B)} \\ \vdash \quad & y = \oplus / ((\times x) \star \texttt{split(B)}) = \overset{+\!\!+}{\curlyvee} / ((\times x) \star \texttt{split(B)}) \end{aligned}$$

This is a skeleton the form of which entails two things: The argument vector x has to be broadcasted over all processors. And the result vector is composed from the piecewise results of the individual processors.

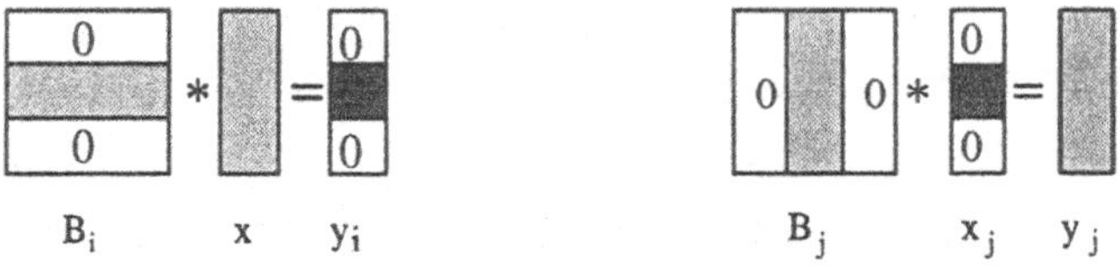

Fig. 8. Block row view: (a) row view (b) column view

Column-oriented view When *split* produces a block column cover, then we obtain the situation in Fig. 8 (b). That is, we obtain y by adding the intermediate vectors $y_j = B_j \times x_j$. (Note that here we also construct a corresponding cover for x.) Using our functionals, this reads

$$y = \oplus/\,(\mathtt{split}(\mathrm{B}) \overset{\times}{\curlyvee} \mathtt{split}(\mathrm{x})) = \overset{+}{\curlyvee}\,/\,(\mathtt{split}(\mathrm{B}) \overset{\times}{\curlyvee} \mathtt{split}(\mathrm{x}))$$

The form of this skeleton entails the following layout: The fragments of B and x are distributed over the local memories. And at the end of the computation we have to perform a multinode accumulation

$$y = y_1 \overset{+}{\curlyvee} y_2 \overset{+}{\curlyvee} \dots \overset{+}{\curlyvee} y_p$$

This skeleton may be implemented e.g. by accumulating the elements along the rows in a staggered manner (see Fig. 9) such that at any point in time all processors can do work. At the end, the i-th processor holds the i-th component of y.

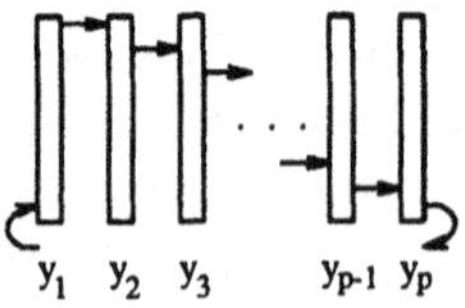

Fig. 9. Multinode accumulation

Tiling The same principles apply to a cover obtained by tiling (see Fig. 10). As a matter of fact, this algorithm can easily be derived from the two previous ones: First we form a block row cover consisting of $\sqrt{p}$ submatrices. And then we build for each of these submatrices a column row cover consisting of $\sqrt{p}$ blocks each. This nicely demonstrates the compositionality of the approach.

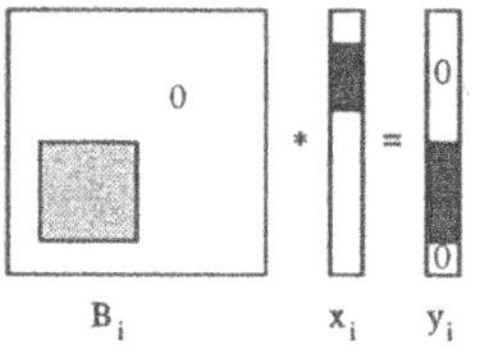

Fig. 10. Tiling

Parallelization of the Gauss-Seidel Iteration We had indicated above that the Gauss-Seidel method is not suited for parallelization. This is actually not quite true. Consider the problematic part of Fig. 7 and imagine a block column technique for this multiplication (see Fig. 11). This means that we obtain an expression of the following form:

$$y = L \times y = (L_1 \times y_1 + \ldots + L_p \times y_p)$$

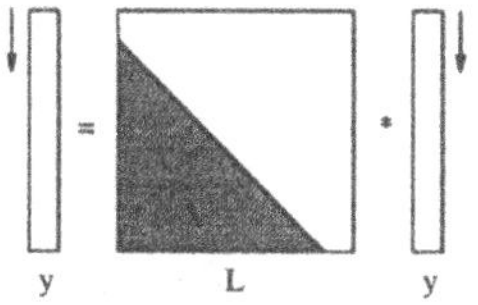

Fig. 11. Triangular systems

The only serious complication now is that the vector y_i depends on the previous vectors, that is, $y_i = \Phi(y_1, \ldots, y_{i-1})$. This leads to a data flow along the processors. If we follow the policy that each processor shall try to communicate data as early as possible to its neighbour, then the stream equations of the above block column multiplication will yield a wavefront algorithm for this triangular matrix. (As a matter of fact, this is the basic principle of the algorithm given by Young-il Choo [YiC93] for solving triangular systems of equations through back-substitution.)

3.3 Partial Differential Equations

Now we will briefly sketch, how our methods can be applied to partial differential equations — one of the most important application areas of parallel programming. As a simple example consider Poisson's equation (with some boundary conditions the details of which we can ignore here)

$$\frac{\partial^2 u(x,y)}{\partial x^2} + \frac{\partial^2 u(x,y)}{\partial y^2} = g(x,y) \qquad (x,y) \in [0,1] \times [0,1]$$

By discretizing the unit square $[0,1] \times [0,1]$ into a $N \times N$- grid we can solve this problem by a Gauss-Seidel iteration of the form (where $h = \frac{1}{N}$ is the grid spacing)

$$U_{i,j}^{(t+1)} = \frac{1}{4}(U_{i,j-1}^{(t+1)} + U_{i+1,j}^{(t+1)} + U_{i,j+1}^{(t)} + U_{i-1,j}^{(t)} - h^2 g(i,j))$$

In order to abstract away the excessive indexing, we introduce more readable shorthand notations (with the obvious meanings)

$$U^+ = \frac{1}{4}(U_{left}^+ + U_{top}^+ + U_{right} + U_{bottom} - h^2 g(i,j))$$

A little analysis shows that this leads to some kind of "wavefront" algorithm. In the sequel we sketch a systematic derivation of such an algorithm. In order to keep the presentation short, we restrict ourselves to block column splitting.

In order to describe the algorithm formally, we define a **cover** by block columns C such that the following properties hold (see Figure 12): Each block C has an *interior* and a *fringe*. This fringe consists of one column on each side, which overlaps with the interior of the neighbouring block column. (That is, any two neighbouring blocks have two columns in common.)

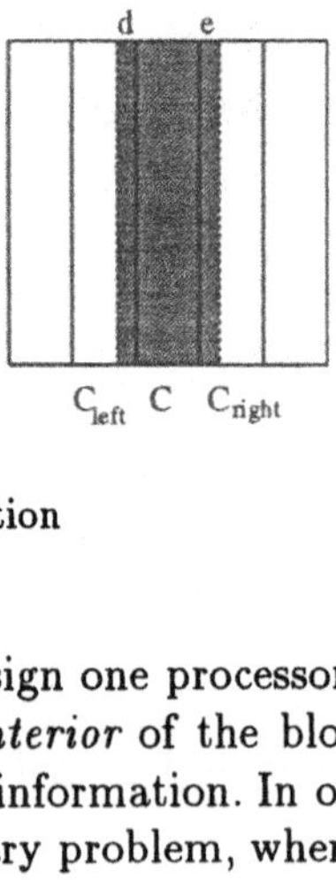

Fig. 12. Cover for Poisson's equation

The idea now is that we assign one processor P_i to each block C_i, and that this processor computes the *interior* of the block. In order to do this, it has to use the fringe as additional information. In other words, each P_i performs a sequential solution of a boundary problem, where the boundary is given by the fringe.

In order to make this cover into a sheaf – such that the local solutions together constitute a global solution – we obviously have to require as the *compatibility criterium* that the fringe of C coincides with the corresponding values of C_{left} and C_{right}, respectively. More precisely, in the iteration $C^{(t+1)}$ the left fringe must coincide with $C_{left}^{(t+1)}$ and the right fringe with $C_{right}^{(t)}$. These observations immediately yield the essence of the algorithm:

- The overall communication pattern for P is depicted in Fig. 13.

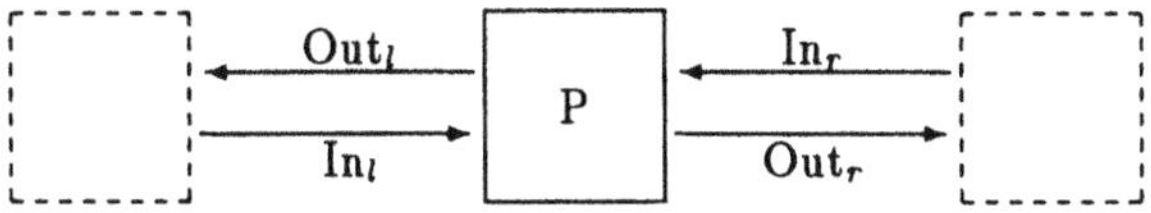

Fig. 13. Communication pattern for Poisson's equation

- In the $(t+1)$st iteration P effects the transition from $C^{(t)}$ to $C^{(t+1)}$ (for the interior).
- The input stream In_l provides the fringe-columns $\langle d^{(1)}, d^{(2)}, \ldots, d^{(t+1)}, \ldots\rangle$ from the left neighbour, and the input stream In_r provides the fringe-columns $\langle e^{(0)}, e^{(1)}, \ldots, e^{(t)}, \ldots\rangle$ from the right neighbour.
- The output stream Out_l receives the leftmost columns of the interior of $\langle C^{(0)}, C^{(1)}, \ldots, C^{(t)}, \ldots\rangle$, and the output stream Out_r receives the rightmost columns of the interior of $\langle C^{(1)}, C^{(2)}, \ldots, C^{(t+1)}, \ldots\rangle$.
  By a simple induction it is immediately seen that this output behaviour ensures the correct input behaviour for the left and right neighbour of P.

One more aspect remains to be considered: The processor P has to send in each iteration a whole column to each of its neighbours. In which order shall the elements of these columns be transmitted? The answer is obvious from the behaviour of the local computation: Due to the causal dependencies in the Gauss-Seidel iteration, the column should be sent from top to bottom. And the local computation should proceed in such a way that the stream elements can be sent at the earliest possible time.

3.4 Reachability in Graphs

Let a graph $\mathcal{G}$ be given by its characteristic function **edge(x, y)**. We are looking for the transitive closure of $\mathcal{G}$, i.e. we want to know if for any two nodes x, y, y can be reached from x.[6]

$$\texttt{FUN}\ \mathbf{reachable} : \mathbf{node} \times \mathbf{node} \rightarrow \mathbf{bool}$$
$$\mathbf{AXM}\ \mathbf{reachable(x, y)} = \{\exists\, \mathbf{p} : \mathbf{seq[node]} \mid \mathbf{ispath(x \cdot p \cdot y)}\}$$

Our proposed solution will be only briefly outlined here. (For a formal development, see, e.g., [PM91]. We blacken the nodes in sequence and only consider paths whose *inner* nodes are black. This idea immediately entails the following recursive solution (where $N =$ **Nodes**($\mathcal{G}$) is the set of all nodes of $\mathcal{G}$).

$$\begin{aligned}
&\texttt{LAW}\ \mathbf{reachable\,(x, y)} = \mathbf{reaches\,(N)(x, y)}\\
&\texttt{LAW}\ \mathbf{reaches\,(\emptyset)(x, y)} = \mathbf{edge\,(x, y)}\\
&\texttt{LAW}\ \mathbf{reaches\,(S \cup \{a\})(x, y)} = (\mathbf{reaches\,(S)(x, y)} \vee\\
&\qquad\qquad (\mathbf{reaches\,(S)(x, a)} \wedge \mathbf{reaches\,(S)(a, y)})
\end{aligned} \tag{1}$$

For the subsequent discussions, the following definitions apply:

[6] The same algorithm can be used to calculate the *length* of the paths between the nodes. We need only substitute $\vee$ by 'min' and $\wedge$ by $+$ (see [PM91]).

- $\mathtt{a}_1, \ldots, \mathtt{a}_n$ refers to the nodes in the order in which they were blackened.
- $\mathtt{S}_0 = \emptyset$, $\mathtt{S}_i = \mathtt{S}_{i-1} \cup \{\mathtt{a}_i\}$ refers to the associated sets.

Partitioning the Data Space Our task is to calculate **reachable**$(\mathbf{x}, \mathbf{y})$ for all node pairs $(\mathbf{x}, \mathbf{y}) \in \mathbf{N} \times \mathbf{N}$, that is **reachable** $*$ $(\mathbf{N} \times \mathbf{N})$. If we observe the entire (dynamic) data space during the calculation process in accordance with the equations (1), the situation portrayed in Fig. 14 is realised. By permuting the arguments of **reaches** we obtain a function **reaches′** such that the equation

$$(\mathbf{reaches} * \langle \mathtt{S}_0, \ldots, \mathtt{S}_n \rangle) * (\mathbf{N} \times \mathbf{N}) \quad = \quad (\mathbf{reaches}' * (\mathbf{N} \times \mathbf{N})) * \langle \mathtt{S}_0, \ldots, \mathtt{S}_n \rangle$$

holds. The left- and right-hand side of this equation induce two different parallelizations (by considering the outer '$*$'-operator):

1. A processor $\mathtt{P}_{\mathtt{x},\mathtt{y}}$ is allocated to each matrix element — i.e. to each node pair $(\mathbf{x}, \mathbf{y})$.
2. A processor $\mathtt{P}_i$ is assigned to each matrix — i.e. to each set $\mathtt{S}_i$.

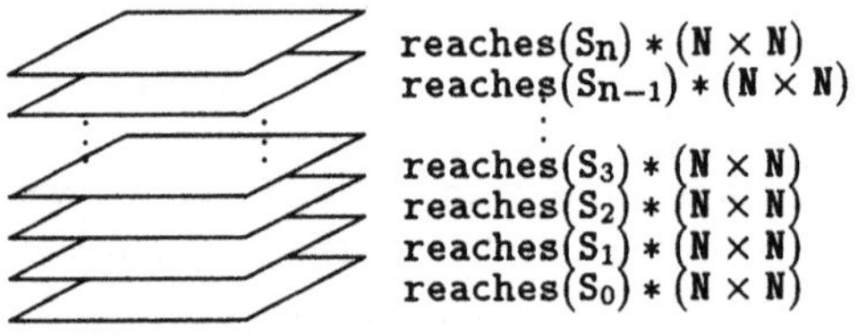

Fig. 14. Data space during calculation

A brief analysis shows that the causal dependencies of Version 2 would permit only negligible parallelism. We therefore concentrate on Version 1. For simplicity, we first make the – unrealistic – assumption, that we actually have n^2 processors at our disposal. (Later, we will revise this assumption.) Here, we formulate this in the ad-hoc notation:

PROCESS $\mathtt{P}_{\mathtt{x},\mathtt{y}}$ COMPUTES
– **reachable**$(\mathbf{x}, \mathbf{y})$

In order to choose a suitable cover, we consider the essential recurrence in (1):

$$\begin{aligned}\mathbf{reaches}\,(\mathtt{S}_i)(\mathbf{x}, \mathbf{y}) = {} & \mathbf{reaches}\,(\mathtt{S}_{i-1})(\mathbf{x}, \mathbf{y}) \; \vee \\ & \mathbf{reaches}\,(\mathtt{S}_{i-1})(\mathbf{x}, \mathtt{a}_i) \wedge \mathbf{reaches}\,(\mathtt{s}_i)(\mathtt{a}_i, \mathbf{y})\end{aligned}$$

Hence, the value at the point (x, y) depends on all values from the row x and from the column y. This motivates the choice of a cover as depicted in Fig. 15. This cover is based on the following rows and columns. (Note that — in the visualization of Fig. 14 — the elements of these rows and columns come from different matrices!)

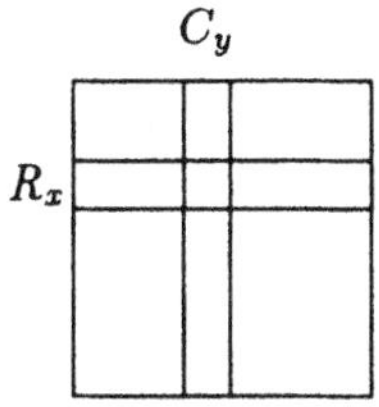

Fig. 15. Covering the matrix for reachability computation

$$\begin{aligned} \mathtt{Rx}[\mathtt{a_i}] &\stackrel{\text{def}}{=} \mathtt{reaches}(\mathtt{S_{i-1}})(\mathtt{x}, \mathtt{a_i}); \\ \mathtt{Cy}[\mathtt{a_i}] &\stackrel{\text{def}}{=} \mathtt{reaches}(\mathtt{S_{i-1}})(\mathtt{a_i}, \mathtt{y}). \end{aligned}$$

Thus the above equation can be changed to

$$\mathtt{reaches}(\mathtt{S_i})(\mathtt{x}, \mathtt{y}) = \mathtt{reaches}(\mathtt{S_{i-1}})(\mathtt{x}, \mathtt{y}) \vee (\mathtt{Rx}[\mathtt{a_i}] \wedge \mathtt{Cy}[\mathtt{a_i}]).$$

Using simple induction, it immediately follows that:

$$\begin{aligned} \mathtt{reachable}(\mathtt{x}, \mathtt{y}) &= \mathtt{edge}(\mathtt{x}, \mathtt{y}) \vee (\mathtt{Rx}[\mathtt{a_1}] \wedge \mathtt{Cy}[\mathtt{a_1}]) \vee \ldots \vee (\mathtt{Rx}[\mathtt{a_n}] \wedge \mathtt{Cy}[\mathtt{a_n}]) \\ &= \mathtt{edge}(\mathtt{x}, \mathtt{y}) \vee (\vee/ (\mathtt{Rx} \stackrel{\wedge}{\Upsilon} \mathtt{Cy})) \end{aligned}$$

Data flow Each processor $P_{x,y}$ computes its own values $\mathtt{reaches}(\mathtt{S_i})(\mathtt{x}, \mathtt{y})$ basing on values from the row $\mathtt{Rx}$ and the column $\mathtt{Cy}$. These "foreign" data have to be transmitted by other processors. The systolic layout of Fig. 16 only necessitates communication between neighbouring processors. (Note the similarity between this derivation and the one from Section 3.3.) If we assume — as in-

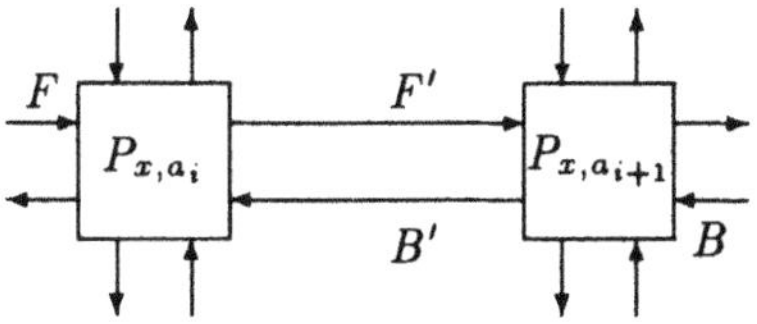

Fig. 16. Systolic data flow

duction hypothesis — that the following relationships hold:

$$\begin{aligned} \mathtt{F} &= \mathtt{Rx}[1 \mathinner{..} \mathtt{i} - 1] \\ \mathtt{B} &= \mathtt{Rx}[\mathtt{i} + 2 \mathinner{..} \mathtt{n}] \end{aligned}$$

then we can pass on the input streams and append the value computed by the processor itself (that is $\mathtt{r_{x,a_i}} = \mathtt{reaches(S_{i-1})(x, a_i)}$):

$$\mathtt{F}' = \mathtt{F} \mathbin{+\!\!+} \mathtt{r_{x,a_i}} = \mathtt{R_x}[1 \,..\, \mathtt{i}]$$
$$\mathtt{B}' = \mathtt{r_{x,a_{i+1}}} \mathbin{+\!\!+} \mathtt{B} = \mathtt{R_x}[\mathtt{i}+1 \,..\, \mathtt{n}]$$

Hence, every processor can reconstruct the row $\mathtt{R_x}$ from its left and right input streams. (Analogously for the column $\mathtt{C_y}$).

$$\mathtt{R_x} = \mathtt{F} \mathbin{+\!\!+} \mathtt{r_{x,a_i}} \mathbin{+\!\!+} \mathtt{B}'$$

Hence, each processor $\mathtt{P_{x,y}}$ has to compute three functions. (We assume that $\mathtt{x}$ will be blackened in the i-th step and $\mathtt{y}$ in the j-th step.)

PROCESS $\mathtt{P_{x,y}}$ COMPUTES

$$-\mathtt{reachable(x,y)} = \mathtt{edge(x,y)} \vee (\vee/\,(\mathtt{R_x} \overset{\wedge}{\curlyvee} \mathtt{C_y}))$$
$$-\mathtt{r(x,y)} = \mathtt{edge(x,y)} \vee (\vee/\,(\mathtt{R_x}[1 \,..\, \mathtt{j}-1] \overset{\wedge}{\curlyvee} \mathtt{C_y}[1 \,..\, \mathtt{j}-1]))$$
$$-\mathtt{c(x,y)} = \mathtt{edge(x,y)} \vee (\vee/\,(\mathtt{R_x}[1 \,..\, \mathtt{i}-1] \overset{\wedge}{\curlyvee} \mathtt{C_y}[1 \,..\, \mathtt{i}-1]))$$

Obviously, any good compiler will find out the common subexpressions in these definitions.

Partitioning revisited A closer analysis shows that the above algorithm does not yield a reasonable amount of parallelization. (It is a nice exercise to visualize the rather odd wave patterns caused by our design.) But if we give up the unrealistic assumption that there is one processor for each pair of nodes, things improve considerably. Let us briefly compare the two covers of Fig. 17. A short back-of-the-envelope calculation shows that tiling requires in the order of $\frac{\sqrt{q}}{2}$ less communication than the block column partitioning (where q is the number of processors).

It is evident, how this modification of the covering changes the above programming concepts.

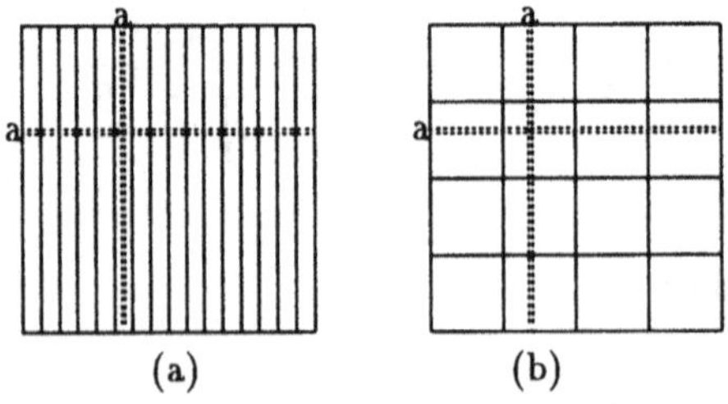

Fig. 17. Two possible partitionings

4 Conclusion

Even though the description was sometimes relatively informal, the above examples should help to elucidate the underlying *principle of skeletons*: Skeletons are functional forms on a very high and abstract level, which are based on appropriate coverings of the data objects under consideration.

There are, however, two aspects (which will be treated in companion papers) that we have completely ignored in this paper:

Architecture mapping To each skeleton we have to associate implementation patterns for the various parallel architectures of interest (see e.g. [DFH+92]). As a matter of fact, the skeletons themselves may be classified into different levels of abstraction such that a program development can proceed through several stages.
In the `reachable`-example this may look as follows (see Fig. 18). The highest level S_1 is given by the recursion equation for `reaches`. Then we transform this to the skeleton S_4, which is characterized by stream equations. And this relatively low-level skeleton can then be implemented on various architectures $A_1, A_2, A_3, \ldots$ (broadcasting, shared-memory, distributed-memory, ...).

Compositionality Realistic programs do not only consist of single algorithms but rather are a combination of a multitude of algorithms. Therefore the issue of *compositionality* plays a major role in the design of parallel programs. Compositionality is straightforward on the high functional level, but as of today it is not quite clear, how this extends to the lower implementation levels.

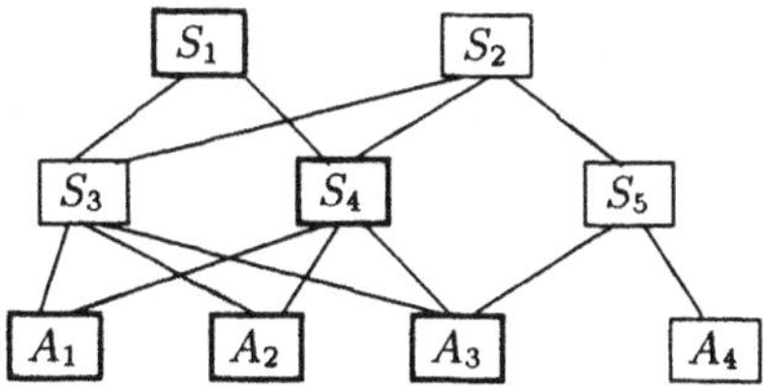

Fig. 18. Hierarchy of skeletons

As sketched in this paper, a general theory of skeletons based on an algebra of data partitioning can be formulated concisely in a category-theoretic framework. While this formalization is too abstract to be of practical use, carrying out parallel programming on the level of index manipulation is also an impractical solution for real problems. We are therefore looking for a methodology which is based on a very abstract formalization, but offers programming means suitable for describing parallel algorithms in a precise but high-level way.

In our experience these goals cannot be achieved by only considering suitable language concepts. What is needed in addition is a calculus for program derivation. With a calculus of this kind we can start from clear problem specifications and move in a step-by-step controlled process to the potentially very complex parallel implementations.

Acknowledgement: We thank our colleagues from the OPAL-project at the Technical University of Berlin for many stimulating discussions. The visit of Young-il Choo also provided us with new and helpful insights. Niamh Warde was of great assistance in the preparation of the manuscript.

References

[Bir87] R. S. Bird. An introduction to the theory of lists. In M. Broy, editor, *Logic of Programming and Calculi of Discrete Design*, volume 36 of *NATO ASI, F*, pages 5–42. Springer Verlag, 1987.

[Bir88] R. S. Bird. Lectures on constructive functional programming. In M. Broy, editor, *Constructive Methods in Computing Science*, volume 55 of *NATO ASI, F*, pages 151–216. Springer Verlag, 1988.

[BW90] M. Barr and C. Wells. *Category Theory for Computing Science*. Prentice Hall, 1990.

[Col89] M. Cole. *Algorithmic Skeletons: Structured Management of Parallel Computation*. MIT Press, 1989.

[DFH+92] J. Darlington, A. J. Field, P. G. Harrison et al. Parallel programming using skeleton functions. Personal Communication. Presented at PARLE '93.

[Gri93] A. S. Grimshaw. Easy-to-use object-oriented parallel processing with Mentat. *IEEE Computer*, 26(5):39–51, May 1993.

[Pep93] P. Pepper. Deductive derivation of parallel programs. In R. Paige, J. Reif, and R. Wachter, editors, *Parallel Algorithm Derivation and Program Transformation*. Kluwer Academic Publishers, 1993. To appear. Also: Technical Report 92-23, Technische Universität Berlin, July 1992.

[PM91] P. Pepper and B. Möller. Programming with (finite) mappings. In M. Broy, editor, *Informatik und Mathematik*, pages 381–405. Springer Verlag, 1991.

[Ski92] D. B. Skillicorn. The Bird-Meertens formalism as a parallel model. Presented at: NATO ARW Software for Parallel Computation, Cosenza, Italy, June 1992, 1992.

[Sri91] Y. V. Srinivas. A sheaf-theoretic approach to pattern matching and related problems. Technical report, Kestrel Institute, 1991.

[vH75] F. W. von Henke. On generating programs from types: An approach to automatic programming. In G. Huet and G. Kahn (eds.) *Construction, Amélioration et Vérification des Programmes*, 57–69. Colloques IRIA, 1975.

[YiC92] J. A. Yang and Y. il Choo. Formal derivation of an efficient parallel 2-d gauss-seidel method. In *Proceedings of the 6th International Processing Symposium*, pages 204–207. IEEE Computer Society Press, March 1992.

[YiC93] Young-il Choo. Data fields as parallel programs. Talk at TU Berlin, 1993.

Observing Some Properties of Event Structures*

I.B.Virbitskaite

Institute of Informatics Systems
Siberian Division of the Russian Academy of Sciences
630090, Novosibirsk, Russia

Abstract. The intention of the paper is to study prime event structures as models of nondeterministic processes. We characterize and examine some properties known as discreteness, density and crossing which allow inconsistency to be avoided between syntactic and semantic representations of processes. A number of close relationships between density and crossing concepts of the chosen model is shown. An algebraic system whose terms are interpreted as dense prime event structures is proposed.

1 Introduction

In concurrency semantics, a proliferation of models, namely: acyclic Petri nets, posets, event structures, etc., has been proposed to represent and study the behaviour of concurrent/distributed systems. However, not all instances of these models are suitable for the purpose of an adequate representation of 'reasonable' concurrent processes. There has been a line of research originating from [9] such as [1, 2, 6] where interest has mainly concentrated on the 'density', 'crossing' and 'discreteness versus continuity' properties of occurrence nets. In [9], a property called K-density has been defined, motivated by intuitive idea that every sequential subprocess of a process should always be in a well-defined state. [1] characterizes K-density in terms of other properties and some consequences of this are proved in [2]. L- and M-density as meaningful properties for acyclic Petri nets with nondeterministic choices have been introduced in [7]. Finally, Plünnecke [10] proved a variety of results on the relationship between K- and N-density in the context of posets in general.

The relative strength and significance of the mentioned above properties are not self-evident for event structures which are reminiscent of many poset models. The advantage of event structures is that the nondeterministic aspects of concurrent processes are explicitly described and the choices can naturally be expressed. From our point of view, an investigation of the K-density and related concepts for event structures is interesting for several reasons. First of all, it is always intriguing to see what are the consequences of small modifications and generalizations of important definitions. Algebraically, these properties lead to

* This work is suppoted in part by the Fundamental Research Fund of Russia (grant No 93-012-986).

elegant and simple laws [4, 5]. Moreover, we expect that event structures with such properties have a nice characterization in terms of temporal logic languages [11].

In this paper, we try to convey an understanding and evaluate the power and limitations of density and crossing properties for prime event structures (here, event structures for the sake of brevity). In Section 2 we shortly recall some basic notions of event structures. A number of density and crossing properties for event structures is treated in Section 3. As could be assumed several finiteness properties guarantee K- and L-crossing. The coincidence of L-density with L-crossing for the chosen model is established. A number of close relationships between different density concepts is also shown. Next, following [4] we describe an algebraic system whose terms are interpreted as finite dense event structures. A complete axiomatization of the interpretation equality is given in Section 4. We end with concluding remarks in Section 5 where the future lines of research are pointed out.

2 Event Structures

Our framework is event structures introduced ten years ago by Nielsen, Plotkin and Winskel in [8] as a fundamental model for computational processes.

Definition 1. An *event structure* is a triple $S = (E, \leq, \#)$, where

- E is a set of events,
- $\leq \subseteq E \times E$ is a partial order (the *causality relation*) satisfying the *principle of finite causes*:
  $\forall e \in E : \{d \in E \mid d \leq e\}$ is finite,
- $\# \subseteq E \times E$ is a symmetric and irreflexive relation (the *conflict relation*) satisfying the *principle of conflict heredity*:
  $\forall e_1, e_2, e_3 \in E : e_1 \leq e_2 \ \& \ e_1 \# e_3 \Rightarrow e_2 \# e_3$.

From now on for an event structure $S = (E, \leq, \#)$, we let $id = \{(e, e) \mid e \in E\}$; $< \, = \, \leq \setminus id$; $\leq^2 \subseteq \leq$ (*transitivity*); $\lhd = < \setminus <^2$; ${}^{\bullet}e = \{e' \in E \mid e' \lhd e\}$ and $e^{\bullet} = \{e' \in E \mid e \lhd e'\}$; $\smile = (E \times E) \setminus (\leq \cup \geq \cup \#)$ (*concurrency*); $e \#_1 d$ iff $(e \# d) \ \& \ (\forall e', d' \in E : (e' \leq e \ \& \ d' \leq d \ \& \ e' \# d' \Rightarrow e' = e \ \& \ d' = d))$ (*immediate conflict*).

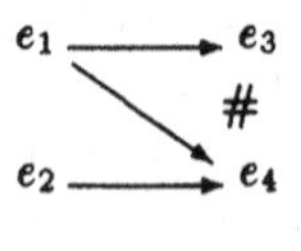

Fig. 1.

In graphic representations only immediate conflicts — not the inherited ones — are pictured. The <-relation is represented by arcs, omitting those derivable by transitivity. Following these conventions, a trivial example of the event structure is shown in Fig. 1.

Next we introduce some auxiliary notions which will be useful throughout the paper. First, however, we need the following. Let $B \subseteq E \times E$ be a relation on E. Then B^ϵ is the reflexive and symmetric closure of B.

Definition 2. Let $S = (E, \leq, \#)$ be an event structure.

- $A \subseteq E$ is a *cf-set* iff $\forall e_1, e_2 \in A : e_1 \#^\epsilon e_2$, A is a *cf-section* iff A is a maximal cf-set. The set of cf-sections in S is denoted by $CF(S)$,
- $C \subseteq E$ is a *co-set* iff $\forall e_1, e_2 \in C : e_1 \smile^\epsilon e_2$, C is a *co-section* iff C is a maximal co-set. The set of co-sections in S is denoted by $CO(S)$,
- $L \subseteq E$ is a *li-set* iff $\forall e_1, e_2 \in L : e_1 <^\epsilon e_2$, L is a *li-section* iff L is a maximal li-set. The set of li-sections in S is denoted by $LI(S)$.

An event structure $S = (E, \leq, \#)$ is *cf-finite* (respectively, *co-finite*, *li-finite*) iff any cf-section in S (respectively, co-section, li-section) is finite, S is *CFF-structure* (respectively, *COF-structure*, *LIF-structure*), if $\# = \emptyset$, (respectively, $\smile = \emptyset$, $< = \emptyset$).

Let us make some informal comments concerning substructures of S. A *CFF-substructure* of S contains only causally and concurrently related events. Hence, the set of maximal CFF-substructures of S characterizes the projection of S to a plane formed by li- and co-axes. A *COF-substructure* (*LIF-substructure*) of S is the counterpart of its CFF-substructure with respect to li- and cf-axes (co- and cf-axes).

Definition 3. Let $S = (E, \leq, \#)$, $S' = (E', \leq', \#')$ be event structures.

- S' is a *substructure* of S ($S' \subseteq S$) iff $E' \subseteq E$, $\leq' \subseteq \leq \cap E'^2$, $\#' \subseteq \# \cap E'^2$,
- S' is a *maximal substructure* of S if for any substructure S'' of S such that $S' \subseteq S''$ is valid that $S' = S''$,
- S' is a *maximal CFF-substructure* (respectively, *maximal COF-substructure*, *maximal LIF-substructure*) of S iff S' is a CFF-structure (respectively, COF-structure, LIF-structure) and a maximal substructure of S.

It is known that in any event structure there is no infinite li-set, be it ascending or descending between any pair of events, i.e. event structures are *discrete* models of processes. The following auxiliary lemma exhibits this fact.

Lemma 4. *Let $S = (E, \leq, \#)$ be an event structure. Then*

$$\forall e_1, e_2 \in E, \forall L \in LI(S) : |\, [e_1, e_2] \cap L \,| < \infty,$$
where $[e_1, e_2] = \{e \in E \mid e_1 \leq e \leq e_2\}$.

In such a way, we have recalled basic terminology of event structures and defined some additional notions needed to introduce the density and crossing concepts for event structures.

3 Density and Crossing Properties of Event Structures

Our aim in this section is to introduce a hierarchy of density and crossing properties which are motivated by the wish to exclude unreasonable processes and to give a few key results pertaining to the properties. First we rephrase the notions of K-, N-density and, the so-called, K-crossing [10] in terms of event structures. In doing so it will be convenient to adopt the following notations.

Let $S = (E, \leq, \#)$ be an event structure and $X \subseteq E$. Then $\downarrow X = \{e' \mid e' \in E \ \&\ \exists e \in X : e' \leq e\}$, and $\uparrow X = \{e' \mid e' \in E \ \&\ \exists e \in X : e \leq e'\}$.

Definition 5. Let $S = (E, \leq, \#)$ be an event structure and $S' = (E', \leq')$ be a maximal CFF-substructure of S. Then

- S' is *K-dense* iff $\forall L \in LI(S'),\ \forall C \in CO(S') : |\, L \cap C \,| = 1$,
- S' is *K-crossing* iff $\forall L \in LI(S'),\ \forall C \in CO(S') : L \cap \downarrow C \neq \emptyset \ \&\ L \cap \uparrow C \neq \emptyset$,
- S is *K-dense* (*K-crossing*) iff any maximal CFF-substructure of S is K-dense (K-crossing),
- S is *N-dense* iff $\forall e_0, e_1, e_2, e_3 \in E$: if $(e_1 < e_0 \ \&\ e_0 \smile e_2)$ and $(e_3 < e_2 \ \&\ e_3 \smile e_1)$ then $e_1 < e_2 \Rightarrow e_3 < e_0$.

An example of the K-, N-dense and K-crossing event structure is shown in Fig. 1. Illustrating further, the event structure in Fig. 2(a) is N-dense, but neither K-dense nor K-crossing, whereas the event structure in Fig. 2(b) is K-crossing but neither K-dense nor N-dense.

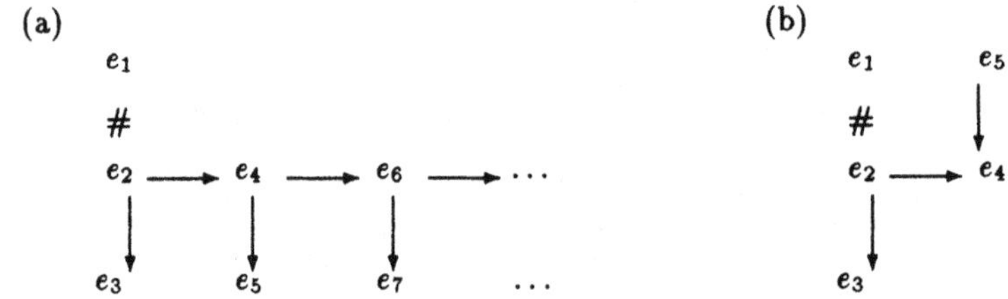

Fig. 2.

The following result states a connection between the properties defined prior to that.

Proposition 6. *Let $S = (E, \leq, \#)$ be an N-dense event structure. Then S is K-dense, iff S is K-crossing.*

Proof. It follows from Theorems 2.2.6, 2.3.11 in [3] and Lemma 4. □

The next proposition establishes some basic relationships between several finiteness and the properties above.

Proposition 7. *Let $S = (E, \leq, \#)$ be a co-finite or li-finite event structure. Then*

(i) S is K-crossing,
(ii) S is K-dense, iff N-dense.

Proof. Let S be co-finite (li-finite). Then
(i) is an immediate generalization of Proposition 3.2 (5) in [10] (Lemma 4 and Proposition 3.2 (6,8) in [10]).
(ii) follows from Theorem 4.3 in [10] (and Proposition 3.2(8) in [10]). □

We now wish to formulate the definitions of the L-density [7] and L-crossing properties of event structures as follows.

Definition 8. Let $S = (E, \leq, \#)$ be an event structure and $S' = (E', \leq', \#')$ be a maximal COF-substructure of S.

- S' is *L-dense* iff $\forall L \in LI(S'),\ \forall A \in CF(S') : |\, L \cap A \,| = 1$,
- S' is *L-crossing* iff $\forall L \in LI(S'), \forall A \in CF(S') : L \cap \downarrow A \neq \emptyset \ \&\ L \cap \uparrow A \neq \emptyset$,
- S is *L-dense* (*L-crossing*) iff any maximal COF-substructure S' of S is L-dense (L-crossing).

According to the above definition, the event structure in Fig. 3(a) is L-dense and L-crossing, when the event structure in Fig. 3(b) is neither L-dense nor L-crossing.

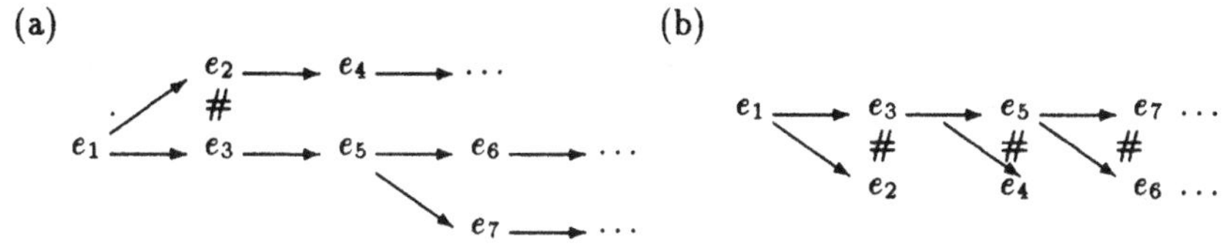

Fig. 3.

Proposition 9. *Let $S = (E, \leq, \#)$ be an event structure. Then S is L-dense iff S is L-crossing.*

Proof. ($\Rightarrow$): This is obvious.
($\Leftarrow$): Let $S' = (E', \leq', \#')$ be a maximal COF-substructure of S. Let $L \in LI(S')$, $A \in CF(S')$. According to the L-crossing property there exists a maximal element $e \in L \cap \downarrow A$ and a minimal element $d \in L \cap \uparrow A$. Assume $e \neq d$, then $e < d$.

- If $e \lhd d$, then there exist $a, a' \in A$ such that $e \leq a$ and $a' \leq d$.
  1. If $e = a = a'$ or $d = a = a'$, then the result is proved.
  2. If $a = a'$ and $e \neq a$, $d \neq a'$, then this contradicts the definition of $\lhd$.

3. If $a \neq a'$, $e \neq a$ and $d \neq a'$, then it contradicts the definition of a COF-structure.

– If $\neg(e \lhd d)$, then there exists $a \in L$ such that $e < a < d$. If $a \in \downarrow A$, then it contradicts the maximality of e. If $a \in \uparrow A$, then it contradicts the minimality of d. But no other cases remain because $\uparrow A \cup \downarrow A = E$. Hence $e = d \in \downarrow A \cap \uparrow A = A$. □

Some finiteness restrictions which imply L-crossing are established by the following proposition.

Proposition 10. *Let $S = (E, \leq, \#)$ be a cf-finite or li-finite event structure. Then S is L-crossing.*

Proof. Let S be cf-finite. Then $\forall L \in LI(S)$, $\forall A \in CF(S) : (L \subseteq \downarrow A \Rightarrow \exists a \in A$ such that $L \subseteq \downarrow \{a\})$ & $(L \subseteq \uparrow A \Rightarrow \exists a \in A$ such that $L \subseteq \uparrow \{a\})$. Using this fact the required result is proved..
Next, let S be li-finite. We assume that S is not L-crossing. Then there exists a maximal non-L-crossing COF-substructure $S' = (E', \leq', \#')$ of S. Let $\emptyset \neq A \in CF(S')$, $\emptyset \neq L \in LI(S')$ such that $L \cap \uparrow A = \emptyset$ or $L \cap \downarrow A = \emptyset$. Then $L \subseteq \downarrow A \setminus A$ or $L \subseteq \uparrow A \setminus A$, respectively. Let $L = \{x_1, ..., x_n\}$.

- If $L \subseteq \downarrow A \setminus A$, then there exists $a \in A$ such that $x_n < a$. Hence L is not an li-section.
- If $L \subseteq \uparrow A \setminus A$, then the proof is analogous. □

One may generalize the density concept still further by introducing the notions of R- and N′-density. Formally, these properties can be defined as follows.

Definition 11. Let $S = (E, \leq, \#)$ be an event structure and $S' = (E', \leq', \#')$ be a maximal LIF-substructure of S. Then

- S' is *R-dense* iff $\forall C \in CO(S')$, $\forall A \in CF(S') : |C \cap A| = 1$,
- S is *R-dense* iff any maximal LIF-substructure S' of S is R-dense,
- S is called *N′-dense* iff $\forall e_0, e_1, e_2, e_3 \in E$: if $(e_0 \# e_1$ & $e_0 \smile e_2)$ and $(e_2 \# e_3$ & $e_1 \smile e_3)$ then $e_1 \# e_2 \Rightarrow e_0 \# e_3$.

Illustrating the concepts, the event structure in Fig. 4(a) is R-dense and N′-dense, whereas the event structure in Fig. 4(b) is neither R-dense nor N-dense.

Fig. 4.

The proposition below states that N′-density implies R-density for event structures satisfying some finiteness constraints.

Proposition 12. *Let $S = (E, \leq, \#)$ be a cf-finite or co-finite event structure. Then S is R-dense if S is N′-dense.*

Proof. We suppose a contrary. This means that there exists a maximal LIF-substructure S' of S in which there are a cf-section A and a co-section C such that $\mid C \cap A \mid \neq 1$. Two cases are admissible.

- If $\mid C \cap A \mid = 0$. Since S is cf-finite or co-finite then we get a contradiction because S is N′-dense.
- If $\mid C \cap A \mid > 1$. This contradicts the definittion of the $\smile$ relation. □

Here we shall introduce a few properties which in a certain sense ensure the coincidence of R- and N′-density. An event structure S is *cf-binary* (*co-binary*) iff $\forall A \in CF(S) : \mid A \mid \leq 2$ ($\forall C \in CO(S) : \mid C \mid \leq 2$).

Proposition 13. *Let $S = (E, \leq, \#)$ be a cf- or co-binary event structure. Then S is R-dense iff S is N′-dense.*

Proof. First we assume R-density of S. Suppose that S is not N′-dense. This means that there exist events $e_0, e_1, e_2, e_3 \in E$ such that ($e_0 \# e_1$ & $e_0 \smile e_2$), ($e_2 \# e_3$ & $e_1 \smile e_3$) and $e_1 \# e_2$. Then $\{e_0, e_3\}$ is a co-set and $\{e_1, e_2\}$ is a cf-set in S. Since S is R-dense then there exists a maximal LIF-substructure S' of S in which there is an event e_4 such that $\{e_0, e_3, e_4\}$ is a co-section and $\{e_1, e_2, e_4\}$ is a cf-section. This contradicts both co- and cf-binarity of S.
Now we suppose that S is N′-dense but not R-dense. This means that there exists a maximal LIF-substructure S' of S in which there are a cf-section A and a co-section C such that $\mid C \cap A \mid \neq 1$. We have arrived at a special case of the situation considered in the proof of Proposition 12. □

We now aim at defining some modifications of the M-density concept [4] for event structures as follows.

Definition 14. Let $S = (E, \leq, \#)$ be an event structure. Then

- S is *M^{cf}-dense* iff the intersection of any maximal LIF-substructure of S with any maximal COF-substructure of S results in some (unique) cf-section of S,
- S is *M^{co}-dense* iff the intersection of any maximal LIF-substructure of S with any maximal CFF-substructure of S results in some (unique) co-section of S,
- S is *M^{li}-dense* iff the intersection of any maximal CFF-substructure of S with any maximal COF-substructure of S results in some (unique) li-section of S,
- S is *M-dense* iff S is M^{cf}-, M^{co}- and M^{li}-dense.

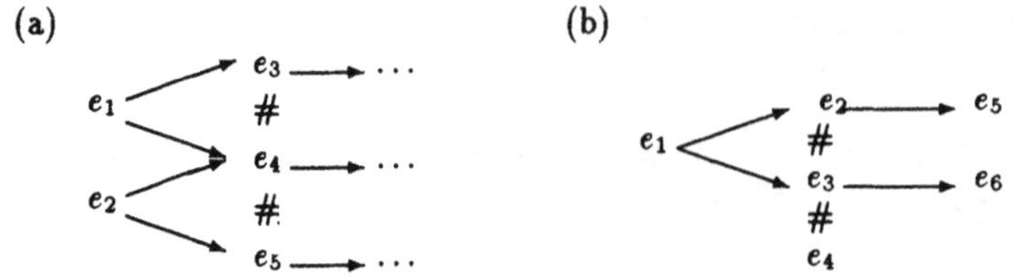

Fig. 5.

As an illustration, we may use the M-dense event structure in Fig. 5(a) but the event structure in Fig. 5(b) is neither M^{cf}-dense nor M^{co}-dense nor M^{li}-dense.

In order to establish the close relationship between different density concepts it is necessary to define another requirement which we may call the triangle-feeness property: an event structure S satisfies this property (referred to as the ∇-*freeness* property) if it does not contain a configuration: $\forall e_0, e_1, e_2 \in E$: $e_0 < e_1 \# e_2 \smile e_0$.

Proposition 15. *Let $S = (E, \leq, \#)$ be an event structure. Then*

(i) If S is K-dense and ∇-free then S is M^{cf}-dense,
(ii) If S is L-dense and ∇-free then S is M^{co}-dense,
(iii) If S is R-dense and ∇-free then S is M^{li}-dense.

Proof. (iii) We suppose that S is not M^{li}-dense. Then there exist its maximal CFF-substucture $S' = (E', \leq', \#')$ and maximal COF-substructure $S'' = (E'', \leq'', \#'')$ such that $E' \cap E'' = L$ is not a li-section in S.

- Let $L = \emptyset$. Two cases are admissible.
  1. There exist at least four events $e_0, e_1, e_2, e_3 \in E$ such that $E' = \{e_0, e_3\}$ and $E'' = \{e_1, e_2\}$, i.e. ($e_0 \# e_1$, $e_2 \# e_3$, $e_0 \smile e_2$, $e_1 \smile e_3$) and $e_1 \# e_2$. Since S is R-dense then there exists a maximal LIF-substructure S''' of S in which there is an event e_4 such that $\{e_0, e_3, e_4\}$ is a co-section and $\{e_1, e_2, e_4\}$ is a cf-section in S'''. This contradicts the maximality of S''.
  2. There exist events $e_0, e_1, e_2, e_3, \ldots \in E$ such that $E' = \{e_0, e_2, \ldots\}$ and $E'' = \{e_1, e_3, \ldots\}$. Then there exists a maximal LIF-substructure S''' of S in which there are a co-section $C = E'$ and a cf-section $R = E''$ such that $| C \cap A | = 0$. We get a contradiction to R-density of S.
- Let $L = \{e_0, ..., e_n\}$ be a li-set.
  1. If there exists an event e' such that $e' < e_0$ then $e', e_0 \in E'$ by definition of a maximal CFF-substructure and there is an event $e'' \in E''$ such that $e'' \# e_0$ and $e'' \smile e'$. This contradicts ∇-freeness of S.
  2. If there exists an event e' such that $e_n < e'$ then the reasoning is analogous to the one in the previous case. We get a contradiction to the principle of conflict heredity .

The proofs of parts *(i)* and *(ii)* are similar to the proof of part *(iii)*. □

4 An Algebraic System for Dense Event Structures

First by labelling the events with the actions taken from some alphabet we get labelled event structures. Then to denote finite labelled event structures we introduce an abstract syntax where the primitive constructs are sequential composition, parallel composition and sum. It follows closely to work [4] but we show that our algebriac system is oriented on the class of dense labelled event structures.

Definition 16. Let A be a nonempty set. A *labelled event structure* (LS) is a structure $(E, \leq, \#, \lambda)$ where $\lambda : E \Rightarrow A$ is the labelling function.

From now on, let $*$ be one of $\leq, \smile, \#$ and S_0, S_1 be LS's. Then $S_0 * S_1$ is the structure we get by binding S_0 and S_1 and setting the $*$-relation between the events of S_0 and S_1. If $*$ is $\leq$, this is called the *sequential composition* of S_0 and S_1 and denoted $S_0; S_1$. If $*$ is $\smile$, this is the *parallel composition* $S_0 \parallel S_1$, and if $* = \#$, this is the *sum* $S_0 + S_1$.

Let $S_i = (E_i, \leq_i, \#_i, \lambda_i)$ for $i \in \{0, 1\}$, then $S_0 * S_1 = (E, \leq, \#, \lambda)$, where
$E = E_0 \uplus E_1$, i.e. $E = \{0x \mid x \in E_0\} \cup \{1x \mid x \in E_1\}$;
$ix \leq jy \Leftrightarrow (i = j \text{ and } x \leq_i y)$ or $(* =\leq,\ i = 0,\ \#_0 = \emptyset, \text{ and } j = 1)$;
$ix \# jy \Leftrightarrow (i = j \text{ and } x \#_i y)$ or $(* = \# \text{ and } i \neq j)$;
$\lambda(ix) = \lambda_i(x)$.

These operations are defined up to isomorphism (notation $\rightleftharpoons$) as follows:

$$S_0 \rightleftharpoons S_0' \text{ and } S_1 \rightleftharpoons S_1' \Rightarrow \begin{cases} S_0; S_1 \rightleftharpoons S_0'; S_1' \\ S_0 + S_1 \rightleftharpoons S_0' + S_1' \\ S_0 \parallel S_1 \rightleftharpoons S_0' \parallel S_1' \end{cases}$$

A labelled event structure $S = (E, \leq, \#, \lambda)$ is *dense* if $(E, \leq, \#)$ is K-, L-, R- and M-dense. Let $\mathcal{D}(A)$ be the set of finite dense LS's.

Lemma 17. *Let $S_0, S_1 \in \mathcal{D}(A)$. Then*

(i) $S_0 + S_1, S_0 \parallel S_1$ are in $\mathcal{D}(A)$, and
(ii) If S_0 is a CFF-structure then $S_0; S_1$ is in $\mathcal{D}(A)$.

Proof. Straightforward. □

An abstract *syntax* for the set of terms $T(A)$ to denote finite LS's is generated by the following production system:

$F ::= P \mid P; F \mid F \parallel F \mid F + F,$
$P ::= \mathbf{1} \mid a \mid P; P \mid P \parallel P$, where $a \in A$.

Let $\mathcal{T}(p)$ be the labelled event structure denoted by the term p, defined as follows:
$\mathcal{T}(\mathbf{1}) = (\emptyset, \emptyset, \emptyset, \emptyset)$ (the empty structure),
$\mathcal{T}(a) = (\{e\}, =, \emptyset, \lambda)$ with $\lambda(e) = a$,

$\mathcal{T}(p;q) = \mathcal{T}(p);\mathcal{T}(q)$,
$\mathcal{T}(p \parallel q) = \mathcal{T}(p) \parallel \mathcal{T}(q)$,
$\mathcal{T}(p+q) = \mathcal{T}(p) + \mathcal{T}(q)$.

For example, the term $(a \parallel b);(c \# d)$ denotes the event structure shown in Fig. 6.

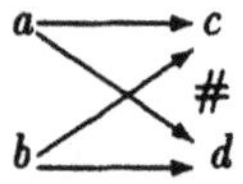

Fig. 6.

Let $[S]$ be the isomorphism class of S. Then the *interpretation* of $p \in T(A)$ is: $\mathcal{F}(p) \overset{\text{def}}{\Leftrightarrow} [\mathcal{T}(p)]$. Let us denote the interpretation equality as follows:
$p =_{\mathcal{F}} q \overset{\text{def}}{\Leftrightarrow} \mathcal{F}(p) = \mathcal{F}(q) \overset{\text{def}}{\Leftrightarrow} \mathcal{T}(p) \rightleftharpoons \mathcal{T}(q)$.

Now, we will define the *axiom system* corresponding to the interpretation equality defined above. In the following equations, $\mathbf{1}, p, q, r \in T(A)$.

(i) $A0 : (p;(q;r)) = ((p;q);r)$
$U0 : (p;\mathbf{1}) = p = (\mathbf{1};p)$
(ii) $A1 : (p \parallel (q \parallel r)) = ((p \parallel q) \parallel r)$
$U1 : (p \parallel \mathbf{1}) = p = (\mathbf{1} \parallel p)$
$C1 : (p \parallel q) = (q \parallel p)$
(iii) $A2 : (p + (q + r)) = ((p + q) + r)$
$U2 : (p + \mathbf{1}) = p = (\mathbf{1} + p)$
$C2 : (p + q) = (q + p)$

Let Θ be the equational system whose axioms are A0 to A2, U0 to U2, C1 and C2 and let $=_\Theta$ be the congruence on $T(A)$ generated by these equations. Then it is obvious that $p =_\Theta q \Rightarrow p =_{\mathcal{F}} q$. We have the system Θ be *sound*.

In order to show that the system Θ is complete we have to describe a *normal form* for a term of $T(A)$. Let $\mathcal{N}(A) = \{\mathbf{1}\} \cup \mathcal{W}(A)$, where $\mathcal{W}(A)$ is the least set of terms built according to the following rules:

(i) every atom $a \in A$ is in $\mathcal{W}(A)$ and has no head operator,
(ii) if $p \in \mathcal{W}(A)$ does not have ; (respectively, $\parallel, +$) as head operator and if $q \in \mathcal{W}(A)$, then $(p;q)$ (respectively, $(p \parallel q), (p+q)$) is in $\mathcal{W}(A)$ and has ; (respectively, $\parallel, +$) as head operator.

Lemma 18. *Let φ be the system whose axioms are A0 to A2 and U0 to U2 and ψ be the system consisting of A0 to A2, C1 and C2. Then*

(i) for each term $p \in T(A)$ there exists a normal form $t \in \mathcal{N}(A)$ such that $p =_\varphi t$,
(ii) for two normal forms $t, t' \in \mathcal{N}(A)$ $t =_\Theta t' \Leftrightarrow t =_\psi t'$.

Proof. Straightforward. □

Proposition 19.

(i) $\forall p \in T(A) \mid \mathcal{T}(p) \rightleftharpoons S : S \in \mathcal{D}(A)$,
(ii) $p =_{\mathcal{F}} q \Leftrightarrow p =_{\Theta} q$.

Proof. (i) K-density of S follows from the definition of the syntax of terms in $T(A)$. L-density of S is an immediate consequence of Propositions 9 and 10. According to the definition of the syntax of terms in $T(A)$ and Theorem 3.4 in [4], S is N′-dense and ∇-free. By Proposition 12, we have R-density of S. Moreover, S is M-dense by Proposition 15. So, the result is proved.
(ii) We must show $t, t' \in \mathcal{N}(A) \Rightarrow \mathcal{T}(t) \rightleftharpoons \mathcal{T}(t') \Leftrightarrow t = \psi t'$. The proof of this point is omitted here. □

5 Concluding Remarks

In these paper, we have tried to present a variety of density and crossing properties of prime event structures and discuss why these might be useful properties. We have done so both by proving some new results and by generalizing old ones. Perhaps the most interesting outcome of our work has been the close relationship between different density concepts. To get this we had to introduce some new notions, namely: R-density and a number of modifications of M-density.

The work presented here is by no means complete. Regarding future works, some lines of research may be pointed out. The first line of research to pursue should be to provide dense and crossing event structures with a concrete interpretation. This would enable us to evaluate the power and the weakness of these properties of event structures. A concrete interpretation would add some insight to the vague explanations of the various concepts which we have provided within the paper. We also plan to enrich the algebraic syntax with recursive definitions and show how to interprete infinite terms as dense event structures. So far we have limited ourselves to prime event structures. We expect that our results may be generalized by the concept of flow event structures introduced in [4].

References

1. Best, E.: The relative strenghth of K-density. Lecture Notes in Computer Science **84** (1980) 261-276
2. Best, E.: A Theorem on the Characteristics of Non-Sequential Processes. Fundamenta Informaticae **3** (1980) 77-94
3. Best, E., Fernandez, C., Plünnecke, H.: Concurrent systems and processes. Final Report on the Foundational Part of the Project BEGRUND, FMP-Studien N 107, GMD, Sankt Augustin, FDR (March 1985)
4. Boudol, G., Castellani, I.: Concurrency and atomicity. Theoretical Computer Science **59** (1988) 25-84

5. Cherkasova, L., Kotov, V.: Descriptive and analytical process algebras. Lecture Notes in Computer Science **424** (1989) 77-104
6. Fernandez, C.: Non-sequential processes. Lecture Notes in Computer Science **254** (1986) 95-116
7. Kotov, V., Cherkasova, L.: On structural properties of generalized processes. Lecture Notes in Computer Science **188** (1984) 288-306
8. Nielsen, M., Plotkin, G., Winskel, G.: Petri nets, event structures and domains. Theoretical Computer Science **13** (1981) 85-108
9. Petri, C.: Concurrency as a basis for system thinking. ISP-Report 78.06, St. Augustin: Gesellschaft für Mathematik und Detenverarbeitung (1978)
10. Plünnecke, H.: K-density, N-density and finiteness properties. Lecture Notes in Computer Science **188** (1984) 392-412
11. Virbitskaite, I., Votintseva A.,: A Temporal Logic Characterization of a Subclass of Flow Event Structures. (in preparation)

The Other Linear Logic
(Extended Abstract)

M.Taitslin, D.Arkhangelsky

Tver State University, Tver, Russia

Abstract. One of computer science motivations for Linear Logic is that proofs in this logic reflect the use of certain resources. However, the problem of availability of resources is not addressed. In this paper we are attempting to fill this gap.
Our approach can be illustrated most clearly by example of a Horn fragment. In this fragment, the left part of each sequent is a conjunction of simple implications and supplies, and the right part is a simple conjunction of supplies. An implication is called simple, if both its succedent and antecedent are conjunctions of supplies. As usual, a supply is simply a letter.
The standard axiom system for this Horn fragment consists of three inference rules, while the axioms are all the sequents with a single letter both for the left and right parts (the letter is the same).
On the other hand, the implications can be thought of as determining relations in a commutative semigroup and then we show that a sequent is provable iff the collection of supplies in its right part can be obtained from the left one by applying all the relations from the left part in appropriate order.
Now, let us require that the left part (i.e. all its supplies) is available. For this case we construct a new axiom system with sligtly more complex axioms and inference rules in which any axiom can be used at most once. Among other corollaries of this result we obtain the NP-completeness of the Horn fragment (known result).
However, for a bounded number of implication types we construct a subexponential algorithm for the derivation search.

We propose a calculus for describing the behaviour of a net with restricted resources. Then we prove the completeness theorem and canonical proof theorem for this calculus.

One of the computer science motivations for Linear Logic is that proofs in this logic reflect the use of certain resources. However, the problem of availability of these resources is not addressed. In this paper we are attempting to fill this gap. We are attempting to investigate the common work of a net constructed from elements with restricted resources. Our approach can be most clearly illustrated by the example of the Horn fragment.

Horn implications are formulas of the form

$$(X \rightarrow Y)$$

where X, Y are conjunctions of letters. Horn sequent is a sequent whose left part is a conjunction of Horn implications and letters and whose right part is simply a conjunction of letters.

A Horn implication $(X \to Y)$ is applicable to a conjunction Z of letters if there exists a conjunction U of letters such that Z can be represented as (XU). The result of the application is (YU).

A sequence $\alpha_1, \ldots, \alpha_n$ of Horn implications is applicable to Z iff there exists a natural i such that α_i is applicable to Z and $\alpha_1, \ldots, \alpha_{i-1}, \alpha_{i+1}, \ldots, \alpha_n$ is applicable to the result. We consider letters as transformable objects and we use implications as tools for transformation. It is convenient to interpret implications as tools for transformation of objects indicated in the premise to objects indicated in the conclusion.

We prefer to consider formulas instead of lists of formulas. We replace a list of formulas by a conjunction of the formulas.

Two formulas are equivalent if one of them can be obtained from another using commutative and associative rules for conjunction.

The calcucus for Horn sequents.

Axioms:

$p \vdash p$ where p is a letter.

Proof rules:

1. $$\frac{A \vdash B \quad D \vdash C}{(AD) \vdash (BC)}$$

2. $$\frac{A \vdash C \quad (BD) \vdash F}{((AB)(C \to D) \vdash F}$$

3. $$\frac{A \vdash C}{B \vdash D}$$ if A is equivalent to B and C is equivalent to D.

A Horn sequent $X\alpha_1 \ldots \alpha_n \vdash Y$ where X is a word and $\alpha_1, \ldots, \alpha_n$ are Horn implications is true iff $\alpha_1 \ldots \alpha_n$ is applicable to X, and the result of applying is Y.

The question: which Horn sequents are provable?

The answer: a Horn sequent is provable if and only if it is true.

An example.

Consider the sequent

$$X^2Y^3(X^3Z \to Y)(Y \to XZ) \vdash Y^3$$

We apply $Y \to XZ$ to X^2Y^3. The result is X^3Y^2Z. Then we apply $X^3Z \to Y$ to the result, and obtain Y^3. The sequent

$$X^2Y^3(X^3Z \to Y)(Y \to XZ) \vdash Y^3$$

is true.

A proof of the sequent:

$$\dfrac{Y\vdash Y \quad \dfrac{\dfrac{\dfrac{\dfrac{X\vdash X \quad \dfrac{X\vdash X \quad X\vdash X}{X^2\vdash X^2}\,1}{X^3\vdash X^3}\,1 \quad Z\vdash Z}{X^3Z\vdash X^3Z}\,1 \qquad \dfrac{Y\vdash Y \quad \dfrac{Y\vdash Y \quad Y\vdash Y}{Y^2\vdash Y^2}\,1}{Y^3\vdash Y^3}\,1}{X^3ZY^2(X^3Z\rightarrow Y)\vdash Y^3}\,2}{X^2Y^2(X^3Z\rightarrow Y)XZ\vdash Y^3}\,3}{\dfrac{YX^2Y^2(X^3Z\rightarrow Y)(Y\rightarrow XZ)\vdash Y^3}{X^2Y^3(X^3Z\rightarrow Y)(Y\rightarrow XZ)\vdash Y^3}\,3}\,2$$

The proof of the answer is simple.

Indeed, let a sequent be provable. If it is an axiom then its truth is obvious. Assume the rule 1 is used:

1. $\dfrac{A\vdash B \quad D\vdash C}{(AD)\vdash(BC)}$.

First we obtain B from A and then we obtain D from C.

Assume the rule 2 is used:

2. $\dfrac{A\vdash C \quad (BD)\vdash F}{((AB)(C\rightarrow D)\vdash F}$

First we obtain C from A, and then we apply $(C\rightarrow D)$ to C, and obtain D. Then we obtain F from (BD).

Let a sequent be true. Suppose the left part is $X\alpha_1\ldots\alpha_n$ where X is a word and $\alpha_1,\ldots,\alpha_n$ are implications. We use induction on n. If $n=0$ the case is obvious. Otherwise we have:

$$\dfrac{A\vdash A \quad (BD)\alpha_2\ldots\alpha_{n+1}\vdash Y}{(A\rightarrow B)\alpha_2\ldots\alpha_{n+1}\vdash Y} \qquad AD$$

The problem of Horn sequent provability in Girard's Linear Logic is NP-complete.

This result was proved by Kanovich in 1991. We propose a simple proof for a stronger statement.

The sequent

$$Y^{9a}(Y^{3a}\rightarrow Z^{a-a_1}Y^{a_1})\ldots(Y^{3a}\rightarrow Z^{a-a_{3m}}Y^{a_{3m}})(Y^aZ^{2a}\rightarrow Y^{9a})^m\vdash Y^{9a}$$

is provable if and only if positive natural numbers $a_1,\ldots,a_{3m}$ having the following properties:

$$a_1,\ldots,a_{3m}<a;$$

$$a_1 + \ldots + a_{3m} = ma$$

can be partitioned into m triples whose sum is equal to a.

Indeed, we can apply only less than 4 implications. We cannot apply the last implication. If we apply less than 3 implications we cannot obtain Z^{2a}, and then we cannot apply the last implication to the result.

If we apply exactly 3 implications, and obtain Z^{2a}, then suppose that we apply implications

$$(Y^{3a} \to Z^{a-a_{i1}}Y^{a_{i1}})(Y^{3a} \to Z^{a-a_{i2}}Y^{a_{i2}})(Y^{3a} \to Z^{a-a_{i3}}Y^{a_{i3}}).$$

Thus $a_{i1} + a_{i2} + a_{i3} \leq a$.

If $a_{i1} + a_{i2} + a_{i3} < a$ then we cannot obtain Y^a and cannot apply the last implication.

If $a_{i1} + a_{i2} + a_{i3} = a$ we use induction on m.

Suppose we fix a list of implications

$$(A_1 \to B_1), \ldots, (A_s \to B_s)$$

and we will use the implications from this list only.

So we will consider sequents of the form

$$A(A_1 \to B_1)^{n_1}, \ldots, (A_s \to B_s)^{n_s} \vdash B$$

where A and B are any conjunctions of $x_1, \ldots, x_m$, and $A_1, B_1, \ldots, A_s, B_s$ are fixed conjunctions of $x_1, \ldots, x_m$.

There is an algorithm with complexity $n^{c*\log(n)}$ where c is a constant which constructs a proof for a given sequent or states that the sequent is not provable.

The idea uses the fact that the result of applying a sequence of implications to a word does not depend on the order which the implications are applied in, if the result exists.

We can partition the sequence of implications into two subsequences of the same length, and try to apply the first of them to the word and second of them to the result.

But to have a possibility to spend resources, it needs before to have these resources.

In Girard's Linear Logic the resource availability condition is not taken into account. If we add the resource availability condition to Girard's Logic, then with checking a sequent truth it is necessary:

a) to check its provability,

b) to check the condition of sufficiency of available resources.

We propose a modified Linear Logic calculus having the following property:

A sequent is provable in our calculus if and only if both it is provable, and it uses available resources only.

We obtain many different calculi depending on available resources.

This idea permits us to study problems of borrowing of supplies.

Suppose we have a fixed set S of supplies.

We fix a finite set P of implications.

We denote the union of S and P by R.

For each $D \in R$ we give a number $\alpha(D)$ which denotes amount of available resources.

The calculus T_α.

Axioms.

We have $\alpha(p)$ copies of the axiom $p \vdash p$ for each $p \in S$ and $\alpha(X \to Y)$ copies of the axiom $(X \to Y) \vdash (X \to Y)$ for each $(X \to Y) \in$ P.

Proof rules.

1. $\dfrac{A \vdash B \quad D \vdash C}{(AD) \vdash (BC)}$
2. $\dfrac{A \vdash (BH) \quad C \vdash ((B \to D)F)}{\vdash ((HD)F)}$ $\qquad AC$
3. $\dfrac{A \vdash C}{B \vdash D}$ if A is equivalent to B and C is equivalent to D.

A sequent is called elementary if its left and right parts are conjunctions of letters and Horn implications.

A proof is a finite tree. Each leaf of the proof is indexed by an axiom. The number of all the leaves indexed by a given axiom is less than or equal to the number of copies of this axiom in T_α. Each node being not a leaf is indexed by a sequent obtained from the sequents by indexing the node successors using one of proof rules 1,2,3.

An elementary sequent is true in α if and only if one can obtain B from A using relations from A and, in addition, for each $D \in R$, the number of D in A is not greater than $\alpha(D)$.

Theorem. An elementary sequent is true in α if and only if it is provable in T_α.

Acknowledgements. We would like to thank S.Adian and S.Artemov and also M.Dehttar and Musikaev for helpful discussions.

Duration Calculi: An Overview
(Extended Abstract)

ZHOU Chaochen*

Abstract. The Duration Calculi are calculi for designing real-time software embedded systems. The Calculi overviewed in the paper include the Duration Calculus, the Extended Duration Calculus, the Mean Value Calculus and the Probabilistic Duration Calculus. They are extensions of an Interval Temporal Logic, and can accommodate some concepts of mathematical analysis such as integrals, mean values, piecewise continuity and differentiability of functions over continuous time. The Calculi can be used to capture and refine real-time requirements for hybrid systems, to define real-time behaviour/semantics of computing systems, and to calculate system dependability in terms of requirement satisfaction probability.

1 Summary

The research of the Duration Calculi was started by the ProCoS project (Provably Correct Systems - Esprit BRA 3104) in 1989, when the project was researching formal techniques for designing time/safety critical software embedded systems. Four calculi have been developed since then. They are the Duration Calculus, the Extended Duration Calculus, the Mean Value Calculus and the Probabilistic Duration Calculus.

The Duration Calculus is a real-time interval logic [14]. It formalises integrals of Boolean functions over intervals, and is used to specify and reason about timing and logical constraints on discrete states of a system. All the other calculi are extensions of the Duration Calculus. The Extended Duration Calculus [16] extends the Duration Calculus with piecewise continuity/differentiability of functions. It can capture properties of continuous states. This calculus is useful for reasoning about hybrid systems with a mixture of continuous and discrete states. The Mean Value Calculus [15] extends the Duration Calculus by replacing integrals of Boolean functions with their mean values, so that it can use δ-functions to represent instant actions such as communications and events. The Mean Value Calculus can be used to refine from state based requirements via mixed state and event specifications to event based specifications or programs. The Probabilistic Duration Calculus [5] [6] provides designers with a set of rules to reason about and calculate whether a given requirement written in the Duration Calculus will hold with a sufficient high probability for given failure

* UNU/IIST, P. O. Box 3058, Macau. E-mail: zcc@iist.unu.edu. On leave of absence from the Software Institute, the Chinese Academy of Sciences, Beijing, China.

probabilities of the components used in a design, where the design with imperfect components is taken to be a finite automaton with history-independent transition probabilities.

The Duration Calculi have been used to capture and refine requirements and designs for a number of examples, including a gas burner [9], a railway crossing [10], a water level controller [1] and an auto pilot [8]. The Calculi have also been used to define real-time semantics for Occam-like languages [12] [4], and to specify real-time behaviour of schedulers [12] and circuits [2].

In the paper we give an overview of the four duration calculi.

2 The Duration Calculus

A case study of the ProCoS project was asked to formulate a requirement for a gas burner:
"The proportion of time when gas is leak is not more than one twentieth of the elapsed time, if the system is observed for more than one minute."

The direct formulation of the requirement can be obtained by applying mathematical analysis. That is

$$(e-b) \geq 60\,sec \Rightarrow 20\int_b^e Leak(t)dt \leq (e-b)$$

where *Leak* is a Boolean function, which represents the leak state of the gas burner. The function is from reals $\mathbf{R}$ (representing *time*) to $\{0,1\}$, where 1 denotes that the system is in the state, and 0 denotes that the system is not in the state. The observation interval is taken to be bounded, and b stands for its start and e stands for its end.

Unfortunately, at that time, no software design calculi were available to express and reason about properties of integrals or differentials of functions, although integrals and differentials have been widely used for ages in control theory to model dynamic systems.

Confining ourselves to the formalisation of integrals of Boolean functions, we developed the Duration Calculus [14]. Integrals can be considered *curried* functionals from state functions and intervals to reals:

$$\int : \mathbf{S} \rightarrow (\mathbf{I} \rightarrow \mathbf{R})$$

where $\mathbf{S}$ stands for states (i.e. Boolean functions), and $\mathbf{I}$ for bounded intervals. Therefore the Interval Temporal Logic [7] (extended with continuous time) is adopted as its base logic, and interval functions $\int S, \int P, ...$ become interval variables of the calculus, where S and P are states. It follows that $\int 1 = e - b$, and we use l as an abbreviation of $\int 1$ to be a mnemonic notation of interval length. Thus the requirement can be encoded more succinctly as *Req*:

$$l \geq 60 \Rightarrow 20\int Leak \leq l$$

With integrals we can also express a lasting presence of a state over an interval by defining a *ceiling* operator $\lceil . \rceil$:

$$\lceil S \rceil \mathrel{\hat{=}} (\textstyle\int S = l \ \wedge \ l > 0)$$

$\lceil S \rceil$ holds in an interval, when the interval is not a point interval, and state S takes value 1 (almost) everywhere in the interval. Then formula $Des1$

$$(\lceil Leak \rceil \Rightarrow l \leq 1)$$

formulates a possible design decision for the gas burner to detect and stop a flame failure within one second.

The modality of the Interval Temporal Logic is "*chop*" (;), semantically defined as

$$A; B[b, e] \mathrel{\hat{=}} \exists m : b \leq m \leq e.\ A[b, m] \ \wedge \ B[m, e]$$

It means that formula $A; B$ holds for interval $[b, e]$ iff the interval can be chopped into two subintervals such that A holds for the first subinterval and B holds for the second one. *Chop* is a contraction operator which enables designers to specify properties of systems over an observational interval by assembling properties over its subintervals. For instance, with *chop* we can formulate another design decision for the gas burner, denoted $Des2$. It specifies 30 seconds as the lower bound of time distance between two leakages, so that we can reject frequent flame failures of the gas burner

$$(\lceil Leak \rceil; \lceil \neg Leak \rceil; \lceil Leak \rceil \ \Rightarrow \ l \geq 30)$$

The conventional modalities can also be defined by *chop*.

$$\Diamond A \mathrel{\hat{=}} true; A; true$$
$$\Box A \mathrel{\hat{=}} \neg \Diamond \neg A$$

$\Diamond A$ is true of an interval in which A is true of some subinterval, and $\Box A$ is true of an interval in which A is true of every subinterval.

By assuming the *finite variability* of states, six axioms were developed, which are similar to the axioms for measures of finite sets of intervals in Measure Theory, and constitute a relatively *complete* calculus [3] for the Duration Calculus. They are

1. $\int 0 = 0$
2. $\int P \geq 0$
3. $\int P + \int Q \ = \ \int (P \vee Q) + \int (P \wedge Q)$
4. $(\int P = r + s) \ \Leftrightarrow \ (\int P = r); (\int P = s)$
5. $(\lceil\,\rceil \vee \lceil P \rceil; true \vee \lceil \neg P \rceil; true)$
6. $(\lceil\,\rceil \vee true; \lceil P \rceil \vee true; \lceil \neg P \rceil)$

where $\lceil\,\rceil \mathrel{\hat{=}} (l = 0)$. Axiom 5 & 6 stipulate the finite variability of state P. That is, any non-point interval can be split into a finite alternation of P and $\neg P$.

With the calculus, a designer can verify the correctness of the previous two design decisions by proving

$$\Box Des1 \land \Box Des2 \Rightarrow Req$$

to be a theorem of the calculus.

3 The Extended Duration Calculus

Piecewise continuity and differentiability are function properties over intervals also. The Interval Temporal Logic can still be used as a logical framework to accommodate those analytic concepts. We generalise the arguments of the ceiling operator $\lceil . \rceil$ in the Duration Calculus from Boolean functions to properties like equality, inequality, continuity, etc. $\lceil H \rceil$ holds in an interval iff the interval is not a point interval and the property H holds everywhere *inside* the interval. Thus we can specify a continuous behaviour of a gas valve of the burner. Let $v : \mathbf{R} \rightarrow [0,1]$ be a function to describe the movement of the valve between fully closed 0 and fully open 1.

1. The range of v is $[0,1]$:

$$(\lceil\,\rceil \lor \lceil 0 \leq v \leq 1 \rceil)$$

2. The valve is either stable, closing or opening for some time

$$(\lceil\,\rceil \lor (\lceil \dot{v} = 0 \rceil ; \mathit{true}) \lor (\lceil \dot{v} = k_{\text{off}} \rceil ; \mathit{true}) \lor (\lceil \dot{v} = k_{\text{on}} \rceil ; \mathit{true}))$$

and

$$(\lceil\,\rceil \lor (\mathit{true} ; \lceil \dot{v} = 0 \rceil) \lor (\mathit{true} ; \lceil \dot{v} = k_{\text{off}} \rceil) \lor (\mathit{true} ; \lceil \dot{v} = k_{\text{on}} \rceil))$$

where $k_{\text{on}} > 0$ and $k_{\text{off}} < 0$ to denote some constant speeds.

3. Once it starts to close, it will complete closing.

$$((\lceil v > 0 \rceil \land (\lceil \dot{v} = k_{\text{off}} \rceil ; \mathit{true})) \Rightarrow \lceil \dot{v} = k_{\text{off}} \rceil)$$

Similarily for opening.

4. The movement v is totally continuous

$$(\lceil\,\rceil \lor \lceil \mathit{Continuous}(v) \rceil)$$

To describe boundary conditions of differential equations, we introduce interval variables $\mathbf{b}.v$ (and $\mathbf{e}.v$) to denote the right limit (and the left limit) of function v at the beginning (and end) point of a non-point interval.

The main axiom and rule for this notation are

1. $\lceil Continuous(f) \rceil \wedge (l > 0 \wedge \mathbf{e}.f = r_1); (\mathbf{b}.f = r_2 \wedge l > 0) \Rightarrow r_1 = r_2$
2. Let $R(\lceil H(f) \rceil,\ \mathbf{b}.f,\ \mathbf{e}.f,\ l)$ be a formula without *chop*.
   If $\vdash_{MT} \forall c < d.R(\forall t \in (c,d).H(f(t)),\ f(c^+),\ f(d^-),\ d-c)$ then

$$\lceil \cdot \rceil \vee R(\lceil H(f) \rceil, \mathbf{b}.f, \mathbf{e}.f, l)$$

   where MT stands for an existing mathematical theory for the property H.

The first axiom defines the relations among $\mathbf{b}.f$ and $\mathbf{e}.f$ of consecutive intervals, when f is continuous. With this axiom, we can reason about *discrete transitions* of controls of *piecewise smoothly* changed functions. The second rule just honestly confesses that we have no way but to use classical mathematical theory to reason about analytic properties in a single interval. Using these two rules, we can present in the calculus, for example, a numerical algorithm for solving a differential equation.

Differential and integral are dual. From differential equations, we can derive integrals.. The calculus for specifying piecewise continuity and differentiability is therefore called the Extended Duration Calculus. To derive integrals of states from differential equations, we introduce for any state P, a function f_P which is defined by:

$$(\lceil\ \rceil \vee \lceil Continuous(f_p) \rceil) \wedge \Box(\lceil P \rceil \Rightarrow (\dot{f}_p = 1)) \wedge \Box(\lceil \neg P \rceil \Rightarrow (\dot{f}_p = 0))$$

Then we define:

$$\int P \mathrel{\widehat{=}} \mathbf{e}.f_p - \mathbf{b}.f_p$$

With these definitions the six axioms of the Duration Calculus are proved to be theorems in the Extended Duration Calculus.

4 The Mean Value Calculus

In order to be able to refine a system requirement in terms of states into a set of system components which coordinate their behaviour by communications, we have also extended the Duration Calculus with point values of Boolean functions such that Boolean δ-functions can be introduced to represent instant actions. This extended calculus is called Mean Value Calculus. The Mean Value calculus uses mean value of Boolean function F (denoted $\overline{F}$) over intervals to replace the integral of F, $\int F$, in the original calculus. The integral over a point interval always degenerates to zero value, while the mean value of a function over a point interval delivers the value of the function at that point

$$\overline{F}[b,e] = \begin{cases} \int_b^e F(t)dt/(e-b) & \text{if } e > b \\ F(b) & \text{if } e = b \end{cases}$$

$\int F$ can be defined by $\overline{F}$

$$\int F \mathrel{\widehat{=}} \overline{F} * l$$

Replacing $\int F$ with $\overline{F} * l$ in the axioms of the Duration Calculus and introduing two more axioms to define point values, we have also established a relatively *complete* inference system for the Mean Value Calculus.

Generalise the ceiling operator $\lceil . \rceil$ to include point intervals and reformulate it as

$$\begin{aligned}
\lceil F \rceil^0 &\,\widehat{=}\, \lceil\,\rceil \wedge (\overline{F} = 1) \\
\lceil F \rceil &\,\widehat{=}\, (l > 0) \wedge \neg((l > 0); (\lceil\,\rceil \wedge \overline{F} = 0); (l > 0)) \\
\lceil F \rceil^* &\,\widehat{=}\, \lceil F \rceil \vee \lceil F \rceil^0
\end{aligned}$$

That is, $\lceil F \rceil^0$ holds for a point interval when the value of F at the point is 1, and $\lceil F \rceil$ holds for a non-point interval when F will not take value 0 anywhere *inside* the interval.The two additional axioms for point values are

$$\begin{aligned}
&\lceil\,\rceil \Rightarrow (\lceil F \rceil^0 \vee \lceil \neg F \rceil^0) \\
&\lceil F \rceil^* \Rightarrow (\overline{F} = 1)
\end{aligned}$$

The first axiom asserts that at any point the value of F is either 1 or 0. The second one asserts that the mean value of F over an interval is 1, if the value of F everywhere inside/in the interval is always 1.

A Boolean δ-function E is a function which satisfies

$$\neg\Diamond\lceil E \rceil$$

That is, any Boolean δ-function will not last its presence for a non-zero period. Instant actions can then be represented by Boolean δ-functions in the Mean Value Calculus.

Events cause state transitions of a system. State transition from S_1 to S_2 determines the system states in a neighbourhood of the transition. That is, right before the transition the system stays at S_1 and right after the transition it stays at S_2. The transition can be expressed by the conjunction of δ-functions

$$\downarrow S_1 \wedge \uparrow S_2$$

where $\downarrow S_1$ represents *falling edge* of S_1 and $\uparrow S_2$ represents the *rising edge* of S_2.

In Topology, neighbourhood properties of functions are characterised by so-called *function germs*. A function germ at a point is an equivalence class of functions which have the same value in an arbitrarily small neighbourhood of the point. States are Boolean functions without *isolated* points, so there are only four interesting state germs at any point. For a point t, they are

$\mathcal{G}_1(t)$: $\{S \mid \exists\delta\forall x : (0,\delta).S(t-x) = 0 \wedge S(t+x) = 1\}$,
$\mathcal{G}_2(t)$: $\{S \mid \exists\delta\forall x : (0,\delta).S(t-x) = 1 \wedge S(t+x) = 0\}$,
$\mathcal{G}_3(t)$: $\{S \mid \exists\delta\forall x : (0,\delta).S(t-x) = 0 \wedge S(t+x) = 0\}$, and
$\mathcal{G}_4(t)$: $\{S \mid \exists\delta\forall x : (0,\delta).S(t-x) = 1 \wedge S(t+x) = 1\}$.

$\uparrow S$ becomes a state 'germship', which discriminates the membership of S for germ $\mathcal{G}_1$: $\uparrow S(t) = 1$ iff $S \in \mathcal{G}_1(t)$. Similarly $\downarrow S(t) = 1$ means $S \in \mathcal{G}_2(t)$. The other two germships corresponding to $\mathcal{G}_3$ and $\mathcal{G}_4$ are denoted $\top S$ (*keep high*) and $\bot S$ (*keep low*). The state transition $\downarrow S_1 \wedge \uparrow S_2$ becomes a *composite* state germship.

The behaviour of an automaton is determined by the correspondence between its events and state transitions. It can be specified in the Mean Value Calculus with formulas over events and state germships. For example, let E_1 and E_2 be δ-functions to stand for two events. Then

$$(\downarrow S_1 \wedge \uparrow S_2) \Rightarrow (E_1 \vee E_2)$$

specifies that E_1 and E_2 are the only events which may drive an automaton from S_1 to S_2.

The state germships can be easily defined in the Mean Value Calculus. The axioms to define $\uparrow S$ and $\downarrow S$ are

$$(l > 0); \uparrow S; (l > 0) \Leftrightarrow \mathit{true}; \lceil \neg S \rceil; \lceil S \rceil; \mathit{true}$$
$$(l > 0); \downarrow S; (l > 0) \Leftrightarrow \mathit{true}; \lceil S \rceil; \lceil \neg S \rceil; \mathit{true}$$

where the heavy notations $\lceil . \rceil^0$ around $\uparrow S$ and $\downarrow S$ are elided for simplification. The axioms for $\top S$ and $\bot S$ are similar.

A calculus for the germships follows. For any point intervals

1. Completeness & Exclusiveness
   $\bigvee_\alpha \alpha P$ $\qquad \alpha \in \{\uparrow, \downarrow, \top, \bot\}$
   $\bigwedge_{\alpha \neq \alpha'} (\alpha P \Rightarrow \neg \alpha' P)$ $\qquad \alpha, \alpha' \in \{\uparrow, \downarrow, \top, \bot\}$
2. Constant Zero
   $\neg \uparrow 0$
   $\neg \downarrow 0$
   $\neg \top 0$
   $\bot 0$
3. Negation
   $\uparrow \neg P \Leftrightarrow \downarrow P$
   $\downarrow \neg P \Leftrightarrow \uparrow P$
   $\top \neg P \Leftrightarrow \bot P$
   $\bot \neg P \Leftrightarrow \top P$
4. Conjunction
   $\uparrow (P \wedge Q) \Leftrightarrow (\uparrow P \wedge \top Q) \vee (\uparrow Q \wedge \top P) \vee (\uparrow P \wedge \uparrow Q)$
   $\downarrow (P \wedge Q) \Leftrightarrow (\downarrow P \wedge \top Q) \vee (\downarrow Q \wedge \top P) \vee (\downarrow P \wedge \downarrow Q)$
   $\top (P \wedge Q) \Leftrightarrow (\top P \wedge \top Q)$
   $\bot (P \wedge Q) \Leftrightarrow \bot P \vee \bot Q \vee (\uparrow P \wedge \downarrow Q) \vee (\downarrow P \wedge \uparrow Q)$

In [15] the germship calculus has been applied in specifying and reasoning about circuits and real-time automata.

5 The Probabilistic Duration Calculus

Satisfaction of a system requirement by an implementation does not only depend on the functional behaviour of the implementation, but also the reliability of the components used in the implementation. The previous three calculi can assist designers to develop a *functionally* correct implementation with respect to a given requirement, but cannot help designers to calculate the satisfaction probability of the requirement by an implementation with unreliable components.

The Probabilistic Duration Calculus has been established to complement the other three calculi for this purpose. The calculus assumes discrete time and an imperfect implementation to be modelled as a finite automaton with history-independent transition probabilities.

For example, an implementation of a gas burner with unreliable flame detector can be modelled by the probabilistic automaton shown in the following figure. In this gas burner, *Leak* and $\neg$*Leak* are the only system states. The probabilities of the system starting in states $\neg$*Leak* and *Leak* are p_1 and p_2 respectively, where $0 \leq p_1, p_2 \leq 1$ and $p_1 + p_2 = 1$. We usually assume that the gas burner starts from the state $\neg$*Leak*, i.e. $p_1 = 1$ and $p_2 = 0$. The probability of the system to stay burning within one time unit is p_{11}. The probability of flame failure within one time unit is p_{12}. So $0 \leq p_{11}, p_{12} \leq 1$ and $p_{11} + p_{12} = 1$. The probability of the detector to detect the leakage (thereby causing re-ignition of the flame) within one time unit is p_{21}. The probability with which the detector fails to detect the leakage within one time unit is p_{22}, where $0 \leq p_{21}, p_{22} \leq 1$ and $p_{21} + p_{22} = 1$. Here all the transition probabilities are independent of the transition history.

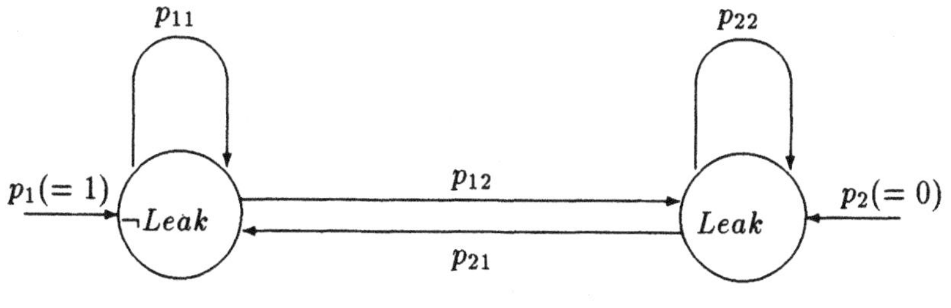

The probability of a behaviour (a state sequence) of the automaton can be obtained by multiplying the initial probability of the first state of the behaviour with the probabilities of all the state transitions of the behaviour. For example, let

$$\sigma^{[3]} : \neg Leak, \neg Leak, Leak$$

$\sigma^{[3]}$ denotes a behaviour of the automaton in the period of first three time units since its starting: start from $\neg Leak$, stay at $\neg Leak$ for one more time unit, and then transit to $Leak$. The probability of $\sigma^{[3]}$ is

$$p_1 * p_{11} * p_{12}$$

The satisfaction probability of the requirement

$$l \geq 60 \Rightarrow 20 \int Leak \leq l$$

within a period can be defined as the probability of the set of behaviours of the automaton (in this period) which satisfy the requirement.

Let G be a probabilistic automaton, D be a duration formula, $[0, t]$ be an interval of discrete time, and $\mu(D)[t]$ denote the ***satisfaction probability*** of D by G within the time interval $[0, t]$. The Probabilistic Duration Calculus provides a set of rules to calculate and reason about $\mu(D)[t]$. They are, for example,

1. $\mu(true)[t] = 1$
2. $\mu(D)[t] + \mu(\neg D)[t] = 1$
3. $\mu(D_1 \vee D_2)[t] + \mu(D_1 \wedge D_2)[t] = \mu(D_1)[t] + \mu(D_2)[t]$
4. If $D_1 \Rightarrow D_2$ holds in the Duration Calculus, then $\mu(D_1)[t] \leq \mu(D_2)[t]$ holds in the Probabilistic Duration Calculus.
5. $\mu(D)[t] = \mu(D \wedge (l = t))[t]$
6. $\mu(D; (l = t'))[t + t'] = \mu(D)[t]$
7. $\mu(\lceil v \rceil^1)[1] = \tau_0(v)$
   where v is a state of the automaton G, and τ_0 is the initial probability mass function of G. $\tau_0(v)$ defines the probability of G starting from v. And $\lceil v \rceil^1 \,\widehat{=}\, \lceil v \rceil \wedge (l = 1)$ (i.e. the automaton stays at v for one time unit).
8. $\mu(D; \lceil v_i \rceil^1; \lceil v_j \rceil^1)[t + 1] = \tau(v_i, v_j) \cdot \mu(D; \lceil v_i^1 \rceil)[t]$
   where v_i, v_j are states of G, and τ is the single-step probabilistic transition function which defines the transition probabilities of G.

[5] contains examples of a gas burner with unreliable failure detector and a protocol over unreliable medium. They illustrate the possible application of the calculus. [6] derives from the calculus numerical algorithms to compute $\mu(D)[t]$.

6 Remarks

1. Without infinite intervals the Duration Calculi cannot describe the so-called *qualitative* fairness nor liveness. An Infinite Duration Calculus is being established.
2. As to mechanical proof/model-checking support for the duration calculi, the decidability and undecidability results of the Duration Calculus has been published [13], and an automatic model checker for a decidable subclass of the Duration Calculus has been implemented in Standard ML [11]. An efficient model checking algorithm for Linear Duration Invariants has been discovered. It employs the technique of Linear Programming. A report of

it will be available soon. Some experiments implementing proof support systems for the Duration Calculus with the Larch Prover have been made.

Acknowledgements: I would like to give my sincere thanks to all the authors and coauthors of the papers which I refer to in this paper, and to Dines Bjørner for his various comments and suggestions.

References

1. M. Engel, M. Kubica, J. Madey, D.J. Parnas, A.P. Ravn, A.J. van Schouwen: A Formal Approach to Computer Systems Requirements Documentation, *presented in the Workshop on Theory of Hybrid Systems*, Lyngby, Denmark, 19-20 Oct. 1992
2. M.R. Hansen, Zhou Chaochen, J. Staunstrup: A Real-Time Duration Semantics for Circuits, *Proc. of the Workshop on Timing Issues in the Specification and Synthesis of Digital Systems, Princeton,* March 1992
3. M.R. Hansen, Zhou Chaochen: Semantics and Completeness of Duration Calculus, *J.W. de Bakker, C. Huizing, W.-P. de Roever, G. Rozenberg, (Eds) Real-Time: Theory in Practice, REX Workshop, LNCS 600,* pp 209-225, 1992
4. He Jifeng, J. Bowen: Time Interval Semantics and Implementation of a Real - Time Programming Language, *Proc. 4th Euromicro Workshop on Real-Time Systems, IEEE Press,* June 1992
5. Liu Zhiming, A.P. Ravn, E.V. Sørensen, Zhou Chaochen: A Probabilistic Duration Calculus, *presented in the 2nd Intl. Workshop on Responsive Computer Systems, Saitama, Japan,* Oct. 1-2, 1992, *published in H. Kopetz and Y. Kakuda (eds), Dependable Computing and Fault-Tolerant Systems Vol. 7 : Responsive Computer Systems, pp 30-52. Springer-Verlag Wien New York*, 1993.
6. Liu Zhiming, A.P. Ravn, E.V. Sørensen, Zhou Chaochen: Towards a Calculus of Systems Dependability, *presented in the Workshop on Theory of Hybrid Systems,* Lyngby, Denmark, 19-20 Oct. 1992. *accepted by High Integrity Systems Journal, Oxford University Press.*
7. B. Moszkowski: A Temporal Logic for Multi-level Reasoning about Hardware. In *IEEE Computer, Vol. 18(2)*, pp 10-19, 1985.
8. A.P. Ravn, H. Rischel: Requirements Capture for Embedded Real-Time Systems, *Proc. IMACS-MCTS'91 Symp. Modelling and Control of Technological Systems, Vol 2,* pp. 147-152, Villeneuve d'Ascq, France, 1991
9. A.P. Ravn, H. Rischel, K.M. Hansen: Specifying and Verifying Requirements of Real-Time Systems, *Proceedings of the ACM SIGSOFT'91 Conference on Software for Critical Systems, New Orleans, December 4-6, 1991, ACM Software Engineering Notes, Vol 15, No 5,* pp 44-54, 1991 (published also in *IEEE Trans. Software Eng., Vol 19, No 1,* pp 41-55, January 1993)
10. J.U. Skakkebæk, A.P. Ravn, H. Rischel, Zhou Chaochen: Specification of Embedded Real-Time Systems, *Proc. 4th Euromicro Workshop on Real-Time Systems, IEEE Press,* pp 116-121, June 1992
11. J.U. Skakkebæk, P. Sestoft: Checking Validity of Duration Calculus Formulas, *submitted to Conference on Computer-Aided Verification, Crete,* June 1993
12. Zhou Chaochen, M.R. Hansen, A.P. Ravn, H. Rischel: Duration Specifications for Shared Processors, *Proc. of the Symposium on Formal Techniques in Real-Time and Fault-Tolerant Systems, Nijmegen,* January 1992, *LNCS 571,* pp 21-32, 1992

13. Zhou Chaochen, M.R. Hansen, P. Sestoft: Decidability and Undecidability Results for Duration Calculus, *Proc. of STACS '93. 10th Symposium on Theoretical Aspects of Computer Science, Würzburg,* Feb. 1993
14. Zhou Chaochen, C.A.R. Hoare, A.P. Ravn: A Calculus of Durations, *Information Processing Letter, 40, 5,* pp. 269-276, 1991
15. Zhou Chaochen, Li Xiaoshan: A Mean-Value Duration Calculus, *UNU/IIST Report No. 5,* March 1993, to be published in the Hoare *Festschrift, Prentice-Hall International*
16. Zhou Chaochen, A.P. Ravn, M.R. Hansen: An Extended Duration Calculus for Hybrid Real-Time Systems, *to be published in the Proc. of the Workshop on Theory of Hybrid Systems*, Lyngby, Denmark, 19-20 Oct. 1992

A Unique Formalism for Specifying and Designing Objects in a Parallel Environment

J.P. BAHSOUN, C. SERVIERES[1]
{bahsoun, serviere}@irit.fr

C. SEGUIN[2]
seguin@tls-cs.cert.fr

Abstract : When objects are put in a parallel environment, new problems appear. Constraints on methods may change, the composition and extension of objects is more difficult. We introduce a concurrent object concept to implement solutions of these problems in object-oriented languages. A formalization of this concept and proofs of properties are built up in a Multimodal Logic of Actions. We terminate this paper by a discussion about the proof of inheritance properties in a parallel environment.

1 Introduction

Object-Oriented concepts are used more and more in several areas of Computer Science, such as programming languages and databases, software engineering and artificial intelligence [22], [29], [32]. Several computer languages currently contain objects as a basic concept ([1], [2], [3], [4], [5], [10], [13], [16], [17], [18], [24], [30]). The usual definition of an object is an identity, a state, and methods which modify its state. Objects interact by means of messages, and they can be defined and implemented in a parallel or distributed environment. The more complex applications using objects are, the more necessary it becomes to have a formal semantics, a proof system and a suitable work environment.

In this paper we will present our formal approach to design objects in a parallel environment. We focus on complex problems as the composition and the extension of objects. In the second section we will first explain our informal and intuitive approach to defining concurrent objects. In the third section we will present an adequate logical framework to formalize our approach, a Multimodal Logic of Actions [27], defined from the Temporal Logic, and a Logic of Actions which has been thoroughly tested in the Artificial Intelligence area of planning. This formalism has shown itself to be expressive, flexible and complete. We will give the description of objects – in particular the bounded buffer one – with this formalism and sketch out the proof of some properties of these objects.

1 IRIT, 118 route de Narbonne, 31062 Toulouse cedex, France

2 ONERA-CERT, 2 av Edouard Belin, bp 4025, 31055 Toulouse cedex, France

2 An Informal Approach to Concurrent Objects

2.1 The Concept of Concurrent Object

In our approach, the definition of a concurrent object is composed of two parts :

- the first being the classical one : a *state* and *methods*,
- the second being *constraints* : these constraints depend on the representation of the object and the parallel environment.

The state and the constraints of an object are hidden and only the methods are visible on the outside of the object. Methods constraints bracket each method by a precondition and a postcondition. To execute a method, its precondition must hold. Moreover, its postcondition must be satisfied just after this execution.

An object interacts with other objects by sending messages. Each message is a request for the execution of a method.

The Object Manager

An object manager — whose role is to receive messages and to check the access to the methods — is associated with each object. Actually, the object manager acts like an interface between the object and its environment. The behavior of the object manager is repetitive:

- It stores received messages.
- It selects a request for a method that can be performed. If there are several messages waiting for different methods whose preconditions hold the object manager chooses one of them in an arbitrary way whilst respecting some fairness criteria. In order to have fairness between requests for the same method, we suppose that the first request arrived at the object will be the first request treated by the object manager.
- After choosing a message, the object manager initiates the method corresponding to that message, and it simultaneously continues working, in parallel with the method. So several methods can be executed concurrently.

A Class

A class describes a collection of objects sharing the same features. So an object is nothing else than a class instance.

2.2 From an Informal Approach to a Formal one

We need to formalize concurrent objects in order to be able to develop programs and to prove them. Correctness means that the program satisfies some desired properties. Programs, properties and satisfaction are three separated concepts. So, in a proof system, three things have to be defined: a programming language, a specification language to express properties, and a satisfaction relation. The complexity of such a system is related to the formalisms used to support the three concepts described above, and their differences.

In [19], [20], the author talks about two methods to specify a program, a constructive one [25] and an axiomatic one [6], [7], [9], [14], [26]. In our opinion, a formal method to develop a program has to be both constructive and axiomatic, following the way of object-oriented design. In order to implement a program the constructive approach is more intuitive but to specify a program the axiomatic one may be better because it is compatible with the hierarchy concept.

In order to avoid the complexity of the use of different formalisms, we propose an easier approach, as in Lamport [21], in which both the program and the property are specified in the same logical formalism. Proving a program amounts to showing that the formula specifying the program implies the formula specifying the property. Our motivation is the rigorous reasoning which is the only way to avoid errors in concurrent programming.

Our choice of formalism is a multimodal logic based on the Logic of Actions [27] and the Temporal Logic. The union of both these logics enables us to express a specification (thanks to the notion of state and the modal operators) as well as a detailed conception (thanks to the actions as in process algebra).

3 Object Formalization

3.1 The Logic of Actions

In this section, we present a logical framework to deal with objects. The formalism uses the logic of actions (see [27]), based on Dynamic Logic [12]. We extend the notion of atomic programs met in Dynamic Logic to construct primitive actions. Each primitive action represents an elementary change that may be applied to objects. Composition of elementary changes is a way to define methods and managers of the objects. This possibility of composition is an important feature of our approach. It allows us to extend or compose easily the objects.

To define our logic of actions, we use the Propositional Deterministic Dynamic Logic (see [12]). Two propositional formulae are associated with each atomic action: the preconditions and postconditions of the action; the preconditions are conditions necessary to trigger an action and the postconditions describe constant effects of the action. We assume that postconditions determine all changes caused by the action. So, if a property is not explicitly affected by an action, we suppose that it remains unchanged after the action: there is a minimal change between the two "situations". The notion of minimal change is classical in the field of artificial intelligence to study changes and to update knowledge bases (see [8], [33], [28], [11]). These features are crucial to deal with the constraints stated to describe the right use of the objects.

On these bases, we introduce the notion of invisible action or not observable action. This notion is useful to describe concurrent behaviors, as we will see later.

3.1.1 Syntax

Let VAR and PRIMACT denote two symbol sets: atomic propositions and primitive deterministic actions, respectively.

Each action of PRIMACT is a couple (precondition of the action , postcondition of the action) where the precondition is a formula of the classical propositional logic and the postcondition is a conjunction of literals[3]. In order to shorten the notations, a name is associated with each action. Generally, an action will be referenced by means of its name instead of the couple pre and postcondition. So if the action α is defined by $\alpha \equiv (pre_\alpha, post_\alpha)$, the identifier α will be employed to designate the action $(pre_\alpha, post_\alpha)$.

We also use several constructors of actions to build complex actions using actions already defined (primitive or not). Intuitively, the compound actions have the following intended meaning :

$\alpha ; \beta$ do α followed by β without any interruption,

$\alpha + \beta$ do either α or β non deterministically,

α^* repeat α a finite time, but non deterministically determined number of times (this number can be zero).

The compound actions so built can be abbreviated by an identifier. For example, if we want to reference the action A;B – where A and B are any already defined actions – by means of a complex action C, we write $C \equiv A; B$. We do not associate preconditions and postconditions to compound actions. Properties of compound actions are deduced from both their definition and preconditions and postconditions of primitive actions used to define them.

Let ATOMACT be the set of atomic actions. By atomic actions, we mean actions whose execution cannot be interrupted. Of course, primitive actions are atomic. Other atomic actions can be defined as a particular composition of atomic actions, using the three constructors quoted above.

In a parallel context, actions may be executed simultaneously. So we need a new action constructor to represent parallel execution. In Dynamic Logic, the parallel execution of atomic actions α and β is defined by the non deterministic interleaving of α and β: $\alpha \| \beta = \alpha;\beta + \beta;\alpha$. However, this construction is not accurate enough to describe constrained behaviors. So we introduce another kind of actions defined as follow. From an action A and a set of atomic actions called *observable actions for A* and noted for example OBS_A, we can build another action representing in fact all the interleavings allowed between A and the atomic actions different from those ones contained in OBS_A. This new action will be noted : $(\backslash OBS_A)\ A$.

Let us now define formally this action construction, from the constructors previously introduced. If A is an action, and OBS_A a subset of ATOMACT then the action $(\backslash OBS_A)\ A$ represents in fact the action $\left(+_{A_i \in (ATOMACT \setminus OBS_A)} A_i\right)^*\ ;\ A$, i.e. the action A preceded by any sequence of actions contained in the complementary set of OBS_A in ATOMACT, which is the set of the ***non observable actions for*** **A**, or the actions that can interleave with A. Such an action will be called ***visible A action***, this expression means that we focus on the A action, and represents all the concurrent executions of A i.e. the executions that end by A what either the non observable actions interleaved before A.

From PRIMACT and VAR, we inductively construct the set of well-formed formulae (wff):

- if p ∈ VAR, then p is a wff;

[3] This restriction was chosen to characterize the deterministic primitive actions.

- if P and Q are wffs and A is any action, then $\neg P$, $P \rightarrow Q$, $P \vee Q$, $P \wedge Q$, $P \leftrightarrow Q$, True, False, [A] P and $< A >$ P are wff.

[A] and $< A >$ are modal operators. They have the following intended meaning:

- $< A >$ P: it is possible to execute A reaching a situation in which P is true.
- [A]P: every possible execution of A leads to a situation in which P is true.

We will give the semantics and an axiomatic for the connectives $\neg$, $\rightarrow$ and the family of modal operators [A]. The meaning of the other connectives and the formulae True and False is given by the definitions above:

Def1 : $True \overset{def}{=} P \rightarrow P$ *Def2* : $False \overset{def}{=} \neg(P \rightarrow P)$

Def3 : $P \wedge Q \overset{def}{=} \neg(P \rightarrow \neg Q)$ *Def4* : $P \vee Q \overset{def}{=} \neg P \rightarrow Q$

Def5 : $P \leftrightarrow Q \overset{def}{=} (P \rightarrow Q) \wedge (Q \rightarrow P)$

Def6 : $< A > P \overset{def}{=} \neg[A]\neg P$ with A any action

Def7 : $< A_1; A_2 > P \overset{def}{=} < A_1 >< A_2 > P$ with A_1, A_2 any actions

Def8 : $< A_1 + A_2 > P \overset{def}{=} < A_1 > P \vee < A_2 > P$ with A_1, A_2 any actions.

It's usually convenient to define a particular action, $\lambda \equiv$ (True,True), to represent the void action.

3.1.2 Semantics

A model M for the logic of actions is a triple (W, R, m) where:

- W is a non empty set of "worlds",
- R is a family of relations R_α over W. Each R_α represent the next-state relation for a primitive action α of PRIMACT. w R_α w' means (w,w') $\in R_\alpha$ or w' is reached from w by performing the α action. The definition of a relation R_α depends on the precondition and the postcondition of the α action, as we will see below.
- m: W $\rightarrow$ Val(VAR) is the meaning function. It associates a valuation with each world of W; a valuation is a mapping from VAR to {0,1}. m(w)(p)=1 means that the proposition p is true in w and m(w)(p)=0 means that p is false in w.

The satisfiability relation $\models$ between pairs (M,w) and formulae is defined by the following rules:

- (M,w) $\models$ p iff m(w)(p)=1 with p $\in$ VAR;
- (M,w) $\models$ P iff non (M, w) $\models$ P;
- (M,w) $\models$ $P \rightarrow Q$ iff non (M, w) $\models$ P or (M, w) $\models$ Q;
- (M,w) $\models$ $[\alpha]P$ iff $\forall\ w' \in W$, if w R_α w', then (M, w') $\models$ P
- (M,w) $\models$ $[\alpha^*]P$ iff $\forall\ w' \in W$, if w R_{α^*} w', then (M, w') $\models$ P with $R_{\alpha^*} = (R_\alpha)^* = \{\ (w_i, w_f)\ /\ \exists k \in N, \exists w_0, w_1, ..., w_k$ *such that* $w_0 = w_i$ *and* $w_k = w_f$ *and* $(\forall\ 1 \leq i \leq k,\ (w_{i-1}, w_i) \in R_\alpha)\}$

From these definitions and the definition of the other connectives (classical as well as modal), the satisfiability relation may be extended to a modal operator for general compound actions in the following way:

(M,w) $\models$ [A]P iff $\forall w' \in W$, if w R_A w', then (M, w') $\models$ A with :

$$R_A = R_{A_1;A_2} = R_{A_1}.R_{A_2} = \{(w_i, w_f)/\exists(w_i, w_k) \in R_{A_1}\ \wedge\ (w_k, w_f) \in R_{A_2}\}$$

$$R_A = R_{A_1+A_2} = R_{A_1} \cup R_{A_2}$$

$R_A = R_{((\backslash OBS_B)B)} = \left(\bigcup_{A_i \in ATOMACT \backslash OBS_B} R_{A_i} \right)^* .R_B$, this relation is deduced from the previous ones.

A formula P is *satisfiable* if there is a model M= (W, R, m) and a world w of W such that (M,w)⊨P. A formula is *valid* if it is satisfied in every world of every model.

These definitions are usual in Dynamic Logic. We introduce new semantic constraints to handle reasoning about actions. In a model M, each R_α characterizing an action α of PRIMACT verifies:

- *Determinism*: R_α is a functional relation over W (i.e. at most one next state is associated with a given one)
- *Necessary conditions for the execution of an action*:
  $\forall w \in W, if\ \exists w' \in W / w R_\alpha w'\ then\ (M, w) \models pre(\alpha)$
- *Effect of the execution of an action*: $\forall w \in W, if\ \exists w' \in W / w R_\alpha w'\ then$
  - *Effects of* α: $(M, w') \models post(\alpha)$
  - *Minimal Change*:
    $\forall l$ literal, if (M,w) ⊨ l and ¬l is not a conjunct of post(α), then (M,w') ⊨ l

Comments: The feature "Minimal change" guarantees that every literal which is compatible with the effects described in the postconditions remains unchanged after the execution of the action. This "inertia" of literals ensures that there is a minimal number of changes between a state and its successor.

3.1.3 Axiomatic

The following system captures the semantics given in the previous section.

Axioms:

Ax0 Axioms of propositional calculus

Ax1 $[\alpha](P \rightarrow Q) \rightarrow ([\alpha]P \rightarrow [\alpha]Q)$

Ax2 $< \alpha^* > P \leftrightarrow (P \vee < \alpha >< \alpha^* > P)$

Ax3 *Induction axiom*: $[\alpha^*](P \rightarrow [\alpha]P) \rightarrow (P \rightarrow [\alpha^*]P)$

Ax4 *Determinism*: $< \alpha > P \rightarrow [\alpha]P$

Ax5 *Necessary condition for the execution of* α: $< \alpha > P \rightarrow pre(\alpha)$

Ax6 *Effects of* α: $[\alpha]post(\alpha)$

Ax7 *Minimal change*: if a literal ¬*l* is not a conjunct of post(α), then l →[α]l,

Inference Rules:

R1 from P, P→Q derive Q (modus ponens)

R2 from P, derive [α]P (necessitation) with α any action

3.2 The Temporal Logic

As we have said previously, the operators of the temporal logic may be embedded in our logic. We can either define operators for branching or linear time. To specify properties of the designed systems more easily, we have chosen to introduce operators of the linear temporal logic (see [23]). From a semantic point of view, we have to constraint models of the logic of actions. Indeed former models were trees which could stand for families of time lines when a unique time line is the required model to give a semantics to temporal operators for linear time. We will state this property by the definition of the operator *next*.

So in this section, we define the *always, eventually, next* operators. We note these new modal operators $\Box$, $\Diamond$ and $\bigcirc$.

By adding new operators to our language, we obtain new well formed formulae. If P is a wff then:

- $\Box P$ is a wff which intended meaning is: in the future, P will always be true.
- $\Diamond P$ is a wff which intended meaning is: P will be true in the future.
- $\bigcirc P$ is a wff which intended meaning is: after the next action, P will be true.

In order to handle the modal operators $\bigcirc$, $\Box$, and its dual $\Diamond$[4], we add two new accessibility relations, $R_{\bigcirc}$ and R_{τ}, built upon the family R of R_α relations, to each model M of the logic of actions. Now a model M is defined by M=(W, R, $R_{\bigcirc}$, R_{τ}, m).

$R_{\bigcirc}$ is $\bigcup_{\alpha \in PRIMACT} R_\alpha$, and the R_α relations are such that $R_{\bigcirc}$ is a total function. So, in each state either an action modifying this state or the void action λ is performed and $R_{\bigcirc}$ is a mapping from W to W, that is to say the family of relations R is such that $\forall \alpha, \beta \in PRIMACT, \forall w, w', w'' \in W$ if $w R_\alpha w'$ and $w' R_\beta w''$ then $w' = w''$ and $\alpha = \beta$.

R_τ is the reflexive, transitive closure of $R_{\bigcirc}$.

The satisfiability relation is extended by:

- (M,w) $\models \bigcirc$P iff $\forall w' \in W$ if $w R_{\bigcirc} w'$ then (M,w') $\models$ P
- (M,w) $\models \Box$P iff $\forall w' \in W$ if $w R_\tau w'$ then (M,w') $\models$ P

To capture these new features, we add the axioms and the inference rule:

Ax1'	$\bigcirc P \leftrightarrow [+_{\alpha_i \in PRIMACT}\ \alpha_i]P$	*Ax2'*	$\bigcirc P \leftrightarrow \neg \bigcirc \neg P$
Ax3'	$\bigcirc(P \rightarrow Q) \rightarrow (\bigcirc P \rightarrow \bigcirc Q)$	*Ax4'*	$\Box(P \rightarrow Q) \rightarrow (\Box P \rightarrow \Box Q)$
Ax5'	$\Box P \rightarrow P \wedge \bigcirc \Box P$	*Ax6'*	$\Box(P \rightarrow \bigcirc P) \rightarrow (P \rightarrow \Box P)$
R1'	from P, derive $\Box$ P		

3.3 Examples

In this section, we formalize the example of a simple buffer using our multimodal logic, in order to be able to formally state and verify its properties. Moreover we give the description of a bounded buffer as a composition of simple buffers.

3.3.1 The Delayed Message Queue

As we have said in the second part, requests for the same method must be treated respecting a fairness criteria. At the implementation level, we have chosen to store delayed requests in a bounded queue. So a bounded queue managed by a FIFO policy is associated with each method of an object. We need first of all to describe such a queue with our formalism. For sake of shortness we won't give the whole description, but only the subset necessary to deal with our examples.

Let $VAR_{QUEUE(X)}$[5] be the set of all propositional variables employed to describe the different states of a queue. In particular this set contains the proposition *empty_queue(X)* which is true when the queue X is empty and false otherwise. Notice

---

4 $\Diamond$ is defined by $\neg\Box\neg$ as usually.

5 This set is finite since we suppose that the queue is bounded.

that we parameterize all the terms of a queue by X, this in order to instantiate them with the name of an object method.

Let $PRIMACT_{QUEUE(X)}$ be the set of all primitive actions of the object QUEUE(X).

Such an object owns two methods defined by two compound actions *enqueue(X)* and *dequeue(X)*, respectively inserting and removing an element of the queue X. Both these actions are atomic, so contained in the set $ATOMACT_{QUEUE(X)}$. We are not concerned with the values inserted in the queue, consequently the methods are not parameterized by a value.

We suppose that any behavior of a QUEUE(X) object satisfies the formula $BEHAV_{QUEUE(X)}$. In particular the initial state of a QUEUE(X) is given by this formula.

If the description of a QUEUE is complete we will be able to show the following property : $\Box(< enqueue(X) > True \rightarrow \bigcirc\neg empty\_queue(X))$ (i.e. if a state is reached by an enqueue(X) action then in this state the proposition empty_queue(X) is false)

3.3.2 The Simple Buffer

We want to describe a memory cell i called SBUF(i) by the methods which can be applied to it. These methods are PUT and GET. So the memory cell is also composed of two queues to stored delayed requests for PUT or GET method, respectively called: QUEUE(P_i) and QUEUE(G_i). We suppose that a GET can only be made if the cell is full and a PUT is possible if the cell is empty. Therefore for us, the cell can only be in one of two states: either empty or not empty. Consequently the propositional variables of a SBUF(i) object are the variable empty(i) and all the variables of QUEUE(P_i) and QUEUE(G_i).

$VAR_{SBUF(i)} = \{empty(i)\} \bigcup VAR_{QUEUE(P_i)} \bigcup VAR_{QUEUE(G_i)}$

We define PUT and GET methods by two atomic actions: put(i) and get(i). This leads to the following definitions :

- $PRIMACT_{SBUF(i)} =$
  $\{put(i)\ ,\ get(i)\} \bigcup PRIMACT_{QUEUE(P_i)} \bigcup PRIMACT_{QUEUE(G_i)}$ with :
  $put(i) \equiv$ (empty(i) , $\neg$ empty(i))
  $get(i) \equiv$ ($\neg$ empty(i) , empty(i))
- $ATOMACT_{SBUF(i)} =$
  $\{get(i), put(i)\} \bigcup ATOMACT_{QUEUE(P_i)} \bigcup ATOMACT_{QUEUE(G_i)}$

In order to simplify, let us first consider the case in which parallelism is not possible between actions of the bounded buffer. At each time, the object manager may either dequeue a request and try to satisfy it or enqueue new access requests. This behavior consists in repeating infinitely often the action: (dequeue(G_i);get(i)) + (dequeue(P_i);put(i)) + enqueue(G_i) + enqueue(P_i) + λ .

It is possible now to take into account more access requests thanks to the queues associated with the buffer. However, in the description of the behavior, we did not define the actions which can be executed concurrently. For example, the action dequeue(G_i) may be performed simultaneously with the sequence dequeue(P_i);put(i). On the contrary, dequeue(G_i) cannot interleave with dequeue(G_i);get(i). If such a thing is authorized, then the second request may be satisfied before the first one. So we need to distinguish between the actions whose execution produces bad interferences and the others.

We describe this fact by the visible action concurget(i) defined as below.
$concurget(i) \equiv (\backslash OBS_{concurget(i)})dequeue(G_i); (\backslash OBS_{concurget(i)})get(i)$
where $OBS_{concurget(i)} = \{dequeue(G_i), get(i)\}$ is the set of observable actions for $dequeue(G_i)$ and $get(i)$.

The symmetric method may be defined in a similar way by:
$concurput(i) \equiv (\backslash OBS_{concurput(i)})dequeue(P_i); (\backslash OBS_{concurput(i)})put(i)$
with $OBS_{concurput(i)} = \{dequeue(P_i), put(i)\}$.

The behavior of our buffer is an infinite repetition of the action (concurget(i) + concurput(i) + enqueue(G_i) + enqueue(P_i) + λ).

This behavior could be directly expressed with our formalism if we introduced a fixpoint operator. But we will not talk about this possibility in this paper, since there is another way to deal with our examples, as we will see below.

The sequence of states reached by actions of $PRIMACT_{SBUF(i)}$ must satisfy the two following formulae:

- $INTERLEAVING_{SBUF(i)} =$
  $\Box(< dequeue(P_i) > True \rightarrow < concurput(i) > True) \wedge$
  $\Box(< dequeue(G_i) > True \rightarrow < concurget(i) > True)$
  which guarantees that the two compound actions $concurget(i)$ and $concurput(i)$ have a correct interleaving saying that if the first action of one of these actions is performed then the complete action is performed.
- $WF_{SBUF(i)} =$
  $\Box(\Diamond < concurput(i) > True \vee \Diamond\neg(empty(i) \wedge \neg empty\_queue(P_i))) \wedge$
  $\Box(\Diamond < concurget(i) > True \vee \Diamond\neg(\neg empty(i) \wedge \neg empty\_queue(G_i)))$
  which guarantees a weak fairness between the two actions concurput(i) and concurget(i). None of these actions can be enabled infinitely often without being performed. Such a property for an action A is expressed by $\Box(\Diamond < A > True \vee \Diamond\neg precondition(A))$.

The initial state of a BSUF(i) must satisfy the following formula :
$INIT_{SBUF(i)} = empty(i)$.

So the behavior of a SBUF(i) is given by:

$$BEHAV_{SBUF(i)} = INIT_{SBUF(i)} \wedge INTERLEAVING_{SBUF(i)} \wedge WF_{SBUF(i)} \wedge BEHAV_{QUEUE(P_i)} \wedge BEHAV_{QUEUE(G_i)}.$$

A sequence of states each of them is reached by an action of $PRIMACT_{SBUF(i)}$, whose first state satisfies the formula $BEHAV_{BSBUF(i)}$, represents a behavior of the SBUF(i) object .

We can summarize the features of an SBUF(i) object by the following tuple, giving respectively the sets of propositional variables, primitive actions, atomic actions, the names of the actions used to access a method, and the behavior of the object,
SBUF(i) = $< VAR_{SBUF(i)}, PRIMACT_{SBUF(i)}, ATOMACT_{SBUF(i)}, \{enqueue(G_i), enqueue(P_i)\}, BEHAV_{SBUF(i)} >$.

3.3.3 Properties of the Simple Buffer

We are going to sketch out the proof (see appendix) of two properties of an object of the SBUF(i) class: one of safety, the other one of liveness.

A Safety Property If we want the SBUF(i) object to work correctly, the action of inserting a value in the buffer must not be performed when the buffer is not empty. This property is expressed by the formula:
$\Box\neg(\neg empty(i) \wedge < put(i) > True)$. We can show that this property leads from the definition of the SBUF(i) object. The particularities of our formalism make this proof very easy.

A Liveness Property As a liveness property we can proove:
$\Box(empty(i) \wedge \Diamond < enqueue(P_i) > True \rightarrow \Diamond < put(i) > True)$. That is to say, if the buffer is empty and in the future, a PUT request arrives at the buffer, then the PUT method will be finally performed. The main interest of proving such a property is to give an illustration of the use of our formalism. More complex properties could be shown from other properties of the delayed request queue object.

3.3.4 Composition: The Bounded Buffer

A bounded buffer (BSBUF) can be defined as a limited number of memory cells. It is used by two methods (*enq* to insert a value and *deq* to remove a value) and managed by a circular distribution of the requests. In order to obtain maximum parallelism the bounded buffer can be created from a composition of SBUF(i). We assume the number of cells is MAX, these cells are indexed by 0, 1, ... MAX-1.

The execution of the *enq* (resp. *deq*) method is ensured by two steps:

- Obtain the number of a cell from the manager of the BSBUF. This number is an index on the bottom (resp. on the top) of the BSBUF,
- Do a PUT (resp. a GET) on this cell.

A request for an *enq* is delayed at a cell and not delayed at the bounded buffer, and thus PUT methods can be executed in parallel on different cells of the BSBUF. The same holds for the GET method. Moreover GET and PUT methods can also be executed in parallel.

To formally describe the bounded buffer we need some other propositional variables, and we obtain

$$VAR_{BSBUF} = \bigcup_{i=0}^{MAX-1} \{val(bottom, i), val(top, i)\} \cup \bigcup_{i=0}^{MAX-1} VAR_{SBUF(i)}.$$

The variable val(top,i) is true when the top of the bounded buffer is the cell i; val(bottom,i) is true when the bottom of the bounded buffer is i.

New primitive actions using the propositions val(bottom,i) and val(top,i) are defined.

$$PRIMACT_{BSUF} = \bigcup_{i=0}^{MAX-1} PRIMACT_{SBUF(i)} \cup$$
$$\bigcup_{i=0}^{MAX-1} \{next(bottom, i), next(top, i), val(bottom, i)?, val(top, i)?\}$$

An action which name is *p?* where p is a propositional variable, is a test action defined

by p ≡ (p,True). The ***next(bottom,i)*** and ***next(top,i)*** actions are defined as follow[6]:
$next(bottom, i) \equiv$ (val(bottom,i), ¬val(bottom,i) ∧ val(bottom,i⊕1))
$next(top, i) \equiv$ (val(top,i), ¬val(top,i) ∧ val(top,i⊕1)).

These primitive actions are composed with a non deterministic choice to give the two following actions that change the value of bottom or top by its successor.
$next(bottom) \equiv +_{i=0..MAX-1}\ next(bottom, i)$
$next(top) \equiv +_{i=0..MAX-1}\ next(top, i)$.

Accessing a cell i is locally solved by methods concurget(i) and concurput(i). So, the methods to access the whole buffer may be described by the actions: $concurget \equiv +_{i=0..MAX-1} concurget(i)$ and $concurput \equiv +_{i=0..MAX-1} concurput(i)$. We notice that these actions are non determinist and this allows the expression of the method parallelism.

Now we have to define how the global object manager puts requests for getting or putting a value in the local queue of the correct cell. A request for putting a value in the composed buffer is enqueued in the queue of the bottom cell. After that, the index of bottom is moved on the next cell. This treatment may be described by:
$request\_enq \equiv +_{i=0..MAX-1}\ val(bottom, i)?; enqueue(P_i); next(bottom)$

In a similar way, how is handled a request for getting a value from the queue is described by: $request\_deq \equiv +_{i=0..MAX-1} val(top, i)?; enqueue(G_i); next(top)$.

If we want the buffer to behave correctly both these actions must be atomic. The resulting set of atomic actions is:

$$ATOMACT_{BSBUF} = \{request\_enq, request\_deq\} \bigcup \bigcup_{i=0}^{MAX-1} ATOMACT_{SBUF(i)}$$

The behavior of the bounded buffer consists in repeating infinitely the action : (concurget + concurput + request-enq + request-deq + λ).

The initial state of the bounded buffer is described by the formula:

$$INIT_{BSBUF} = val(bottom, 0) \wedge \bigwedge_{i=1..MAX-1} \neg val(bottom, i) \wedge$$
$$val(top, 0) \wedge \bigwedge_{i=1..MAX-1} \neg val(top, i)\ .$$

Moreover each request_enq or request_deq action which is begun must be fully performed. The following formula ensures this property:

$$FP_{BSBUF} =$$
$$\Box(\bigvee_{i=0..MAX-1} < val(bottom, i)? > True \rightarrow < request\_enq > True)$$
$$\Box(\bigvee_{i=0..MAX-1} < val(top, i)? > True \rightarrow < request\_deq > True)$$

A sequence of states whose first state satisfies the following formula is a correct behavior of the bounded buffer:

$$BEHAV_{BSBUF} = INIT_{BSBUF} \ \wedge \ FP_{BSBUF} \ \wedge \ \bigwedge_{i=0..MAX-1} BEHAV_{SBUF(i)}.$$

So this composition of objects can easily be expressed with our formalism. The three sets VAR, PRIMACT and ATOMACT are defined from the corresponding sets of the composed objects by adding new propositional variables or new actions. The formula BEHAV also reuses formulae expressing features of the composed objects.

---

[6] ⊕ is the addition modulo MAX.

4 Conclusion

In this paper, we have introduced a logic to design objects in a parallel environment. This logic is adapted to specify properties of concurrence by its capability to define temporal connectives (as shown in section 3.2) and programs behavior by its capability to define complex actions as a composition of primitive ones. Moreover the system given in 3.1.3 is complete but we did not study this problem in this paper. This framework is comparable with Temporal Logic of Actions [21] in the sense that both programs and their properties are described within the same formalism. The difference with TLA is that our formalism supports complex actions. In TLA we didn't have compositional operators and that leads the developer to describe the complex actions entirely with atomic actions and control predicates. The concept of minimal change states what is changed and what remains unchanged after performing an action, this property is important in a refinement process to develop programs. Actually, the concept of the effect of an action under the minimal change hypothesis gives a lot of interest to our approach. It allows reasoning with both transitions and state properties since it specifies how states and actions are linked.

In a future work we will enhance our logic with a fixpoint operator to be able to describe the infinite behavior of the manager.

Defining a class by inheritance may in general introduce additional local variables, define new methods, redefine the behavior of the concurrent object for previously defined methods, or refine the initialization condition. All these changes correspond well to syntactical operations on the formulae defining the concurrent object behavior if these are written in the style proposed in the preceding sections. Moreover proofs about new object's behavior may rely on properties proven for the ancestors.

However, we feel that for a more adequate style of reasoning about the behavior of concurrent objects defined in an object-oriented framework, new logical operators expressing the inheritance relation should be introduced and studied in their own right. One particular goal of a logical analysis of the inheritance relation would be to support the necessary adaptation of preconditions expressing synchronization constraints caused by the introduction of new methods, using deductive methods.

References

[1] G. Agha and C. Hewitt Concurrent Programming Using Actors, In Object-Oriented Concurrent Programming, (ed) A. Yonezawa and M. Tokoro, MIT Press, 1987, 37-53.
[2] P. America, F. Vander Linden A Parallel Object Oriented Language with Inheritance and Subtyping", Proc. ECOOP'90. pp161-168.
[3] J. P. Bahsoun, L. Feraud, C. Btourn "A two degrees of freedom approach for parallel programming", Proc. IEEE International Conference on Computer Languages.March 12-15 1990.New Orleans.pp261-270.
[4] P. Cointe "Meta-classes are first class: the ObjVlisp model", Proc. 2nd OOPSLA Orlando Florida, pp 156-167.
[5] O. J. Dahl, K. Nygaard "SIMULA - an ALGOL based simulation language", CACM 9(9) 1966 pp 671-678.
[6] E.W. Dijkstra. Hierarchical ordering of sequential processes. Acta Informatica. 1 - 1971.

[7] E.W. Dijkstra. A Discipline of Programming. Prentice Hall, Englwood Cliffs, New Jersey, 1976.
[8] L. Farias del Cerro, A. Herzig "An Automated Modal Logic of Elementary Changes", in "Non-standard Logics for Automated Reasoning", Mandani, Smets, Dubois and Prade Eds, Academic Press, 1988.
[9] R.W. FLOYD Assigning meanings to programs. Proc Symp in AMS vol 19 Provinceton RI 1967.
[10] A. Golberg, A. Robson "SMALLTALK: the language and its implentation", Addison -Wesley 1980.
[11] S. Hanks, D. Mac Dermott Default Reasoning, Nonmonotonic Logic and the Frame Problem, Proc. of AAAI, 1986.
[12] D. Harel, "Dynamic Logic", Handbook of Philosophical Logic, D. Gabbay and F. Guenthner Eds., D. Reidel, 1984.
[13] C. Hewitt, R. Atkinson "Specification and proof technique for serializers", IEEE Trans. Soft. Eng. 1979 pp 10-13.
[14] C.A.R. Hoare An axiomatic basis for computer programming. CACM 12 1969.
[15] D. Kafura, K. Lee "Inheritance in actor based concurrent object-oriented languages", TR 88-53 Virginia Polytechnic Institue and state university. 1988.
[16] K. Kahn, E. D. Tribble, M. Smiller, D. G. Bobrow "VULCAN: logical concurrent objects".In Research directions in object oriented programming.B.Shriver,P.Wegner Editors 112 MIT Press 1987.
[17] S. E. Keene "Object oriented programming in common Lisp" A progammer's guide to CLOS", Addison Wesley 1989.
[18] S. Krakoviak, M. Meysembourg, H. Nguyen Van, M. Riveill, X. Rousset De Pina "Design and implementation of an object oriented strongly typed language for distributed applications", JOOP Sept/Oct. 1990 pp 11- 22.
[19] L. Lamport Specifying concurrent program modules. ACM Transactions on Programming Languages and systems, 5(2) 190-222 March 1977.
[20] L. Lamport What good is temporal logic? Proceeding of IFIP 9th World Congress September 1983.
[21] L. Lamport The Temporal Logic of Actions SRC DEC December 1991.
[22] B. Liskov, J. Guttag "Abstraction and specification in program developments", MIT Press/McGraw Hill 1986.
[23] Z. Manna and A. Pnueli The temporal logic of reactive and concurrent systems, Springer Verlag, 1992
[24] B. Meyer "Object oriented software construction", Prentice Hall 1988.
[25] R. Milner "Operational and Algebraic Semantics of Concurrent Processes", Handbook of Theorical Computer Science, J. van Leeuwen Ed., Elsevier Science, 1990.
[26] S. Owicki, D. Gries An Axiomatic proof technique for parallel programs. Acta Informatica. Vol 19 May 1976.
[27] C. Seguin "De l'action l'intention: vers une caractrisation formelle des agents", Ph.D. thesis, Universit Paul sabatier, Toulouse, France, 1992.
[28] Y. Shoham Reasoning about Change, MIT Press, 1987, Boston.
[29] A.Snyder "Inheritance and the development of encapsulated software systems", In Research directions in objetc-oriented programming. A.Shriver, P.Wegner editors MIT Press pp 165-188.
[30] B. Stroustup The C++ programming language Addison & Wesley 1986.

[31] T. Tomlison, V. Singh "Inheritance and synchronization with enabled sets", Proc. of OOPSLA'89.Oct.1-6.1989 pp 103-112.
[32] P. Wegner Concepts and paradigms of object-oriented programming", Expansion of Oct. 4 OOPSLA-89 keynote talk.OOPS Messenger. Vol.1 n1 Aug.1990 pp-84.
[33] M. Winslett "Updating Logical Databases", Cambridge University Press, 1990.

Appendix : Proofs

A Safety Property

The following paragraph is a proof of : $\Box(\neg(\neg empty(i) \wedge < put(i) > True))$. In parentheses we indicate how we have obtained each theorem from the previous one, or from an axiom.

$< put(i) > True \rightarrow empty(i)$	(Ax5)
$\neg(\neg empty(i) \wedge < put(i) > True)$	(Propositional Logic)
$\Box(\neg(\neg empty(i) \wedge < put(i) > True))$	(Inference Rule : from P derive $\Box P$)

A Liveness Property

In this section, we are going to sketch out the proof of the formula : $\Box(empty(i) \wedge \Diamond < enqueue(P_i) > True \rightarrow \Diamond < put(i) > True)$. Using propositional logic and the inference rule R1', this formula can easily be deduced from the following one :
$\Diamond < enqueue(P_i) > True \wedge empty(i) \wedge \neg\Diamond < put(i) > True \rightarrow False$.

Let us prove it.

$$\Diamond < enqueue(P_i) > True \wedge empty(i) \wedge \neg\Diamond < put(i) > True \rightarrow False$$
$$\overset{step1}{\rightarrow} \Box empty(i) \wedge \Diamond\neg empty\_queue(P_i) \wedge \neg\Diamond < put(i) > True$$
$$\overset{step2}{\rightarrow} \Box empty(i) \wedge \Diamond\Box\neg empty\_queue(P_i) \wedge \neg\Diamond < concurput(i) > True$$
$$\overset{step3}{\rightarrow} \Box(\Box empty(i) \wedge \neg\Diamond < concurput(i) > True) \wedge \Diamond\Box\neg empty\_queue(P_i)$$
$$\overset{step4}{\rightarrow} \Box\Diamond empty\_queue(P_i) \wedge \neg\Box\Diamond empty\_queue(P_i)$$
$$\overset{step5}{\rightarrow} False$$

Step1 follows from the property of delayed message queue (see 3.3.1) and the property *prop1* : $empty(i) \wedge \neg\Diamond < put(i) > True \rightarrow \Box empty(i)$.

Step2 is justifyied by both the following properties :
prop2 : $\neg\Diamond < put(i) > True \rightarrow \neg\Diamond < concurput(i) > True$
prop3 : $\Diamond\neg empty\_queue(P_i) \wedge \neg\Diamond < put(i) > True \rightarrow \Diamond\Box\neg empty\_queue(P_i)$.

We achieve Step3 using theorems of temporal logic.

Step4 uses the formula $WF_{SBUF(i)}$ expressing weak fairness (see 3.3.2).

The proof of *prop1* and *prop3* is based on the following proof rule: if l is a literal, if A is a set of primitve actions such that ¬l doesn't appear in the postcondition of any action of PRIMACT \ A (i.e. the complementary set of A) then we have:
$l \wedge \Box\neg < +_{\alpha\in A}\alpha > True \rightarrow \Box l$.

On the Smooth Calculation of Relational Recursive Expressions out of First-Order Non-Constructive Specifications Involving Quantifiers

Armando M. Haeberer ♠
Gabriel A. Baum ♣
Gunther Schmidt ❄

ABSTRACT
The work presented here has its focus on the formal construction of programs out of non-constructive specifications involving quantifiers. This is accomplished by means of an extended abstract algebra of relations whose expressive power is shown to encompass that of first-order logic. Our extension was devised for tackling the classic issue of lack of expressiveness of abstract relational algebras first stated by Tarski and later formally treated by Maddux, Németi, etc. First we compare our extension with classic approaches to expressiveness and our axiomatization with modern approaches to products. Then, we introduce some non-fundamental operations. One of them, the *relational implication*, is shown to have heavy heuristic significance both in the statement of Galois connections for expressing relational counterparts for universally quantified sentences and for dealing with them. In the last sections we present two smooth program derivations based on the theoretical framework introduced previously.

1 Introduction

The last few years have witnessed a renewed interest of the computing science community in relational programming calculi. The main reason for such interest is the postulated advantage of such calculi for dealing with non-determinism. Also, the claim about its fitness for expressing what-to-do instead of how-to-do specifications should not be ignored.

In the meantime, another major issue of whether or not a programming calculus should encompass the whole path from non-constructive specifications to programs has been almost completely neglected. In the case of some methods, as CIP, such disregard simply leads to a bias in the research work favouring algorithm optimisation to the detriment of algorithm construction. However, looking at program calculi in general, and at relational ones in particular, the situation is worse because any decision about this issue biases harmfully the choice of an underlying theoretical framework. In the particular case of relational programming calculi, the question is as

---

♠ Pontifícia Universidade Católica do Rio de Janeiro (PUC-Rio). Departamento de Informática, Av. Marquês de São Vicente 225, 22453 Rio de Janeiro, RJ, Brazil.

♣ Universidad Nacional de La Plata, Facultad de Ciencias Exactas, Departamento de Informática, 1 y 50 La Plata, Buenos Aires, Argentina.

❄ Universität der Bundeswehr, München, D-8014 Neubiberg.

crucial as whether or not Abstract Relational Algebras (in the sequel ARAs) qualify as an appropriate basis for them.

Along this paper we first argue, in Section 2, that, under the light of the restricted expressiveness of ARAs we should abandon them as a basis of relational programming calculi. We also examine the implications of the so-called representability problem over programming calculi development. In the following section, we first analyse a well-known solution to representability and subsequently some classical solutions to the expressiveness problem, namely, Cylindric and Polyadic algebras. Later on, in Section 4, we propose a formal extension to ARAs, prove that it has the same expressive power than first-order logic with equality, discuss its representability and compare it with both classical algebras for first-order logic and other extensions to ARA under use, for supporting relational programming calculi.

In the second part of the paper, i.e., in Section 5, we first introduce in Subsection 5.1 some results about partial identities as a tool for dealing with types within an homogeneous framework as well as a refinement order relation among relations. In the following subsection we introduce an example of the calculation of a recursive expression out of a first-order non-constructive specification involving an existential quantifier. In Subsection 5.4 we recall the concept of residuals and introduce a convenient heuristic way for considering them. Finally, in Subsection 5.5 we illustrate with an example how to derive recursive expressions out of non-constructive specifications involving universal quantifiers[1].

2 Are Abstract Relational Algebras an Appropriate Support for Programming Calculi?

One of the various presentations (axiomatizations) for ARAs is that due to L. Chin and A. Tarski [Chi51]. The version we choose here is that presented by G. Schmidt and T. Ströhlein [Sch85].

Definition 1 An *Abstract Relational Algebra* is a structure $A=\langle \mathfrak{R},+,\bullet,^{-},;,^{\mathsf{T}}\rangle$ over a non-empty set $\mathfrak{R}$, such that .

(1) $\langle \mathfrak{R},+,\bullet,^{-}\rangle$ is a complete atomic Boolean Algebra (in the sequel BA). Its *zero element* will be denoted by *0*, and its *unit element* by ∞. The symbol $\subset$ is used for the ordering with respect to the lattice structure and is called *inclusion*

(2) $\langle \mathfrak{R},;\rangle$ is a semigroup with exactly one *identity element* which is denoted by *1*, i.e. $(r;s);t=r;(s;t)$ & $r;1=1;r=r$

(3) $r;s\subset t \leftrightarrow r^{\mathsf{T}};\bar{t}\subset\bar{s}\leftrightarrow \bar{t};s^{\mathsf{T}}\subset\bar{r}$ *Schröder rule*

(4) $r\neq 0\rightarrow \infty;r;\infty=\infty$ *Tarski rule*

---

1 The properties of the "classical" relational operations are well-known and can be found elsewhere [Sch89].

The standard models for ARAs so defined are called Proper Relational Algebras (in the sequel PRAs). A PRA [Tar41] is a first-order theory satisfying the following extralogical axioms,

i.	$\forall x \forall y\,(x \infty y)$	*vii.*	$\forall x \forall y\,(x r^{\mathsf{T}} y \leftrightarrow y r x)$
ii.	$\forall x \forall y\,(\neg\, x 0 y)$	*viii.*	$\forall x \forall y\,(x r + s\, y \leftrightarrow x r y \vee x s y)$
iii.	$\forall x\,(x\, 1\, x))$	*ix.*	$\forall x \forall y\,(x r \bullet s\, y \leftrightarrow x r y \wedge x s y)$
iv.	$\forall x \forall y \forall z\,((x r y \wedge y\, 1\, z) \rightarrow x r z)$	*x.*	$\forall x \forall y\,(x r \ddagger s\, y \leftrightarrow (\forall z)(x r z \vee z s y)$
v.	$\forall x \forall y\,(x \wp y \leftrightarrow \neg\, x\, 1\, y)$	*xi.*	$\forall x \forall y\,(x r\, ;\, s\, y \leftrightarrow (\exists z)(x r z \wedge z s y)$
vi.	$\forall x \forall y\,(x \bar{r} y \leftrightarrow \neg\, x r y)$	*xii.*	$r = s \leftrightarrow \forall x \forall y\,(x r y \leftrightarrow x s y)$

This is a two-sorted theory, i.e., its variables are of two kinds, one ranging over *individuals* (denoted here by x, y, z, . . .) while the other ranges over relations (denoted here by italic letters $r, s, t, \ldots$). The atomic sentences are of the form $x r y$ (with intended meaning "x is in relation r with y") and $r = s$ (where the symbol = denotes equality over relations). Variables on individuals range over a fixed set A, while variables on relations range over some subsets of $A \times A$ –this *simplicity* is introduced by the inclusion of the so-called *Tarski rule* within the axiomatization of ARAs.

The symbols introduced by Tarski are, in our notation, ∞ (for the universal relation), 0 (for the null relation), 1 (for the identity relation) and $\wp$ (for the diversity relation), as relational constants, together with the following operations on relations $\bar{}$ (complement), ^{T} (converse), + (union), • (intersection), ‡ (relative addition), and ; (relative product).

Tarski [Tar41] posed two questions that are not only fundamental for the fields of Relational Algebras and Algebraic Logic but that we consider of primordial importance for relational programming calculi. The first was, can every property of relations, relations among relations, etc., that can be defined in a PRA be expressed in an ARA?. While the second can be stated as, is every model of (1) to (4) of Definition 1 isomorphic to a PRA (its standard model)? In other words, is every ARA representable?

It has been known for many years that the answer to both questions is negative.

The first one was answered by Tarski himself by showing two simple first-order expressions [Tar41] that cannot be expressed within his relational calculus (ARA). For instance, no sentence of an ARA is satisfied by exactly the same relations that satisfy expression $(\forall x)(\forall y)(\forall z)(\exists u)(x r u \wedge y r u \wedge z r u)$. Some years later Henkin and Monk [Hen74] and Maddux [Mad83, Mad91] stated that ARAs correspond to predicate logic in which only three individual variables are allowed and all predicates are at most binary.

The *Representation Problem* (i.e., whether every ARA is representable) received a negative answer from Roger Lyndon [Lyn50]. Lyndon constructed a finite nontrivial simple relational algebra which is non-representable.

Our view of Programming Calculi is, on one hand, that such calculi should be unifying formalisms for writing either non-constructive or constructive specifications and for calculating programs out of them. On the other hand, we claim that *actual formal programming methodologies* will contain, together with specification

construction and validation tools, not one but various interacting programming calculi (then, the possibility of inter-translation of terms between them is a must). Thus, from our viewpoint, we should provide not only programming calculi exhibiting –at least[2]– the expressive power of first-order logic with equality, but also calculi for translating back and forth first-order expressions to terms of such programming calculi. Moreover, if we do not want to fall into a *linguistic trap* resulting in divorce from reality (i.e., practical software engineering) as had happened in the case of logical empiricists with respect to the explication of the construction of scientific theories, we should bear in mind that while calculating a program we can often interpret terms as binary (semantic) relations relating pairs of *states of affairs*.
So, it seems we should abandon ARAs as the basis for programming calculi, especially if we consider that it's expressive power do not march neither that of first-order logic nor that of the language of general recursive functions (i.e., the constructive first-order logic).

3 "Classical" Solutions to Representability and Expressiveness Problems

3.1 Point Axiom and the Solution of the Representability Problem

The solution to the problem of the non-representability of ARAs is well known, requiring the inclusion of an extra axiom. For example G. Schmidt and T. Ströhlein have proposed [Sch85] the introduction of an axiom that they have called *point axiom*. Let us analyse succinctly their proposal. We call a relation x a *vector* iff $x = x\,;\infty$ and a *point* iff x is a vector, $x \neq 0$, $x\,; x^{\mathsf{T}} \subset 1$.
Notice that the interpretation of a vector in a PRA is a relation $B \times A \subset A \times A$. Then, since A, the fixed set we have taken in the standard model we are dealing with, is invariant for every vector, such vector is one of the ways of representing sets within an Algebra of Relations. In fact this is the way chosen by Hoare and He [Hoa86]. Along the same line of thinking, notice that points are then the representation of unary sets.
Continuing with the representation problem we will say that an ARA satisfies the *point axiom* if $r \neq 0 \rightarrow (\exists x)(\exists y)(x, y \text{ points} \wedge x\,; y^{\mathsf{T}} \subset r)$. Now we state the following

Theorem 4 An ARA satisfying the point axiom is a PRA. More precisely it is representable.

3.2 Classical Algebras for First-order Logic

Since BA has the expressive power of propositional logic, it seems natural to try to obtain an algebra for first-order logic (in particular for first-order logic with equality) by extending BA with some operations so as to cope with quantifiers (and the equality predicate).

---

2 In [Mad83] it is studied with detail the problem of how many variables are needed to do a prove within Proper Relational Algebra, providing a criteria for deciding whether or no a give proof can be accomplished in Abstract Relational Algebra. If we want to talk about termination within the programming calculus itself, we should require that such calculus have an expressive power which encompasses that of first-order logic.

Boolean Algebras with Operators. B. Jónsson and A. Tarski defined the concept of *Boolean Algebras with Operators* (in the sequel BAOs) [Jón51], i.e., an algebra resulting of extending a BA with a number, not necessarily finite, of operations satisfying certain conditions. They also show [Jón51] that ARA is a BAO -unfortunately with not enough expressive power to be an algebra of first-order logic. Speaking informally (see [Jón51] for a formal treatment of the subject), given a non-empty set $\mathfrak{A}$ and the BA $\mathcal{B} = \langle \mathfrak{A}, +, 0, \bullet, \infty \rangle$, then, a function f on $\mathfrak{A}^m$ to $\mathfrak{A}$ is called *additive* if it distributes over + and *completely additive* if it distributes over *supremum* (i.e. Boolean summation).

Now, by a *Boolean Algebra with Operators* we shall mean an algebra $\mathcal{O} = \langle \mathfrak{A}, +, 0, \bullet, \infty, f_0, f_1, \cdots, f_\xi, \cdots \rangle$, such that $\langle \mathfrak{A}, +, 0, \bullet, \infty \rangle$ is a BA and the functions f_ξ are additive. By an *atom* of $\mathcal{O}$ we mean an atom of the BA $\langle \mathfrak{A}, +, 0, \bullet, \infty \rangle$. We say that $\mathcal{O}$ is *atomistic* if the BA $\langle \mathfrak{A}, +, 0, \bullet, \infty \rangle$ is atomistic. We call $\mathcal{O}$ *complete* if the BA $\langle \mathfrak{A}, +, 0, \bullet, \infty \rangle$ is complete and if each of the operations f_ξ is completely additive.

Then, it is evident that every algebra satisfying the axioms (1) to (4) in Definition 1 –i.e., every ARA–, is a complete atomic BAO, the operators being **;** and ^{T}.

Two Classical Solutions for the Expressiveness Problem. In defining ARA, A. Tarski et al. extended BA with a *finite set* of operators –a way we will call *Peircean*–. The efforts of L. Chin, A. Tarski and F. B. Thompson [Chi48, Tar52] and later of Henkin, Monk and Tarski [Hen74], for defining an algebra for first-order logic with equality led to the introduction of *Cylindric Algebras*. But, in doing this, they abandon the Peircean way of extending BA in favour of extending it with *infinitely many operators*, thus obtaining algebras of *sets of infinitary sequences*, and likewise for Halmos's *Polyadic Algebras* [Hal62].

Before we begin with a succinct analysis of both algebras let us recall that the *Lindenbaun Algebra of Formulae* of a theory in a first-order language $\mathcal{L}$ is a structure $\mathfrak{F}_{\mathcal{L}} = \langle \Phi_{\mathcal{L}}, \vee, \wedge, \neg, \mathsf{F}, \mathsf{T}, \exists v_\zeta, \iota v_\zeta v_\xi \rangle_{\zeta, \xi < \omega}$, where $\Phi_{\mathcal{L}}$ is the set of formulae of $\mathcal{L}$ and $\iota v_\zeta v_\xi$ is the equivalence class of the *equation* $v_\zeta \approx v_\xi$ for every $\zeta, \xi < \omega$. We will denote by Γ the set of sentences of $\mathcal{L}$ and by $\mathfrak{F}_{\mathcal{L}}/\!\equiv_\Gamma$ the quotient algebra with respect to the equivalence relation $\equiv_\Gamma$ defined as, $\varphi \equiv_\Gamma \psi$ holds iff $\varphi \leftrightarrow \psi$ is a consequence of Γ. This quotient algebra is the so-called *Tarski algebra* associated with $\mathcal{L}$ and Γ.

Cylindric Algebras. Let Y be a non-empty set. Subsets of ${}^{\alpha}Y$ may be thought of α-ary relations, and we may perform on them the usual Boolean operations $\cup$, $\cap$ and $-$ (the latter, complementation, being performed with respect to ${}^{\alpha}Y$). Now for each $\kappa < \alpha$ we introduce a unary operation $\mathfrak{C}_\kappa^{({}^{\alpha}Y)}$, or simply $\mathfrak{C}_\kappa$, as follows. For any $\mathfrak{r} \subset {}^{\alpha}Y$,

$$\mathfrak{C}_\kappa^{({}^{\alpha}Y)}\mathfrak{r} = \left\{ x \in {}^{\alpha}Y\text{: there is a } y \in \mathfrak{r} \text{ with } x_\lambda = y_\lambda \text{ for all } \lambda < \alpha \text{ with } \lambda \neq \kappa \right\} \quad (5)$$

Also we consider special sets $\mathfrak{D}_{\kappa\lambda}^{({}^{\alpha}Y)}$:

$$\mathfrak{D}_{\kappa\lambda}^{({}^{\alpha}Y)} = \left\{ x \in {}^{\alpha}Y\text{: } x_\kappa = x_\lambda, \text{ for any } \kappa, \lambda < \alpha \right\} \quad (6)$$

Now, a *Cylindric Set Algebra of Dimension* α *with base* Y *and unit set* ${}^{\alpha}Y$, for brevity $\mathcal{CS}_\alpha$, is a structure

$$\mathcal{CS}_\alpha = \left\langle \mathcal{P}({}^{\alpha}Y), \cup, \cap, -, \varnothing, {}^{\alpha}Y, \mathfrak{C}_\kappa^{({}^{\alpha}Y)}, \mathfrak{D}_{\kappa\lambda}^{({}^{\alpha}Y)} \right\rangle_{\kappa, \lambda < \alpha}$$

An idea of these notions can be easily captured by considering the case $\kappa = 0, \alpha = 3$. A subset $\mathfrak{r} \subset {}^3Y$ can then be pictured as a point set in three dimensional space, $\mathfrak{C}_0\mathfrak{r}$ being the cylinder generated by moving $\mathfrak{r}$ parallel to the 0-axis (this is the reason why operations $\mathfrak{C}_\kappa$ are called *cylindrifications*). $\mathfrak{D}_{02}$ consists of all points equidistant from the 0 and 2 axis, and is thus a diagonal plane (this is the reason why elements $\mathfrak{D}_{\kappa\lambda}$ are called *diagonal elements*).

The connection of $\mathcal{CS}_\alpha$'s with first-order logic with equality can be seen as follows. Let $\mathcal{L}$ be a first-order language with equality, $\Phi_{\mathcal{L}}$ the set of formulas of $\mathcal{L}$, and $\mathcal{M} = \langle Y, \mathfrak{r}_i \rangle_{i \in I}$ a model for $\Phi_{\mathcal{L}}$. For each formula φ of $\mathcal{L}$ let $S_{\mathcal{M}}\varphi$ be the collection of all $x \in {}^{\omega}Y$ which satisfy φ in $\mathcal{M}$. Sets $S_{\mathcal{M}}$ induce a homomorphism $\hat{S}_{\mathcal{M}} : \mathfrak{F}_{\mathcal{L}}/_{\equiv_\Gamma} \to \mathcal{P}({}^{\omega}Y)$ with base Y. In particular, for any formula φ of $\mathcal{L}$ and any $\kappa < \omega$, we have

(7) $\hat{S}_{\mathcal{M}}[v_\kappa = v_\lambda] = \mathfrak{D}_{\kappa\lambda}$

(8) $\hat{S}_{\mathcal{M}}[\neg\varphi] = \overline{\hat{S}_{\mathcal{M}}[\varphi]}$

(9) $\hat{S}_{\mathcal{M}}[\varphi \vee \psi] = \hat{S}_{\mathcal{M}}[\varphi] \cup \hat{S}_{\mathcal{M}}[\psi]$

(10) $\hat{S}_{\mathcal{M}}[\varphi \wedge \psi] = \hat{S}_{\mathcal{M}}[\varphi] \cap \hat{S}_{\mathcal{M}}[\psi]$

(11) $\hat{S}_{\mathcal{M}}[\varphi \to \psi] = \overline{\hat{S}_{\mathcal{M}}[\varphi]} \cup \hat{S}_{\mathcal{M}}[\psi]$

(12) $\hat{S}_{\mathcal{M}}[(\exists v_\kappa)\varphi] = \mathfrak{C}_\kappa \hat{S}_{\mathcal{M}}[\varphi]$

By abstraction from this set-theoretic version we define a *Cylindric Algebra of dimension* α is a structure $\mathcal{CA}_\alpha = \langle \mathbf{A}, +, \bullet, -, 0, \infty, c_\kappa, \delta_{\kappa\lambda} \rangle_{\kappa,\lambda \le \alpha}$ satisfying the following axioms for all $\kappa, \lambda, \mu < \alpha$ and $x, y \in \mathbf{A}$:

C_0 $\langle \mathbf{A}, +, \bullet, -, 0, \infty \rangle$ is a BA, $c_\kappa : \mathbf{A} \to \mathbf{A}$ and $\delta_{\kappa\lambda} \in \mathbf{A}$

C_1 $c_\kappa 0 = 0$

C_2 $x \prec c_\kappa x$[3]

C_3 $c_\kappa(x \bullet c_\kappa y) = c_\kappa x \bullet c_\kappa y$

C_4 $c_\kappa c_\lambda x = c_\lambda c_\kappa x$

C_5 $\delta_{\kappa\kappa} = \infty$

C_6 if $\kappa \neq \lambda \neq \mu$ then $\delta_{\lambda\mu} = c_\kappa(\delta_{\kappa\lambda} \bullet \delta_{\kappa\mu})$

C_7 if $\kappa \neq \lambda$ then $c_\kappa(\delta_{\kappa\lambda} \bullet x) \bullet c_\kappa(\delta_{\kappa\lambda} \bullet \bar{x}) = 0$

Notice that $\mathcal{CA}_\alpha$ is a BAO and that $\mathfrak{F}_{\mathcal{L}}/_{\equiv_\Gamma}$ is a cylindric algebra of dimension ω, which satisfies the additional condition of *locally finite dimension*:

C_8 for any $x \in \mathbf{A}$ there are only finitely many $\kappa \le \lambda$ such that $c_\kappa x \neq x$

Hence, the eight axioms defining a *Cylindric Algebra of dimension* ω (for brevity CA) furnish a *theory presentation* for the *Tarski algebra* associated with $\mathcal{L}$ and Γ.

Polyadic Algebras. Let us, now, recall the concept of *Polyadic Algebra*, via *Quantifier* and *Transformation Algebras* [Hal62].

First, a *Quantifier Algebra of degree* χ (where χ is a cardinal number) is a triple $\langle \mathcal{B}, J, \exists \rangle$, where $\mathcal{B}$ is a BA, J is a set with cardinality χ, and $\exists$ is a function from subsets of J to quantifiers (endomorphisms) on $\mathcal{B}$, such that for all $t \in \mathcal{B}$, $\exists(\emptyset)t = t$, and $\exists(H)\exists(K)t = \exists(H \cup K)t$ whenever H, K $\subseteq$ J. The set J is referred to as the index set for the variables. We are interested in algebras of degree ω actually in those whose elements correspond to formulas with finitely many variables; these are the so-called *locally finite Quantifier Algebras of degree* ω, i.e., to every $t \in \mathcal{B}$ there corresponds a finite subset H of J such that $\exists(J - H)t = t$.

3 Where $\prec$ represents the natural order within the Boolean reduct.

Now, a *Transformation Algebra* is a triple $\langle \mathcal{B}, J, S\rangle$, where $\mathcal{B}$ is a BA, J is a set, and S is a function from transformations on J (i.e., mappings of J into itself) to endomorphisms on $\mathcal{B}$, such that for all $t \in \mathcal{B}$, $S(\delta)t = t$ where δ is the identity transformation on J, and $S(\sigma)S(\tau)p = S(\sigma\tau)p$, whenever σ and τ are transformations on J.

Finally, a *Polyadic Algebra of degree* ω (for brevity PA) is a quadruple $\mathfrak{P} = \langle \mathcal{B}, J, S, \exists\rangle$, where $\langle \mathcal{B}, J, \exists\rangle$ is a *locally finite Quantifier Algebra of degree* ω and $\langle \mathcal{B}, J, S\rangle$ is a *Transformation Algebra*, such that for all $t \in \mathcal{A}$, $S(\sigma)\exists(H)t = S(\rho)\exists(H)t$ whenever $H \subseteq J$ and σ and ρ are transformations on J agreeing on the set $J - H$, and $\exists(H)S(\rho)t = S(\rho)\exists(\rho^{-1}H)t$ whenever $H \subseteq J$ and ρ is a transformation on J never mapping any two distinct elements of J into the same element of H. We can have a better grasp of the idea behind PAs by considering that PA provides just one function $\exists$ from subsets of J to quantifiers and infinitely many transformations on J (each one allowing the construction of one of the infinitely many alphabetical arrangements of the variables). PA is then an algebra of pure first-order logic. Notice again that a PA $\mathfrak{P} = \langle \mathcal{B}, J, S, \exists\rangle$ is a BAO.

The Fitness of Abstract Relational Algebras, Cylindric and Polyadic Algebras to Underlying Programming Calculi. Thus, as we have seen above both CAs and PAs are BAOs, the former being an algebra for first-order logic –due to their diagonal elements– while the latter is just an algebra for pure first-order logic.

The attractive aspect of ARA for underlying a programming calculus resides in its absence of variables over individuals, an aspect that led *functional languages* and functional based programming calculi to deserve so much attention. This absence of variables is due to the fact that the extension to BA necessary for constructing ARAs is finitary. In fact, ARA is a BA extended with just two operators, namely, **;** and ^{T}.

We should notice that both Tarski et al., with CAs, and Halmos, with PAs abandon the Peircean-like way of extending BA in favour of a more direct attack to the problem. Both approaches maintain a BA over a universe of elements whose internal structure is immaterial. By imposing an infinite arity, they produce theories over sequences of infinite length restricted by a local finiteness condition. Tarski et al. choose to represent the existential quantifier by infinitely many cylindrifications c_ζ (one for each variable v_ζ), and equations $v_\zeta = v_\xi$ by means of a doubly infinite sequence of diagonal elements $\delta_{\zeta\xi}$. Halmos, on the other hand, represents the existential quantifier by means of a single function $\exists$ but introduces infinitely many transformations on the index set J. So both are models laden with a syntactical artefact, in that CAs, by means of each of the "ζ-axis", and PAs, by means of the index set J, have hidden "names" for variables over individuals.

4 ∇-Extended Abstract Relational Algebra

Let us analyse here, a way for constructing a finitary algebra of first-order logic with equality introduced by A. M. Haeberer, P. A. S. Veloso, G. A. Baum and P. Elustondo [Hae90, 91, 93, Vel91, 92].

4.1 Tackling the Expressiveness Problem as Posed By Tarski, Henkin and Monk

First of all we will analyse and solve the expressiveness problem as it was posed by Tarski [Tar41]. So, we will try to express, by means of ∇ operation –in addition to classical relational operations– one of the first-order expressions stated by Tarski, say the four-variables one $(\forall x)(\forall y)(\forall z)(\exists u)(x r u \wedge y r u \wedge z r u)$. First, notice that there is no difficulty in expressing the simpler Three-variables version $(\forall x)(\forall y)(\exists u)(x r u \wedge y r u)$ since it is equivalent to $(\forall x)(\forall y)(\exists u)(x r u \wedge u r^{\mathsf{T}} y)$ it can be expressed by $r ; r^{\mathsf{T}} = \infty$. The key idea here is the fact that the existential quantifier $(\exists u)$ can be simulated by the *relative* product. If we would try to apply this simple idea to the four-variables expression, we would be led to something like $(\forall x)(\forall y)(\forall z)(\exists u)(x r u \wedge u r^{\mathsf{T}} y \wedge u r^{\mathsf{T}} z)$. Unfortunately cannot simulate here the effect of $(\exists u)$ by the relative product. The reason for this is the fact that variable u now occurs three times in the matrix of the formula. Tarski's relational calculus has no variables over individuals, and the relative product $r ; r^{\mathsf{T}}$ "consumes", so to speak, variable u.

Let us extend PRA with a new operation ∇ –which we will call *fork* – defined as

$$(\forall x)(\forall y)(\forall z)(x\, r \nabla s\, [y,z] \leftrightarrow x r y \wedge x s z) \tag{13}$$

Where [ ] is a non-commutative pair formation operation. Notice that in doing so, variables over individuals will no more range over a fixed set A without any noticeable structure but over a set A^* of all finite binary trees made out of elements of a fixed set A, i.e., a free groupoid over A.

Another useful operation which can be defined from ∇ is $r \otimes s = \left((1 \nabla \infty)^{\mathsf{T}} ; r\right) \nabla \left((\infty \nabla 1)^{\mathsf{T}} ; s\right)$. It is easy to see that the Extended PRA expression for such operation will be,

$$(\forall x)(\forall y)(\forall z)(\forall w)([x,z]\, r \otimes s\, [y,w] \leftrightarrow x r y \wedge z s w)$$

Then, by finitely extend the PRA with the operation ∇ we obtain a Proper Algebra over locally finite trees.

Shifting back our attention to the problem of expressing relationally Tarski's four-variables first-order formula, it is easy to see that in the same way that we were able to express the three variables expression as $r ; r^{\mathsf{T}} = \infty$, we can now express Tarski's one as $r ; (r^{\mathsf{T}} \nabla r^{\mathsf{T}}) = \infty \nabla \infty$.

4.2 An Axiomatization of ∇-Extended Abstract Relational Algebra

Let us now introduce an axiomatization for ∇-Extended Abstract Relational Algebra.

Definition 10 A *∇-Extended Abstract Relational Algebra* (for brevity ∇ARA) is a structure $A\nabla = \langle \Re, +, \bullet, ^{-}, ;, ^{\mathsf{T}}, \nabla \rangle$ over a non-empty set $\Re$, such that

(14) $\langle \Re, +, \bullet, ^{-}, ;, ^{\mathsf{T}}, \nabla \rangle$ is an ARA. Its *zero element* will be denoted by *0*, its *unit element* by ∞, and its *identity element* by *1*. The symbol $\subset$ is used for the ordering with respect to the lattice structure and is called *inclusion*

(15) $r \nabla s = (r ; (1 \nabla \infty)) \bullet (s ; (\infty \nabla 1))$

(16) $(r \nabla s) ; (t \nabla q)^{\mathsf{T}} = (r ; t^{\mathsf{T}}) \bullet (s ; q^{\mathsf{T}})$

(17) $\left((1 \nabla \infty)^{\mathsf{T}} \nabla (\infty \nabla 1)^{\mathsf{T}}\right) + 1 = 1$

Notice that in the preceding section we call, for the sake of compactness, $\pi = (1 \nabla \infty)^{\mathsf{T}}$ and $\rho = (\infty \nabla 1)^{\mathsf{T}}$. The intention behind the choice of symbols π and ρ is to connect $(1 \nabla \infty)^{\mathsf{T}}$ and $(\infty \nabla 1)^{\mathsf{T}}$ with the first and second projections respectively. Formally speaking we have extended algebra $A\nabla$ by definition with π, ρ and $\otimes$.

The concept of *projections* and of some kind of *product* in connection with algebras of first-order logic and, in special, with RAs appears as early as 1946 with the seminal work of Everett and Ulam on *Projective Algebra* [Eve46]. Later on, De Roever [Roe72], the Munich group, i.e., Schmidt, Ströhlein, Berghammer, Zierer along the 80's[4], and Backhouse et al. [Bac92][5], introduced the product and projections either as data types or as operations. However, each one of them failed to formulate that a RA extended with a product of this kind has enough expressive power. Everett and Ulam (and, later, Bednarek and Ulam [Bed78]) when defining Projective Algebra, extended BA with three fundamental operations: two projections p_1 and p_2 and a product $\square$. In doing so, they did not discuss the power of their combination with Peircean operations for the construction of a RA over locally finite trees. Those researchers involved in computer science failed because they had a categorical viewpoint that "masked" the non fundamental character of projections.

4.3 The Expressiveness of ∇-Extended Abstract Relational Algebra

Encouraged by the result presented in Subsection 4.1, one is tempted to conjecture that a BA extended with operations ; and ^{T} as above and an abstract operation ∇ –resulting from the axiomatization of ∇ from its properties derived within the above extended PRA– has the expressive power of first-order logic. Due to lack of space, we will only give here a convincing argument for showing that this is the case (a formal and detailed prove can be found in [Vel91]).

Consider a *first-order language* $\mathcal{L}$ with equality $\approx$ and a structure $\mathfrak{B}$ for it assigning to each formula $\varphi \in \Phi_{\mathcal{L}}$ with n free variables a set $\varphi^{\mathfrak{B}} = \{\langle [b_1, [\ldots, [b_{n-1}, b_n]], [b_1, [\ldots, [b_{n-1}, b_n]] \rangle : \mathfrak{B} \models \varphi(b_1, \ldots, b_n)\}$ of identities on finite trees over a basic set B. Let us call $\mathfrak{I}$ the set of terms of $A\nabla$ denoting identities, then, it is possible to construct a function $\mathcal{T}\colon \Phi_{\mathcal{L}} \to \mathfrak{I}$ such that $\varphi^{\mathfrak{B}} = \mathfrak{B}[\mathcal{T}(\varphi)]$. Where $\mathfrak{B}[t]$ is the interpretation by structure $\mathfrak{B}$ of the relational term t (see [Vel91]).

Notice that for every $n \geq 0$, there exist $A\nabla$-relational terms d_n and e_m, such that, for every formula φ with n free variables, if φ' is the formula $\varphi \wedge v_1 \approx v_1 \wedge \ldots \wedge v_{m+n} \approx v_{m+n}$, one can have $\mathcal{T}(\varphi') = \mathcal{T}(\varphi) \otimes e_m$ and $\mathcal{T}(\varphi) = d_n^{\mathsf{T}} ; \mathcal{T}(\varphi') ; d_n$. Also, notice that given a substitution σ on $\{1, \ldots, n\}$, there exists a term $s(\sigma)$ (easily constructed by means of 1, ∇ and ∞), such that for any formula φ with n free variables, if $\sigma(\varphi)$ is the formula obtained by applying σ to φ, then one can take $\mathcal{T}(\sigma(\varphi)) = s(\sigma)^{\mathsf{T}} ; \mathcal{T}(\varphi) ; s(\sigma)$.

For the two variables case, function $\mathcal{T}$ will be as follows,

$$\mathcal{T}(p(v_1, \ldots, v_n)) = (p ; p^{\mathsf{T}}) \bullet 1 \qquad \mathcal{T}(\sigma(\alpha(v_1, \ldots, v_n))) = s^{\mathsf{T}} ; \mathcal{T}(\alpha(v_1, \ldots, v_n)) ; s$$

---

4 See the bibliography included in [Sch89].

5 The idea of direct product and direct sum presented in this paper by Backhouse et al. was borrowed from the works of the Munich group.

$\mathcal{T}(v_1 \approx v_2) = (1 \nabla 1)^{\mathsf{T}} ; 1_B ; (1 \nabla 1)$ $\quad \mathcal{T}(\neg\psi) = \overline{\mathcal{T}(\psi)} \bullet e_n$

$\mathcal{T}(\psi \vee \theta) = \mathcal{T}(\psi') + \mathcal{T}(\theta')$ $\quad \mathcal{T}((\exists v_n)\psi') = (1 \nabla \infty) ; \mathcal{T}(\psi') ; (1 \nabla \infty)^{\mathsf{T}}$

Notice that we define the translation $\mathcal{T}$ for the existential quantifier just for the case of the quantification on variable v_n. For generalising the quantification on other variables we relay on terms d_n and $s(\sigma)$ for respectively adjusting the number of variables and producing an alphabetical transformation of the original formula. This resembles the way Polyadic Algebras treat the same problem, the difference being that, in case of $A\nabla$, this trick is an artefact of the proof, while in the former it is inherent to the algebra itself. Notice also that both d_n and $s(\sigma)$ are constructed with the constants and operations of the $A\nabla$.

By calling $\pi = (1 \nabla \infty)^{\mathsf{T}}$ and $\rho = (\infty \nabla 1)^{\mathsf{T}}$ and using a somewhat two-dimensional notation we can express, for instance, $\mathcal{T}((\exists v_1)(\exists v_3)\psi(v_0, v_1, v_2, v_3, v_4))$ as,

$$\underbrace{\begin{pmatrix} \pi;\pi;\begin{pmatrix}\pi;\pi\\ \nabla\\ \rho\end{pmatrix} \\ \nabla \\ \rho \\ \\ \end{pmatrix}^{\mathsf{T}}}_{ds} ; \mathcal{T}\big(\psi(v_0,v_1,v_2,v_3,v_4)\big) ; \underbrace{\begin{pmatrix} \pi;\pi;\begin{pmatrix}\pi;\pi\\ \nabla\\ \rho\end{pmatrix} \\ \nabla \\ \rho \\ \\ \end{pmatrix}}_{ds^{\mathsf{T}}}$$

where the pair of terms $\langle ds, ds^{\mathsf{T}} \rangle$ is the construction, within ∇-Extended Relational Algebra, of the cylindrification $c_1 c_3\big(\psi^H(v_0,v_1,v_2,v_3,v_4)\big)$.

4.4 Comparing our Axiomatization with Other Approaches to Projections and Product.

We are able now to compare our axiomatization with both, the Munich one and that due to Backhouse et al.

First let us prove that the Munich group axiomatization can be derived from our one. The Munich axiomatization is, in our own notation,

$M_0 \quad \pi^{\mathsf{T}} ; \pi = 1$

$M_1 \quad \rho^{\mathsf{T}} ; \rho = 1$

$M_2 \quad (\pi ; \pi^{\mathsf{T}}) \bullet (\rho ; \rho^{\mathsf{T}}) = 1$

$M_3 \quad \pi^{\mathsf{T}} ; \rho = \infty$ and $\rho^{\mathsf{T}} ; \pi = \infty$ [6]

$M_4 \quad r \nabla s = (r ; \pi^{\mathsf{T}}) \bullet (s ; \rho^{\mathsf{T}})$

M_0 can be derived as follows, by applying the definition of π, whereby we can write $(1 \nabla \infty);(1 \nabla \infty)^{\mathsf{T}}$, which by axiom (16) equals $(1;1^{\mathsf{T}}) \bullet (\infty;\infty^{\mathsf{T}}) = 1$. M_1 can be proved similarly. To derive M_2 notice that from the definition of $\otimes$ one has $1 \otimes 1 = \big((1 \nabla \infty)^{\mathsf{T}} ; 1\big) \nabla \big((\infty \nabla 1)^{\mathsf{T}} ; 1\big) = (1 \nabla \infty)^{\mathsf{T}} \nabla (\infty \nabla 1)^{\mathsf{T}}$ which by applying axiom (15) can be rewritten as $\big((1 \nabla \infty)^{\mathsf{T}} ; (1 \nabla \infty)\big) \bullet \big((\infty \nabla 1)^{\mathsf{T}} ; (\infty \nabla 1)\big)$ which by the definitions of π and ρ is $(\pi ; \pi^{\mathsf{T}}) \bullet (\rho ; \rho^{\mathsf{T}})$ – notice that we proved equality with $1 \otimes 1 \subset 1$ because we are working with homogeneous relations. M_3 can be derived in a

---

6 Actually one of them suffices for this axiomatization, we include both just for symmetry.

straightforward way from the definitions of π and ρ and (16). Finally, after the definitions of π and ρ M_4 is equal to (15). Equality in axiom (16) is not derivable from Munich axioms. Without a *point-like* axiom, only strict inclusion is derivable.
As for the axiomatization due to Backhouse et al. we can show that it exhibits more axioms than are necessary. As presented in [Bac92] it looks like,

B_0 $\quad r \nabla s = (r ; \pi^{\mathsf{T}}) \bullet (s ; \rho^{\mathsf{T}})$ $\qquad$ B_2 $\quad (r \nabla s);(t \nabla q)^{\mathsf{T}} = (r ; t^{\mathsf{T}}) \bullet (s ; q^{\mathsf{T}})$

B_1 $\quad (\pi ; \pi^{\mathsf{T}}) \bullet (\rho ; \rho^{\mathsf{T}}) \subset 1$ $\qquad$ B_3 $\quad \mathcal{D}om(\pi) = \mathcal{D}om(\rho)$

Notice that B_0 equals (15), B_1 equals M_2 in the homogeneous case, B_2 equals (16) and B_1 equals (17).
As for B_3 let us introduce as definition $1_{\mathcal{D}om(r)} = (r ; r^{\mathsf{T}}) \bullet 1$, then, B_3 can be derived immediately from M_2 and the preceding definition provided the equivalence $\mathcal{D}om(r) = \mathcal{D}om(s) \leftrightarrow 1_{\mathcal{D}om(r)} = 1_{\mathcal{D}om(s)}$ holds, which in turn, is obvious, since 1_X is the equivalence relation restricted to set X. Here, we are using the usual meaning for $\mathcal{D}om(r)$, i.e., $(\forall x)((x \in \mathcal{D}om(r)) \leftrightarrow (\exists y)(xry))$.

4.5 On the Representability of ∇-Extended Abstract Relational Algebra

The representability of ∇ARA as defined by axioms (14) to (17) is still an open problem, being the object of ongoing research. However, as was discussed in [Bau92], the relational reduct of the ∇-Extended Abstract Relational Algebra is representable if we add $t \neq 0 \wedge t \subset \infty \nabla \infty \rightarrow (\exists v)(\exists w)(0 \neq v \nabla w \subset t)$ to the above mentioned axiom.
Then, denoting by $\mathcal{A}$ the set of atoms of A∇, representability can be proved by first proving that $\alpha \in \mathcal{A} \rightarrow \alpha \nabla \alpha \in \mathcal{A}$ and then using this result for proving $\alpha \in \mathcal{A} \rightarrow \alpha$ *is functional*. Hence, by a well-known result [Jón51], since its atoms are functional, the relational reduct of the ∇ARA extended with the new axiom is representable, i.e., all its models are isomorphic to the relational reduct of the Proper ∇-Extended Relational Algebra.

5 On the Calculation of Recursive Expressions for Quantified Relational Terms

5.1 Some Useful Results on Partial Identities

As it should be noticed, in constructing the ∇ARA we chose to work in a homogeneous framework (for a comparative discussion about both homogeneous and heterogeneous approaches see [Ber93]). Thus, we are forced to have some kind of type relativization mechanism. It is obvious that such mechanism depends on the way we have chosen for internalising sets in the relational framework. Some authors, as Hoare and He [Hoa86] use vectors (which they call *conditions*) and, therefore, intersection as relativizing operation. Others, as ourselves, chose partial identities, therefore using left multiplication (;) for relativizing[7] relations. All the results here introduced, without proof due to lack of space, can be proved within the abstract framework (proofs can be found in [Hae93]).

---

7 For a discussion about different ways of internalising sets in a relational framework see [Vel92).

First we should verify the equivalence between both ways of relativization, i.e., $(1 \bullet (s ; \infty)) ; r = r \bullet (s ; \infty)$. After that we can show that in restricting partial identities with partial identities, left multiplication and intersection are interchangeable, i.e., $1_X \bullet 1_Y = 1_X ; 1_Y$ and $(1_X ; r) \bullet (1_Y ; r) = (1_X \bullet 1_Y) ; r$. Now, we introduce some useful and necessary results for dealing with partial identities, the equality $(r ; r^{\mathsf{T}}) \bullet 1 = (r ; \infty) \bullet 1 = (1 \nabla r) ; \pi$ gives different ways of expressing $1_{\mathcal{D}om(r)}$, while $d \subset 1 \rightarrow 1 \bullet (d ; \infty) = d$ shows in the abstract environment, the obvious proper algebraic result that any part of the identity relation is itself a partial identity. The next two results show that for a given relation r, the relation $1_{\mathcal{D}om(r)}$ is exactly its left identity. The equality $1_{\mathcal{D}om(r)} ; r = r$ states that given any relation, the partial identity over its domain contains its left identity; while $1_X ; r = r \rightarrow (r ; \infty) \bullet 1 \subset 1_X$ shows that any part of an identity relation which behaves as left identity of a given relation, contains the identity over its domain.

5.2 On Refinements

It is noticeable in calculating programs that equality is a too narrow substitution criterion while inclusion is obviously too wide. What we need is to introduce a *refinement order relation* among relations. So, let us state the following,

Definition 4 $r \sqsubseteq s$ iff $1_{\mathcal{D}om(s)} ; r \subset s$ and $1_{\mathcal{D}om(s)} \subset 1_{\mathcal{D}om(r)}$

Its "programming" meaning is obvious, given two specifications (or programs) Σ_1 and Σ_2, we will say that Σ_2 is totally correct with respect to Σ_1 iff within its precondition, Σ_1 satisfies Σ_2 –not necessarily with the same degree of non-determinism–, and the set denoted by the precondition of Σ_2 is included in that denoted by the precondition of Σ_1

The equality $\mathcal{D}om(r) ; \infty = r ; \infty$, which is obviously valid in PRA, is also a theorem in the abstract framework.

For relativizing via fork and projections we can use the identity $(r \nabla s) ; \pi = 1_{\mathcal{D}om(s)} ; r$, which is also valid in the abstract algebra.

Finally, $\mathcal{D}om(r ; s) = \mathcal{D}om(r ; \mathcal{D}om(s))$ and $\mathcal{D}om(r \nabla s) = \mathcal{D}om(r) \bullet \mathcal{D}om(s)$ are an example of refinements valid under certain conditions (this result will be used in the derivation presented in Subsection 5.5). We can also prove without variables over individuals that if $\mathcal{D}om(s) = \mathcal{D}om(t)$ then, $\mathcal{D}om(s \nabla t) = \mathcal{D}om(s)$, $\mathcal{D}om(r ; s) = \mathcal{D}om(r ; t)$ and $\mathcal{D}om(r ; s) = \mathcal{D}om(r ; s \nabla r ; t) = \mathcal{D}om(r ; (s \nabla t))$. Along the same line we can prove that if $\mathcal{D}om(r ; s) = \mathcal{D}om(r ; t)$ then $r ; (s \nabla t) \sqsubseteq (r ; s) \nabla (r ; t)$.

5.3 Calculating Algorithms From First-Order Specifications Involving Existential Quantifiers

One of the ways of translating first-order specifications onto terms of the ∇ARA is by the simple and straightforward application of the expressiveness result of Subsection 4.3. However, by doing so we obtain expressions that, from the view point of algorithmic complexity, are extremely inefficient. The ability of expressing quantifiers not by fundamental operations but by terms written from such operations, in this cases, helps dealing with such complexity.

To illustrate this claim, let us calculate a very simple and "classic" program from a non constructive first-order specification, i.e., a program for deciding whether a natural number n is of the form $2^i - 1$ for some $i \in Nat$, where Nat is the type *natural numbers.*

Let us first introduce some relational constants. Thus, for x, y ∈ Nat we will write x *zero* 0, x *suc* x+1, x *pred* x-1 (whenever x≠0), x *pot2* 2^x, x *true* T, x *false* F, [x, y] *and* (x ∧ y).

Let us now give a first-order specification of our problem,

$$\varphi(n) = (\exists i)\left(\underbrace{i \in Nat \wedge n = 2^i - 1}_{\alpha(n,i)}\right)$$

Recall that, $\mathcal{T}((\exists y)(\psi(x,y))) = \pi^T ; \mathcal{T}(\psi(x,y)) ; \pi$. Since, as it is obvious, $\mathcal{T}(\alpha(n,i)) = (1 \otimes (pot2 ; pred)) ; (1 \nabla 1)^T ; (1 \nabla 1)$, we can write 1_{prop} (i.e., the identity relation which interprets φ(n)) as

$$1_{prop} = 1_{Nat} ; \pi^T ; (1 \otimes (pot2 ; pred)) ; (1 \nabla 1)^T ; (1 \nabla 1) ; \pi$$

But $(1 \nabla 1)^T ; (1 \nabla 1) ; \pi = (1 \nabla 1)^T ; (1 \nabla 1) ; (1 \nabla \infty)^T$ and by applying axiom (16), we can write $(1 \nabla 1)^T ; 1 \bullet \infty = (1 \nabla 1)^T ; 1 = (1 \nabla 1)^T$. Thus, we have

$$1_{prop} = 1_{Nat} ; \pi^T ; (1 \otimes (pot2 ; pred)) ; (1 \nabla 1)^T$$

From this partial identity, by using $r[\![s]\!] = (r \nabla s) ; \rho + (r \nabla \overline{s ; \infty}) ; \pi$, we can calculate a *total predicate* [8]

$$\ddot{1}_{prop} = (1_{Nat} ; false)[\![1_{prop}]\!] = ((1_{Nat} ; false) \nabla 1_{prop}) ; \rho + \left((1_{Nat} ; false) \nabla \overline{1_{prop} ; \infty}\right) ; \pi$$

which can be written

$$\ddot{1}_{prop} = 1_{Nat} ; \pi^T ; (1 \otimes (pot2 ; pred)) ; (1 \nabla 1)^T ; true + 1_{\overline{prop}} ; false$$

Now, by taking into account that $(\exists i)(n = 2^i - 1) \leftrightarrow (\exists i)(i \le n \wedge n = 2^i - 1)$ and defining the new relational constant $\le n = \{\langle x, y\rangle : x, y \in Nat \wedge y \le x\}$, it is easy to obtain the following equality

$$(1 \otimes (pot2 ; pred)) ; (1 \nabla 1)^T = (1 \otimes \le n) ; (1 \otimes (pot2 ; pred)) ; (1 \nabla 1)^T$$

Substituting and recalling the definition of π we have

$$\ddot{1}_{prop} = 1_{Nat} ; (1 \nabla \infty) ; (1 \otimes \le n) ; (1 \otimes (pot2 ; pred)) ; (1 \nabla 1)^T ; true + 1_{\overline{prop}} ; false$$

Notice that here we are in the crucial point of the derivation. Due to the fact that projections –and as a consequence cylindrifications– are not fundamental operations, we can write

$$(1 \nabla \infty) ; (1 \otimes \le n) = (1 \nabla \le n)$$

thus restricting the *combinatorial explosion* produced by π^T. Then, substituting and distributing ∇ over ⊗, we have[9]

$$\ddot{1}_{prop} = 1_{Nat} ; (1 \nabla (\le n ; pot2 ; pred)) ; (1 \nabla 1)^T ; true + 1_{\overline{prop}} ; false$$

---

8 Notice that $r[\![s]\!]$ equals s within $\mathcal{D}om(r) \bullet \mathcal{D}om(s)$ and r. within $\mathcal{D}om(r) - \mathcal{D}om(s)$

9 General properties of ∇ and ⊗, can be found in [Hae91].

But notice that $\leq n = pred^{;*}$, where $pred^{;*}$ is the transitive closure of *pred*. Then

$$\ddot{1}_{prop} = 1_{Nat} ; \left(1 \nabla \left(pred^{;*} ; pot2 ; pred\right)\right) ; \left(1 \nabla 1\right)^{\mathsf{T}} ; true + 1_{\overline{prop}} ; false$$

which is an iterative algorithm.

5.4 Some Considerations About Residuals and Relational Implication

For the subsequent derivation, some discussion about residuals will be useful.
Let $s \rfloor r$ be the *left residual* and $s \llcorner r$ the *right residual*.. Their relational expressions can be calculated by means of Galois connections from the equations $x ; r \subset s$ and $r ; x \subset s$, respectively, by using Schröder rule, i.e.,

$$x \subset s \rfloor r \leftrightarrow x ; r \subset s \qquad (18) \qquad\qquad x \subset s \llcorner r \leftrightarrow r ; x \subset s \qquad (19)$$

Since $s \rfloor r$ and $s \llcorner r$ are the weakest and the strongest solutions of equations $x ; r \subset s$ and $r ; x \subset s$, respectively, we can write $s \rfloor r = \overline{\overline{s} ; r^{\mathsf{T}}}$ and $s \llcorner r = \overline{r^{\mathsf{T}} ; \overline{s}}$ (recall that $r \dagger s = \overline{\overline{r} ; \overline{s}}$).

Now, notice that $\left(s \rfloor r\right)^{\mathsf{T}} = \left(\overline{\overline{s} ; r^{\mathsf{T}}}\right)^{\mathsf{T}} = \overline{\left(\overline{s} ; r^{\mathsf{T}}\right)^{\mathsf{T}}} = \overline{r ; (\overline{s})^{\mathsf{T}}} = \overline{r} \dagger s^{\mathsf{T}}$ then, recalling that $\overline{\alpha} \vee \beta \leftrightarrow \alpha \rightarrow \beta$ and the definition of $r \dagger s$ we can write

$$(\forall x)(\forall y)\left(x\left(s \rfloor r\right)^{\mathsf{T}} y \leftrightarrow (\forall z)(x r z \rightarrow y s z)\right)$$

This definition lead us to introduce a new symbol –which we call *relational implication*–, namely, $r \twoheadrightarrow s = \left(s \rfloor r\right)^{\mathsf{T}}$. Notice also that

$$r^{\mathsf{T}} \twoheadrightarrow s^{\mathsf{T}} = \left(s^{\mathsf{T}} \rfloor r^{\mathsf{T}}\right)^{\mathsf{T}} = \left(\overline{\overline{s^{\mathsf{T}}} ; r^{\mathsf{T}^{\mathsf{T}}}}\right)^{\mathsf{T}} = \overline{r^{\mathsf{T}} ; \overline{s}} = s \llcorner r \qquad (20)$$

Then, $r^{\mathsf{T}} \twoheadrightarrow s^{\mathsf{T}} = \sup\{x : r ; x \subset s\} = \overline{r^{\mathsf{T}}} \dagger s$, thus

$$(\forall x)(\forall y)\left(x r^{\mathsf{T}} \twoheadrightarrow s^{\mathsf{T}} y \leftrightarrow (\forall z)(z r x \rightarrow z s y)\right) \qquad (21)$$

5.5 Specifications Involving Universal Quantifiers - Formal Development of a Constructive Specification of Quicksort from the Partial Identity over Ordered Lists

As was done in the preceding section, translation from first-order formulae to terms in ∇ARA can be accomplished by the straightforward application of the constructive proof of the expressiveness outlined in Subsection 4.3. However, here we will begin our formal development by doing another kind of translation, which we will call "by *pattern matching*".

Let ℓ, ℓ_1 and ℓ_2 be lists, *Ord* (ℓ) a predicate whose truth-value is true iff ℓ is an ordered list and last and head the usual operations on lists. A first-order definition of *Ord* could be,

$$(\forall \ell)\left(\mathit{Ord}(\ell) \leftrightarrow (\forall \ell_1)(\forall \ell_2)\left(\ell \in \mathcal{L}^1 \vee \ell = \ell_1 \# \ell_2 \rightarrow \mathrm{last}(\ell_1) \leq \mathrm{head}(\ell_2)\right)\right)$$

which can be written in PRA as

$$(\forall \ell)\left(\ell \ \mathit{Ord} \ \mathsf{T} \leftrightarrow (\forall [\ell_1, \ell_2])\left(\ell \in \mathcal{L}^1 \vee [\ell_1, \ell_2] \ cnc \ \ell \rightarrow [\ell_1, \ell_2] \left(lst \otimes hd\right) ; 1_{\leq}\right) \mathrm{pair}\right) \qquad (22)$$

where, $\mathcal{L}^1$ is the set of lists of length 1, $\mathcal{L}^*$ is the set of all lists, $1_{\mathcal{L}^*}$ is the identity relation restricted to the set of all lists, $1_{\mathcal{L}^i}$ is the identity restricted to the set of lists of length i or less, *lst* is a relation which for each list gives its `last` element, *hd* is a relation giving for each list its `head`, *cnc* is a relation which given two lists of length

1 or greater concatenates them to form a new list[10] and $1_{\leq}$ is the partial identity over pairs pair of elements satisfying relation $\leq$.
Comparing expressions (22) and (21) it is evident that both definiens match.

$$\underbrace{(\forall \ell)} \quad (\underbrace{\ell\, \mathit{Ord}\, \mathsf{T}} \leftrightarrow (\forall\underbrace{[\ell_1,\ell_2]})(\underbrace{\ell\in\mathcal{L}^1 \vee [\ell_1,\ell_2]\, conc\, \ell} \rightarrow \underbrace{[\ell_1,\ell_2]\, ((lst\otimes hd);1)\, \text{pair}})$$

$$(\forall x)(\forall y)\ (x\, r^{\mathsf{T}} \twoheadrightarrow s^{\mathsf{T}}\, y \leftrightarrow (\forall z) \quad (\quad z\, r\, x \quad \rightarrow \quad z\, s\, y \quad)$$

Then, replacing them by their definienda we can write

$$(\forall \ell)\left(\ell\, \mathit{Ord}\, \mathsf{T} \leftrightarrow 1_{\mathcal{L}^1} + \ell\left(cnc^{\mathsf{T}} \twoheadrightarrow ((lst\otimes hd);1_{\leq})^{\mathsf{T}}\right)\text{pair}\right) \tag{23}$$

recalling that $1\bullet(r;\infty)$ is the identity over the domain of r and distributing intersection over sum we can write

$$1_{\mathit{Ord}} = 1_{\mathcal{L}^*}\bullet(\mathit{Ord};\infty) = \underbrace{1_{\mathcal{L}^*}\bullet 1_{\mathcal{L}^1}}_{1_{\mathcal{L}^1}} + 1_{\mathcal{L}^*}\bullet\left(\left(cnc^{\mathsf{T}} \twoheadrightarrow ((lst\otimes hd);1_{\leq})^{\mathsf{T}}\right);\infty\right)$$

Notice that the expression $\left(\left(cnc^{\mathsf{T}} \twoheadrightarrow ((lst\otimes hd);1_{\leq})^{\mathsf{T}}\right);\infty\right)$ is not defined over $\mathcal{L}^1$.

Then, assigning to operation $+$ angelic non determinism, we can write the following deterministic expression

$$1_{\mathit{Ord}} = 1_{\mathcal{L}^1} + 1_{\mathcal{L}^*-\mathcal{L}^1}\bullet\left(\left(cnc^{\mathsf{T}} \twoheadrightarrow ((lst\otimes hd);1_{\leq})^{\mathsf{T}}\right);\infty\right) \tag{24}$$

Let us forget the trivial part $1_{\mathcal{L}^1}$ and concentrate ourselves in the second non trivial part of (36). Thus, calling $1_{\mathit{Ord}\,2} = 1_{\mathit{Ord}} - 1_{\mathcal{L}^1}$ and recalling (20) we can introduce the right residual

$$1_{\mathit{Ord}\,2} = 1_{\mathcal{L}^*-\mathcal{L}^1}\bullet\left(\left((lst\otimes hd);1_{\leq}\,\lfloor cnc\right);\infty\right) \tag{25}$$

which is, by a Galois connection, equivalent to

$$1_{\mathit{Ord}\,2} = 1_{\mathcal{L}^*-\mathcal{L}^1}\bullet\left(\sup\{x : cnc\,;x \subset (lst\otimes hd);1_{\leq}\};\infty\right) \tag{26}$$

Notice that we must solve the equation $\sup\{x : cnc\,;x \subset (lst\otimes hd);1_{\leq}\}$ over the set of ordered lists. For doing so, we can introduce the following two lemmas[11]:

Lemma 5 If $\mathcal{Dom}(x)$ is the set of lists ℓ satisfying $\ell\, \mathit{Ord}\, \mathsf{T}$ then $cnc\,;x = (1_{\mathit{Ord}}\otimes 1_{\mathit{Ord}});cnc\,;x$.

Lemma 6 $\sup\{x : 1_{\mathit{Ord}} = 1\bullet(x;\infty) \wedge cnc\,;x \subset (lst\otimes hd);1_{\leq}\} = cnc^{\mathsf{T}};(1_{\mathit{Ord}}\otimes 1_{\mathit{Ord}});(lst\otimes hd);1_{\leq}$

Then, we can rewrite (26) as

$$1_{\mathit{Ord}\,2} = 1_{\mathcal{L}^*-\mathcal{L}^1}\bullet\left(\left(cnc^{\mathsf{T}};(1_{\mathit{Ord}}\otimes 1_{\mathit{Ord}});(lst\otimes hd);1_{\leq}\right);\infty\right)$$

which obviously equals

$$1_{\mathit{Ord}\,2} = 1_{\mathit{Ord}}\bullet\left(\left(cnc^{\mathsf{T}};(1_{\mathit{Ord}}\otimes 1_{\mathit{Ord}});(lst\otimes hd);1_{\leq}\right);\infty\right)$$

---

10 Notice that *cnc* is a partial relation not defined on empty lists.

11 The proofs of these lemmas can be found elsewhere [Hae93]

Now, recalling $(r;r^{\mathsf{T}})\bullet 1=(r;\infty)\bullet 1=(1\nabla r);\pi$ we can write the above expression as

$1_{Ord2}=\left(\left(1_{Ord};cnc^{\mathsf{T}};(1_{Ord}\otimes 1_{Ord});cnc\right)\nabla\left(cnc^{\mathsf{T}};(1_{Ord}\otimes 1_{Ord});(lst\otimes hd);1_{\leq}\right)\right);\pi$

But, by Lemma 6, the domain of $cnc^{\mathsf{T}};(1_{Ord}\otimes 1_{Ord});(lst\otimes hd);1_{\leq}$ is the set of ordered lists. Then, from the definition of ∇ we have

$$1_{Ord2}=\left(\left(cnc^{\mathsf{T}};(1_{Ord}\otimes 1_{Ord});cnc\right)\nabla\left(cnc^{\mathsf{T}};(1_{Ord}\otimes 1_{Ord});(lst\otimes hd);1_{\leq}\right)\right);\pi$$

But from the results presented in Subsection 5.2 and Definition 4 we can easily derive $1\bullet(r;s;\infty)=1\bullet(r;t;\infty)\rightarrow r;(s\nabla t)\Subset((r;s)\nabla(r;t))$, and since 1_{Ord} is univalent, $\Subset$ is equivalent to $=$. Then, we can write

$$1_{Ord2}=cnc^{\mathsf{T}};\left((1_{Ord}\otimes 1_{Ord});cnc\,\nabla\,(lst\otimes hd);1_{\leq}\right);\pi \qquad (27)$$

It is important to notice that what we have carried out was the construction of a recursive refinement (27) for the universal quantified relational expression $1_{L^*-L^1}\bullet\left(\left(cnc^{\mathsf{T}}\rightarrow((lst\otimes hd);1_{\leq})^{\mathsf{T}}\right);\infty\right)$. In doing so, we could resort earlier to a Galois connection, i.e.,

$$x\subset((lst\otimes hd);1_{\leq}\lfloor cnc)\leftrightarrow cnc;x\subset(lst\otimes hd);1_{\leq}$$

However. the intuition needed for stating equation $cnc;x\subset(lst\otimes hd);1_{\leq}$ is not trivial. Notice also that the solution on x of this equation must be a relation with the same domain as relation $\mathcal{O}rd$. Then, to guaranteeing the inclusion $x\subset cnc^{\mathsf{T}};(1_{Ord}\otimes 1_{Ord});(lst\otimes hd);1_{\leq}$ the domain of x must be the domain of ordered lists. What we has done is to replace "clever intuition" by mere calculation.

Now, by folding, unfolding and the intensive use of the results presented along the previous sections, we can derive (see [Hae93]) from the obvious specification $sort=perm;1_{Ord}$ the expression

$$sort=1_{L^1}+1_{L^*-L^1};perm;cnc^{\mathsf{T}};\left(1\ \nabla\left((max\otimes min);1_{\leq}\right)\right);\pi;(sort\otimes sort);cnc$$

Where $max=sort;lst$ and $min=sort;hd$. Notice that if we do $partition=perm;cnc^{\mathsf{T}};\left(1\ \nabla\left((max\otimes min);1_{\leq}\right)\right);\pi$ and fold we have

$$sort=1_{L^1}+1_{L^*-L^1};partition;(sort\otimes sort);cnc$$

which is obviously a specification of Quicksort. A sort algorithm can be derived by calculating out of the specification of *partition*..

6 Conclusions

We have presented an alternative basis for relational programming calculi by extending ARAs with a fork operator. This operator, as well as the direct product and the projections, has being used by other researchers. However, they fail to formulate that the new algebra has the expressive power of first-order logic We introduce an axiomatization for the ∇-Extended Abstract Relational Algebra, compare it to "classical" algebras for first-order logic and, later, proved that alternative axiomatizations for fork and projections can be derived from our one.

At this point we show, by formally calculating algorithms out of first-order specifications of two "classical" examples, the advantage of having quantifiers expressed by terms. We also show two ways of translating first-order expressions onto relational terms. Two main points of this second part of the paper are the examples of how to cope with the algorithmic complexity of the relational term expressing existential quantifier and the proposed heuristic for deriving the equation on residuals expressing universal quantifier.

A point to be emphasised is that we will have another indication of the fitness of this algebra to underlay relational programming calculi, if it turns out by subsequent research that the ∇-Extended Relational Algebra is representable without the inclusion of an extra axiom.

Future research will be carried out along two main directions. The first one is the exhaustive analysis of the ∇-Extended Relational Algebra as an algebraic structure. The second one is the full development of a programming calculus. This involves answering the question of whether or not we should deal not trivially with partiality (as discussed in Section 4 of [Hae91]). After a conclusive answer to this question, we will develop a typing system for our evolving relational programming calculus.

References

[Bac92] Backhouse, R. and Tormenta, P., "Polynomial Relators" in *State-of-the-Art Seminar on Formal Program Development,* IFIP - WG 2.1 Algorithmic Languages and Calculi, 1992.

[Bau92] Baum, G., Haeberer, A.M., and Veloso, P.A.S.., "On the Representability of the ∇-Abstract Relational Algebra" *IGPL Newsletter*, vol. 1, no. 3, European Foundation for Logic, Language and Information, Interest Group on Programming Logic.

[Bed78] Bednarek, A. R. and Ulam, S. M., "Projective Algebra and the Calculus of Relations" *Journal of Symbolic Logic*, vol. 43, no. 1, 56 - 64, March 1978.

[Ber93] Berghammer, R., Haeberer, A.M., Schmidt, G., and Veloso, P.A.S., "Comparing two Different Approaches to Products in Abstract Relation Algebras", Proceedings of the Third International Conference on Algebraic Methodology and Software Technology, AMAST'93, 1993.

[Chi48] Chin, L. and Tarski, A., "Remarks on Projective Algebras" *Bulletin of the American Mathematical Society,* vol. 54, 80 - 81, 1948.

[Chi51] Chin, L. H. C. and Tarski, A., "Distributive and Modular Laws in the Arithmetic of Relation Algebras" in *University of California Publications in Mathematics*, of California, U., Ed. University of California, 1951, pp. 341 - 384.

[Eve46] Everett, C. J. and Ulam, S. M., "Projective Algebra," *American Journal of Mathematics*, vol. 68, 77 - 88, 1946.

[Hae90] Haeberer, A.M., Veloso, P.A.S., and Elustondo, P., "Towards a Relational Calculus for Software Construction" in *41st Meeting of the,* vol. Document IFIP Working Group 2.1 Algorithmic Languages and Calculi, Chester - England, 1990.

[Hae91] Haeberer, A.M. and Veloso, P.A.S., "Partial Relations for Program Derivation: Adequacy, Inevitability and Expressiveness" in *Constructing Programs From Specifications - Proceedings of the IFIP TC2 Working*

Conference on Constructing Programs From Specifications, N. H., IFIP WG 2.1, Bernhard Möller, 1991, pp. 319 - 371.

[Hae93] Haeberer, A.M., Baum, G., and Schmidt, G. "Dealing with Non-Constuctive Specifications Involving Quantifiers" Res. Rept. MCC 4, 1993.

[Hal62] Halmos, P. R., *Algebraic Logic*. New York: Chelsea Publishing Company, 1962.

[Hen74] Henkin, L. and Monk, J. D., "Cylindric Algebras and Related Structures" in *Proceedings of the Tarski Symposium,* vol. 25 American Mathematical Society, 1974, pp. 105 - 121.

[Hoa86] Hoare, C. A. R. and He, J., "The Weakest Presepecification, Part I" *Fundamenta Informatica*, vol. 4, no. 9, 51 - 54, "Part II" vol. 4, no. 9, 217 - 252, 1986.

[Jón51] Jónsson, B. and Tarski, A., "Boolean Algebras With Operators PART I" *American Journal of Mathematics*, vol. 73, 891 - 939, 1951. "PART II", vol. 74, 127 - 162, 1952.

[Lyn50] Lyndon, R., "The Representation of Relational Algebras" *Annals of Mathematics (series 2)*, vol. 51, 707 - 729, 1950.

[Mad83] Maddux, R., "A Sequent Calculus for Relation Algebras" *Annals of Pure and Apply Logic*, vol. 25, 73 - 101, 1983.

[Mad91] Maddux, R., "The Origin of Relation Algebras in the Development and Axiomatization of the Calculus of Relations" *Studia Logica*, vol. L, no. 3-4, 412 - 455, 1991.

[Ném91] Németi, I., "Algebraization of Quantifier Logics, an Introductory Overview" *Studia Logica*, vol. L, no. 3-4, 485 - 569, 1991.

[Roe72] Roever, W. P. de, "A Formalization of Various Prameter Mechanisms as Products of Relations Within a Calculus of Recursive Program Schemes" in *Theorie des Algorithmes, des Languages et de la Programation,* 1972, pp. 55 - 88.

[Sch85] Schmidt, G. and Ströhlein, T., "Relation Algebras: Concept of Points and Representability" *Discrete Mathematics*, vol. 54, 83 - 92, 1985.

[Sch89] Schmidt, G. and Ströhlein, T., *Relationen und Graphen*. Mathematik für Informatiker, Springer-Verlag, 1989. Republished in English in EATCS Monographs on Theoretical Computer Science, Springer-Verlag, 1993.

[Tar41] Tarski, A., "On the Calculus of Relations" *Journal of Symbolic Logic*, vol. 6, 73 - 89, 1941.

[Tar52] Tarski, A. and Thompson, F. B., "Some General Properties of Cylindric Algebras" *Bulletin of the American Mathematical Society*, vol. 58, 65, 1952.

[Vel91] Veloso, P.A.S. and Haeberer, A.M., "A Finitary Relational Algebra For Classical First-Order Logic" *Bulletin of the Section of Logic of the Polish Academy of Sciences*, vol. 20, no. 2, 52 - 62, June 1991.

[Vel92] Veloso, P.A.S., Haeberer, A.M., and Baum, G., "Formal Program Construction Within an Extended Calculus of Binary Relations" Res. Rept. MCC 19, 1992.

Saturation Replaces Induction for a Miniscoped Linear Temporal Logic

Regimantas Pliuškevičius

Institute of Mathematics and Informatics,
Akademijos 4, Vilnius 2600, LITHUANIA

Abstract. A new type of deductive principle (named the saturation one) is introduced for a complete class (called miniscoped) of the first order linear temporal logic with $\bigcirc$("next") and $\Box$("always"). The saturation replaces induction-like postulates and intuitively corresponds to a certain type of regularity in the derivations for the logic. Non-logical axioms in "saturated calculi" are some sequents, indicating the saturation of the derivation process in these calculi. The saturation suggests that "nothing new" can be obtained continuing the derivation process. The non-logical axioms of the saturated calculus are constructed dependent on specific peculiarities of the given sequent. Therefore, for each given sequent (whose derivability requires the application of the induction-like postulate) a concrete saturated calculus is constructed. This property makes saturated calculi more effective than the traditional ones (based on the fixed induction-like postulate).

1 Introduction

It is known that in some cases the first order linear temporal logic is finitary complete (see e.g. [2]). The main rule of inference in this case is the following one (see [3]):

$$\frac{\Gamma \to \Delta, R;\ R \to \bigcirc R;\ R \to A}{\Gamma \to \Delta, \Box A} (\to \Box).$$

This rule of inference is called the induction one because it corresponds to the induction-like axiom $A \wedge \Box(A \supset \bigcirc A) \supset \Box A$. The formula R ia called an invariant formula. How can one find the invariant formula? Maybe there is another way of construction of derivations without using $(\to \Box)$? The answer to both questions is: let us construct a saturated calculus. Non-logical axioms in a saturated calculus are some sequents, indicating the saturation of the derivation process in the calculus. The saturation intuitively corresponds to a certain type of regularity in the derivations for the logic. The saturation suggests that "nothing new" can be obtained continuing the derivation process. The derivability in a finitary saturated calculus serves as a finitary completeness criterion for the first order linear temporal logic [4]. The "finitary saturation" replaces the invariant rule and is more "computer-aided" than the latter. The non-logical axioms of the saturated calculus are constructed dependent on specific peculiarities of the given sequent. Therefore, for each given sequent (whose derivability requires the application of the induction-like postulate) a concrete saturated calculus is constructed. This property makes saturated calculi more effective than the traditional ones (based on the fixed induction-like postulate). In [5, 6] the saturated calculus mostly were described and founded for the so-called class of Horn miniscoped formulas (in short: MH-formulas) with "next", "always" and "next", "unless", respectively. MH-formulas have the form $((A \supset)^{\circ}\Box^{\circ}B)$, where A does not contain a positive occurrence of $\Box, (A \supset)^{\circ} \in \{\varnothing, A \supset\}, \Box^{\circ} \in \{\varnothing, \Box\}, B$ does not contain a positive occurrence of $\Box$. Besides, in their temporal parts quantifiers enter only the formulas of the form $Q\overline{x}E(Q \in \{\forall, \exists\})$, where E is an elementary formula (i.e. $E = P(t_1, \ldots, t_n)$, P is an n-place ($n \geqslant 0$) predicate symbol). The purpose of this paper is to present the saturation for non-Horn miniscoped formulas (in short: M-formulas) with "next" and "always". The saturation for this cases is based on the so-called "saturated" disjunction property and on some new technical details (see section 4).

2 Description of the Infinitary Sequential Calculus $G_{L\omega}$

The foundation of a saturated calculus is carried out with the help of the infinitary calculus $G_{L\omega}$ containing the ω-type rule of inference. In $G_{L\omega}$ we consider arbitrary formulas (i.e. not only (M-formulas) which are determined by means of logical symbols and a temporal operator $\Box$ as usual. Throughout the paper $\omega := \{0, 1, \ldots, n, \ldots\}$, the formula $\circ \ldots \circ A$ (k-times next A, $k \geqslant 1$) will be abbreviated as A^k (i.e. as a formula with the index k) more precisely: 1) if E is an elementary formula, $i, k \in \omega$, $k \neq 0$, then $(E^i)^k := E^{i+k}$ $(E^0 := E)$; E^l $(l \geqslant 0)$ will be called an atomic formula (if $l = 0$ then E^l becomes an elementary one); 2) $(A \odot B)^k := A^k \odot B^k$ if $\odot \in \{\supset, \vee, \wedge\}$; 3) $(\sigma A)^k := \sigma A^k$ if $\sigma \in \{\urcorner, \Box, \forall x, \exists x\}$.

A sequent is an expression of the form $\Gamma \to \Delta$, where Γ, Δ are arbitrary finite sets (not sequences or multisets) of formulas.

The calculus $G_{L\omega}$ is defined by the following postulates.

Axiom: $A \to A$.

Rules of inference:

1) temporal rules:

$$\frac{A, \Box A^1, \Gamma \to \Delta}{\Box A, \Gamma \to \Delta}(\Box \to) \qquad \frac{\{\Gamma \to \Delta, A^k\}_{k \in \omega}}{\Gamma \to \Delta, \Box A}(\to \Box_\omega),$$

where $k \in \omega$; here and below Γ^1 means $A_1^{k_1+1}, \ldots, A_n^{k_n+1}$, if $\Gamma = A_1^{k_1}, \ldots, A_n^{k_n}$, $n \geqslant 1$, $k_i \geqslant 0$, $1 \leqslant i \leqslant n$;

2) logical rules of inference consist of traditional invertible rules of inference;

3) structural rules: (W) (weakening); from the definition of a sequent it follows that $G_{L\omega}$ implicitly contains the structural rules "contraction" and "exchange".

Derivations in $G_{L\omega}$ are built up in a usual way (for the calculi with the ω-rule), i.e. in the form of an infinite tree (with finite branches); the height of a derivation D is an ordinal (defined in a traditional way) denoted by $O(D)$. Let I be some calculus, then $I \vdash S$ means that the sequent S is derivable in S.

A derivation D in $G_{L\omega}$ will be called atomic if all axioms occurring in D have the form $E \to E$, where E is an atomic formula.

Lemma 2.1. An arbitrary derivation in $G_{L\omega}$ may be transformed into an atomic one.

Proof: using the rules of inference of $G_{L\omega}$.

Remark 2.1 All derivations in $G_{L\omega}$ will be regarded as atomic ones.

Lemma 2.2 (invertibility of the rules of inference of $G_{L\omega}$). If S_1 is a premise and S is the conclusion of any rule of inference of $G_{L\omega}$, different from (W)), then $G_{L\omega} \vdash S \Longrightarrow G_{L\omega} \vdash S_1$.

Proof: by induction on $O(D)$, where D is the given atomic derivation.

3. Completeness of $G_{L\omega}$

A model M over which a formula of temporal logic under consideration is interpreted as a pair $< N, V >$, where N is a triple $< D, \omega, \leqslant >$ called frame, V is a valuation function, defined in a traditional way, $\leqslant$ is the usual order relation on ω.

The concept "A is valid in $M =< N, V >$ at time point $k \in \omega$" (in symbols $M, k \vDash A$) is defined as follows:

1) $M, k \vDash P^l(t_1, \ldots, t_m) \iff < V(t_1, \ldots, t_m),\ k + l > \in V(P)$;

2) $M, k \vDash \Box A \iff \forall l(l \in k) M,\ k + l \vDash A$.

Other cases are defined as in first order logic. Using this concept we can define (in a traditional way) the concept of universally valid sequent.

Theorem 3.1 (soundness of $I \in \{G_{L\omega}, G_{L\omega} + \text{cut}\}$). If $I \vdash S$, then S is universally valid.

Proof: by induction on $O(D)$.

To prove completeness of $G_{L\omega}$ let us introduce the "symmetric" calculus $G_{L\omega k}$ obtained from $G_{L\omega}$ replacing $(\Box \to)$ by the following one:

$$\frac{A^k, \Box A, \Gamma \to \Delta}{\Box A, \Gamma \to \Delta}\ (\Box^k \to)\ (k \in \omega)$$

Theorem 3.2 (completeness of $G_{L\omega k}$). If a sequent S is universally valid, then $G_{L\omega k} \vdash S$.

Proof: applying method of Shutte (see [1]).

Lemma 3.1. $G_{L\omega k} \vdash S \Longrightarrow G_{L\omega} \vdash S$.

Proof. At first, let us prove (by induction on k) the admissibility in $G_{L\omega}$ the following rule of inference:

$$\frac{\Box A^k, \Box A, \Gamma \to \Delta}{\Box A, \Gamma \to \Delta}\ (\Box^{k*} \to)\ (k \in \omega).$$

The case when $k = 0$ is obvious. Let $G_{L\omega} \vdash S = \Box A^{k+1}, \Box A, \Gamma \to \Delta$, then applying $(\mathcal{W})$ to S we get $G_{L\omega} \vdash S' = A^k, \Box A^{k+1}, \Box A, \Gamma \to \Delta$. Applying $(\Box \to)$ to S' and using induction hypothesis we get $\Box A, \Gamma \to \Delta$. Now let us prove admissibility of $(\Box^k \to)$ in $G_{L\omega}$. Let us prove admissibility of $(\Box^k \to)$ in $G_{L\omega}$. Let $G_{L\omega} \vdash S = A^k, \Box A, \Gamma \to \Delta$, applying $(\mathcal{W})$ to S we get $G_{L\omega} \vdash S' = A^k, \Box A^{k+1}, \Box A, \Gamma \to \Delta$. Applying $(\Box \to)$, $(\Box^{k*} \to)$ to S' we get $G_{L\omega} \vdash \Box A, \Gamma \to \Delta$. Therefore $G_{L\omega k} \vdash S \Longrightarrow G_{L\omega} \vdash S$.

Theorem 3.3 (completeness of $G_{L\omega}$). If a sequent S is universally valid, then $G_{L\omega} \vdash S$.

Proof: follows from Theorem 3.2 and Lemma 3.1.

Theorem 3.4 (admissibility of cut in $G_{L\omega}$). $G_{L\omega} + \text{cut} \vdash S \Longrightarrow G_{L\omega} \vdash S$.

Proof: follows from Theorems 3.1, 3.3.

4 Description and Investigation of the Sequential Saturated Calculus Sat

Let $S = A_1, \ldots, A_n \to B_1, \ldots, B_m$, then $S^F = \bigwedge_{i=1}^{n} A_i \supset \bigvee_{i=1}^{m} B_i$. The sequent S will be called M-sequent, if S^F is the M-formula. First let us define the canonical form of M-sequents (simply: sequents). A sequent S will be called primary, if $S = \Sigma_1, \Pi_1^1, \Box\Omega^1 \to \Sigma_2, \Pi_2^1, \Box\Delta^1$, where $\Sigma_i = \varnothing$ $(i = 1, 2)$ or consists of logical formulas without indices; $\Pi_i^1 = \varnothing$ $(i = 1, 2)$ or consists of logical formulas with indices; $\Box\Omega^1, \Box\Delta^1 = \varnothing$ or consists of arbitrary M-formulas of the form $\Box A^1$. If $S = \Sigma_1, \Pi_1^1, \Box\Omega \to \Sigma_2, \Pi_2^1, \Box\Delta$ (i.e., it is not necessary that $\Box\Omega = \Box\Omega_1^1, \Box\Delta = \Box\Delta_1^1$) then such a sequent S will be called quasiprimary. It is clear that every primary sequent is the quasiprimary one. The sequent S will be called ordinary, if S contains both negative and positive occurrences of $\Box$. The sequent S will be called singular (simple) if S does not contain negative (positive, respectively) occurrences of $\Box$. An ordinary primary (quasiprimary) sequent will be called proper, if $S \neq \Box\Omega^1 \to \Box\Delta^1$ $(S \neq \Box\Omega \to \Box\Delta$, respectively).

At first, let us introduce the notion of the blanked saturated calculus, denoted by bSat. The rules of inference of the calculus bSat consist of the rules of inference of $G_{L\omega}$, different from $(\to \Box_\omega)$, and the following two rules of inference:

$$\frac{\Pi_1 \to \Pi_2, \Box A^1}{\Pi_1 \to \Pi_2, \Box A} (\to \Box_b^1) \qquad \frac{S_i^*}{S^*} (A) \ (i \in \{1, 2\}),$$

S^* is a primary sequent, i.e. $S^* = \Sigma_1, \Pi_1^1, \Box\Omega^1 \to \Sigma_2, \Pi_2^1, \Box\Delta^1$; $S_1^* = \Sigma_1 \to \Sigma_2$; $S_2^* = \Pi_1, \Box\Omega \to \Pi_2, \Box\Delta$.

Lemma 4.1 (invertibility of the rules of inference of bSat). All rules of inference of bSat, different from (W) and (A), are invertible in $G_{L\omega}$.

Proof. The invertibility of rules of inference, except $(\to \Box^1), (A), (W)$. follows from invertibility of the rules of inference of $G_{L\omega}$. The invertibility of $(\to \Box^1)$ follows from the fact that $G_{L\omega} \vdash \Box A \to A^1$ and the admissibility of cut in $G_{L\omega}$.

Lemma 4.2 (disjunctional invertibility of (A)). Let S^* be the conclusion of (A), i.e. $S^* = \Sigma_1, \Pi_1^1, \Box\Omega^1 \to \Sigma_2, \Pi_2^1, \Box\Delta^1$, then $G_{L\omega} \vdash S^* \Longrightarrow \text{Log} \vdash \Sigma_1 \to \Sigma_2$

or $G_{L\omega} \vdash \Pi_1, \Box\Omega \rightarrow \Pi_2, \Box\Delta$, where Log is the calculus obtained from $G_{L\omega}$ by dropping $(\rightarrow \Box_\omega), (\Box \rightarrow)$.

Proof: by induction on the height of a given atomic derivation of the sequent of the form $\Sigma_1, \Gamma^1 \rightarrow \Sigma_2, \Delta^1$.

A derivation in *b*Sat is constructed bottom-up applying the rules of inference of *b*Sat. As follows from Lemmas 4.1, 4.2 all sequents from the derivation in *b*Sat are derivable in $G_{L\omega}$. Some sequents which indicate the saturation of a derivation process in *b*Sat play the role of non-logical axioms in *b*Sat. To define the class of these sequents let us define the tactic of constructing a derivation in *b*Sat. First let us define the notion of reduction of a sequent. Let $\{i\}$ denote the set of rules of inference of calculus I, let $S, S_1, \dots, S_n$ denote sequents. The $\{i\}$-reduction (or briefly: reduction) of S to $S_1, \dots, S_n$, denoted by $R(S)\{i\} \Longrightarrow \{S_1, \dots S_n\}$ or briefly by $R(S)$, is defined to be a tree of sequents with the root S and leaves $S_1, \dots, S_n$, and possibly some logical axioms such that every sequent in $R(S)$ different from S is an "upper sequent" of the rule of inference in $\{i\}$ whose "lower sequent" also belongs to $R(S)$.

Lemma 4.3. For each sequent S one can construct $R(S)\{i\} \Longrightarrow \{S_1, \dots, S_n\}$,where $\forall i$ $(1 \leqslant i \leqslant)$ S_i is a primary (quasiprimary) sequent; $\{i\}$ is the set of rules of inference of *b*Sat, different from (A) and the rules of inference for quantifiers (and $(\Box \rightarrow), (\rightarrow \Box_b^1)$, respectively); besides $G_{L\omega} \vdash S \Longrightarrow G_{L\omega} \vdash S_i$ $(i = 1, \dots, n)$.

Proof: follows from Lemma 4.1.

The set of the primary (quasiprimary) sequents from Lemma 4.3 will be denoted by $P(S)$ ($QP(S)$, respectively). Let I_1 be the calculus obtained from $G_{L\omega}$ by dropping $(\rightarrow \Box_\omega)$.

Lemma 4.4. Let $G_{L\omega} \vdash S$, then $I_1 \vdash S$, if S is simple.

Proof: applying Lemma 4.1.

Let S be a quasiprimary sequent, i.e. $S = \Sigma_1, \Pi_1^1, \ \Box\Omega \rightarrow \Sigma_2, \Pi_2^1, \Box A$ let us define the notion of resolvent of the sequent S, denoted by $Re(S)$. Let us construct $P(S)$ and let $S_j^* \in P(S)$ $(j = 1, \dots, n)$; let us apply (A) bottom-up to S_j^* and let S_{j2}^* be the "temporal" conclusion of this bottom-up application of (A) (i.e. $S_{j2}^* = \varnothing$, if $i = 1$ in (A)), then $Re(S) = \{S_{12}^*, \dots, S_{n2}^*\} = \{S_1, \dots, S_k\}$ $(k \leqslant n)$. We say that a sequent S absorbs the sequent S' (in symbols: $S' \prec S$ or $S \succ S'$), if S' can be obtained from S with the help of the structural rule of inference (W). Let us introduce the notion of the k-th resolvent of S, denoted by $Re^k(S)$ $(k \in \omega)$. $Re^k(S)$ is constructed as a tree of sequents, defined as follows $Re^0(S) = S$; $Re^1(S) = Re(S)$. Let $S_k \in Re^k(S)$, $S_l \in Re^l(S)$ $(k < l)$ and S_k, S_l belong to the same branch; we say that S_k is blanked saturated (in short: b-saturated), if $S_k \succcurlyeq S_l$ ($S' \preccurlyeq S''$ means $S' = S''$ or $S' \prec S''$); the sequent S_l is called the absorbed one. Let $S_k \in Re^k(S)$, we say that the sequent S_k is blocked if S_k is absorbed and $\forall j$ $(f - 1 \leqslant j < k - 1)$ $S_j \in Re^j(S)$ is either b-saturated, or absorbed, and S_f is b-saturated; f is some natural number and $S_f, \ S_{f+1}, \dots, S_k$ belong to the same branch; otherwise $S_k \in Re^k(S)$ will be called non-blocked. Let $S_i \in Re^k(S)$ and S_i be non-blocked, then $Re^{k+1}(S) = \bigcup_i Re(S_i)$. A branch B in the derivation D will be called b-saturated if the leaf of B is blocked. A branch B in the derivation D in bSat will be called closed if the leaf of B is either a logical axiom or B is b-saturated. The derivation D in bSat of the sequent S will be called closed (in symbols: $b\text{Sat} \vdash^D S$) if all the branches of D are closed. Let $S_i \in Re^k(S)$, then the sequent S_i is called the resolvent one. We say that the derivation D in bSat is constructed using the resolvent tactic, if D is constructed by generating $Re^k(S)$ ($k \in \omega$). Let $b\text{Sat} \vdash^D S$, then the set of saturated sequents from D will be denoted by $b\text{Sat}^* \{S\}$. Let $b\text{Sat}\{S\}$ be the set obtained from the set $b\text{Sat}^*\{S\}$ by dropping those elements which are absorbed by some elements of $b\text{Sat}^*\{S\}$. The set $b\text{Sat}\{S\}$ will be called the minimal saturted set of S.

Example 4.1. (a). Let $S = P, \Box\Omega \to \Box A$, where $\Omega = (P \supset \neg\Box\neg P^1)$, $A = \Box\neg\Box\neg P$, then it is easy to verify that $b\text{Sat}\{S\} = \Box\Omega \to \Box A, \Box P$ and $b\text{Sat} \vdash S$.

(b) Let $S = \Box\Omega \to A, \Box\neg\Box A$, where $\Omega = \neg(\neg A \wedge \Box(A \supset \Box A))$, then it is easy to verify that $b\text{Sat} = \Box\Omega \to \Box(A \supset \Box A), \Box\neg\Box A$.

Remark. 4.1. The trivial example P, $\Box(P \supset P^1) \to \Box P_1$ shows that the implication $b\text{Sat}\{S\} \Rightarrow G_{L\omega} \vdash S$ is invalid. More complicated situation reflects the following example, let $S = P, \Box\Omega \to \Box Q_1^1$, $\Box Q_2^1$, where $\Box\Omega = \Box(P \supset (Q_1 \wedge Q_2)), \Box(\neg P \supset (Q_1 \vee Q_2)), \Box(P \equiv \neg P^1), \Box(\Box Q_1^1 \vee \Box Q_2^1) \supset \Box Q_1)$ (The construction of this example belongs to N.Bjorner). Then $b\text{Sat} \vdash S$, $b\text{Sat}\{S\} = \{P, \Box\Omega \to \Box Q_1, \Box Q_2;\ \Box\Omega \to P, \Box Q_1, \Box Q_2\}$, but $G_{L\omega} \nvdash S$. Indeed, let us bottom-up apply ($\Box \to$), logical rules and ($\to \Box_\omega$) to S. Let us consider the premise of ($\to \Box_\omega$) of the form $S_0^* = P, \Box\Omega \to Q_1^1, Q_2^3$. Again, let us bottom-up apply ($\Box \to$), logical rules and ($\to \Box_\omega$) to S. Now, let us consider the premise of ($\to \Box_\omega$) of the form $S^* = \Box\Omega^1, P \to P^1, Q_1^1, Q_2^3, Q_1^5$. Let us bottom-up applly (A) to S^* (using Lemma 4.2) getting the sequent $S' = \Box\Omega \to P, Q_1, Q_2^2, Q_1^4$. Repeating we get the sequent $S_1^* = \Box\Omega, P \to Q_2^1, Q_1^3$, again repeating we get the sequent S_0^*. This "degenerate saturation" implies that $G_{L\omega} \nvdash S$. A simple criterion for "degenerate saturation" will be indicated below.

Let us define the notion of a subformula and of a resolvent subformula of an arbitrary formula of the M-sequent. The set of subformulas (resolvent subformulas) of formula A will be denoted by Sub(A) (RSub(A), respectively) To define the set RSub(A) an auxiliary set R^*Sub(A) is introduced. 1. $\text{Sub}(A) = R^*\text{Sub}(A) = A$, if A is an elementary formula. 2. $\text{Sub}(A \odot B) = \{A \odot B\} \cup \text{Sub}(A) \cup \text{Sub}(B)$; $R^*\text{Sub}(A \odot B) = \{A \odot B\} \cup R^*\text{Sub}(A) \cup R^*\text{Sub}(B)$; where $\odot \in \{\supset, \wedge, \vee\}$; 3. $\text{Sub}(\neg A) = \{\neg A\} \cup \text{Sub}(A)$; $R^*\text{Sub}(\neg A) = \{\neg A\} \cup R^*\text{Sub}(A)$. 4. $\text{Sub}(QxA(x)) = \{QxA(x)\} \cup \text{Sub}\{A(t)\}$ ($Q \in \{\forall, \exists\}$, t is some term); $R^*\text{Sub}(QxA(x)) = \{QxA(x)\}$, if $A(x) = Q_1x_1, \dots, Q_nx_nE(x, x_1, \dots, x_n)$ ($Q_i \in \{\forall, \exists\}$, E is elementary), otherwise $R^*\text{Sub}(QxA(x)) = \varnothing$. 5. $\text{Sub}(A^1) = \{A^1\} \cup \{\text{Sub}(A)\}$; $R^*\text{Sub}(A^1) = \{A^1\} \cup R^*\text{Sub}(A)$. 6. $\text{Sub}(\Box A) = \{\Box A^k \mid k \in \omega\} \cup \text{Sub}(A)$; $R^*\text{Sub}(\Box A) = \{\Box A\} \cup R^*\text{Sub}(A)$. Let A be a formula, then A^{-1} will mean the formula B^{k-1}, if $A = B^k$ ($k > 0$) and the empty word, otherwise. Let $R^*\text{Sub}(A) = \{A_1, \dots, A_n\}$, then $R\text{Sub}(A) = \{A_1^{-1}, \dots, A_n^{-1}\}$. Therefore, Sub$(A)$ (where $\Box \in A$) is infinite, whereas RSub(A) is finite, even in the non-propositional case.

Example 4.2. Let $A = \Box(\exists xP(x) \supset \forall xP^1(x))$, then $R\text{Sub}(A) = \{A, \forall xP(x)\}$.

Lemma 4.5. Let S be any ordinary quasiprimary sequent and let $b\mathrm{Sat} \vdash^D S$, Then either $\forall i\ (1 \leqslant i \leqslant n)\ S_i \in b\mathrm{Sat}\{S\} \Longrightarrow S_i = \Pi_i, \Box\Omega \to \Delta_i, \Box\nabla$, where $\Pi_i, \Delta_i \in R\mathrm{Sub}(\Omega)$ or $S_i = \Box\Omega \to \Box\nabla$; besides $G_{L\omega} \vdash S \Longrightarrow G_{L\omega} \vdash S_i$ $(i = 1, \ldots, n)$ (a).

Proof. The form of S_i follows from the construction of D; the point (a) follows from Lemmas 4.1, 4.2.

Now we shall prove a "saturated" disjunctive property for $G_{L\omega}$ with respect to a sequent $S_i = \Pi_i, \Box\Omega \to \Delta_i, \Box\nabla \in b\mathrm{Sat}$ $(i = 1, \ldots, n, \Pi_i, \Delta_i \neq \varnothing)$. First we shall state some auxiliary Lemmas. Following by S.Kleene we shall introduce the notion of product of sequents S_1, S_2. Let $S_1 = \Gamma \to \Delta, S_2 = \Pi \to \Theta$, then $S_1 \cdot S_2 = \Gamma, \Pi \to \Delta, \Theta$.

Lemma 4.6. Let $S_0 = \Pi_i, \Box\Omega^\sigma \to \Delta_i$, where $\Pi_i, \Delta_i \in R\mathrm{Sub}(\Omega^\sigma)$, $\sigma \in \{\varnothing, \omega\}$, $S_k^{i_k} = \Gamma^{i_k} \to \Delta^{i_k}$ $(k = 1, \ldots, n)$, where $i_k \in \omega$, $i_1 \neq \ldots \neq i_n \neq j$, j is any index from S_0, then $G_{L\omega} \vdash S_0 \cdot S_1^{i_1} \cdot \ldots \cdot S_n^{i_n} \Rightarrow G_{L\omega} \vdash S_0 \cdot S_k^{i_k}$ $(1 \leqslant k \leqslant n)$ or $\Box\Omega,\ \Gamma^{i_k} \to \Delta^{i_k}$.

Proof: by induction on $O(D)$, where D is the given atomic derivation.

Lemma 4.7. $G_{L\omega} \vdash S_0 \cdot S_k^{i_k} \Rightarrow G_{L\omega} \vdash S_0 \cdot S_k^{i}$, where $S_0, S_k^{i_k}$ are the same as in Lemma 4.6; $i \in \omega$.

Proof: by repeating the bottom-up applications of (A) (which are admissible in $G_{L\omega}$ by Lemma 4.2).

Lemma 4.8. $G_{L\omega} \vdash S_0 \cdot S_k^i \Rightarrow G_{L\omega} \vdash S_0 \cdot S_k'$ where $S_k' = \to \Box A_k$.

Proof: follows from Lemma 4.7 by application of $(\to \Box_\omega)$.

Theorem 4.1 (saturated disjunctive property for $G_{L\omega}$). $\forall i (1 \leqslant i \leqslant n)$ $(G_{L\omega} \vdash S_i = \Pi_i, \Box\Omega \to \Delta_i, \Box A_1, \ldots, \Box A_n \Rightarrow G_{L\omega} \vdash \Pi_i, \Box\Omega \to \Delta_i, \Box A_k)$, where $S_i \in b\mathrm{Sat}\{S\}, \Pi_i, \Delta_i \in R\mathrm{Sub}(\Omega)$, $1 \leqslant k \leqslant n$ or $G_{L\Omega} \vdash \Box\Omega \to \Box A_1, \ldots, \Box A_n$.

Proof: follows from Lemmas 4.6, 4.8.

Remark 4.2. As seen from the quasiprimary sequents $\to (\Box A \supset B), \Box(\Box B \supset A)$; $\Box\neg(A \wedge B), \Box\neg(A \wedge \Diamond B), \Box\neg(B \wedge \Diamond A) \to \Box\neg A, \Box\neg B$, $\Box(\Diamond A \vee \Diamond B) \to \Box\Diamond A, \Box\Diamond B$, $P,\ \Box(P \supset (O^1 \wedge \Box(C \vee D) \wedge \Box(C \supset c^1) \wedge \Box(D \supset D^1))) \to \Box C,\ \Box D$ and $P,\ \Box(P \supset (*P^1 \vee \Box(C \vee \neg A) \wedge \Box(C \supset C^1) \wedge \Box(\Box(A \supset \Box A) \supset A))) \to \Box C,\ \Box\neg\Box A$ the ordinary disjunctive property fails for $G_{L\omega}$.

A derivation D in bSat will be called normal, if all b-saturated sequents in D either have the form $S_i \Pi_i, \Box\Omega \to \Delta_i,\ \Box A$ (where $\Pi_i, \Delta \in R\mathrm{Sub}(\Omega), \Pi_i, \Delta_i \neq \varnothing$) or $S_i = \Box\Omega \to \Box\nabla$. The notation $b\mathrm{Sat}(n) \vdash S$ means that given derivation of S in bSat is normal.

Theorem 4.2 (normalization theorem for bSat). Let $G_{L\omega} \vdash S$, then $b\mathrm{Sat} \vdash S \Rightarrow b\mathrm{Sat}(n) \vdash S$.

Proof: follows from Lemma 4.5 and Theorem 4.1.

Remark 4.3. Further all derivations in bSat will be regarded as normal ones.

Lemma 4.9. Let S be a proper quasiprimary sequent, i.e. $S = \Sigma_1, \Pi_1^1, \Box\Omega \to \Sigma_2, \Pi_2^1, \Box A_1, \ldots, \Box A_n$, let $G_{L\omega} \vdash S$ and $b\mathrm{Sat} \vdash S$, then $b\mathrm{Sat}(n) \vdash S' = \Sigma_1, \Pi_1^1, \Box\Omega \to \Sigma_2, \Pi_2^1, \Box A_k$ $(1 \leqslant k \leqslant n)$ or $G_{L\Omega} \vdash \Box\Omega \to \Box A_n$.

Proof: follows from Theorem 4.2.

Now let us describe the saturated calculus Sat. The rules of inference of Sat are obtained from the rules of inference of bSat byreplacing the rule of inference $(\to \Box_b^1)$ by the following ones:

$$\frac{\Pi_1 \to \Pi_2, A;\ \Pi_1 \to \Pi_2, \Box A^1}{\Pi_1 \to \Pi_2, \Box A}(\to \Box^1) \qquad \frac{S_1; \ldots; S_n}{\Box\Gamma \to \Box A_1, \ldots, \Box A_n}(\Box_n),$$

where $S_1 = \Box\Gamma \to A_1, \Box A_2, \ldots, \Box A_n; \ldots; S_n = \Box\Gamma \to \Box A_1, \ldots, \Box A_{n-1}, A_n$.

Remark 4.4. The rule of inference $(\Box_n)$ was introduced independently in [3, 7].

Let us describe the construction of the derivation in Sat. of a sequent S. Let $b\text{Sat} \vdash^D S$ and $b\text{Sat}\{S\} = \Box\Gamma \to \Box\nabla$, then let us bottom-up apply $(\Box_n)$ to S and take $S_i = S$ $(i = 1 \ldots n)$, where S_i is a premise of $\Box_n$). Let $b\text{Sat}\,\{S\}$ consists of a sequent of a form $S_i = \Gamma_i, \Box\Omega \to \Delta_i, \Box A_1, \ldots, \Box A_p$, where $\Gamma_i, \Delta_i \in R\text{Sub}\,(\Omega)$and $\Box A_j (1 \leqslant j \leqslant p)$ is not a descendant of formulas from Ω in creating $b\text{Sat}\{S\}$; formulas $\Box A_1, \ldots, \Box A_p$ will be called induction formulas. Let $G_{L\omega} \vdash S$ then (as follows from Lemma 4.9) instead af S we can consider the sequent $S' = \Sigma_1, \Pi^1, \Box\Omega \to \Sigma_2, \Pi_2^1, \Box A_j$ $(1 \leqslant j \leqslant p)$. Let us denote $b\text{Sat}\,\{S\}$ by $b_1^0\text{Sat}\,\{S\}$ and let us consider the lowest application of $(\to \Box_b^1)$ in D. Instead of this application let us consider the application of $(\to \Box^1)$. If the left premise of this application of $(\to \Box^1)$ is simple sequent, then we get the correct application of $(\to \Box^1)$. In opposite case let us consider the normal derivation of the sequent $S^* = QP(\Gamma_i, \Box\Omega \to \Delta_i, A_j)$ in bSat and let us denote $b\text{Sat}\,\{S^*\}$ by $b_2^0\text{Sat}\,\{S\}$. Now let us consider the remaider applications of $(\to \Box_b^1)$ in D and instead of these applications let us apply $(\to \Box^1)$. The saturation process in the obtained tree is definmed in the same way as in the case of bSat, (now the rule of inference $(\Box_n)$ may by applied in the left premise of $(\to \Box^1)$), besides in obtained tree in creating the set of saturated sequents the saturated sequents from $b_1^0\text{Sat}\,\{S\}$ is used. Let us denoted saturated sequents by $b_1\text{Sat}\,\{S\}$. Analogously we get $b_2\text{Sat}\,\{S\}, \ldots b_n\text{Sat}\,\{S\}$. Therefore $\text{Sat}\,\{S\} = \bigcup_{k=1}^{n} b_k\,\{S\}$, where $S = \Sigma_1, \Pi_1^1, \Box\Omega \to \Sigma_2 \Pi_2^1, \Box A_j$ and n is the number of positive occurrences of $\Box$ in $\Box A_j$ $(1 \leqslant j \leqslant p)$. *Caution.* Let $S_i \in \text{Sat}\{S\}$ and $S_i = \Gamma_i,\ \Box\Omega \to \Delta_i$, where Γ_i, Δ_i consists only of descendants of formulas from Ω in creating of $\text{Sat}\{S\}$. In this case we collide with the degenerate saturation and in this case $\text{Sat} \nvdash S$. Therefore we the simple criterion for degenerate saturation: all subformulas of a induction formula disappear in creating of $\text{Sat}\{S\}$.

Example 4.2. (a) Let S be the same as in Example 4.1 (a). Then easy to verify that $\text{Sat}\,\{S\} = b_1\text{Sat}\,\{S\} = b\text{Sat}\,\{S\}$ and $\text{Sat} \vdash S$.

(b) Let S be the same as in Example 4.1 (b) Then one can to verify that $\text{Sat}\,\{S\} = b_1\text{Sat}\,\{S\} = \{S_1, S_2\}$, where $S_1 = \Box\Omega \to \Box(A \supset \Box A), \Box\urcorner\Box A$; $S_2 = \Box\Omega \to \Box A, \Box\urcorner\Box A$;

(c) Let $S = P,\ \Box(P \supset \neg\Box\neg P^1) \to \Box\neg P^1$. Then it easy to verify that in searching of $\text{Sat}\,\{S\}$ we collide with the degenerate saturation, therefore $\text{Sat} \nvdash S$. The same picture we notice in the case of the sequent S from Remark 4.1.

Lemma 4.10. Let $\text{Sat} \vdash^D S$ and $S' \in b_k\text{Sat}\,\{S\}$, let $Re(S') = \{S_1, \ldots, S_n\}$, then $\forall i\ (1 \leqslant i \leqslant n)\ S_i \preccurlyeq S^* \in b_k\text{Sat}\,\{S\}$.

Proof: follows from the definition of $b_k\text{Sat}\,\{S\}$.

Example 4.3. Let S, S_1, S_2 be the same as in Example 4.2 (b) then $Re(S_2) = \{S_1, S_2, S_3\}$, where $S_3 = \Box\Omega \rightarrow \Box A, \Box(A \supset \Box A), \Box\neg\Box A$ and $S_2 \succ S_3$; $Re(S_1) = \{S_1, S_2, S_3\}$.

5 Description of an Invariant Calculus IN and the Equivalence of $G_{L\omega}$, Sat and IN for M-sequents

To clarify the role of saturated sequents let us introduce an "invariant" calculus IN. The postulates of the calculus IN are obtained from the calculus $G_{L\omega}$ replacing $(\rightarrow \Box_\omega)$ by $(\rightarrow \Box^1), (\Box_1)$ and the following one:

$$\frac{\Gamma, \Box\Omega \rightarrow \Delta, R_k; \quad R_k \rightarrow R_k^1; \quad R_k \rightarrow A}{\Gamma, \Box\Omega \rightarrow \Delta, \Box A} (\rightarrow \Box),$$

where (1) $\Gamma, \Box\Omega \rightarrow \Delta, \Box A \in b_k\text{Sat}\{S\} = \{\Pi_1, \Box\Omega \rightarrow \Delta_1, \Box A, \ldots, \Pi_n, \Box\Omega \rightarrow \Delta_n, \Box A\}$ and S is a proper ordinary quasiprimary sequent such that $b\text{Sat} \vdash S$; (2) $R_k = \bigvee_{i=1}^{n} (\Pi_i^\wedge \wedge \neg\Delta_i^\vee) \wedge \Box\Omega$ (where $\Gamma^\wedge (\Gamma^\vee)$ means the conjunction (disjunction) of formulas from Γ); the formula R_k is called an invariant formula.

Remark 5.1. It is clear that the rule of inference $(\rightarrow \Box)$ destroys the subformula property (it becomes an analytic cut-like rule of inference after adding the conditions (1), (2), see above), which is restored by means of saturation principle. The rule of inference $(\rightarrow \Box)$ corresponds to the induction-like axiom $A \wedge \Box(A \supset A^1) \supset \Box A$.

Lemma 5.1. In the calculi $G_{L\omega}$, IN the following rule of inference: $\frac{\Gamma \rightarrow \Delta}{\Gamma^1 \rightarrow \Delta^1} (+1)$ is admissible.

Proof: by induction on the height of the derivation of the sequent $\Gamma \rightarrow \Delta$.

Lemma 5.2. In the calculus $G_{L\omega}$ the rule of inference $(\rightarrow \Box)$ is admissible.

Proof. Using the premises of $(\rightarrow \Box)$, admisibility of cut and $(+1)$ in $G_{L\omega}$ and by induction on k we get $G_{L\omega} \vdash \Gamma \rightarrow \Delta, A^k$ $(k \in \omega)$. Hence, by $(\rightarrow \Box_\omega)$ we get $G_{L\omega} \vdash \Gamma \rightarrow \Delta, \Box A$.

Lemma 5.3. The rule of inference $(\rightarrow \Box^1)$ is admissible in $G_{L\omega}$.

Proof. Using the fact that $G_{L\omega} \vdash A \wedge \Box A^1 \rightarrow \Box A$ and the admissibility of cut in $G_{L\omega}$.

Lemma 5.4. The rule of inference $(\Box_n)$ is admissible in $G_{L\omega}$.

Proof. First, by induction on n (and using Lemma 5.3) one can prove the admissibility in $G_{L\omega}$ the following rule of inference:

$$\frac{S_1; \; \ldots; \; S_n; \; S_{n+1}}{\Gamma \rightarrow \Delta, \Box A_1, \ldots, \Box A_n} (\Box_n^*),$$

where $S_1 = \Gamma \rightarrow \Delta, A_1, \Box A_2, \ldots, \Box A_n; \ldots; S_n = \Gamma \rightarrow \Delta, \Box A_1, \ldots, \Box A_{n-1}, A_n$; $S_{n+1} = \Gamma \rightarrow \Delta, \Box A_1^1, \ldots, \Box A_n^1$. Now we prove (using induction on k) the admissibility in $G_{L\omega}$ the following rule of inference:

$$\frac{S_1; \; \ldots; \; S_n}{\Box\Gamma \rightarrow \Box A_1, \ldots, \Box A_n} (\Box_n^k),$$

where $S_1, \dots S_n$ are the same as in $(\Box_n)$. If $k = 0$ then S_1 coincides with the conclusion of $(\Box_n^k)$. Let $k > 0$, then by induction hypothesis we have $G_{L\omega} \vdash \Box\Gamma \to A_1^{k-1}, \Box A_2, \dots, \Box A_n$. Using Lemma 5.1, (W) and $(\Box \to)$ we get $G_{L\omega} \vdash S_n' = \Box\Gamma \to A_1^k, \Box A_2^1, \dots, \Box A_n^1$. Using the fact that $G_{L\omega} \vdash \Box A_1 \to A_1^k$ from the premises $S_2, \dots, S_n$ of $(\Box_n^k)$ by cut we get $G_{L\omega} \vdash S_1' = \Box\Gamma \to A_1^k, A_2, \Box A_3, \dots, \Box A_n; \dots;$ $G_{L\omega} \vdash S_{n-1}' = \Box\Gamma \to A_1^k, \Box A_2, \dots, \Box A_{n-1}, A_n$. Applying $(\Box_n^*)$ to $S_1', \dots, S_{n-1}'$, S_n' we get $G_{L\omega} \vdash S_k^* = \Box\Gamma \to A_1^k, \Box A_2, \dots, \Box A_n$ $(k \in \omega)$, i.e., the conclusion of $(\Box_n^k)$. Applying $(\to \Box_\omega)$ to S_k^* we obtain the conclusion of $(\Box_n)$.

Theorem 5.1. $IN \vdash S \Longrightarrow G_{L\omega} \vdash S$.

Proof: follows from Lemmas 5.2, 5.3, 5.4.

Lemma 5.5. Let Sat $\vdash S$ and $S_i = \Pi_i, \Box\Omega \to \Delta_i, \Box A \in b_k\mathrm{Sat}\{S\}$ $(i = 1, \dots, n; k$ is some natural number) then $\forall i$ $(1 \leqslant i \leqslant n)$ $I \vdash S_i = \Pi_i, \Box\Omega \to \Delta_i, B \Longrightarrow I \vdash S_i' = \Pi_i, \Box\Omega \to \Delta_i, B^1$, where I is some calculus in which $(+1)$ is admissible; besides the part of the resulting derivation from the sequent S_i' contains the application of the postulates obtained from the postulates of $G_{L\omega}$ by replacing $(\to \Box_\omega)$ by $(\to \Box^1)$.

Proof. As follows from Lemma 4.10 $\forall i$ $(1 \leqslant i \leqslant n)$ there exists $Re(S_i) = \{S_{i_1}, \dots, S_{i_m}\}$ such that $\forall j$ $(1 \leqslant j \leqslant m)$ $S_{i_j} \in \mathrm{Sat}\{S\}$ or $S_{i_j} \preccurlyeq S' \in \mathrm{Sat}\{S\}$. Let us consider $R(S_i)\{k\} \Longrightarrow \{S_{i_1}, \dots, S_{i_m}\}$, i.e. reduction of S_i to the sequents $S_{i_1}, \dots, S_{i_m}$. In each branch of this reduction there is only one bottom-up application of (A) :

$$\frac{S^* = \Pi_j^*, \Box\Omega \to \Delta_j^*, \Box A}{\Sigma_1, \Pi_j^{*1}, \Box\Omega \to \Sigma_2, \Delta_j^{*1}, \Box A^1}\,(A)$$

$$\vdots$$

$$\Pi_i, \Box\Omega \to \Delta_i, \Box A,$$

where $\Pi_j^* = \Pi_j$; $\Delta_j^* = \Delta_j$, if $S^* \in \mathrm{Sat}\{S\}$ and $\Pi_j^* = \Pi_j', \Pi_j$; $\Delta_j^* = \Delta_j', \Delta_j$, if $S^* \prec S' \in b_k\mathrm{Sat}\{S\}$. Let us perform in $R(S_i)$ (for all i) the following operations: (1) if $S^* \prec S' \in b_k\mathrm{Sat}\{S\}$, then let us bottom-up apply (W) to S^* (with the sequent S' as the premise of (W)); (2) above (A) let us replace $\Box A$ by B; (3) below (A) let us replace $\Box A^1$ by B^1. Then instead of S^* we get the sequent $\Pi_j, \Box\Omega \to \Delta_j, B$, which is derivable in I by assumption of the Lemma (and by (W), if $S^* \prec S' \in b_k\mathrm{Sat}\{S\}$). Instead of bottom-up applications of (A) we get (W) and $(+1)$ (which is admissible in I). All bottom-up application of rule of inference (k), except (A), in the reduction $R(S_i)$ we get the application of the same rule of inference (k). Therefore, instead of the reduction $R(S_i)$ we get $I \vdash \Pi_i, \Box\Omega \to \Delta_i B^1$, for all i.

Now, we prove that Sat $\vdash S \Longleftrightarrow IN \vdash S$.

Lemma 5.6. Let Sat $\vdash S$, then $I \vdash R_k \to R_k^1$, where R_k is the invariant formula, as indicated above, I be the calculus obtained from $G_{L\omega}$ replacing $(\to \Box_\omega)$ by $(\to \Box^1)$.

Proof: Taking R_k instead of B in Lemma 5.5 and using logical rules of inference we get $I \vdash \Pi_i, \Box\Omega \to \Delta_i, R_k$ (for each i), therefore, by Lemma 5.5 we get $I \vdash S_i^* = \Pi_i, \Box\Omega \to \Delta_i, R_k^1$. Applying logical rules of inference to S_i^* we get $I \vdash R_k \to R_k^1$.

Lemma 5.7. Let Sat $\vdash^D S$, $S_{i_k} = \Pi_{i_k}, \Box\Omega_k \to \Delta_{i_k}, \Box A_k \in b_k\mathrm{Sat}\{S\}$, $(1 \leqslant i \leqslant l)$, $(1 \leqslant k \leqslant n)$ then $\forall ik\ (IN \vdash S_{i_k})$.

Proof: by induction of $\|S_{i_k}\|$, where $\|S_{i_k}\|$ means the number of positive occurrences of $\Box$ in S_{i_k}. From the form of R_k we get Prop $\vdash \Pi_{i_k}, \Box\Omega_k \to \Delta_{i_k}, R_k$ (1). ¿From Lemma 5.6 we get $I \vdash R_k \to R_k^1$ (2). Let us consider the sequent $S'_{i_k} = \Pi_{i_k}, \Box\Omega_k \to \Delta_{i_k}, A$. If $\|S_{i_k}\| = 0$, then from the construction of D it follows, that $I \vdash S'_{i_k} = \Pi_{i_k}, \Box\Omega_k \to \Delta_{i_k}, A$. Let $\|S_{i_k}\| > 0$, then let us consider a sequent $dS^* \in QP(S'_{i_k})$. As $\|S'_{i_k}\| < \|S_{i_k}\|$, then $IN \vdash S'_{i_k}$ ($3'_i$). Using the logical rules of inference we get $IN \vdash R \to A$ (3). Applying $(\to \Box)$ to (1), (2), (3) we get $IN \vdash S_{i_k}$ for $\forall ik$.

Now, we prove that Sat $\vdash S \Longrightarrow G_{L\omega} \vdash S$.

Lemma 5.8. Let Sat $\vdash^D S$ and $S_p \in D$ $(p = 1, 2, \ldots)$ then $G_{L\omega} \vdash S_p$ for all p.

Proof. Let $S_i^* \in b_k\mathrm{Sat}\{S\}$, and $S_i^* = \Pi_i,\ \Box\Omega \to \Delta_i,\ \Box A$. Let us prove (by induction on k) that $\forall k (k \in \omega)\quad G_{L\omega} \vdash \Pi_i,\ \Box\Omega \to \Delta_i, A^k$. Let $k = 0$, then from the construction of D it follows that $S_{i_0} = \Pi_i,\ \Box\Omega \to \Delta_i,\ A \in D$. Let us prove that $G_{L\omega} \vdash S_{i_0}$. Let $\|S_{i_0}\|$ denotes the number of positive occurences of $\Box$ in A. Let $\|S_{i_0}\| = 0$, then as follows from the construction of D we get $I \vdash S_{i_0}$, where I is obtained from $G_{L\omega}$ replacing $(\to \Box_\omega)$ by $(\to \Box^1)$. Let $\|S_{i_0}\| > 0$, then analogously as in proof of Lemma 5.7 (instead of using the induction assumption on $\|S_{i_0}\|$ we apply the fact that Sat $\vdash S_{i_0}$) we get $IN \vdash S_{i_0}$. Using Theorem 5.1 we get $G_{L\omega} \vdash S_{i_0}$. Let $G_{L\omega} \vdash S_{i_k} = \Pi_i,\ \Box\Omega \to \Delta_i, A^k$ (for some $k \geqslant 1$). Applying Lemma 5.5 we get $G_{L\omega} \vdash S_{i_{k+1}} = \Pi_i,\ \Box\Omega \to \Delta_i,\ A^{k+1}$. Therefore $\forall k (G_{L\omega} \vdash S_{i_k} = \Pi_i,\ \Box\Omega \to \Delta_i,\ A^k)$ and by $(\to \Box_\omega)$ we get $G_{L\omega} \vdash S_i^* = \Pi_i,\ \Box\Omega \to \Delta_i,\ \Box A$. The same we get when $S_i^* = \Box\Omega \to \Box\nabla$. Therefore all saturated sequents from D are provable in $G_{L\omega}$. A bottom-up applications of rule of inference (i), except (A), $(\to \Box^1)$, may be replaced by applications the same rule of inference (i). Bottom-up applications of (A) may be replaced by (W), $(+1)$ (which is admissible in $G_{L\omega}$), bottom-up applications of $(\to \Box^1)$ may be replaced (using Lemma 4.3) by applications of rules of inference of $G_{L\omega}$. Therefore if $S_p \in D$ $(p = 1, 2, \ldots)$ then $G_{L\omega} \vdash S_p$ for all p.

Theorem 5.2. Sat $\vdash S \Longrightarrow G_{L\omega} \vdash S$.

Proof: follows from Lemma 5.8.

Lemma 5.9. $IN \vdash S \Longrightarrow$ Sat $\vdash S$.

Proof. Let V be the given derivation in IN. All rules of inference, except $(\to \Box)$, and axioms of IN include in the set of postulates of Sat. Therefore, any application in V of a rule of inference (i) can be modelled by bottom-up application of the same rule of inference (i). Let us consider any application of $(\to \Box)$ in V. As follows from the restriction on $(\to \Box)$, the conclusion of $(\to \Box)$ is derivable in Sat. Therefore Sat $\vdash^D S$.

Theorem 5.3. Sat $\vdash S \Longleftrightarrow IN \vdash S$.

Proof. The part $\Longrightarrow$ follows from Lemma 5.7 the part $\Longleftarrow$ follows from Lemma 5.9.

Lemma 5.10 Let $S = \Pi, \Box\Omega \to \Delta, \Box\nabla$ be a proper ordinary quasiprimary sequent, then $G_{L\omega} \vdash^D S \Longrightarrow$ Sat $\vdash^{D_1} S$.

Proof. Let us consider any two applications of $(\rightarrow \Box_\omega)$ in the given atomic derivation D of S. These application will be called different if the principal formulas are either diffferent (formulas distingushed by indices are considered as coincident) or coincide but have different descendants. Therefore, different applications of $(\rightarrow \Box_\omega)$ in D are finite and let $(\rightarrow \Box_\omega)[D]$ means this number. The proof of the Lemma is carried out by induction on $(\rightarrow \Box_\omega)[D]$. If $(\rightarrow \Box_\omega)[D] = 0$, then $D = D_1$. Let $(\rightarrow \Box_\omega)[D] > 0$. Let us consider the lowest application of $(\rightarrow \Box_\omega)$ in D. Without loosing of generality we can consider that the conclusion of this application is quasiprimary sequent $S^0 = \Sigma_1, \Pi_1^1, \Box\Omega \rightarrow \Sigma_2, \Pi_2^1, \Box\nabla$. Let us consider two cases. (1) $S^0 = \Box\Omega \rightarrow \Box A_2, \ldots, \Box A_n$. Using the invertability of $(\rightarrow \Box_\omega)$ we get $G_{L\omega} \vdash S_1 = \Box\Omega \rightarrow A_1, \Box A_2, \ldots, \Box A_n; \ldots; G_{L\omega} \vdash S_n = \Box\Omega \rightarrow \Box A_1, \ldots, \Box A_{n-1}, A_n$. Applying to $S_1, \ldots, S_n$ the induction assumption and $(\Box_n)$ we get Sat $\vdash S$. (2) S^0 is a proper quasiprimary sequent, then starting from S let us construct a tree D of sequents, applying the resolvent tactic. Since $G_{L\omega} \vdash S$, (as follows from Lemmas 4.1, 4.2), invertibility of $(\rightarrow \Box_\omega)$ all the sequents from D are derivable in $G_{L\omega}$. Let us consider the bottom-up applications of $(\rightarrow \Box^1)$ in D. As $G_{L\omega} \vdash S$, then using induction hypothesis and invertibility of $(\rightarrow \Box_\Omega)$ we conclude that the left premise of $(\rightarrow \Box^1)$ is derivable in Sat. Next, let us consider the bottom-up applications of (A) in D. Since $G_{L\omega} \vdash S$, (as follows from Lemma 4.2) the "logical premise' of (A) is derivable in Log (where Log is obtained from $G_{L\omega}$ by dropping $(\rightarrow \Box_\omega), (\Box \rightarrow)$). Relative to the "temporal" applications of (A) we conclude that the "temporal" premise of (A) is of the form $\Gamma, \Box\Omega \rightarrow \Theta, \Box\nabla$. Therefore, applying the resolvent tactic to S^0 (and using the fact that $G_{L\omega} \vdash S$) we can reduce S either to logical axioms or to the sequents $S_i = \Pi_i, \Box\Omega \rightarrow \Delta_i, \Box\nabla$, (in separate case $\Pi_i, \Delta_i = \varnothing$) where $\Pi_i, \Delta_i \in R\mathrm{Sub}(\Omega)$. Since the set $R\mathrm{Sub}(\Omega)$ is finite, starting from some k-th resolvent $Re^k(S)$ we shall generate the resolvent sequents which were obtained previously in a branch of the tree D and each resolvent sequent is either saturated or absorbed in each branch of the tree D. Therefore each "resolvent branch" of D is saturated and (each "resolvent branch") can be cut off getting the blocked leaf. The leaves of other branches of D are logical axioms. Thus, we get Sat $\vdash^D S^0$. Let us note that we described the worst way of obtaining the derivation in Sat of the proper quasiprimary sequent S^0. Instead of getting the equality $S^{**} = S^*$ (which yields that S^* is saturated and S^{**} is absorbed) we can get (in some cases earlier) the inequality $S^{**} \prec S^*$. Therefore it is not necessary to generate resolvent sequents, containing all possible resolvent subformulas of the formulas from Ω.

Theorem 5.4. Let S be an M-sequent, then $G_{L\omega} \vdash S \Longleftrightarrow \mathrm{Sat} \vdash S$.

Proof. The part $\Longrightarrow$ follows from Lemma 5.10, the part $\Longleftarrow$ follows from Theorem 5.2.

Theorem 5.5. Let S be an M-sequent, then $G_{L\omega} \vdash S \Longleftrightarrow IN \vdash S$.

Proof: follows from Theorems 5.3, 5.4.

References

1. H.Kawai: Sequential calculus for a first order infinitary temporal logic. Zeitshr. für Math. Logik und Grundlagen der Math. 33, 423–432 (1987).
2. S.Merz: Decidability and incompleteness results for first-order temporal logic of linear time. Journal of Applied Non-classical Logics 2, 139–156 (1992).
3. R.Pliuškevičius, Investigation of finitary calculi for temporal logics by means of infinitary calculi. In B.Rovan (ed.): Mathematical foundations of computer science 1990. Lecture Notes in Computer Science 452, Berlin: Springer 1990, pp.464–469.
4. R.Pliuškevičius: Completeness criterion for the Horn-like first order linear temporal logic. Proceedings of conference on applied logic. Logic at Work, Amsterdam 1992.
5. R.Pliuškevičius: On saturated calculi for a linear temporal logic, Proceedings of MFCS-93, Gdansk (in print).
6. R.Pliuškevičius: On the saturation principle for a linear temporal Logic. In Proceedings of KGC-93, Brno (in print).
7. T.Shimura: Cut-free systems for the modal logic S4.3 and S4.3GRZ. Reports on Mathematical Logic 25, 57–73 (1991).

A Formal Approach to Requirements Engineering*

Friederike Nickl, Martin Wirsing
Institut für Informatik
Ludwig-Maximilians-Universität München
Germany

Abstract

In this paper the algebraic/axiomatic approach to software specification is proposed as an integrating formal model for requirement descriptions. It is argued that formal specificaton techniques are complementary to informal ones and that they can safely be integrated in an iterative requirements development process. Three results from the German National Project Korso are presented: entity-relationship diagrams, data flow diagrams and access rights are modelled by algebraic specifications in the language SPECTRUM. These schematic techniques are illustrated by a small part of a Korso case study on the requirements specification for a patient information system of a hospital.

1. Introduction

Requirements engineering is the discipline for developing a complete consistent unambiguous specification - which can serve as basis for a common agreement between all parties concerned - describing what the software product will do [Boehm 79]. Requirements engineering takes place in the beginning of the software development. It describes the process leading in a systematic way from the informal objectives to a requirements specification. This specification describes the requirements for the automatization of a system including its functional and performance capabilities as well as requirements for hardware, software and the development process. The requirements specification serves as a contract for the consecutive software development.

Current requirements engineering methods, such as SSADM [Yourdon 89], Object-Oriented Analysis [Coad, Yourdon 89] or SDL use a combination of notations including entity-relation diagrams, data flow and control flow diagrams, object-oriented diagrams and finite state machines. Other, mostly academic methods, such as CIP (cf. e.g. [CIP 85], [van Diepen, Partsch 90]), ERAE (cf. e.g. [Dubois et al. 86]), NATURAL [Jarke et al. 93] propose fully formal descriptions for the functional requirements. Both approaches have their advantages and disadvantages: the informal diagrammatic methods are easier to understand and to apply but they can be ambiguous and due to the different nature of the employed diagrams and descriptions it is often difficult to get a comprehensive view of all requirements. On the other hand, the formal

* This research has been partially sponsored by the German National Project Korso.

approaches are more difficult to learn and to communicate to the user. But they provide mathematical rigour for analysis and prototyping of requirements.
To close partly this gap we propose the algebraic/axiomatic approach to software specification as integrating formal model for requirement description. Algebraic/axiomatic specifications are known to be a suitable tool for the formal description of all functional aspects of a software system including data structure, control and reactive behaviour. In this paper we will show that they are also suitable for formally describing entity-relationship diagrams, data flow diagrams and access rights. Moreover, the notion of refinement between specifications provides a formal tool for checking the correctness of requirements. Therefore algebraic/axiomatic specifications are a universal tool for giving a coherent view of all formalizible requirements and for formally guiding the requirements development process. Formal specification techniques are complementary to pre-formal and informal ones. The integration of both leads to an improved requirement analysis.

In this paper we present ongoing work on a method for the development of requirements specification leading from informal requirements to an algebraic/axiomatic specification (of the functional requirements). The method instantiates the Korso approach to correct software development [Wirsing et al. 92], [Pepper et al. 93]; it is based on many ideas from SSADM and ERAE. Requirements analysis is seen as an iterative process leading from the actual state of the system (if such a system exists already) to the requirement description of the intended system. Each step of this evolutionary process consists of four activities: problem identification, modelling, analysis of requirements and conversion into a user friendly notation. In our approach formal methods play a major role mainly in the modelling and analysis phase. In the modelling phase a domain model is constructed and expressed in the specification language SPECTRUM [Broy et al. 92]. The construction starts with the pre-formal diagrams and descriptions from the problem identification phase. Then the diagrams are schematically translated to specifications written in SPECTRUM and detailed constraints are formally expressed in this language. In the analysis phase the formal specifications are validated using techniques such as theorem proving and rapid prototyping. As examples we describe in this paper how entity-relationship diagrams and data flow diagrams can be schematically translated to SPECTRUM (for details see [Hettler 93], [Nickl 93]) and outline how access rights can be formalized [Renzel 93].

The paper is organized as follows: In section 2 we present our general approach to requirement engineering. In the section 3 we elaborate a small example that is taken from a Korso case study on the requirement specification for a patient information system [Hußmann et al 93]. In the concluding remarks the formalization of access rights is outlined. In the appendix some features of the language SPECTRUM are summarized and a formal specification of the basic data structure of STREAMS is given.

2. Method for the development of a requirement specification

In this section we describe first the process model for requirements engineering with algebraic/axiomatic specifications. It is based on the "development graph" describing the established documents and their interrelationships. As examples for the interplay between diagrams and formal specifications the construction of the requirements for the data model and the dynamic behaviour is outlined.

2.1 Requirement life cycle

Requirement engineering comprises three main activities:
Problem identification, modelling and analysis of requirements.

Problem identification includes the discussion of functional requirements between the customer and the requirements engineer, the agreement on the quality attributes and the exploration of the environment for the system and its development. Often the analysis of the actual system is important; it helps the engineer to understand better the application and enhances the communication between user and engineer. Pre-formal notation, such as entity-relationship or data flow diagrams are used in this phase for the communication between user and engineer. The goal is to approve, correct or complete the material gathered at that point.

In the **modelling activity** the informal input material is converted into its formal counterpart in the requirement specification language. In Korso this conversion consists of schematic translations of diagrams into SPECTRUM specifications. The data model is described by extended entity-relationship diagrams, the dynamic behaviour by extended data flow diagrams. The translation makes incomplete or ambiguous statements of the identification phase appearent. Moreover, the translation leads to systematic structuring and classification of the requirements; possible relationships between the individual requirements are investigated.

The **analysis activity** includes checking the new material for qualities, such as consistency and completeness, and testing and validating the new material using theorem proving and prototyping. In particular, many annotations of the diagrams can now be proven using the formal translations. Then the existing document is updated with the new material and the integration of the new material is checked, tested and validated. In the analysis phase also new information for removing incompleteness for structuring and simplifying the document and for proposing solutions to satisfy the objectives is conjectured. Then that part of the document is selected and prepared which will be used in the next identification step. This includes selecting inconsistencies deduced from the new material or the updated document, warnings of potential incompletions and surprising implications.

Often a fourth activity, called **conversion** is necessary: If the customers do not understand the requirements engineering language it is necessary to convert the specifications into different representations. For customers this results in natural language, drawings, diagrams, pre-formal language or a prototype.

Thus, the development of a requirement specification consists of a series of the following four steps [Dubois et al. 86]:

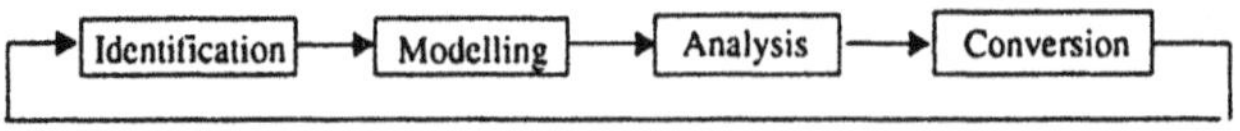

2.2 Development graph

In general, many iterations of this process are necessary making management and version control to another important activity. In Korso all documents as well as all results of activities of the required engineering process are reported in a **development graph**. The elements of this graph are called "units" which are connected by certain relations. A unit is understood to be any software object involved in the development process.

In requirements engineering units are diagrams, informal comments and formal specifications. There are three kinds of relations: Syntactic relations indicate the syntactic dependencies between the requirements units (denoted by ·······▸), modifications report changes a unit has to undergo in the analysis activity; semantic relations are subject to verification: the refinement relations (denoted by ~~~>) states that the new document is a correct implementation of the preceding one and a translation (denoted by ⟶) gives different correct views on the same unit. Each development step in Korso represents a local or a global change of the system of units. It depends on the actual context of the system and of its environment. In the case of the establishment of a semantic relation it has a well defined semantics and comes as a set of verification conditions which have to be verified in order to guarantee the correctness of the step. As much as possible a software library is involved which stores specification schemata and transformations that help to reduce the amount of verification conditions.

The following development graph shows the construction of the requirements of the example "blood test system". The (hyper-) node "BloodTest-Req" describes the state of the actual non-computerized system, the node "BloodTest-System" describes the requirements for the inclusion of a database. Nodes marked E/R represent entitiy-relationship diagrams, labels DF denote data flow diagrams.

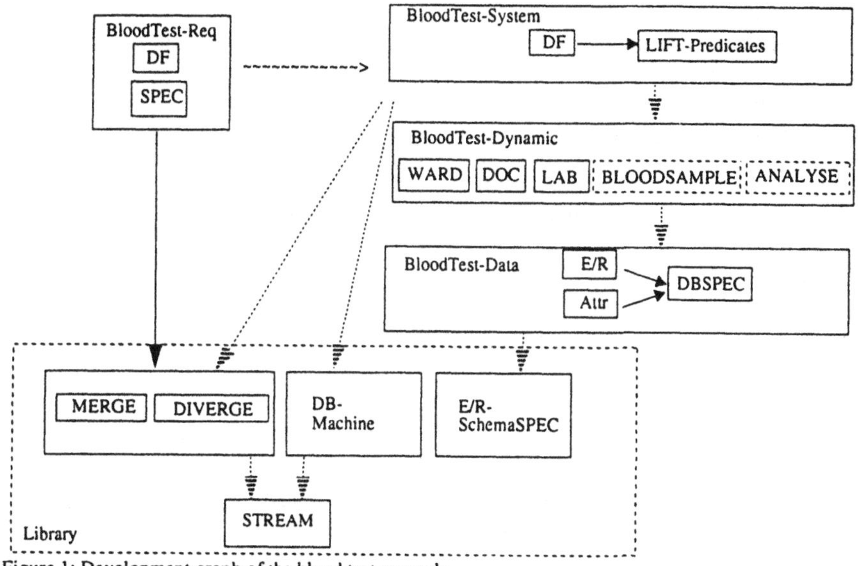

Figure 1: Development graph of the blood test example

2.3 Data modelling

The data model of the intended system is described by entity-relationship diagrams with attributes and integrity conditions. An entity type E is given together with a set of attributes a_i of type T_i $(i = 1, \ldots, n)$ in the following form:

$\underline{E}$
$a_1 : T_1;$ **a_1 primary key;**
$\vdots$
$a_n : T_n;$

where one of the attributes is selected as primary key. The relationship R between two entity types E_1 and E_2 is denoted diagrams written in SSADM-notation [Downs et al. 92]: E.g.

denotes a n:1 relationship, n ≥ 0. These two forms of descriptions are schematically translated into a SPECTRUM specification of an entity-relationship database (for details see [Hettler 93]). In this specification a type `Db` for the defined states of the database is introduced. Each state consists of a collection of finite sets of entities and a finite number of relations which are characterized by the functions

```
entE: Db → Set E;
R : Db → ((E1 × E2) → Bool);
```

For each entity type E there are operations

```
putE, delE : E × Db → Db;
getE : KeyE × Db → E;
```

(where `KeyE` denotes the primary keytype of the entity type E) to insert, delete or retrieve entities from the database. For each relation R there are operations

```
estR, relR : Db × E1 × E2 → Db;
```

to establish and release relationships.
From the properties of R expressed in the entity-relationship diagram (for instance, 1:1 or 1:n relations) an integrity condition

```
CR : Db → Bool
```

can be synthesized in a schematic way. Additional integrity conditions may involve conditions on the attribute values of the involved entites and therefore cannot be derived directly for the entity-relationship diagram. These are specified by an additional predicate `C : Db → Bool` directly in SPECTRUM. The overall integrity condition on the state of the database is then given by the predicate

```
Ok : Db → Bool
```

with

```
Ok(db) ⇔ C(db) ∧ CR1(db) ∧ ... ∧ CRm(db)
```

where $R_1, \ldots, R_m$ are the relations of the database.

2.4 Modelling of dynamic behaviour

The dynamic behaviour of the system is described by data flow diagrams extended by axiomatic descriptions of pre- and postconditions. To indicate the difference between (trans-)actions of the system and actions performed by the environment we use dotted lines surrounding environment actions. The data flow diagrams have three kinds of nodes:

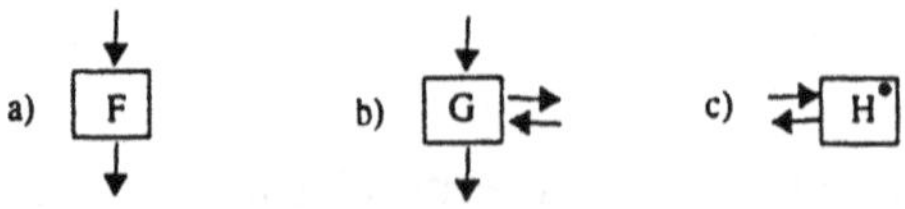

Vertical lines denote input and output lines, horizontal ones communication with a repository. A node F of type (a) is associated with a function symbol

$f : In \to Out.$

A node G ot type (b) corresponds to a function symbol

$g: In \times State \to Out \times State$

with an additional State parameter, where State denotes the state of a repository. A node H of type (c) denotes a repository represented by a function symbol

$h\_mach: Call \times State \to Answer \times State$

that receives a message and produces an answer and an update of the state.

We formalize the data flow diagram in three steps:

Step 1:

To each node G of type (b) a specification of the behaviour of g and of its precondition

$pre_g : In \times State \to Bool$

is constructed which ensures that g can be safely executed. Similarly for nodes of type (a) and (c).

Step 2:

For each node a predicate on streams

(a) $LIFT_f: In^{\omega} \times Out^{\omega}$ to Bool

(b) $LIFT_g: (In^{\omega} \times Answer^{\omega}) \times (Out^{\omega} \times Call^{\omega})$ to Bool

(c) $LIFT_h: Call^{\omega} \times Answer^{\omega}$ to Bool

is systematically constructed.

Here, for any sort T, T^{ω} denotes the set of all finite or infinite streams $<a_n>_{n<\omega}$ with elements $a_n : T$ (for details see Appendix 2). Each predicate is subject to a condition expressing the connection with f. For instance, if in (a) f acts in a sequential way the output is just the sequence of the values of the inputs:

$LIFT_f(in, out) \Leftrightarrow out = <f(in_j)>_{j<\omega}$ (where $in = <in_j>_{j<\omega}$)

Step 3:

All node specifications are put together to a specification of the diagram. This step is fully schematic. In case, a node receives several streams of data on the same input line, a "Merge"-node is inserted: The subdiagram (a) is translated to (b)

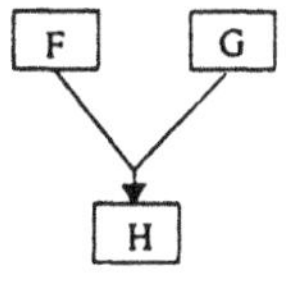

(a) implicit merge

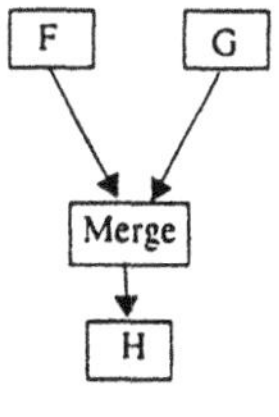

(b) explicit merge

Similarly, for forking outputs additional "Diverge" nodes are introduced.

3. The blood test example

As an example we consider a blood test in a hospital with n wards and a common laboratory. For the sake of simplicity, we assume that there is just one type of blood test (for instance the determination of a particular blood composition). The goal is to derive a requirements specification for the installation of an information system monitoring the blood tests. In section 3.1

the requirements for the actual system are described (as obtained from the problem identification phase). In sections 3.2, 3.3 and 3.4 the modelling for the requirements of the actual system is done. Entity-relationship and data flow diagrams are translated to algebraic specifications. In section 3.5 the requirements of the intended system are given and it is proven that the requirements specification implements a crucial global requirement from the problem identification phase.

3.1 Requirements on the communication model

As a result from the problem identification phase we get the following informal description of the blood test: in each ward, a blood sample is taken from the patient to be examined and the blood sample is sent to the lab. Since all wards use the same lab, the "streams" of blood samples coming from the different wards are merged at the entrance to the lab.
In the lab the incoming blood samples are analyzed, and then the resulting blood pictures - correlated to the patients who gave the samples - are sent back to the ward.
In the case of two wards, this situation is represented by the following dataflow diagram:

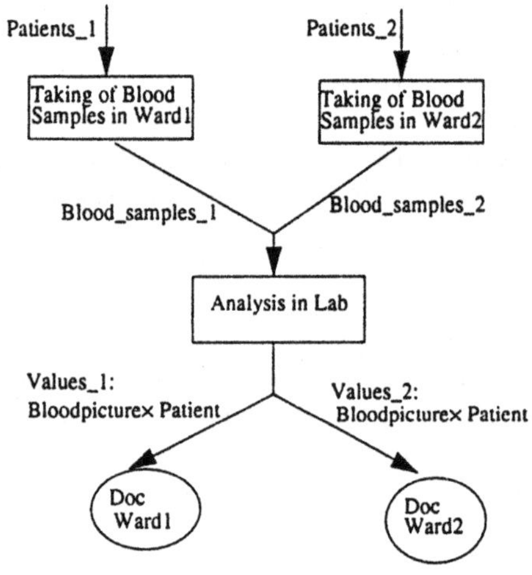

<u>Diagram 3.1: Data flow in the blood test</u>

Here the lines in the diagram are marked with streams of data: for instance, Values_1 is a stream of blood pictures correlated with patients. Streams join at the entrance of the lab and fork at the exit of the lab (where the values are distributed to the wards).

A crucial requirement for the appropriateness of the blood test process in the hospital is that it can be guaranteed that no confusion is possible: if a blood picture `bp` is related to a patient `pat` in the `Values_n` stream, then there must be a blood sample in the stream `Blood_samples_n` containing the blood of patient `pat`, such that the result of the analysis of this blood sample is `bp`. In a first modelling step we express this requirement as follows (with reference to the above diagram): Let the analysis of blood samples be described by a function

```
analyse : BloodSample → BloodPicture;
```

and let the fact that a blood sample `bl_sample` consists of the blood of a patient `pat` be denoted by

```
blood_from(bl_sample, pat)
```

Let furthermore

```
bl_sample in Blood_samples_n
```

stand for the fact that the blood sample `bl_sample` is contained in the stream `Blood_samples_n` of blood samples sent to the laboratory from ward n and

```
(bp,pat) in Values_n
```

mean that the bloodpicture bp correlated to the patient `pat` is an element of the stream `Values_n` of values which is sent from the lab to ward n.
Then the process of the blood testing in the hospital has to meet the following requirement:

(No confusion)

```
∀ bp: BloodPicture, pat:Patient.(bp,pat) in Values_n ⇒
        ∃ bl_sample: BloodSample.
            (bl_sample in Blood_samples_n ∧
            blood_from(bl_sample,pat) ∧
            analyse(bl_sample) = bp)
```

In particular, it follows from the above requirement that it must be possible to retrieve from the blood sample the patient from which the blood has been taken. This association, however, should not be visible to the staff in the laboratory, since only the doctor responsible for a patient should have access to the results of blood tests concerning this patient.

3.2 Specification of the Data Model

In this section we continue the modelling phase by constructing the requirements on the data model. Since each blood sample must be related to a unique patient, but this association should not be visible in the lab, the idea is to associate each blood sample with a unique order to the lab which is related exactly to one patient. Hence the "real" object "blood sample" is supplemented by the abstract entity "LabOrder". The relation between lab-orders, values obtained by the analysis of blood samples corresponding to lab-orders, patients and doctors is specified by the following entity/relationship (E/R)- diagram.

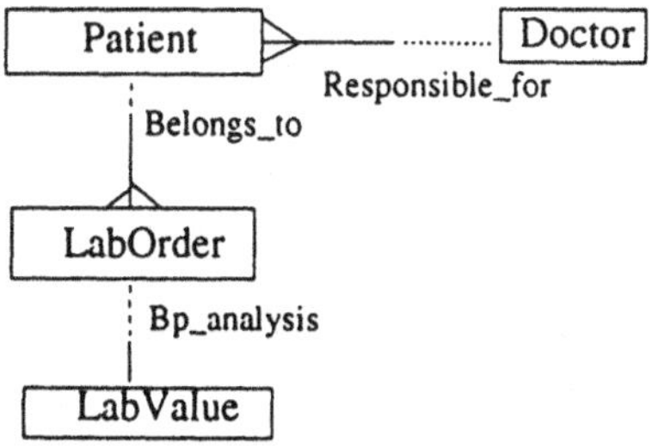

Diagram 3.2: E/R diagram specifiving the data model

This E/R diagram is the basis for the specification of a database supporting the process of blood testing in the hospital. In the next step a number of attributes are associated with each entity type.

Entity Patient

```
PatId : PatId; PatId primary key;
Name : Name;
Date_of_Birth : Date;
Ward : WardNr;
```

Entity Doctor

```
DocId : DocId; DocId primary key;
Name : Name;
Ward : WardNr;
```

The types of the attributes here are the obvious ones: Name is a datatype of strings, `Date` is a datatype for dates, `WardNr` is a number type.

Entity LabOrder

```
Nr : OrderNr; Nr primary key;
Created : Date_with_time;
```

Entity LabValue

```
Nr : OrderNr; Nr primary key;
Value : BloodPicture;
Established: Date_with_time;
```

Here `OrderNr` is an (infinite) number type. Notice that `LabValue` uses the same key type as `LabOrder`. `Date_with_time` is a datatype for dates together with the time of day and `BloodPicture` is a compound datatype, representing all the values that are determined by a standard blood picture analysis.

For each entity type E there is a constructor

createE: $A_1 \times \ldots \times A_n \rightarrow E$

where n is the number of attributes of E and $A_1, \ldots, A_n$ are the types of these attributes. For instance, we have

```
createLabOrder: OrderNr × Date_with_time → LabOrder;
```

The above E/R diagram together with the specification of the attributes of the different entities can be schematically translated into a SPECTRUM specification DBSPEC for a E/R -database (see section 2). A type `Db` standing for defined states of the database is introduced and the states of the database are characterized by the functions

```
entPatient: Db → Set Patient;
entDoc: Db → Set Doc;
entLabOrder: Db → Set LabOrder;
entLabValue: Db → Set LabValue;

Responsible_for: Db → ((Doctor × Patient) → Bool);
Belongs_to : Db → ( (Patient × LabOrder)) → Bool);
Bp_analysis : Db → ( (LabOrder × LabValue)) → Bool);
```

In every defined state of the database, different entities are required to have different primary keys. Hence in our example, the following holds:

```
∀ db: Db, order1, order2: LabOrder.
        (order1 ∈ entLabOrder(db) ∧ order2 ∈ entLabOrder(db)
        ∧ Nr(order1) = Nr(order2) ) ⇒ order1 = order2
```

It is also required that in the database relations only such entities occur which belong to the database.

From the properties of the relationships expressed in the E/R diagram, an integrity condition on the state of the database is synthesized in a schematic way. For details we again refer to

[Hettler 93]. For instance, in the above diagram `Belongs_to` is specified as a 1: n relationship, which is mandatory on the LabOrder- side. This means that, for every entity of type LabOrder, there is a unique patient, to whom this order is related. The dots on the Patient-side indicate that there may be patients in the database with no associated orders. This condition on the relationship Belongs_to can be expressed by the following predicate on database-states:

$C_{Belongs\_to}$: `Db → Bool;` $C_{Belongs\_to}$ **`strict`**`;`

specified by

```
∀ db: Db. C_Belongs_to (db) ⇔
∀ order: LabOrder. order ∈ entLabOrder(db) ⇒
  (∃pat:Patient. Belongs_to db (pat, order) ∧
  ∀pat': Patient. Belongs_to db (pat', order) ⇒ pat = pat')
```

In a similar way specifications for (strict) predicates

$C_{Responsible\_for}$, $C_{Bp\_analyse}$: `Db → Bool;`

are given. Moreover, there are additional integrity conditions on the database which should be preserved, but which involve conditions on the attribute values of the involved entities and therefore cannot be derived directly from the E/R- diagram. In our case, there are two such conditions:

- If a doctor is responsible for a patient, then the doctor's ward must coincide with the patient's ward
- A lab-value is related to a lab-order, if and only if they have the same key.

These additional requirements can be specified by the following predicate

```
C: Db → Bool; C strict;

C(db) ⇔
(∀ doc: Doctor;∀ pat: Patient.
 Responsible_for db (doc, pat) ⇒ Ward(doc) = Ward(pat) )
∧
(∀ order: LabOrder; ∀ val: LabValue.
Bp_analysis db (order, val) ⇔
order ∈ db ∧ val ∈ db ∧ Nr(order) = Nr(val))
```

Notice, that we used the shorthand `order ∈ db` instead of `order ∈ entLabOrder(db)` (`and val ∈ db` instead of `val ∈ entLabValue(db)`). In the sequel we shall always use this shorthand in order to make the specifications more readable.

The overall integrity condition on the state of the database now can be specified by the predicate

`OK : Db → Bool; OK` **`strict`**`;`

with

`OK(db) ⇔` $C_{Belongs\_to}$ `(db) ∧` $C_{Responsible\_for}$ `(db) ∧` $C_{Bp\_analyse}$ `(db) ∧ C(db).`

3. 3 Specification of the dynamic behaviour of the database

By the E/R diagram the static model of our database has been specified. Now we return to the data flow diagram 3.1 and specify in which way the processes of taking blood and analyzing blood samples (the nodes of diagram 3.1) are realized in cooperation with a database-system. To this aim, we refine these processes into elementary transactions which are offered by the database and actions that are performed by the staff of the hospital or by machines in the hospital - in short by the environment of the database system. The notion "elementary transactions" is used since we consider a database machine for which these transactions are elementary in the sense that they are executed as a whole or not executed at all. The states of the database machine are the states of the database (i.e. the elements of sort Db as described in the previous section), and the actions of the database machine are determined by the specification of a finite set ETA of elementary transactions. The standard type of a transaction f in ETA is of the form

```
f: Input × Db → Output × Db
```

If on the right side of the arrow the Output-type is missing, then f only changes the state of the database (i.e. f is a write-transaction) and if the Db-part is missing, then f is a mere read.
In the sequel, the specification of the transactions for our example-system are given. For every such transaction, a precondition

$$\mathtt{pre_f} : \mathtt{Input} \times \mathtt{Db} \rightarrow \mathtt{Bool};$$

is specified (as a total predicate), which ensures that the state of the database is such, that f can be executed safely. The database machine only executes f, provided the precondition pre_f holds.
The set of actions of the database machine are calls to the database to perform elementary transactions with given input data: We consider a sort `Call` for calls of elementary transactions and for every elementary transaction f with parameter type `Input` × `Db` a function

$$\mathtt{f^{call}} : \mathtt{Input} \rightarrow \mathtt{Call}$$

Hence f^{call} is obtained from f by suppressing the Db- parameter. The set of actions of the Db-Machine is

$$\{\mathtt{f^{call}(d)}; \mathtt{f\ is\ an\ elementary\ transaction\ in\ ETA\ with\ input\ type\ Input\ and\ d: Input}\}$$

Given a database state and an element in `Call` , the transition function `db_mach` of the database machine yields some output and a new state. Hence we specify

$$\mathtt{db\_mach} : \mathtt{Call} \times \mathtt{Db} \rightarrow \mathtt{Answer} \times \mathtt{Db}$$

where `Answer` is the union of all data types that are the output of elementary transactions and additionally of a type consisting of two elements `ok` and `error`. The element `error` is considered as an error-message. For every elementary transaction f, the execution of the action $\mathtt{f^{call}(d)}$ in the state `db` is specified by

$$\mathtt{db\_mach(\ f^{call}(d),\ db)} = \textbf{if}\ \neg \mathtt{pre_f(d,\ db)}\ \textbf{then}\ \mathtt{(error,\ db)}$$
$$\textbf{else}\ \mathtt{([v]_o,\ db')}\ \textbf{where}\ \mathtt{(v,\ db')} = \mathtt{f(d,\ db)}$$

Here $[.]_o$ denotes the inclusion of the Output-type of f into the "universe" `Answer`.

For a mere write-transaction f we specify

$$\mathtt{db\_mach(f^{call}(d),\ db)} = \textbf{if}\ \neg\mathtt{pre_f(d,db)}\ \textbf{then}\ \mathtt{(error,\ db)}\ \textbf{else}\ \mathtt{(ok,\ f(d,db)\ )}$$

and in case f is a mere read-transaction

$$\mathtt{db\_mach(f^{call}(d),\ db)} = \textbf{if}\ \neg\mathtt{pre_f(d,db)}\ \textbf{then}\ \mathtt{(error,\ db)}\ \textbf{else}\ \mathtt{([f(d,db)]_o,\ db)}$$

A run of the database machine consists in the evaluation of a stream $\langle c_n \rangle_{n < \omega}$ of actions starting from an initial state `init: Db`. More precisely, `db_mach` can be extended to a stream-processing function

$DB\_mach : Call^{\omega} \times Db \rightarrow Answer^{\omega}$

with an databasestate parameter. DB_mach is specified by the following axioms:

```
DB_mach ( ⊥, db) = ⊥;                                  -- ⊥ is the empty stream
DB_mach (c & calls, db) = v & DB_mach(calls, db')
                              where (v, db') = db_mach(c, db);
```

(For the specification of the stream-operator & we refer to Appendix 2).
It is important that the integrity condition OK on database states is an invariant of the states taken by the database-machine. Hence if DB_mach "starts" in an initial state `init: Db` satisfying the integrity condition `OK(init)`, then all successor states have to satisfy this condition as well. Therefore, every elementary transaction f has to be specified in such a way that the following holds:

$pre_f(d, db) \wedge OK(db) \Rightarrow OK(db')$ **where** $(v, db') = f(d, db)$

(*or* $pre_f(d, db) \wedge OK(db) \Rightarrow OK(f(d, db))$ *iff is a write-transaction*)

3.4 Specification of the elementary transactions of system and environment

In the sequel we now specify elementary transactions for our example-system separately for the different parts of the hospital-organisation: first a transaction to assist the process of blood taking in the ward, then a transaction in the laboratory to write the results of blood analysis into the database, and finally a transaction for the doctors to retrieve the results for their patients.

3.4.1 Specification of the Process of Taking Blood in the Ward

We start by specifying the process of taking blood (at an arbitrary station). The corresponding nodes in Diagram 3.1 are refined to the following data flow diagram.

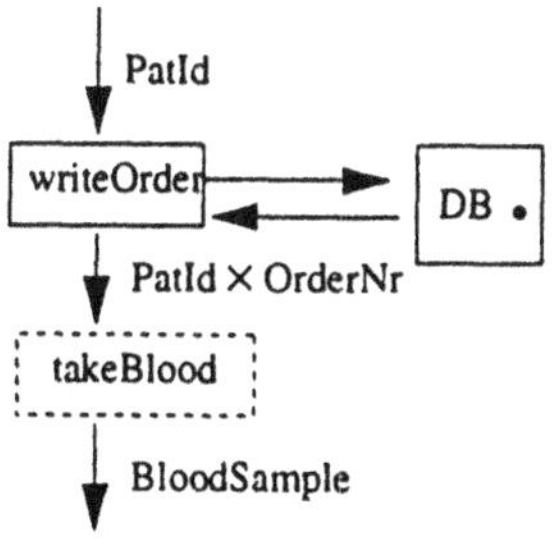

Diagram 3.3: Data Flow for Taking Blood in a Ward

Here writeOrder is an elementary transaction and takeBlood is an action outside the database system. To indicate this difference, we use the dotted lines surrounding takeBlood.
The transaction writeOrder can be precisely specified using the operations on the database given in section 2. For the action takeBlood only rough conditions are given, which ensure that takeBlood cooperates with the system in the right way.
Informally, for a patient identifier `patid` writeOrder creates a new order in the database for the patient with the key `patid` and yields the number (i.e. the key) of this order together with the original patient identifier. This may be conceived as the preparation of an empty blood tube

on which the order number is marked and which also contains an intermediate marker with the patient identifier. Now the action takeBlood consists in filling the blood of the patient with identification `patid` into the prepared tube. After this action, however, the intermediate `patid`- marker must be removed. Hence we specify the type BloodSample and the corresponding function takeBlood roughly as

BLOODSAMPLE = { **enriches** DBSPEC; **strict**; **total**;
    sort BloodSample;
    Nr: BloodSample → OrderNr;
    takeBlood: PatId × OrderNr → BloodSample;

    axioms ∀ pid: PatId; nr: OrderNr; bs: BloodSample **in**
    takeBlood(pid, nr) = bs ⇒ Nr(bs) = nr **endaxioms** }

The specification of writeOrder is split in two parts: First the precondition $\mathtt{pre}_{\mathtt{writeOrder}}$ is specified, which ensures that the state of the database is such, that writeOrder can be executed safely.Then writeOrder is specified on such states of the database on which this precondition holds. Hence writeOrder is loosely specified. The underspecification of writeOrder on db-states which do not satisy the precondition does not influence the behaviour of the db-machine, since, by definition, db_mach yields an error-message in case the precondition does not hold. Since writeOrder creates a new lab-order in the database, we also specify an auxiliary function `newOrderNr` which provides a new key for this order.

```
WARD = { enriches DBSPEC; strict;

    writeOrder: PatId × Db → PatId × OrderNr × Db;
    pre_writeOrder : PatId × Db → Bool;

    newOrderNr: Db → OrderNr;
    newOrderNr total;

    axioms ∀pid: PatId, db: Db, nr: OrderNr in
    pre_writeOrder (pid, db) ⇔
    ( ∃ patient: Patient. patient ∈ db ∧ PatId(patient) = pid);

    newOrderNr(db) = nr ⇒
    ¬∃ order: LabOrder. order ∈ db ∧ Nr(order) = nr;

    pre_writeOrder (pid, db)⇒writeOrder(pid, db) = (pid, nr, db')
    where
      nr = newOrderNr(db) and
      patient = getPatient(pid, db) and
      order = createLabOrder(nr, time(db)) and
      db' =estBelongs_to(putLabOrder(order,db),patient,order);
    endaxioms }
```

Here we assume that each database state contains a date with daytime and that consequently there is a function `time : Db → Date_with_time`.
From the above specification it can be proved that the predicate OK is an invariant for `writeOrder`. More precisely, we have

```
WARD + DBSPEC ⊢
OK(db) ∧ pre_writeOrder (pid,db) ⇒ OK(db')
where (pid',nr,db') = writeOrder(pid,db)
```

The specification of the elementary transaction `writeOrder` can be used to obtain a specification of a stream processing function (an agent) which transforms streams of Patient identifiers into streams of pairs of Ordernumbers and Patient identifiers by interaction with the database. This interaction consists in sending messages to the database machine to perform the transaction writeOrder for the patient identifiers in the input stream and to pass the answers (different from error) of the database to the output stream. This "lifting" of transactions to the stream-level is explained in detail in [Nickl 93] and will be discussed more closely in section 3.5. Now on each ward, there is such an agent interacting with the same database. In the case of 2 wards this situation can be described by the following data flow diagram.

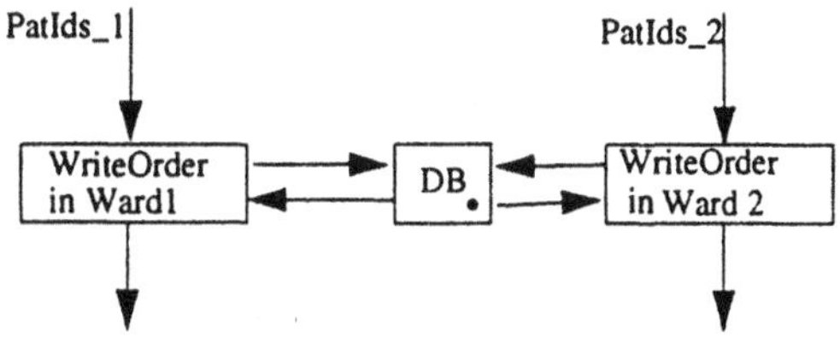

Diagram 3.4 : Ward1 and Ward 2 interacting with the same database

In particular, it is important that orders on Ward1 and Ward2 obtain different numbers. Therefore, the database-machine has to perform `writeOrder` as an atomic action.

3.4.2 Specification of the Analysis of BloodSamples in the Lab

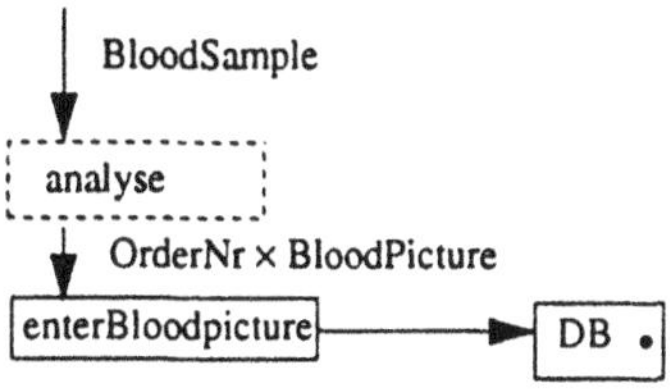

Diagram 3.5 : Data flow in the Lab

Here again `analyse` is a function of the environment which is only roughly specified by

ANALYSE = { **enriches** BLOODSAMPLE ; **strict; total;**
analyse: BloodSample → OrderNr × Bloodpicture;
axioms ∀ bs: BloodSample; bp: BloodPicture; nr: OrderNr **in**
analyse(bs) = (nr, bp) ⇒ nr = Nr(bs) **endaxioms**; }

`enterBloodpicture` is an elementary transaction which is precisely specified over the database model specified in section 3.2.

```
LAB = {enriches DBSPEC;strict;
```

```
enterBloodPicture : OrderNr × BloodPicture × Db → Db;
pre_enterBloodPicture : OrderNr × BloodPicture × Db → Bool;

axioms ∀ db: Db;nr: OrderNr; bp: BloodPicture in
pre_enterBloodPicture(nr, bp, db) ⇔
(∃ order: LabOrder. order ∈ db ∧ Nr(order) = nr ∧
¬∃ lw: LabValue. Bp_analysis db ( order, lw));

pre_enterBloodPicture(nr, bp, db) ⇒ enterBloodPicture( nr, bp, db) = db'
where
   order= getLabOrder( nr, db) and
   lw = createLabValue( nr, bp, time(db) ) and
   db' = estBp_analysis ( put(lw, db), order, lw)
endaxioms; }
```

Again it can be proved that OK is an invariant of `enterBloodPicture`:

DBSPEC + LAB ⊢
OK(db) ∧ $pre_{enterBloodPicture}$(nr,bp,db) ⇒ OK(enterBloodPicture(nr,bp,db))

Notice, that now in the laboratory the values of the analysis are not sent directly to the ward, but written into the database. Therefore, the doctors have to retrieve these results from the database.

3.4.3 Specification of a retrieval-function for the doctor

Here, we specify an elementary read-transaction
`getActual : DocId × PatId × Db → PatId × BloodPicture;`
by which a doctor can retrieve the most actual Bloodpicture for a patient, provided the doctor is responsible for the patient and provided there is a result from the laboratory for that patient in the database. In order to test, whether there is such a result for the patient, we first specify an auxiliary function
`value_for : LabValue × Patient × Db → Bool;`
which yields true on a value `val` and a patient `pat`, if there is some laborder `order` in the database related both to `val` and `pat`.

```
DOC = { enriches DBSPEC; strict;
    getActual : DocId × PatId × Db → PatId × BloodPicture;
    pre_getActual: DocId × PatId × Db → Bool;
    value_for: LabValue × Patient × Db → Bool;

    axioms ∀ db: Db; val: LabValue; pat: Patient; did: DocId;
    pid: PatId; bp: BloodPicture.

    value_for( val, pat, db) ⇔ ∃ order:LabOrder.
    Belongs_to db (pat, order) ∧ Bp_analysis db (order, val);

    pre_getActual(did, pid, db) ⇔
    (∃ doc: Doctor, pat: Pat, val: LabValue.
    doc = getDoctor(did, db) ∧ pat = getPatient(pid, db) ∧
    Responsible_for db(doc, pat) ∧ value_for(val, pat, db));
```

```
pre_getActual(did, pid, db) ⇒
let  (pid', bp) = getActual(did, pid, db) in
  pid = pid' ∧
  (∃ val: LabValue; pat: Patient. pat = getPatient(pid, db)
  ∧ value_for( val, pat, db) ∧ Value(val) = bp
  ∧(∀ val' : LabValue. value_for(val', pat, db) ⇒
  Established(val') before Established(val)) );
endaxioms }
```

Here "before" denotes the linear ordering on Date_with_time.

3.5 Specification of the communication model

The specification of the communication between the wards, the laboratory and the database is obtained by composing the data flow diagrams given in the previous section and interpreting them as a system of streams. To indicate the difference we write the labels of the nodes with capital letters. In the case of two wards, we obtain the following diagram, where, for the moment, we do not yet consider the communication with the doctors.

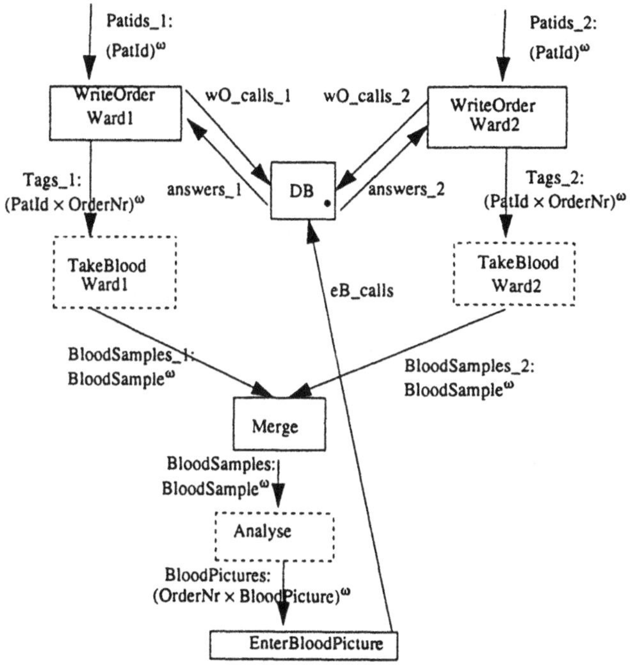

Diagram 3.6 : Specification of the Communication model (without communication with docs)

Here as defined in Appendix 2, for a sort Data, Data$^\omega$ denotes the sort of Streams over Data. Following the approach of M. Broy (see e.g. [FOCUS 92]), the above diagram can be translated systematically in a constraint on the system of those streams by which the edges are marked. With each node of the diagram a constraint on the incoming and outcoming streams is specified. In case of the Analyse-node, for instance, we require that in the outcoming stream `BloodPictures` exactly the values of the function `analyse` applied to all elements in the incoming stream BloodSamples occur. It is not required that the values occur in the same order as the bloodsamples in the incoming stream, since possibly during the analysis overtakings may occur. Hence the following constraint $\mathtt{LIFT}_{\mathtt{analyse}}$(BloodSamples, BloodPictures) states that if BloodSamples and BloodPictures are abstracted to multisets abs(BloodSamples) and abs(BloodPictures), then abs(BloodPictures) is obtained by applying the multiset lifting of the function analyse to the multiset abs(BloodSamples). This can be captured by the following definition (where for a set U and a stream s, #(U©s) denotes the number of occurrences of elements of U in s (see the appendix for a more precise definition) and where == denotes the strict equality.

```
LIFT_analyse (BloodSamples, Bloodpictures) ⇔
∀ nr : OrderNr; bp:BloodPicture.
   #({ (nr,bp) }© BloodPictures) =
   #({bs:BloodSample. analyse(bs)==(nr,bp)} © BloodSamples)
```

In exactly the same way, a constraint $\mathtt{LIFT}_{\mathtt{takeBlood}}$ is derived, where in the above diagram $\mathtt{LIFT}_{\mathtt{takeBlood}}$(Tags_1,Bloodsamples_1) and $\mathtt{LIFT}_{\mathtt{takeBlood}}$(Tags_2, Bloodsamples_2) has to hold.

For the Merge-node, we require

```
MERGE( (Bloodsamples_1, Bloodsamples_2), BloodSamples)
```

with

```
MERGE( (Bloodsamples_1, Bloodsamples_2), BloodSamples) ⇔
∀ bs: BloodSample.
   #({bs} © BloodSamples) =
   #({bs} © BloodSamples_1) + #({bs} © BloodSamples_2)
```

The DB- node consists of the database machine (see section 3.3), extended by a Merge and Diverge-unit, which merges the incoming streams of calls to the database and distributes the answers of the database machine to the nodes, where the calls came from:

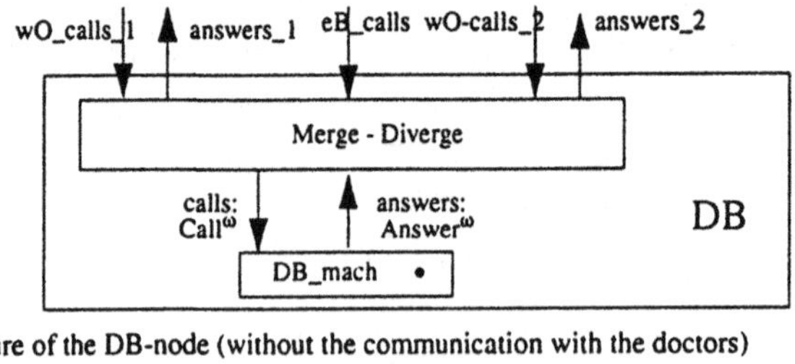

Diagram 3.7: Structure of the DB-node (without the communication with the doctors)

In order to ensure that the distribution works correctly, the calls to the database have to be refined by additional information on the locations of the callers. Such a refinement is considered in [Nickl 93]. The constraint associated with the DB_mach- node is

```
answers = DB_mach(calls,init)
```

where DB_mach is specified as in section 3.3 and init: Db is an initial state of the database.

The constraint $\mathtt{LIFT}_{\mathtt{writeOrder}}$ associated with the WriteOrder-nodes, specifies that the calls in wO_calls_1 (or wO_calls_2 resp.) are obtained by producing calls for all the

patient identifiers in `Patids_1` (`Patids_2` respectively) and with the same multiplicity as these patient identifiers occur in `Patids_i`, i.e. `wO_calls_i` is obtained from `Patids_i` by a multiset-lifting of the function $\texttt{writeOrder}^{\texttt{call}}$. Furthermore it is required that `Tags_1` (and `Tags_2` resp.) contain exactly that values in the stream `Answers_1` (and `Answers_2` resp.) which are different from "`error`".
The constraint $\texttt{LIFT}_{\texttt{enterBloodPicture}}$ specifies that `eB_calls` is obtained from `Bloodpictures` by a multiset-lifting of $\texttt{enterBloodPicture}^{\texttt{call}}$.

Hence by the above described translation, the communication between the different nodes in diagram 3.6 is specified by a constraint on the system of streams in the diagram. This constraint is obtained as the conjunction of the constraints associated with the different nodes. Notice that in this way, the communication of blood samples between the wards and the lab is specified, as well as the interplay between the database system and the environment. Moreover, by the constraints imposed on the dotted nodes, requirements on the behaviour of the environment are imposed on which the system has to rely.

So far, in contrast to the original requirement diagram we started from (diagram 3.1), the values of the blood analysis are only sent to the database, but there are no streams of values sent back to the wards. The communication with the wards now is established via the database by means of the retrieval-function `getActual` for the doctors. Therefore we supplement the above data flow diagram by streams of calls to `getActual` sent from doctors in Ward1 and Ward2 respectively and by respective answer streams, sent back to the doctors. Moreover, since all patients for whom lab-orders are written have to be registered in the database, we assume that there is a further elementary transaction `registerPatient` to insert new patients into the database. We do not specify this transaction here, but asssume that this transaction does not manipulate the relations `Belongs_to` and `Bp_analysis` and that it is specified in such a way that it preserves the OK-predicate. For sake of simplicity, we make the additional assumption that if a patient is registered with a particular station, then during his stay in the hospital he does not change the station.In our diagram, we now also introduce a stream of registerPatient-calls from the registration. Hence the DB-node now becomes

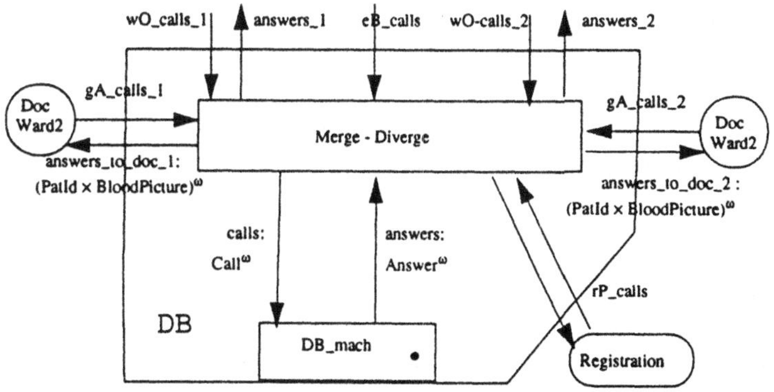

Diagram 3.8 : Revised structure of the DB-node (including Communication with docs and Registration)

Hence, the stream `calls` is required to be a merge of the streams `wO_calls_1`, `wO_calls_2`, `eB_calls`, `gA_calls_1`, `gA_Calls` and `rP_calls` and it is furthermore required that it holds

```
answers= DB_mach(calls,init)
```

where we assume that `init:Db` is an initial state of the Database, in which for every ward at least one doctor is inserted, but no patients, lab-orders and lab-values are inserted. In particular, `OK(init)` holds.

Let

```
C_D(PadIds_1,PatIds_2, wO_calls_1, wO_calls_2,answers_1,answers_2,
Tags_1,Tags2,Bloodsamples_1,Bloodsamples_2,Bloodsamples,Bloodpictures,
eB_calls,gA_calls_1, gA_Calls_2, answers_to_doc_1, answers_to_doc_2,
rP_calls, calls, answers)
```

be the condition on the system of streams obtained by the translation of the above diagram. Let us abbreviate this condition simply by C_D. The verification condition "no confusion" in the context of this diagram is the following (let $i \in \{1,2\}$):

BPS |=

```
C_D ⇒ (∀ bp: BloodPicture; pid: PatId.
    (pid, bp) in answers_to_doc_i ⇒
      (∃bs: BloodSample. ∃nr: OrderNr.
      bs in BloodSamples_i ∧ bs = takeBlood(pid,nr) ∧ analyse(bs) = (nr,bp) )
```

where BPS denotes the union of all specifications given in the paper (i.e. DBSPEC, BLOOD-SAMPLE, WARD, ANALYSE, LAB, DOC, the specification STREAM in Appendix 2 streams (including the specification of the predicate `in`), the specification of the database machine `DB_mach` and the specification of the elementary transaction `registerPatient` and the initial state `init: Db`.

Proof

Let c_n (for $n \geq 1$) denote the n-th element of the stream `calls` and v_n denote the n-th element of the stream `answers`. Let db_0 = `init` and let db_n denote the internal state of the database machine reached after the evaluation of the first n elements of the calls- stream (i.e.

$$db\_mach(c_1, init) = (v_1, db_1) \text{ and } db\_mach(c_{n+1}, db_n) = (v_{n+1}, db_{n+1})).$$

If `(pid, bp)` occurs in the stream `answers_to_doc_i`, then there is an index $n \in N$ such that

$$c_n = Get\_actual^{call}(did, pid)$$

where $Station(getDoctor(did, db_n)) = i$ and such that `(pid, bp)` is the resulting answer of the database machine, i.e.

$$(pid, bp) = v_n.$$

But in this case it follows from the specification of `Get_actual` that the following holds for the database state db_n:

```
∃ val:LabValue,∃ order: LabOrder,∃ pat: Patient.
pat = getPatient(pid, db_n) ∧
Responsible_for db_n (getDoctor(did, db_n), pat) ∧
Belongs_to db_n (pat, order) ∧
Bp_analysis db_n (order, val) ∧ Value(val) = bp;
```

Since $OK(db_n)$ holds, it follows that $Station(getPatient(pid, db_n)) = i$ and that `Nr(order) = Nr(val)`. Let us abbreviate this number by `nr`.

Since only by the transaction enterBloodPicture entities of type LabValue are inserted into the database and since init does not contain Labvalues, there exists some index m<n such that

c_m = enterBloodPicture^{call}(nr, bp)

where the output v_m is different from error.

Now, from the constraint $LIFT_{enterBloodBicture}$(BloodPictures, eB_calls) it follows that (nr, bp) occurs in BloodPictures, and hence by $LIFT_{analyse}$(BloodSamples, BloodPictures) it follows that

$\exists$bs: BloodSample. bs in BloodSamples $\wedge$ analyse(bs) = (nr, bp).

From ANALYSE it follows that Nr(bs) = nr.

Now from MERGE((Bloodsamples_1, Bloodsamples_2), Bloodsamples) together with $LIFT_{takeBlood}$(Tags_1, BloodSamples_1) and $LIFT_{takeBlood}$(Tags_2, BloodSamples_2) we conclude that

($\exists$ pid', nr'. (pid', nr') in Tags_1 $\wedge$ bs = takeblood(pid', nr'))
$\vee$($\exists$ pid', nr'. (pid', nr') in Tags_2 $\wedge$ bs = takeblood(pid', nr'))

By the specification BLOODSAMPLE it follows in both cases that nr' = nr.

Now $LIFT_{writeOrder}$ requires that the elements in Tags_1 (and Tags_2 resp.) are answers of the database-machine to writeOder-calls. Hence there exists some k $\in$ N such that

c_k = WriteOrder^{call}(pid') and v_k = (pid', nr).

From the specification of WriteOrder it follows that in the database state db_{k-1} there is no lab-order with the number nr, and that in the resulting database db_k the lab-order getLabOrder(nr, db_k) is related to the Patient getPatient(patid', db_k). Now since lab-orders are not deleted by the transactions in the system,and since

$\exists$ order: LabOrder. order $\in$ db_n $\wedge$ Nr(order)=nr

it follows that k < n. Since the relationship Belongs_to is not released by the transactions considered in the system, it holds

Belongs_to db_n(getPatient(patid', db_n), getLabOrder(nr, db_n)) $\wedge$
Belongs_to db_n(getPatient(patid, db_n), getLabOrder(nr, db_n)) $\wedge$
OK(db_n)

from which it follows

patid = patid'.

Furthermore, from Station(getPatient(pid, db_n)) = i it follows

takeBlood(pid, nr) in BloodSamples_i

under the above asssumption that patients stay in the station in which they are registered. ◇

4. Concluding remarks

In the preceding sections we have shown how entity-relationship and data flow diagrams can be schematically translated to algebraic/axiomatic specifications providing an integrated view of dynamic and static requirements. Properties that are not expressible by the diagrams can be

formalized and verified by formal techniques. Moreover, structural results from specification theory, process algebra and category theory can help to improve the requirements design (for the entity-relationship model see e.g. [Johnson, Dampney 93]).
Authentication requirements and access rights are another important class of requirements that can be formalized by algebraic/axiomatic specifications. In [Renzel 93] an access control model similar to the approach in [Abadi et al. 91] and [Lampson et al. 92] has been integrated in our requirement process model.
To give an idea we shortly describe (a simple instance of) Renzel's approach. First, an access matrix is defined which determines for each prinicpal p (e.g. a person, a role, a class of terminals) and for each top level function $f : Db \times In \rightarrow Out$ the access rights such that the predicate "p maydo f" is true if p has access to f. Then the application of f is restricted to "secure applications" by introducing a new apply-operator

$$.sapp. : Principal \times (\alpha \rightarrow \beta) \rightarrow \alpha \rightarrow \beta$$

defined by

$$(p \text{ sapp } f)\ x = f(x) \Leftarrow pre_{sapp}(p, f)\ x.$$

The secure application of f to x by the principal p is executable only if p maydo f:

$$pre_{sapp}(p, f)\ x \Leftrightarrow (p \text{ maydo } f) \wedge pre_f(x).$$

In this way access control can be added in a modular way to the requirements specification.

Acknowledgement

This work was highly influenced by stimulating discussions with Manfred Broy and by the work of our colleagues Heinrich Hußmann, Rudi Hettler, Michael Löwe, Stephan Merz, Klaus Renzel and Oskar Slotosch in the HDMS-A- case study group. Thanks of MW to Heidrun Walker for accurate typing during the preparation of the final version of the paper.

Appendix 1

Some remarks on the specification language SPECTRUM

The SPECTRUM language uses the concept of loose semantics. It supports the specification of partial functions. These may also be non-strict. A function f is *strict* iff f yields an undefined result whenever f is applied to at least one undefined argument. A function f is *total* iff f yields a defined result whenever all arguments are defined. If a specification starts with the keyword **strict**, then all functions in this specification are required to be strict. The built-in equality .=. is the strong equality, i.e. it always yields *true* or *false*, considering all "undefined" values of identical sort to be equal (to an undefined value denoted by $\perp$) and all defined values to be different from $\perp$. The quantification operator $\forall$ ranges over all defined values of a sort, whereas $\forall^{\perp}$ also ranges over the undefined value $\perp$.
In the SPECTRUM specifications in this paper we often use the following notational variant: for an expression `e` and definitions `defs`, we write `e` **where** `defs` instead of the SPECTRUM-expression **`let`** `defs` **`in`** `e` **`endlet`**.

Appendix 2

The basic data structure of streams

Let `Data` denote a sort of Data-elements. Then $\texttt{Data}^{\omega}$ denotes the set of streams over `Data`. It is the union of the set of finite and infinite sequences of elements in Data. $\texttt{Data}^{\omega}$ is continuously generated by the non-strict function .&. : $\texttt{Data} \times \texttt{Data}^{\omega} \rightarrow \texttt{Data}^{\omega}$ starting from the undefined (empty) stream $\perp$. The function .&. is strict in its left argument. Hence the specification of $\mathtt{Data}^{\omega}$ is given by

```
STREAM = {
     sort   Dataω;
     .&.:   Data × Dataω → Dataω ;
     ft:    Dataω → Data;
     rt:    Dataω → Dataω;
     .&. strict in 1; ft , rt strict; ft total;
     Dataω freely generated by . &.;

     axioms ∀ d: Data, ∀⊥ s: Dataω in
     ft(d & s) = d; rt(d & s) = s endaxioms }
```

Hence $(.)^{\omega}$ is a sort constructor (where we take some more liberty for the syntax than is allowed in SPECTRUM). The quantification $\forall^{\perp}$ s ranges also over the undefined element of $\mathtt{Data}^{\omega}$, i.e. the empty stream. Due to the phrase **freely generated by** the approximation ordering on $\mathtt{Data}^{\omega}$ is completely determined by the approximation ordering on Data, which for all the non-stream sorts specified so far in our paper is the flat ordering. The sorts endowed with the flat ordering are the elements of the sort class EQ (of sorts with a monotonic strict equality ==). For all sorts Data: EQ the approximation ordering on Data^{ω} is the prefix ordering.

The above specification STREAM can be enriched by the definition of an indexing function

$\cdot_{\cdot} : \mathtt{Data}^{\omega} \times \mathtt{Nat} \rightarrow \mathtt{Data};$ specified by

$s_1 = ft(s) \wedge (n \geq 1 \Rightarrow s_{n+1} = rt(s)_n);$

and by a continuous length function # : $\mathtt{Data}^{\omega} \rightarrow \mathtt{INat}$ specified by

$\#\perp = 0 \wedge \#(d \,\&\, s) = (\#s)+1$

where INat denotes the set of natural numbers with the infinite element ∞, where the approximation ordering is the non-flat ordering $\leq$. Moreover, there is a continuous filter-function .©. : $(\mathtt{Data} \rightarrow \mathtt{Bool}) \times \mathtt{Data}^{\omega} \rightarrow \mathtt{Data}^{\omega}$ specified by

p © ⊥ = ⊥ ∧ p © (d & s) = **if** p d **then** d & (p © s) **else** p © s;

In case, Data is in the sort class EQ (of sorts with a strict equality predicate ==) we write {d} © s for (λx: Data. x == d) © s. The in - predicate on streams (see the specification of no-confusion) now can be specified as

.in. : Data × $\mathtt{Data}^{\omega}$ **to** Bool; .in. **total**;

∀ d: Data; $\forall^{\perp}$ s: $\mathtt{Data}^{\omega}$.(d in s ⇔ #({d}©s) ≠ 0)

Notice, that in is not monotonic (w.r.t. the prefix ordering). It is only used for specification purposes.

References

[Abadi et al. 91] M. Abadi, M. Burrows, B. Lampson, G. Plotkin: A calculus for access control in distributed systems. Techn. Report No. 70, DEC Systems Research Center, 1991.

[Boehm 79] B.W. Boehm: Guidelines for verifying and validating software requirements and design specifications. In: EURO IFIP 79, Amsterdam: North-Holland 1979, 711-719.

[Broy et al. 92] M. Broy, C. Facchi, R. Grosu, R. Hettler, H. Hußmann, D. Nazareth, F. Regensburger, O. Slotosch, K. Stolen: The Requirements and Design Specification Language SPECTRUM- An Informal Introduction, Version 1.0, Tech. Rep., TU München, 1993.

[CIP 85] The CIP Language Group: The Munich Project CIP. Vol 1: The Wide Spectrum Language CIP-L. LNCS 183, Springer 1985.

[Coad, Yourdon 89] P. Coad, E. Yourdon: OOA - Object-Oriented Analysis. Englewood Cliffs, N.J.: Prentice Hall 1989.

[van Diepen, Partsch 90] N.W.P. van Diepen, H.A. Partsch: Formalizing informal requirements: Some aspects. University of Nijmegen, Techn. Report 90-1, 1990.

[Downs et al. 92] E. Downs, P. Clare, I. Coe: Structured Systems Analysis and Design Method - Application and Context. Englewood Cliffs, N.J.: Prentice Hall 1992.

[Dubois et al. 86] E. Dubois, J. Hagelstein. E. Lahou, F. Ponsaert, A. Rifaut, E. Stephens, F. Williams: Model components for requirements engineering. Final Report of the ESPRIT1 "Meteor" Project, Task 1, 1986.

[Focus 92] M. Broy, F. Dederichs, C. Dendorfer, M. Fuchs, T.G. Gritzner, R. Weber: The design of distributed systems - An introduction to FOCUS. TU München, Tech. Rep. TUM I9202, SFB-Bericht Nr. 342.

[Hagelstein 88] J. Hagelstein: Declarative approach to information systems requirements. Knowledge Based Systems 1, 1988, 211-220.

[Hettler 93] R. Hettler: Übersetzung von E/R-Modellen nach SPECTRUM. Internal Report, Korso Project, Institut für Informatik,Technische Universität München, 1993.

[Hußmann et al. 93] H. Hußmann, R. Hettler, S. Merz, F. Nickl, O. Slotosch: Zur formalen Beschreibung der funktionalen Anforderungen an ein Informationssystem. Internal Report Korso Project, TU München & LMU München, 1993.

[Jarke et al. 93] M. Jarke et al.: Requirements Engineering: An integrated view of representation, process and domain. To appear in Proc. ESEC 1993.

[Johnson, Dampney 93] M. Johnson, C.N.G. Dampney: Category theory and information system engineering. In: P. Scollo (ed.): Proc. AMAST 93, Preprint, 1993, 95-103.

[Lampson et al. 92] B. Lampson, M. Abadi, M. Burrows, E. Wobber: Authentication in distributed systems: theory and practice. Tech. Report 83, DEC Systems Res. Center, 1992.

[McMenamin, Palmer 84] S. McMenamin, Y. Palmer: Essential System Analysis. Englewood Cliffs, N.J.: Prentice Hall, 1984.

[Nickl 93] F. Nickl: Ablaufspezifikation durch Datenflußmodellierung und stromverarbeitende Funktionen. Internal Report, Korso Project, Institut für Informatik, LMU München.

[Pepper et al. 93] P. Pepper et al.: Korso: A methodology for the development of correct software. Internal Report, Korso Project, to appear.

[Renzel 93] K. Renzel: Einige Überlegungen zur formalen Beschreibung von Zugriffsrechten und deren Verwaltung. Internal Report, Korso Project, LMU München, 1993.

[Wirsing et al. 92] M. Wirsing, H. Brix, M. Broy, X. Conrad. S. Gastinger, M. Gogolla, F. von Henke, H. Hußmann, M. Jatzeck, B. Krieg-Brückner, J. Lin, P. Pepper, H. Peterreins, A. Poigné, W. Reif, B. Reus, G. Schellhorn, B. Wolff: A framework for software development in Korso. Tech. Report, Korso Project, Universität München, 1992.

[Yourdon 89] E. Yourdon: Modern Structured Anlaysis. Englewood Cliffs, N.J.: Yourdon Press, 1989.

A Two-Phase Approach to Reverse Engineering Using Formal Methods

Gerald C. Gannod and Betty H. C. Cheng

Computer Science Department
Michigan State University
East Lansing, Michigan 48824
USA

Abstract. Reverse engineering of program code is the process of constructing a higher level abstraction of an implementation in order to facilitate the understanding of a system that may be in a "legacy" or "geriatric" state. Changing architectures and improvements in programming methods, including formal methods in software development and object-oriented programming, have prompted a need to reverse engineer and re-engineer program code. This paper presents a two-phase approach to reverse engineering, the results of which can be used to guide the re-implementation of an object-oriented version of the system. The first phase abstracts formal specifications from program code, while the second phase constructs candidate objects from the formal specifications obtained from the first phase.

1 Introduction

Software maintenance has long been a problem facing software professionals, where the average age of software is between 10 to 15 years old [19]. With the development of new architectures and improvements in programming methods and languages, including formal methods in software development and object-oriented programming, there is a strong motivation to reverse engineer and re-engineer existing program code in order to preserve functionality, while exploiting the latest technology.

Reverse engineering of program code is the process of constructing a higher level abstraction of an implementation in order to facilitate the understanding of a system that may be in a "legacy" or "geriatric" state. Re-engineering is the process of examination, understanding, and alteration of a system with the intent of implementing the system in a new form [7]. The benefits offered by re-engineering versus developing software from the original requirements is considered to be a solution for handling legacy code because much of the functionality of the existing software has been achieved over a period of time and must be preserved for many reasons, including providing continuity to current users of the software [2].

One of the most difficult aspects of re-engineering is the recognition of the functionality of existing programs. This step in re-engineering is known as reverse engineering. Identifying design decisions, intended use, and domain specific details are often the main obstacles to successfully re-engineering a system.

The following terms are frequently used in the discussion of re-engineering [7].

Forward Engineering: The process of developing a system by moving from high level abstract specifications to detailed, implementation-specific manifestations.

Reverse Engineering: The process of analyzing a system in order to identify system components, component relationships, and intended behavior. These representations are then used to create higher level abstractions of the system.

Restructuring: The process of creating a logically equivalent system at the same level of abstraction. This process does not require semantic understanding of the system and is best characterized by the act of transforming "spaghetti" (unstructured) code into structured code.

Re-Engineering The examination and alteration of a system to reconstitute it in a new form and the subsequent implementation of the new form.

A diagram depicting the relationships between each of the terms is shownin Figure 1.

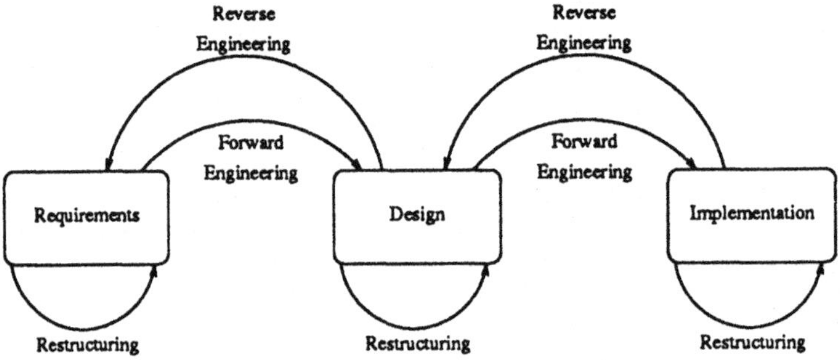

Fig. 1. Relationship between re-engineering terms

Common reverse engineering methods used by software maintenance engineers are observation (i.e., test case analysis) and examination of source code. These techniques are often tedious and error-prone.

Formal methods in software development provide many benefits in the forward engineering aspect of software development [22, 9, 17, 14, 11]. One of the advantages to using formal methods in software development is that the formal notations are precise, verifiable, and facilitate automated processing [3]. For any specification, there may be several implementations that satisfy the specification [10]. There have been recent investigations into reverse engineering that focus on the use of rigorous mathematical methods for extracting formal specifications from existing code [4, 20, 16]. Specifically, Lano and Breuer developed a framework for abstracting Z specifications from programs [16]. This approach involves the translation of imperative programming constructs into a higher level mathematical representation based on category and monad theory. Liu and Wilde have proposed an approach to identifying objects in procedural code [18], where the characterization of candidate objects is based on recognizing common routines, operations, data types, and data items through the examination of global data and major data types. Finally, Haughton and Lano have investigated the identification of objects in procedural code as well as specifications [12]. Recognition is based on translation to an intermediate language, program slicing and heuristic identification of objects.

This paper presents a two-phase approach to reverse engineering that integrates a process for abstracting formal specifications from program code with a technique for identifying candidate objects in program code. One of the difficulties in automating the abstraction of a formal specification from program code is that the specification can often be too tightly bound to the implementation. Ultimately, this coupling necessitates user interaction in order to correctly obtain an accurate high level specification that is free of implementation bias. By taking full advantage of the logical properties of programming constructs, a precise determination of a program's purpose can be represented at a higher level of abstraction, as compared to program code. The constructed formal specification, along with information provided by a domain expert (someone who is knowledgeable about the specific domain, implementation details, and functionality requirements), facilitates determining program correctness using automated reasoning techniques.

Phase 1 of the approach is the process of abstracting formal specifications from program code. In earlier investigations, the AUTOSPEC tool [4] was developed to support the Phase 1 techniques for translating basic programming constructs into equivalent formal representations [4]. In order to illustrate the abstraction process, consider the following sequence of Pascal code.

```
for i := 0 to n
   if a[i] <= a[i+1]
      then
         m := a[i+1]
```

We can reverse engineer this sequence of code to obtain a formal specification that states that variable m is greater than all elements in the array a. Notationally, this specification is expressed as

$$(\forall i : 0 \leq i \leq n : m \geq a[i]).$$

This specification can be used to guide the re-implementation of the original routine in another language [6].

The use of object-oriented analysis, design, and programming has grown significantly in popularity over recent years [8]. Information hiding, abstraction, dynamic binding, and inheritance have made the paradigm appealing. As a result, there is a growing demand for transforming existing software into object-oriented software. The second phase of the approach provides a framework for identifying candidate objects in a system. Information found in formal specifications constructed by Phase 1 provides essential details that are used to identify candidate objects embedded in the implementation of a system. Reverse engineering program code into a representation that identifies candidate objects can provide a view of the subject system that facilitates the forward engineering of the design into an object-oriented implementation.

The remainder of this paper is organized as follows. Section 2 describes the techniques used to abstract formal specifications from program code. The process used to determine candidate objects in formal specifications is discussed in Section 3. Section 4 gives concluding remarks and briefly discusses future investigations.

2 Translation of Source Code into Formal Specifications

Reverse engineering program code into a formal specification facilitates the utilization of the benefits of formal methods for the maintenance and re-implementation of legacy code [4]. This section describes a translational approach that uses the logical properties of programs to abstract formal specifications from program code.

2.1 Review of Formal Methods

A *precondition* describes the initial state of a program, and a *postcondition* describes the final state. For a given postcondition R and a statement S, the *weakest precondition wp(S,R)* describes the set of states in which the statement S can begin execution and terminate with R *true* [10].

2.2 Abstraction of Formal Specifications from Programs

This section describes the techniques used to abstract formal specifications of programming constructs (i.e., assignments, alternatives, iteratives, and procedures).

Assignments. Given an assignment statement of the form `x:= e;` and a precondition U, where U is a logical expression, the following annotated code is constructed

```
{U}                /* precondition */
x := e;
{x = e ∧ U}        /* postcondition */
```

The *wp* of an assignment statement is expressed as $wp(\texttt{x:=expr}, R) = R^x_{expr}$, which represents the postcondition R with every occurrence of x replaced by the expression *expr*. This type of replacement is termed a textual substitution of x by *expr* in expression R. If x corresponds to a vector $\overline{y}$ of variables and *expr* represents a vector $\overline{E}$ of expressions, then the *wp* of the assignment is of the form $R^{\overline{y}}_{\overline{E}}$, where each y_i is replaced by E_i, respectively, in expression R. In the above example, the *wp* of the assignment is U, which is satisfied by the original precondition.

Alternatives. Given an alternative statement of the form

$$
\begin{array}{l}
\textit{if} \\
\quad B_1 \rightarrow S_1; \\
\quad B_2 \rightarrow S_2; \\
\quad \ldots \\
\quad B_n \rightarrow S_n; \\
\textit{fi};
\end{array}
$$

and a precondition U, where $B_i \rightarrow S_i$ is a guarded command, and statement S_i is executed only if the boolean expression (guard) B_i is *true*, then a specification is constructed of the form

$$((B_1 \wedge post(S_1)) \vee \ldots \vee (B_n \wedge post(S_n))) \wedge U,$$

where *post*(S_i) is the postcondition of statement S_i. For each guarded command, one disjunct is abstracted, where the guard B_i is directly included as part of the disjunct. The abstraction algorithms are applied to statement list S_i in order to obtain the remaining logical expressions for the disjunct, $B_i \wedge post(S_i)$.

The *wp* of the alternative statement is: at least one guard B_i must be *true* and every guard must logically imply the *wp* of its corresponding statement list S_i with respect to the postcondition R [10]. Symbolically, the *wp* is expressed as

$$(\exists i :: B_i) \wedge (\forall i :: B_i \rightarrow wp(S_i, R)),$$

where '::' indicates that the range of the quantified variable i is not used in the current context.

Iteratives. An iterative statement takes the form

$$\begin{array}{ll} do & \\ & B_1 \rightarrow S_1; \\ & B_2 \rightarrow S_2; \\ & \ldots \\ & B_n \rightarrow S_n; \\ od; & \end{array}$$

Gries defines a number of guidelines for developing loops through the identification of loop invariants [10]. The methods of *deleting a conjunct*, *replacing a constant by a variable*, *enlarging the range of a variable*, and *adding a disjunct* can provide insight into the automated construction of a specification from program code. For instance, a loop written using the method of *replacing a constant by a variable* has properties that allow for the identification of the upper (lower) bound of an incremented (decremented) variable. Furthermore, determining the statements that ensure progress towards termination is facilitated by the properties associated with this class of loops. Figure 2 gives the steps for constructing a specification for a loop that was developed using the replace a constant by a variable strategy for the loop invariant.

In the cases where no automated heuristic can be applied to a loop or the constructed specification is incorrect, the domain expert is prompted for the proper specification of the statement. The following items are then identified in order to confirm that the specification of the loop is complete:

- *invariant (P):* an expression describing the conditions prior to entry and upon exit of the iterative structure.
- *guards (B):* Boolean expressions that restrict the entry into the loop. Execution of each guarded command, $B_i \rightarrow S_i$ terminates with P *true*, so that P is an invariant of the loop.

  $$\{P \wedge B_i\}S_i\{P\}, \text{ for } 1 \leq i \leq n$$

  When none of the guards is *true* and the invariant is *true*, then the postcondition of the loop should be satisfied ($P \wedge \neg BB \rightarrow R$, where $BB = B_1 \vee \ldots \vee B_n$ and R is the postcondition).

- *bound function (t):* an integer expression representing the bound on the number of iterations. If at least one of the guards is true and the invariant is true, then the number of iterations is bounded below by t ($P \wedge BB \rightarrow (t > 0)$).
- *statements that make progress towards termination (S_i):* these statements must decrease the bound function after each iteration. Each loop iteration is guaranteed to decrease the bound function. Formally, this condition is:

$$\{P \wedge B_i\} t_1 := t; S_i \{t < t_1\}, \text{ for } 1 \leq i \leq n,$$

where $P \wedge B_i$ indicates that the invariant P and guard B_i are *true*, the assignment to t_1 represents the statement making progress towards termination, and S_i represents statements necessary to ensure that the invariant P is still *true* after one iteration.

---

1. The abstraction algorithm begins with the template for a quantified expression of the form
$$(Qi : range(i) : expression(i)),$$
where Q represents one of the quantifier symbols $\forall, \exists, \Sigma$.
2. The quantified variable(s) are determined by examining the identifiers occurring in guards B_j.
3. The ranges of the quantified variables are determined by finding statements occurring prior to entry into the loop that assign values to incremented (decremented) variables and their occurrences in the guards.
4. For each guarded command, the corresponding statement list includes statements that ensure progress towards termination; the postcondition for the remaining statements constitute *expression*(i).
5. The bound function becomes the difference between the upper (lower) bound for a variable that is being incremented (decremented) and its value during loop iterations.

Fig. 2. Steps for abstracting the effect of iterative statements

---

Procedure Calls. A procedure *declaration* has the form

$$\textbf{proc}\ p\ (\ \textbf{value}\ \overline{x};\ \textbf{value-result}\ \overline{y};\ \textbf{result}\ \overline{z}\ );$$
$$\{P\}\langle\ \text{body}\ \rangle\{Q\}$$

where $\overline{x}$ represents all the **value** parameters, $\overline{y}$ represents all the **value-result** parameters, and $\overline{z}$ represents all the **result** parameters for the procedure. $\langle$ body $\rangle$ is one or more statements making up the "procedure", while $\{P\}$ and $\{Q\}$ are the precondition and postcondition, respectively. The syntactic *signature* of a procedure appears as

$$\textbf{proc}\ \text{p} : (\text{input\_type\_name})^* \rightarrow (\textit{output\_type\_name})^*$$

where the Kleene star (*) indicates zero or more repetitions of the preceding unit, ***input_type_name*** denotes the name of an input parameter to the procedure p, and ***output_type_name*** denotes the name of an output parameter to procedure p. A specification of a procedure can be constructed of the form

{ **pre:** U }
{ **post:** *post(body)* $\wedge$ U }
proc ⟨ identifier ⟩ : $E_0 \rightarrow E_1$

where E_0 is one or more input parameter types with attribute **value** or **value-result**, and E_1 is one or more output parameter types with attribute **value-result** or **result**. The postcondition for the body of the procedure, *post(body)*, is constructed using the previously defined guidelines for assignments, alternatives, and iteratives as applied to the statements of the procedure body.

Gries defines a theorem for specifying the effects of a procedure call [10]. Given a procedure declaration of the above form, the following condition holds

$$\{PR : P^{\overline{x},\overline{y}}_{\overline{a},\overline{b}} \wedge (\forall \overline{u}, \overline{v} : Q^{\overline{y},\overline{z}}_{\overline{u},\overline{v}} \Rightarrow R^{\overline{b},\overline{c}}_{\overline{u},\overline{v}}\}\ p(\overline{a},\overline{b},\overline{c})\ \{R\}$$

for a procedure call $p(\overline{a},\overline{b},\overline{c})$, where $\overline{a}$, $\overline{b}$, and $\overline{c}$ represent the actual parameters of type **value**, **value-result**, and **result**, respectively. PR must hold before the execution of procedure p in order to satisfy R. PR states that the precondition to procedure p must hold for the parameters passed to the procedure and that the postcondition for procedure p implies R for each **value-result** and **result** parameter. Using this theorem for the procedure call, an abstraction of the effects of a procedure call can be derived using a specification of the procedure declaration.

2.3 Specification Language

Predicate logic is the target language for the abstraction algorithms described in this section because its syntax and semantics are well-defined [15]. In addition, it provides a good basis for migration to some other specification language, if desired. Figure 3 gives the format of the specification of procedures where, again, the Kleene star (*) denotes zero or more repetitions of the previous unit, **in** identifies the formal input parameters, **out** identifies the formal output parameters, **local** identifies the local variables of the procedure, {**pre**:*expression*} is the precondition of the procedure, and {**post**:*expression*} is the postcondition of the procedure.

2.4 Example

Consider the following code sequence written in the Pascal language [1]:

proc function_name : (type_name)* → *type_name**
in((variable: type_name)*)
out((variable: type_name)*)
local ((variable: type_name)*)
{ **pre:** *expression* }
{ **post:** *expression* }

Fig. 3. Function specification format

```
St = record
        t : integer;
        e : array[1..maxlength] of elementtype
     end;

...

function mts ( S : St ) : boolean;
   begin
      if S.t > maxlength then
         mts := true
      else
         mts := false
   end;
```

The alternative statement contained within the `begin-end` block can be translated into the following logical expression

$$(((S.t > maxlength) \wedge (mts = true)) \vee (\neg(S.t > maxlength) \wedge (mts = false))) \wedge U,$$

where U represents the precondition for function `mts`. A formal specification for the `mts` function is given in Figure 5, where $domain(S)$ is a predicate that describes the set of all states in which S is defined. The specification of `mts` identifies a single input and a single output based on the function header. The *precondition* asserts that the **value** and **value-result** parameters of the procedure are well-defined with respect to their domain [10]. The *postcondition* is constructed using the abstraction algorithms, where U represents the conditions prior to the current segment of code being processed. The abstracted formal specifications for each procedure contained in Figure 4 were derived using the previously defined guidelines and can be found in Appendix A.

3 Identification of Classes Using Formal Specifications

The formal specifications constructed in Phase 1 have properties that are amenable to the semi-automated identification of objects. An object is a self-contained module

```
program QeS(output);

const
  maxlength = 50;

type
  St  = record
          t : integer;
          e : array[1..maxlength] of elementtype
        end;
  Qu = record
          e : array[1..maxlength] of elementtype;
          f, r : integer
        end;
  elementtype : Qu;

procedure mns ( var S : St );
  begin
    S.t := maxlength + 1;
  end;

function mts ( S : St ) : boolean;
  begin
    if S.t > maxlength then
      mts := true
    else
      mts := false
  end;

function tp ( var S : St ) : elementtype;
  begin
    if mts(S) then
      writeln('error');
    else
      tp := S.e[S.top]
  end;

procedure po ( var S : St );
  begin
    if mts(S) then
      writeln('error');
    else
      S.t := S.t + 1;
  end;

procedure pu ( x : elementtype; var S : St );
  begin
    if S.t = 1 then
      writeln('error');
    else
      begin
        S.t := S.t - 1;
        S.e[S.t] := x
      end;
  end;

procedure mnq ( var Q : Qu );
  begin
    Q.f := maxlength;
    Q.r := maxlength;
  end;

function mtq ( var Q : Qu ) : boolean;
  begin
    if Q.r = Q.f then
      mtq := true
    else
      mtq := false
  end;

function fr ( var Q : Qu ) : elementtype;
  begin
    if mtq(Q) then
      writeln('error')
    else
      fr := Q.e[Q.f];
  end;

procedure en ( x : elementtype; var Q : Qu );
  begin
    if Q.r = maxlength then
      Q.r := 1
    else
      Q.r := Q.r + 1;
    if Q.r = Q.f then
      writeln('error')
    else
      Q.e[Q.r] := x;
  end;

procedure de ( var Q : Qu );
  begin
    if mtq(Q) then
      writeln('error')
    else
      if Q.f = maxlength then
        Q.f := 1
      else
        Q.f := Q.f + 1;
  end;

var
  ex_s : St;
  ex_q : Qu;

begin

(* QeS Body * )

end
```

Fig. 4. Example Pascal Code

proc mts : St $\rightarrow$ **boolean**
in(S : St)
out(mts : boolean)
{ **pre:** $domain(S)$ }
{ **post:** $(((S.t > maxlength) \wedge (mts = true)) \vee$
$(\neg(S.t > maxlength) \wedge (mts = false))) \wedge domain(S)$ }

Fig. 5. Formal Specification of mts

that includes both the data and procedures that operate on that data. An object can be considered to be an abstract data type (ADT). A class is a collection of objects that have common use [21].

3.1 Guidelines

Using the above definition of an object, a set of guidelines for identifying objects is as follows:

1. Construct a list of all data structures contained within the system. These data structures should not include primitive types.
2. For each data structure contained in the list of system data structures, group together the operations that refer to the data structure as input in the syntactic signature specification of the procedure with the data structure.
3. In the case of conflicts, (i.e., a procedure contains two non-primitive data structures as input in the signature) one of three actions can be taken
   (a) Determine if one data structure is composed of one or more occurrences of the other data structure. This step is performed by checking the definitions of data structures.
   (b) Determine whether the output of the procedure excludes either data structure. If it can be shown that the procedure does not modify a data structure then associate the procedure with the data structure that is modified.
   (c) In the cases where no determination can be made as to how to associate a procedure with a respective data structure, query the domain expert on the appropriate association.

The process of identifying candidate objects is facilitated by the format of the formal specifications for procedures. It is emphasized that the objects identified by this technique are only *candidate* objects. However, once the object definition has been constructed, the formal nature of the specification facilitates formal reasoning about the objects and can aid in the verification of candidate objects as true objects.

3.2 Specification Language

An object specification is constructed by collecting the procedure specifications associated to a data structure and declaring the data structure to be a candidate object. The BNF grammar for the language used to specify potential object classes is given in Figure 6. The definition uses the roman font to describe non-terminals; bold type defines keywords; the Kleene star (*) is used to denote one or more repetitions of the preceding unit; square brackets ([ ]) indicate optional items; parentheses ('()') indicate groupings. The non-terminal, *expression*, represents a predicate logic expression.

```
component = type type_name: has ( method )*
has = has ( data_description )*
data_description =
          variable : type_name
method = method method_name: (type_name)* → (type_name)*
                    in((variable: type_name)*)
                    local((variable: type_name)*)
                    out((variable: type_name)*)
                    { pre: expression }
                    { post: expression }
expression = true
           | false
           | ( expression )
           | ¬ expression
           | expression ∧ expression
           | expression ∨ expression
           | expression ⇒ expression
           | expression ⟺ expression
           | ( ∀ variable : type :: expression )
           | ( ∃ variable : type :: expression )
           | predicate_name [( term (, term)*)]
                  def
           | term  =  expression

term = variable
           | function_name [( term (, term)* )]
```

Fig. 6. Grammar for object specifications

3.3 Example Identification of Candidate Objects

The first step of the analysis is to identify the non-primitive data structures of the system. This process is performed by examining the signatures of the procedure specifications and extracting the unique non-primitive data structure names. For example, analysis of the formal specifications in Appendix A for the program **QeS** (shown in Figure 4) identifies non-primitive data structures named **St**, **Qu**, and **elementtype**. In continuing the analysis of the data structure **St**, the procedures **mns**, **mts**, **tp**, and po are grouped with **St** through examination of the specification signatures. Procedure pu is initially grouped with **St** and **elementtype**, but further analysis leads to a strict association of pu to **St** since the input **elementtype** is not modified by pu. The subsequent object definition for **St** appears in Figure 7.

---

type St:
has
  t : integer;
  e : array of elementtype;

method mns : St → St
  in(S : St)
  out(S : St)
  { **pre:** *true* }
  { **post:** $S.t = maxlength + 1 \wedge true$ }
method mts : St → boolean
  in(S : St)
  out(mts : boolean)
  { **pre:** *domain(S)* }
  { **post:**$(((S.t > maxlength) \wedge (mts = true)) \vee$
    $(\neg(S.t > maxlength) \wedge (mts = false))) \wedge domain(S)$ }
method tp : St → St × elementtype
  in(S : St)
  out(S : St, tp : elementtype)
  { **pre:** *domain(S)* }
  { **post:** $(((S.t > maxlength) \wedge (mts = true)) \wedge post(writeln('error'))) \vee$
    $((\neg(S.t > maxlength) \wedge (mts = false)) \wedge (tp = S.e[S.t])) \wedge domain(S)$ }
method po : St → St
  in(S : St)
  out(S : St)
  { **pre:** *domain(S)* }
  { **post:** $(((S.t > maxlength) \wedge (mts = true)) \wedge post(writeln('error'))) \vee$
    $((\neg(S.t > maxlength) \wedge (mts = false)) \wedge (S.t_1 = S.t_0 + 1)) \wedge domain(S)$ }
method pu : St × elementtype → St
  in(S : St, x : elementtype)
  out(S : St)
  { **pre:** $domain(S) \wedge domain(x)$ }
  { **post:** $(S.t = 1 \wedge post(writeln('error'))) \vee$
    $(\neg(S.t = 1) \wedge (S.t_1 = S.t_0 \wedge S.e[S.t_1 = x)) \wedge domain(S)$ }

Fig. 7. Specification of Object *St*

---

4 Conclusions and Future Investigations

A requirement to achieving the same functionality of existing software on new hardware platforms and in new programming paradigms has prompted a need for reverse engineering techniques that take advantage of rigorous mathematical methods. Automating the process for abstracting formal specifications from program code is sought but, unfortunately, not completely realizable as of yet. Providing maintenance engineers with as much information as possible, whether it is formal or informal, can aid in the successful construction of specifications that accurately describe legacy systems and thus, facilitate its understanding and re-implementation.

As mentioned earlier, our previous investigations led to the development of AutoSpec, a system that abstracts formal specifications from program code using a translational approach to recognizing the effect of basic programming constructs [4]. Future investigations include extending AUTOSPEC to fully support the two phase approach described in this paper. Additional investigations will address the processing and refinement of candidate objects in order to determine true object definitions. In addition, our investigations into the construction of software component libraries [5, 13] will be used to assist in the identification of object instantiations and class hierarchies. Finally, graphical support for Phase 2 will be used to represent message passing between objects identified in systems.

A Specifications of Example

proc mns : St → St
in(S : St)
out(S : St)
{ **pre:** $true$ }
{ **post:** $S.t = maxlength + 1 \wedge true$ }

proc tp : St → St × elementtype
in(S : St)
out(S : St, tp : elementtype)
{ **pre:** $domain(S)$ }
{ **post:** $(((S.t > maxlength) \wedge (mts = true)) \wedge$
$post(writeln('error'))) \vee$
$((\neg(S.t > maxlength) \wedge (mts = false)) \wedge$
$(tp = S.e[S.t])) \wedge domain(S)$ }

proc po : St → St
in(S : St)
out(S : St)
{ **pre:** $domain(S)$ }
{ **post:** $(((S.t > maxlength) \wedge (mts = true)) \wedge$
$post(writeln('error'))) \vee$
$((\neg(S.t > maxlength) \wedge (mts = false)) \wedge$
$(S.t_1 = S.t_0 + 1)) \wedge$
$domain(S)$ }

proc pu : St × elementtype → St
in(S : St, x : elementtype)
out(S : St)
{ **pre:** $domain(S) \wedge domain(x)$ }
{ **post:** $(S.t = 1 \wedge post(writeln('error'))) \vee$
$(\neg(S.t = 1) \wedge (S.t_1 = S.t_0 \wedge S.e[S.t_1 = x))$
$\wedge domain(S) \wedge domain(x)$ }

proc mnq : Qu → Qu
in(Q : Qu)
out(Q : Qu)
{ **pre:** $true$ }
{ **post:** $Q.f = maxlength \wedge$
$Q.r = maxlength \wedge true$ }

proc mtq : Qu → Qu × boolean
in(Q : Qu)
out(Q : Qu, mtq : boolean)
{ **pre:** $domain(Q)$ }
{ **post:** $(Q.r = Q.f \wedge mtq = true) \vee$
$(\neg(Q.r = Q.f) \wedge mtq = false)$
$\wedge domain(Q)$ }

proc fr : Qu → Qu × elementtype
in(Q : Qu)
out(Q : Qu , fr : elementtype)
{ **pre:** $domain(Q)$ }
{ **post:** $((Q.r = Q.f \wedge mtq = true) \wedge$
$(post(writeln('error')))) \vee$
$((Q.r = Q.f \wedge mtq = true) \wedge$
$(fr = Q.e[Q.f])) \wedge domain(Q)$ }

proc en : elementtype × Qu → Qu
in(x : elementtype , Q : Qu)
out(Q : Qu)
{ **pre:** $domain(Q) \wedge domain(x)$ }
{ **post:** $((Q.r_0 = maxlength \wedge Q.r_1 = 1) \vee$
$(\neg(Q.r_0 = maxlength) \wedge$
$(Q.r_1 = Q.r_0 + 1))) \wedge$
$(((Q.r = Q.f) \wedge post(writeln('error'))) \vee$
$(\neg(Q.r = Q.f) \wedge (Q.e[Q.r] = x)))$
$\wedge domain(Q)$ }

proc de : Qu → Qu
in(Q : Qu)
out(Q : Qu)
{ **pre:** $domain(Q)$ }
{ **post:** $((Q.r = Q.f \wedge mtq = true) \wedge$
$(post(writeln('error')))) \vee$
$((\neg(Q.r = Q.f) \wedge mtq = false) \wedge$
$(((Q.f = maxlength) \wedge Q.f = 1) \vee$
$(\neg(Q.f_0 = maxlength) \wedge$
$(Q.f_1 = Q.f_0 + 1)))) \wedge domain(Q)$ }

References

1. Alfred V. Aho, John E. Hopcroft, and Jeffrey D. Ullman. *Data Structures and Algorithms.* Addision-Wesley, 1983.
2. Eric J. Byrne and David A. Gustafson. A Software Re-engineering Process Model. In *COMPSAC.* ACM, 1992.
3. Betty H.C. Cheng. Synthesis of Procedural Abstractions from Formal Specifications. In *Proceedings of COMPSAC'91: Computer Software and Applications Conference,* September 1991.
4. Betty H.C. Cheng and Gerald C. Gannod. Abstraction of Formal Specifications from Program Code. In *Proceedings for the IEEE 3rd International Conference on Tools for Artificial Intelligence.* IEEE, 1991.
5. Betty H.C. Cheng and Jun jang Jeng. Formal methods applied to reuse. In *Proceedings of the Fifth Workshop in Software Reuse,* 1992.
6. Betty Hsiao-Chih Cheng. *Synthesis of Procedural and Data Abstractions.* PhD thesis, University of Illinois at Urbana-Champaign, 1304 West Springfield, Urbana, Illinois 61801, August 1990. Tech Report UIUCDCS-R-90-1631.
7. Elliot J. Chikofsky and James H. Cross II. Reverse Engineering and Design Recovery: A Taxonomy. *IEEE Software,* January 1990.
8. Robert G. Fichman and Chris F. Kemerer. Object-Oriented and Conventional Analysis and Design Methodologies : Comparison and Critique. *IEEE Computer,* October 1992.
9. Susan L. Gerhart. Applications of formal methods: Developing virtuoso software. *IEEE Software,* pages 7–10, September 1990.
10. David Gries. *The Science of Programming.* Springer-Verlag, 1981.
11. Anthony Hall. Seven myths of formal methods. *IEEE Software,* pages 11–19, September 1990.
12. H.P. Haughton and K. Lano. Objects Revisited. In *Conference on Software Maintenance.* IEEE, 1991.
13. Jun jang Jeng and Betty H.C. Cheng. Using Automated Reasoning to Determine Software Reuse. *International Journal of Software Engineering and Knowledge Engineering,* 2(4):523–546, December 1992.
14. Richard A. Kemmerer. Integrating Formal Methods into the Development Process. *IEEE Software,* pages 37–50, September 1990.
15. Robert Kowalski. Predicate logic as a programming language. *Information Processing '74, Proceedings of the IFIP Congress,* pages 569–574, 1974.
16. K. Lano and P.T. Breuer. From Programs to Z Specifications. In *Z User Workshop.* Springer-Verlag, 1989.
17. Nancy G. Leveson. Formal Methods in Software Engineering. *IEEE Transactions on Software Engineering,* 16(9):929–930, September 1990.
18. Sying-Syang Liu and Norman Wilde. Identifying Objects in a Conventional Procedural Language: An Example of Data Design Recovery. In *Conference on Software Maintenance.* IEEE, 1990.
19. Wilma M. Osborne and Elliot J. Chikofsky. Fitting pieces to the maintenance puzzle. *IEEE Software,* January 1990.
20. M. Ward, F.W. Calliss, and M. Munro. The maintainer's assistant. In *Proceedings Conference on Software Maintenance,* pages 307–315, Miami, Florida, October 1989. IEEE.
21. A. Winblad, S. Edwards, and D. King. *Object-Oriented Software.* Addison-Wesley, 1990.
22. Jeannette M. Wing. A Specifier's Introduction to Formal Methods. *IEEE Computer,* September 1990.

Algebraically Provable Specification of Optimized Compilations

Vladimir Levin

Keldysh Institute of Applied Mathematics
Miusskaya sq., 4, 125047 Moscow
alex@alex.srcc.msu.su (V.A. Levin)

Abstract. Hoare's theory of compilation based on refinement algebra of programs is generalized to provide verifiable specification of optimized compilers that produce effective object codes. The compilation relation is modified by including a register table into data representation. The definition of this (modified) compilation relation is given in the framework of the justification scheme for data refinement. A compiler is specified by a set of theorems. They describe target object code for each construct of a source language and state that this partial compiling is correct, i.e. fits to the compilation relation. The proofs of compiler specification theorems are developed in algebraic style of the original theory with additional lemmas that justify alteration of a register table in the process of compilation.

1 Introduction

The key point of the theory [1] is the (basic) compilation relation $CP(p,t,c)$. It means that a source program p can be in principle compiled into an object code c, provided that a symbol table t fixes the data representation; p is in a high level language H, c is in a machine language ML, t maps the global variables of p to the data storage of the target machine.

The refinement algebra of H introduces a partial order <= called improvement. It relates programs: $p<=q$, if q is more predictable than or as good as p, and serves the same purpose; formally

$$p<=q \textit{ iff } p=(p\,|\,q)$$

where | is a symbol of non-deterministic choice [3].

The compilation relation is defined semantically by improvement. As the last holds in H only, the meaning of object code c should be expressed by interpretation in H.

There is the only requirement for p and c: c must be as good as or better than p w.r.t. improvement; so,

$$DEF: CP(p,t,c) = t';F';p <= Ic;t';F'$$

where *t'* implements *t*, *Ic* interprets *c*, *F'* forgets special variables that simulate components of the target machine; *t'*, *Ic* and *F'* are programs in *H*. Forgetting is needed to compare *p* and *Ic* by <=.

The specification method developed in [1,2] uses the predicate *CP(p,t,c)* to specify a particular compilation that translates constructs of a source language to object code compositions. The proof strategy for such a compiler specification is clear: one should prove that the particular compiling meets the definition of *CP(p,t,c)*.

Since users use and designers design a compiler as an accomplished tool, a complete compiling specification that specifies final object code is required.

When optimization is on, final object code produced by a compiler has rather context-dependent structure. The basic compilation relation concentrates context information into a symbol table *t*. But this is a poor structure, and it gives nothing for optimization of object code. Correspondence between fragments of final (optimized) object code and fragments of the source program cannot be reduced to predicate *CP(p,t,c)* in many important cases which occur in practice [5,6].

A more advanced compilation relation should be added to Hoare's algebraic theory of compilation. Optimized compilation of a source program *p* into an object code *c* involves context-sensitive data representation *(t,r)*. It includes the basic component *t* and the alterable component *r*; *r* is a register table that maps a subset of the global variables of *p* to registers of the target machine. The contents of *r* may be different at the starting and finishing points of *c*.

The modified compilation relation is abbreviated as a predicate *CP(p,t,r1,c,r2)*

A similar compilation relation with a register table is introduced for the same goal in the work [4] which was being done in parallel with the present work. The work [4] defines a register table as a mapping from registers to variables (or expressions, in the general case). This simplifies a formal model, but ignores some optimization technique for handling with equal values of different variables. Another (and more important) peculiarity of our approach is the justification scheme for data refinement. It is introduced below as a framework for formal definition of the modified compilation relation.

The definition of *CP(p,t,c)* is treated in [1] as an instance of the refinement formula

$$d';p <= q;d'$$

where *d'* implements the mapping *d* which translates from data space of *q* into that of *p*; *d* is called data refinement.

The modified compilation relation *CP(p,t,r1,c,r2)* involves more sophisticated data refinement, a pair of mappings *(t,r)*. It contains an alterable mapping *r*, which is transformed in the process of execution of program c. The complete interpretation of data refinement (t,r) should also involve some transfer between t and r. So, $CP(p,t,r1,c,r2)$ is not reduced to the simple refinement formula given above, unlike basic compilation $CP(p,t,c)$. Nevertheless, general refinement concept is a good base for algebraic approach in computing [7]. Therefore, we trend to expand it on the modified compilation relation too.

We consider here the general situation: alterable data refinement d is transformed in the process of execution of a program q from input refinement $d1$ to output refinement $d2$ in accordance with some relation J. We treat J as justification of that refinement transformation and extend a refinement formula to the next one:

$$d1';p <= q;d2' \ \& \ J(p,d1,q,d2)$$

This gives the following justification scheme for formal definition of the modified compilation relation:

$$DEF{:}\ CP(p,t,r1,c,r2) = \ t';r1';F';p <= Ic;t';r2';F' \ \& \ J(p,t,r1,c,r2)$$

In order to get the exact definition from this scheme, we determine below both a justification relation J and a transfer of values between basic and alterable components of data refinement (t,r).

2 Interfaces

We use the languages H and ML from [1].

Some details are reminded below. An assertion is defined as follows:

$$\{e\} = \text{IF } e \text{ THEN SKIP ELSE ABORT}$$

The command ABORT is defined as the least predictable program, the bottom of the improvement relation. The commands VAR and END determine scopes of variables. We consider only well-defined expressions that always yield a value.

The basic algebraic laws for H are introduced in [1,2,3]. Some of them are:

$\text{ABORT} <= p$

$p;\text{SKIP} = \text{SKIP};p = p$

$\text{IF } b \text{ THEN } p \text{ ELSE } q;\ w = \text{IF } b \text{ THEN}(p;w) \text{ ELSE}(q;w)$

If x does not appear in e, then

$x{:}{=}e = x{:}{=}e;\{x{=}e\}$

$\{x{=}e\} = \{x{=}e\};x{:}{=}e$

$\text{END } v;\ \text{VAR } v <= \text{SKIP} = \text{VAR } v;\ \text{END } v$

$v{:}{=}e;\text{END } v = \text{END } v$

$\{e\} <= \text{SKIP}$

The loop WHILE b DO p is defined as the least fixed point of the next equation

X = IF b THEN$(p;X)$ ELSE SKIP

ML uses an architecture with

m - the store for machine instructions,

M - the main store for data,

P - the current instruction pointer,

A - the general-purpose register.

We will consider four instructions, which are defined as follows:

$effect($LOAD $n) = (A{:=}M[n];\ P{:=}P+1)$

$effect($STORE $n) = (M[n]{:=}A;\ P{:=}P+1)$

$effect($JUMP $j) = P{:=}j$

$effect($COND $j)$ = IF A THEN $P{:=}P+1$ ELSE $P{:=}j$

Object code c is the machine instruction sequence held in m between locations s and f, except location f. Formally, c is represented by the triple (m,s,f).

Interpretation of c in H is defined as follows:

$Ic = P{:=}s;$ WHILE $P<f$ DO $effect(m[P]);\ \{P{=}f\}$

3 Formal Model

Below we consider the simplest transfer between basic and alterable components of data representation (t,r). The table t is restored each time before execution of the next statement: the value of variable x loaded on register A is stored into location $M[tx]$, even if it is required immediately afterthen. This transfer allows the simplest optimization: if the value of x is required in the following evaluation, it may be taken from A, until A keeps this value.

As ML has the only general register A, we can simplify the formalism. Below a register table r is treated as the set of variables whose current values are loaded on A.

A register table r may be bound to the starting or finishing point of object code c. If $r{=}\{u,...w\}$, this means that at the regarded point of c register A keeps contents of storage locations $M[tu],...M[tw]$. Thus, semantics of r can be expressed by the sequence of assertions

$$\{A{=}M[tu]\};...\{A{=}M[tw]\}$$

This sequence is denoted by r'. If $r{=}\{\}$, we state that $r' {=}$ SKIP; so, SKIP is used here as the empty sequence of assertions.

On the other hand, the register table r may be considered w.r.t. the corresponding (starting or finishing) point of a source program p translated to c. Then we can assert, that all variables from r must be equal to each other at the regarded point of p. Therefore, if $r=\{u,..,v,w\}$, then at this point of p the sequence of assertions

$$\{u=v\};...\{v=w\}$$

is held, or the only assertion $\{u=w\}$ is held, if $r=\{u,w\}$. If $r=\{u\}$, or $r=\{\}$, SKIP is used as the empty sequence of assertions. In all these cases, $r\hat{}$ stands for the corresponding sequence of assertions.

The compilation predicate $CP(p,t,r1,c,r2)$ binds register tables $r1$ and $r2$ to the starting and finishing points of object code c, respectively. This means the following: if register A contains the current values of all variables from $r1$ before execution of c, then A will contain the current values of all variables from $r2$ after execution of c. The necessary condition (justification) for this situation must be provided by a source program p: the last should ensure equality of all variables from $r2$ as the post-condition, if the pre-condition is equality of all variables from $r1$. Formally, this justification is expressed by the equality

$$(r1\hat{};p) = (r1\hat{};p;r2\hat{}) \qquad (J)$$

If it is held, p may be compiled into c provided that the following improvement is held:

$$t';r1';F';p <= Ic;t';r2';F' \qquad (I)$$

Programs t' and F' are defined as follows:

$$DEF: t' \leftharpoondown (x,...z):=(M[tx],...M[tz])$$

where $x,...z$ are the global variables of p ;

$$DEF: F' \leftharpoondown \text{END } A,P,M[FirstAddr],...M[LastAddr]$$

In the whole, the compilation relation is defined by conjunction of improvement (I) and justification (J):

$$DEF: CP(p,t,r1,c,r2) = \; t';r1';F';p <= Ic;t';r2';F' \;\&\; (r1\hat{};p) = (r1\hat{};p;r2\hat{})$$

4 Notation

Below are a few abbreviations.

(1) If the antecedent of a theorem contains some (meta-)variables which do not appear in the conclusion of the theorem, it is meant by default they are bound in the antecedent by the existence quantification.

(2) If the antecedent of a theorem states that $c=(m,s,f)$, then in the formulation and proof of the theorem it is meant that

(2a) $s<=f$;

(2b) $c.M=m$, $c.S=s$, $c.F=f$;

(2c) the predicate

$$c:> c1,c2,c3$$

(for instance, with three components on the right hand side) abbreviates the condition

$c1=(m,s1,f1)$ & $c2=(m,s2,f2)$ & $c3=(m,s3,f3)$

& $s=s1$ & $s1<=f1$ & $s2<=f2$ & $s3<=f3$ & $f3=f$;

(2d) in the predicate above every component *ci* may be substituted by a machine instruction *instr*; then the condition

$$fi=(si+1) \ \& \ m[si]=(instr)$$

is added by & to the abbreviated condition above.

5 The Compiler Specification

Compiler is specified by a set of formulas that define correspondence between constructs of a source language and some object code compositions. Each formula relates a construction *p* to an object code *c* provided that *(t,r1)* and *(t,r2)* represents input and output data for *c*. To prove correctness of the specified particular compilation, the relation *CP(p,t,r1,c,r2)* must be derived from the given structure of *p*, *t*, *r1*, *c*, *r2* by using algebraic laws of the source language.

So, these formulas are treated as compiling theorems that say about correct object codes for constructions of the source language. There are also two auxiliary theorems that say about widening an input register table *r1* and shortening an output register table *r2*, and the general theorem about transition from the modified compilation to the basic one and vice versa. The specification of a compiler from *H* into *ML* consists of twelve theorems.

Below, *c* = *(m,s,f)* everywhere.

Theorem 0 (transitions between compilations).

CP(p,t,{},c,{}) = *CP(p,t,c)*

Theorem 1 (SKIP).

If *s* = *f*

then *CP(*SKIP*,t,r,c,r)*

Theorem 2a (assignment)

If *c:>* *(*LOAD *ty),(*STORE *tx)*

then *CP(x:=y,t,{},c,{x,y})*

Theorem 2b (assignment)

If *c:>* *(*STORE *tx)*

and *y is_in r* and *x is_not_in r*

then *CP(x:=y,t,r,c,r*U*{x})*

Theorem 2c (assignment)

If *s* = *f*

and *r includes {x,y}*

then *CP(x:=y,t,r,c,r)*

Theorem 3 (sequential composition)

If *c:> c1,c2*

and *CP(p,t,r1,c1,r)* and *CP(q,t,r,c2,r2)*

then *CP((p;q),t,r1,c,r2)*

Theorem 4a (conditional)

If *c:> (*LOAD *tb),(*COND *j),c1,(*JUMP *f),c2* and *j=c2*.s

and *CP(p,t,{b},c1,r)* and *CP(q,t,{b},c2,r)*

then *CP(*IF *b* THEN *p* ELSE *q,t,{},c,r)*

Theorem 4b (conditional)

If *c:> (*COND *j),c1,(*JUMP *f),c2* and *j=c2*.s

and *b is_in r1*

and *CP(p,t,r1,c1,r2)* and *CP(q,t,r1,c2,r2)*

then *CP(*IF *b* THEN *p* ELSE *q,t,r1,c,r2)*

Theorem 5a (while)

If *c:> (*LOAD *tb),(*COND *f),c0,(*JUMP *s)*

and *CP(p,t,{},c0,{})*

then *CP(*WHILE *b* DO *p,t,{},c,{})*

Theorem 5b (while)

If *c:> (*COND *f),c0,(*JUMP *s)*

and *b is_in r*

and *CP(p,t,r,c0,r)*

then *CP(*WHILE *b* DO *p,t,r,c,r)*

Theorem 6 (widening an input register table)

If *r includes r1*

and *CP(p,t,r1,c,r2)*

then *CP(p,t,r,c,r2)*

Theorem 7 (shortening an output register table)

If *r2 includes r*

and *CP(p,t,r1,c,r2)*

then *CP(p,t,r1,c,r)*

6 Proof of Compilation Correctness

6.1 Lemmas About Simple Assignments

Lemma 1.

If neither e nor p nor g contains x, then

$(x{:=}e;p;x{:=}g) = (p;x{:=}g)$

Lemma 2.

If p does not contain x and p contains no variable from e, then

$(x{:=}e;p) = (p;x{:=}e)$

Lemma 3.

If neither e nor p nor b contains x, then

$(\texttt{WHILE } b \texttt{ DO } p;\ x{:=}e) = \texttt{WHILE } b \texttt{ DO}(p;x{:=}e)$

Lemma 4.

$x{:=}y;\texttt{WHILE } b \texttt{ DO}(p;y{:=}x) = \texttt{WHILE } b[y/x] \texttt{ DO}(x{:=}y;p;y{:=}x)$

where $b[y/x]$ is produced from b by substitution of x by y.

6.2 Lemmas About Some Specific Programs

Variables P, A, and $M[i]$ are referred to as special ones.
The "inverse" program for F' is introduced as follows:

$$DEF{:}\ \tilde{F} = \texttt{VAR } P,A,M[FirstAddr],...M[LastAddr]$$

Lemma 5.

(5.1) If p contains no special variable, then

$F';p = p;F'$

(5.2) On the same condition:

$\tilde{F};p = p;\tilde{F}$

(5.3) $\tilde{F};F' = \texttt{SKIP}$

(5.4) $F';\tilde{F} <= \texttt{SKIP}$

(5.5) If S is a special variable, then

$F' = (S{:=}\ e;F')$

(5.6) If b does not contain any special variable, then

$F';\ \texttt{IF } b \texttt{ THEN } p \texttt{ ELSE } q = \texttt{IF } b \texttt{ THEN}(F';p) \texttt{ ELSE}(F';q)$

(5.7) On the same condition:

$\tilde{F};\ \texttt{WHILE } b \texttt{ DO}(i;p);F' = \texttt{WHILE } b \texttt{ DO } (\tilde{F};i;p;F')$

where i is an assignment that initializes the special variables of p by non-special variables or constants.

Lemma 6 (assertions)

Let *a*, *a1*, *a2* are sequences of assertions with equalities of variables, and *u*, *v*, and *w* are variables. Then

(6.1) $a = (a;a)$

(6.2) $(a1;a2) = (a2;a1)$

(6.3) $\{u=u\}$ = SKIP

(6.4) $\{u=v\} = \{v=u\}$

(6.5) $(\{u=v\};\{u=w\}) = (\{v=u\};\{v=w\})$

(6.6) if *a* does not contain *x*, then

$(a;x:=e) = (x:=e;a)$

(6.7) $(a;$ IF b THEN p ELSE $q) =$ IF b THEN$(a;p)$ ELSE$(a;q)$

(6.8) WHILE b DO $(p;a)$ <= WHILE b DO $p;\ a$

(6.9) WHILE b DO $(a;p)$ <= $a;$ WHILE b DO p

Operation A<-$\{u,...w\}$ abbreviates writing of non-deterministic choice $A:=u|...|A:=w$.

"Inverse" programs for *t'* and *r'* are introduced:

DEF: $\tilde{t} = (M[tx],...M[tz]):=(x,...z)$

DEF: $\tilde{r} = r\hat{};A$<-r (if *r* is not empty)

DEF: $\tilde{\{\}}$ = SKIP

Lemma 7.

If the domain of *t* includes *r*, then

(7.1) $t';r' = r';t'$

(7.2) $\tilde{t};\tilde{r} = \tilde{t};A$<-$r;r'$

If *u is_in r*, then the following four equalities hold:

(7.3) $\tilde{r} = r\hat{};A:=u$

(7.4) $\tilde{t};\tilde{r} = \tilde{t};A:=u;r'$

(7.5) $t';r' = t';\{A=u\};(r\backslash\{u\})'$

(7.6) $t';r' = t';r';A:=u = t';r';u:=A$

Lemma 8.

If the domain of *t* includes *r*, then

(8.1) $\tilde{F};\tilde{t};\tilde{r};t';r';F' = r\hat{}$

(8.2) $t';r';F';\tilde{F};\tilde{t};\tilde{r}$ <= $t';r'$

Lemma 9.

$CP(p,t,r1,c,r2) = (r1\hat{};p$ <= $\tilde{F};\tilde{t};\tilde{r1};Ic;t';r2';F'\ \&\ (r1\hat{};p)=(r1\hat{};p;r2\hat{}))$

Lemma 10.

$t';r';F';\tilde{F};\tilde{t};\tilde{r};Ic;t'$ <= $Ic;t'$

6.3 Some Properties of Justification

Lemmas J1-J5 say how post-condition *r2^* is related to pre-condition *r1^* for each construct of the source language. Lemmas J6 and J7 say that one can always strengthen pre-condition *r1^* and weaken post-condition *r2^*.

Below, *x*, *y*, *z* are variables, *b* is a boolean variable, *p* and *q* are programs, *r*, *r1*, *r2* are finite sets of variables.

Lemma J1 (SKIP).

*(r^;*SKIP*) = (r^;*SKIP*;r^)*

Lemma J2 (assignment).

If *y is_in r*,

then *(r^; x:=y) = (r^; x:=y; (r* U *{x})^)*

Lemma J3 (sequential composition).

If *(r1^; p) = (r1^; p; r^)*

and *(r^; q) = (r^; q; r2^)*

then *(r1^; p; q) = (r1^; p; q; r2^)*

Lemma J4 (conditional).

If *(r1^; p) = (r1^; p; r2^)*

and *(r1^; q) = (r1^; q; r2^)*

then

(r1^; IF *b* THEN *p* ELSE *q) = (r1^;* IF *b* THEN *p* ELSE *q; r2^)*

Lemma J5 (while).

If *(r^; p) = (r^; p; r^)* then

(J5a) *(r^;*WHILE *b* DO *p) = (r^;* WHILE *b* DO *p; r^)*

(J5b) *(r^;*WHILE *b* DO *p)* = WHILE *b* DO*(r^;p)*

Lemma J6 (strengthening pre-condition).

If *r includes r1*

and *(r1^;p) = (r1^;p;r2^)*

then *(r^;p) = (r^;p;r2^)*

Lemma J7 (weakening post-condition).

If *r2 includes r*

and *p = (p;r2^)*

then *p = (p;r^)*

6.4 The Proofs of the Compilation Theorems

Theorem 0.
The proof directly follows from the definitions of the relations.

Theorem 1.
(Improvement)

$Ic; t'; r'; F'$

(definition of Ic, arithmetic grouping)

$= P:=s; \{P=s\}; t'; r'; F'$

(the laws from [1,2,3] -- see above too, lemma 2)

$= t'; r'; P:=s; F'$

(lemma 5.5)

$= t'; r'; F'$

(the laws)

$= t'; r'; F'; \text{SKIP}$

(Justification) Follows from lemma J1.

Theorem 2a.

(I) $Ic; t'; \{x,y\}'; F'$

(definition of Ic, arithmetic grouping)

$= P:=s; A:=M[ty]; P:=P+1;\ M[tx]:=A; P:=P+1;\ \ \{P=s+2\}; t'; \{x,y\}'; F'$

(the laws)

$>= P:=s; A:=M[ty]; P:=P+1; M[tx]:=A; P:=P+1; t'; F'$

(lemma 2, the laws)

$= A:=M[ty]; M[tx]:=A; t'; P:=s+2; F'$

(lemma 5.5, the laws)

$= M[tx]:=M[ty]; A:=M[ty]; t'; F'$

(the laws)

$= t'; x:=y; M[tx]:=y; A:=y; F'$

(lemmas 5.1, 5.5, definition of $\{\}'$)

$= t'; \{\}'; F'; x:=y$

(J) Follows from the laws.

Theorem 2b.

(I) $Ic; t'; (r\mathrm{U}\{x\})'; F'$

(definition of Ic, arithmetic grouping)

$= P:=s; M[tx]:=A; P:=P+1; \{P=s+1\}; t'; (r\mathrm{U}\{x\})'; F'$

(lemma 2, the laws, lemma 5.5)

$>= M[tx]:=A; t'; (r\mathrm{U}\{x\})'; F'$

(the laws, lemma 7.5)

$= t'; x:=A; M[tx]:=x; \{A=x\}; r'; F'$

(lemmas 6.6, 5.5)

$= t';r';x{:=}A;\{A{=}x\};F'$

(lemma 6.4, the laws, lemma 7.6)

$= t';r';A{:=}y;x{:=}A;F'$

(the laws, lemmas 5.5, 5.1)

$= t';r';F';x{:=}y$

(J) Follows from lemma J2.

Theorem 2c.

(I) $Ic;t';r';F'$

(theorem 1, part I)

$= t';r';F';$ SKIP

(the laws, lemma 7.6)

$= t';r';y{:=}A;x{:=}A;F'$

(the laws, lemmas 5.5, 5.1)

$= t';r';F';x{:=}y$

(J) Follows from lemma J2.

Theorem 3.

(I) The proof is based on lemma 9.

(the antecedents)

$r1^{\wedge};p <= F^{\sim};t^{\sim};r1^{\sim};Ic1;t';r';F'$

$r^{\wedge};q <= F^{\sim};t^{\sim};r^{\sim};Ic2;t';r2';F'$

(the antecedent)

$r1^{\wedge};p;q = r1^{\wedge};p;r^{\wedge};q$

(the laws, the inequalities above)

$<= F^{\sim};t^{\sim};r1^{\sim};Ic1;t';r';F';F^{\sim};t^{\sim};r^{\sim};Ic2;t';r2';F'$

(lemma 10)

$<= F^{\sim};t^{\sim};r1^{\sim};Ic1;Ic2;t';r2';F'$

(lemma 3 from [1])

$<= F^{\sim};t^{\sim};r1^{\sim};Ic;t';r2';F'$

(J) Follows from lemma J3.

Theorem 4a.

The proof in the main repeats the proof of theorem 3 from [1] and uses lemma J4.

Theorem 4b.

(I) $Ic;t';r2';F'$

(lemma 4 from [1])

= IF A THEN$(Ic1;P{:=}f)$ ELSE $Ic2;\quad t';r2';F'$

(the laws, lemmas 2, 5.5, the antecedents)

>= IF A THEN$(t';r1';F';p)$ ELSE$(t';r1';F';q)$

(the laws, lemma 6.7)

$= t';r1';\text{IF } A \text{ THEN}(F';p) \text{ ELSE}(F';q)$

(lemma 7.6)

$= t';r1';A{:=}b;\text{IF } A \text{ THEN}(F';p) \text{ ELSE}(F';q)$

(the laws)

$= t';r1';\text{IF } b \text{ THEN}(A{:=}b;F';p) \text{ ELSE}(A{:=}b;F';q)$

(lemmas 5.5, 5.6)

$= t';r1';F'; \text{ IF } b \text{ THEN } p \text{ ELSE } q$

(J) Follows from lemma J4.

Theorem 5a.
The proof immediately follows from theorem 0, theorem 4 from [1] and lemma J5.

Lemma 11.

If $c{:>} (\text{COND } f),c0,(\text{JUMP } s)$

then $Ic >= \text{WHILE } A \text{ DO } Ic0;\ P{:=}f$

This is near to the lemma 5 from [1].

Theorem 5b.
(I) The proof is based on lemma 9.

$\tilde{F};\tilde{t};\tilde{r};Ic;t';r';F'$

(lemma 11)

$>= \tilde{F};\tilde{t};\tilde{r};\text{WHILE } A \text{ DO } Ic0;\ P{:=}f;\ t';r';F'$

(lemmas 2, 5.5)

$= \tilde{F};\tilde{t};\tilde{r};\text{WHILE } A \text{ DO } Ic0;\ t';r';F'$

(lemmas 7.1, 7.4)

$= \tilde{F};\tilde{t};A{:=}b;r';\text{WHILE } A \text{ DO } Ic0;\ r';t';F'$

(lemmas 6.8, 6.9)

$>= \tilde{F};\tilde{t};A{:=}b;\text{WHILE } A \text{ DO}(r';Ic0;r');\ t';F'$

(lemmas 3, 7.1)

$= \tilde{F};\tilde{t};A{:=}b;\text{WHILE } A \text{ DO}(r';Ic0;t';r');\ F'$

(lemmas 7.6, 4)

$= \tilde{F};\tilde{t};\text{WHILE } b \text{ DO}(A{:=}b;r';Ic0;t';r';b{:=}A);\ F'$

(lemmas 7.6, 7.1, 4, the laws)

$= \tilde{F};\text{WHILE } b \text{ DO}(\tilde{t};A{:=}b;r';Ic0;r';t');\ F'$

(definition of $Ic0$, the laws, lemmas 6.6, 5.7, 7.1)

$= \text{WHILE } b \text{ DO}(\tilde{F};\tilde{t};A{:=}b;P{:=}c0.\text{s};r';Ic0;t';r';F')$

(lemma 6.6, the laws, lemma 7.4)

$= \text{WHILE } b \text{ DO}(\tilde{F};\tilde{t};\tilde{r};Ic0;t';r';F')$

(the antecedent, lemma 9)

$>= \text{WHILE } b \text{ DO}(\hat{r};p)$

(lemma J5b)

$= r\hat{};$WHILE b DO(p)

(J) Follows from lemma J5a.

Theorem 6.
The proof immediately follows from the laws and lemma J6.

Theorem 7.
The proof immediately follows from the laws and lemma J7.

7 Acknowledgements

I am obliged to C.A.R. Hoare, Jonathan Bowen, and He Jifeng for making available to me ProCoS project materials, and Jonathan Bowen and He JiFeng - for useful discussions in January 1992 at Dagstuhl Seminar. The discussions initiated my interest in algebraic approach to compiler specification.

References

[1] C.A.R. Hoare, Refinement Algebra Proves Correctness of Compiling Specifications, in: C.C. Morgan and J.C.P. Woodcock (Eds.), 3rd Refinement Workshop. Springer-Verlag, Workshops in Computing, 33-48, 1991.

[2] He JiFeng and E.-R. Olderog (eds.), Interfaces between Languages for Concurrent Systems, volume II. ESPRIT BRA ProCoS Project Deliverable, 1991.

[3] C.A.R. Hoare, J.M. Spivey and others, Laws of Programming. Comm. ACM, 30, 8 (1987), 672-686.

[4] He JiFeng, Jonathan Bowen, Specification, Verification and Prototyping of an Optimized Compiler (submitted to Formal Aspects of Computing in 1992).

[5] A.V. Aho and J.D. Ullman, Principles of Compiler Design. Addison-Wisley, Series in Computer Science and Information Processing, 1977.

[6] I.V. Pottosin, Analysis of Program Optimization and Further Development. Theoretical Computer Science, 90 (1991), 17-36.

[7] W.-P. de Roever, G. Rozenberg (eds.), Stepwise Refinement of Distributed Systems - Models, Formalisms, Correctness. Lecture Notes in Computer Science, 430, Springer-Verlag, 1990.

FORMAL DERIVATION of an ERROR-DETECTING DISTRIBUTED DATA SCHEDULER USING CHANGELING†

Hanan Lutfiyya
hanan@csd.uwo.ca
Department of Computer Science
University of Western Ontario
London, ONTARIO N6A 5B7
CANADA

Bruce McMillin and Alan Su
ff@cs.umr.edu
Department of Computer Science
University of Missouri-Rolla
Rolla, MO 65401
USA

Abstract. This paper focuses on being able to detect component errors which can lead to system failures in the scheduling part of the lock manager portion of the distributed database system by using embedded executable assertions. The technique used to generate the executable assertions is based on the mathematical model of program verification.

Key words: Distributed Databases, Executable Assertions, Formal Methods, Concurrent Program Verification, Fault Tolerance, Transformation, Changeling.

1. INTRODUCTION

A distributed system is a collection of processors interconnected by a communications network. Processes of a distributed application execute on the processors of the distributed system and cooperate/communicate via message passing over the communications network. Resources in a distributed system include the processors, data/disks, and the communications network.

Distributed database applications are a wide use of distributed systems. Fundamentally, processes execute *transactions* which perform *Lock*, *Read*, *Write*, and *Commit* operations on *entities* stored in the database [BeGo82]. The interleaving of operations from the various transactions, or *schedule* of operations, performed concurrently must be equivalent to the effects of executing the transactions serially [EGLT76]. This is most commonly achieved by the *Two-Phase Locking* protocol (transactions finish locking entities before unlocking any). In the system model, m transaction manager (TM) processes execute transactions which access data uniquely held by n data manager processes each controlled by lock manager (LM) processes. All locks by transaction T_{ij} (jth transaction on ith transaction manager) are released, atomically, with the commit of T_{ij}'s updates to the entities.

† This work was supported in part by the National Science Foundation under Grant Numbers MSS-9216479 and CDA-9222827, and, in part, from the Air Force Office of Scientific Research under contract number F49620-92-J-0546 and the National Sciences and Engineering Research Council of Canada (NSERC) under contract number OGP0138180-S365A2, and in part, from the University of Western Ontario NSERC internal funding under contract number Z001A8-S365A1.

One of the major advantages of distributed database systems is the potential for achieving high availability in the presence of faults. Faults must be handled so that the system still operates or operates in a degraded mode. Many methodologies in the past have attempted to provide fault tolerance efficiently, but have never been successful at eliminating explicit time and space redundancy.

Application-Oriented fault tolerance [McNi92], by contrast, is a promising approach to providing run-time assurance for the distributed environment. In this approach, testing is comprised of executable assertions which are embedded in the program code to ensure that at each testable stage of a calculation, all tested processors conform to the program's specification. These executable assertions in the form of source language statements, are inserted into a program for monitoring the run-time execution behavior of the program. The general form is as follows:

if ¬ *ASSERTION* **then** *ERROR*

Executable assertions are used to ensure that the program state, in the actual run-time environment, is consistent with the logical state specified in the assertion; if not, then an error has occurred and a reliable communication of this diagnostic information is provided to the system such that reconfiguration and recovery can take place.

This paper describes the application of *Changeling* as a formal method to show fault tolerance through software specified executable assertions to distributed databases such that the detection of component errors which can lead to system failures in the scheduling part of the lock manager portion of the distributed database system is possible.

Changeling uses the axiomatic approach to program verification to provide a framework for generating executable assertions. The basic approach is to utilize proofs of the form:

$$<P>S<Q>$$

where P and Q are assertions, and S is a statement or sequence of statements of the language. The interpretation of the theorem is as follows: if P is true before the execution of S and if the execution of S terminates, then Q is true after the execution of S. P is said to be the *precondition* and Q the *postcondition* [Hoar69]. P and Q aare also referred to as a program *specification*. In program verification, logical assertions are determined that describe the effect of each statement that comprises S. Each of these logical assertions must hold if the cumulative effect of these statements results in the execution of S satisfying the logical assertions Q. This collection of assertions is normally referred to as the "verification proof outline".

This paper is organized as follows. Section 2 describes application-oriented fault tolerance and *Changeling*. Section 3 presents a verification proof for the lock manager. Sections 4 and 5 focus on implementation details, fault coverage, and run-time efficiency. Section 6 presents a summary.

2. APPROACH

Application-oriented fault tolerance works on the principle of testing at run time the intermediate logical assertions from the verification proof outline i.e. application-oriented fault tolerance works on the following principle:

> *If we test and ensure intermediate results of a program's computation meet its specification, the end solution meets its specification if the intermediate results meet their specification. If processor errors occur that do not affect the solution, then they are not errors of interest. Program verification provides these tests.*

The above principle yields a formal statement of application-oriented fault tolerance; we generate the executable assertions from the logical assertions used in the verification proof outline of $<P>S<Q>$. The executable assertion generated corresponding to any logical assertion Q_i from the verification proof outline is the following:

$$\text{if } \neg\, Q_i \textbf{ then } ERROR$$

Formally, this ensures that if P is true before the concurrent program S begins execution, S tests at run time that S satisfies the specification as defined by P and Q, by using the embedded executable assertions generated from the assertions of the verification proof. Conversely, the assertions of the verification proof represent the properties that must be satisfied by the run-time environment; an error that causes the execution of the program not to satisfy the specified assertions will be flagged as an error by the executable assertions.

The reader may be suspicious that some program S may be changed into a program S' by an error that satisfies the specification as defined by P and Q. Consider, as an example, a program S computing some value x with postassertion $<Q> \equiv <x \geq 0>$. Suppose that S should compute $x = 3$. A program S' may actually compute $x = 4$. The postcondition is still satisfied, although, the value is not what was intended. This is not a problem with the validity of the postassertion, it is a weakness of the specification. If $x = 3$ was what was really intended, then the proper postassertion should have been $<Q> \equiv <x = 3>$. If $<Q> \equiv <x \geq 0>$ is a sufficient specification for the application at hand, then there is no problem.

The transformation of an algorithm to an error-detecting algorithm involves using the assertions of the verification proof as executable assertions that are to be embedded into the algorithm. To provide for detection of faults in the sequential environment, each Q can be tested at run time. As discussed below, not every Q from the proof outline can be tested at run time.

A proof outline employing global auxiliary variables (GAA) [LeGr81]† as a

† The choice of verification proof system is not important in terms of power [LuMc91]. The choice is based on convenience. Since many people prefer shared memory programming, we chose the global auxiliary system as a starting point.

proof aid is, perhaps, the easiest way to reason about concurrent systems. In this proof system, we reason about processes in isolation. For a full concurrent proof of the appropriate properties of individual processes, we require assumptions to be made about the effect of the communication commands. A "satisfaction proof" is then used to show that these assumptions are "legitimate". Hence, a parallel inference rule for processes indexed by ρ_i is as follows:

$$\frac{(\forall i\colon <P_i> S_i <Q_i>) \text{ and } \textit{satisfied} \text{ and } \textit{interference-free}}{<(\forall i\colon P_i)> [\|_{i=1:n}\ \rho_i\colon S_i] <(\forall i\colon Q_i)>}$$

The parallel rule implies that construction of the proof of a parallel program can be derived from the partial correctness properties of the sequential programs that it comprises. In order to relate the different interleavings of processes via interleaved communication histories requires a proof of "non-interference", or a proof that executions in one process do not affect assertions on another and a "satisfaction" to show soundness.

The GAA system does not reflect the operational environment of a distributed computing system well. It is this difficulty that inspired the development of *Changeling* [LuSM92a].

Changeling consists of four distinct components:

- The GAA Proof System
- An HAA proof system which mimics closely the distributed operational environment
- Formal conversion from GAA to HAA
- Formal translation of assertions in the HAA proof system to executable assertions.

These components are described in the following paragraphs.

2.1 HAA System and Formal Conversion The logical assertions from the GAA verification environment cannot be directly used as executable assertions in the distributed environment. In the distributed environment, there are no global variables, yet the verification proof requires the use of global auxiliary variables. Thus, to evaluate, at run time, logical assertions containing global auxiliary variables, an explicit updating mechanism must be created. We have developed the verification proof system (HAA) in which updates of global auxiliary variables are exchanged at communication time. This matches, more closely, the operational environment. It has been shown [LuSM92a] that every verification proof outline in the GAA proof system has the same properties in the HAA proof system i.e. satisfaction and non-interference. This implies that every verification proof outline of the GAA system is a verification proof outline of the HAA proof system. The existence of the HAA proof system allows for proofs that can be directly transformed to executable assertions in the run-time environment.

Developing the HAA system requires us to keep track of which processes communicate with which other processes. Each process needs to record its global auxiliary variable updates with respect to all other processes. When communication occurs between two processes, they need to exchange the updates and locally apply the updates. The following definition formalizes how the updates are communicated. Details of how the updates are applied can be found in [LuSM92a].

Definition 2.1: *For a process ρ_i, h_i denotes the sequence of all communications that process ρ_i has so far participated in as the receiving process. Thus, h_i is a list consisting of tuples representing matching communication pairs of the form*

$$[\rho, (Var, Val), T, C]$$

where ρ is a process from which ρ_i receives from, Var is the variable that ρ is transmitting to ρ_i with formal parameter Val. T denotes the time at which the value Val was assigned to variable Var and C denotes the communication path.

Since we have several processes running in parallel and there exists no concept of a global time, the time T is a local time represented by an instantiation counter that is incremented by one after every execution of a statement. This permits an ordering (time-stamping) for all updates of the GAVs within each process.

To be able to account for the different operations performed on the auxiliary variables, each process has to keep a history of variable updates with respect to the last communication with the other processes. These variable sets are described using the subscript of the corresponding process, g_{ij}; i.e. g_{ij} contains the changes that were made to the GAVs in ρ_i since the last communication with ρ_j. When two processes ρ_i and ρ_j communicate, where ρ_j is the sender, ρ_j will augment the communication by sending the values of global auxiliary variables that ρ_j updated or received updates of between the last and current communications between ρ_i and ρ_j. We batch the changes made to the local copies of the global auxiliary variables by p_j since the last communication (with any other processor) in g_{jj}. Before a communication, a function ψ applies changes to all g_{jk}'s and g_{jj} is reset to null to collect future changes. For each process ρ_i after an augmented exchange with ρ_j, ρ_i uses a function ϕ to update its set of GAVs, therefore when processes ρ_i and ρ_j communicate, all old values in the set g_{ij} will be replaced by the new variables. Details of this are in [LuSM92a]. paper.

It can be seen that the so-called "global auxiliary variables" in the HAA system are not really global in the sense that all processes have the same values of the variables at all times. Indeed, it is likely that at the end of the process execution some processes that ran in parallel will have different values within their set of GAVs. We have shown [LuSM92a] that because of non-interference, this is not a problem with respect to the proof system.

2.2 Formal Translation of Assertions in the HAA Proof System to Executable Assertions The HAA proof system provides for direct transformation of assertions from the verification environment into executable assertions for the non-faulty distributed operational environment, However, we are concerned with the distributed faulty environment. Thus, it is necessary to ensure that faulty processors cannot fool executable assertions by incorrect augmented communication of g_{ij}'s through sending inconsistent messages to different processors. It is necessary for this to be detected. This is the purpose of *consistency* executable assertions. Mathematically, this can be described as follows:

Definition 2.2: *For a non-faulty process ρ_i, if there exists any two tuples t_1, $t_2 \in h_i$ such that*

$$t_1 = [j, (Var, Val_1), T, C_1]$$

$$t_2 = [j, (Var, Val_2), T, C_2]$$

then if $Val_1 \circ Val_2$ the system is said to be inconsistent otherwise the system is said to be consistent. $\circ$ is defined as a set of functions such that each $\circ' \in \circ$ is of functionality $dt \rightarrow \{T, F\}$ where dt is an abstract data type. Examples of $\circ'$ are $\neq$, $\subseteq$, $\neg prefix$, or some other operator appropriate to the choice of the data type of Var. Where no ambiguity results, we will refer to a particular $\circ'$ simply as $\circ$.

The strongest motivation for the consistency condition is to supplement the power of the executable assertions derived from the HAA system. When the value of a variable computed in time T is communicated to a set of processors on more than one path, there will be two or more tuples in h_i that satisfy the precondition. Under a bounded number of faults, the consistency definition of 2.2 ensures that a non-faulty processor receives a consistent set of input values for its executable assertions, otherwise, $Val_1 \circ Val_2$, and an inconsistent system can be detected. The degree of fault tolerance is based on standard network flow arguments. It should be noted that all faults in communication links are mapped to a processor, thus it is enough to assume only faulty processors.

3. VERIFICATION OF A LOCK MANAGER PROCESS in the GAA SYSTEM

This section highlights the relevant aspects of the lock manager algorithm and the proof outline. Complete details can be found in [LuSM92b]. This section discusses one invariant assertion labelled I. In the proof rules of the previous section, processes ρ_i are either lock manager processes LM_i or transation manager processes TM_i.

The following auxiliary variables are used in the verification proof: $schedule_i$, SG, and TSG. $schedule_i$ denotes the schedule of operations as ordered by the lock manager process LM_i on processor i. SG denotes the partial order between transaction and TSG denotes the total order among transactions.

A transaction T_{kl} finishes before a transaction T_{st}, if all the operations in T_{kl} occur before all the operations in T_{st}. We can define a relation "<" on the set of transactions as the smallest relation satisfying the following condition: If T_{st} requests a lock on entity e that has been previously locked by T_{kl} then $T_{kl} < T_{st}$. Since, T_{st} must wait for T_{kl} to release the lock and since, locks are not released until the termination of the transaction, then it follows that that T_{kl} finishes all its operations before T_{st} finishes all its operations.

It does not make sense for T_{kl} to finish before T_{st} and that T_{st} finishes before T_{kl}. Therefore, "$<$" is antisymmetric. If T_{kl} finishes before T_{st} and T_{st} finishes before T_{yz} then T_{kl} finishes before T_{yz}. Therefore, "$<$" is transitive. Also, since, it doesn't make sense to say that a transaction finishes before itself. This implies that the relation "$<$" is irreflexive. Since, "$<$" is transitive, antisymmetric and irreflexive then "$<$" is an irreflexive partial order. If two transactions do not operate on common entities then it is impossible to say which transaction completes before the other. Therefore, "$<$" is not a total order.

The partial order denoted SG is represented by a graph. $SG = (T, E)$, where T is the set of transactions, and the set E represent the arcs between the transactions. An arc between T_{kl} and T_{st} implies $T_{kl} < T_{st}$. Since, SG is a graphical representation of a partial order, then SG does not contain cycles, i.e. it is acyclic.

The following functions on $schedule_i$ are used in the verification proof:

$Last\_Op_i(e, j)$: This represents the j'th to the last operation on entity e.

$Last\_Trans_i(e, j)$: This represents the j'th to the last transaction to operate on entity e.

The precondition for each lock manager process LM_i assumes that no scheduling of any of the transactions has taken place. In other words, Pre_i is as follows:

$$Pre_i\!: \quad < SG = (T, E), \ where\ E\ is\ \varnothing \wedge schedule_i = the\ null\ sequence >$$

At the termination of the lock manager process LM_i, we want SG to denote an irreflexive partial order, i.e. SG must be acyclic and that the schedule produced by a lock manager on process LM_i is legal. This is represented by the following postcondition $Post_i$:

$$Post_i\!: \ <SG\ is\ acyclic \wedge \neg\exists e, j(Last\_op_i(e, j) = lock \wedge Last\_op_i(e, j+1) = lock)>$$

Let the precondition of the lock management system be the union of Pre_i and let the postcondition, $Post$, of the lock management system as a whole be the union of $Post_i$, where $0 \le i \le N - 1$.

It can be shown that if SG is acyclic and if SG is extended to a total order, denoted by TSG, then TSG will denote a schedule of transactions that is equivalent to a serial schedule of transactions. The proof results from SG representing a partial order of transaction execution. This order of transaction execution represents a dependency among transactions that indicates which transactions must be completed before others. The partial order represented by SG can be shown to be equivalent to the partial order defined in [EGLT76]. [EGLT76] shows that if a set of transactions is well-formed†, two-phase, then any legal schedule for T is consistent and that the partial order can be extended to a total order that defines a serial schedule that is equivalent to the run-time schedule. This total schedule corresponds to TSG. It is obvious that each lock manager must produce only legal schedules, because, otherwise, two transactions may have a simultaneous lock on the same entity.

† A well-formed transaction does not attempt to lock an entity that has already been locked nor does it read or modify an entity unless it has already locked that entity.

The best way of ensuring that *SG* is acyclic at termination of the lock managers is to ensure that *SG* is acyclic is invariantly true throughout program execution and to show that when an operation for a transaction is scheduled that it does not violate the conditions of a legal schedule. This can be formally represented as follows:

$$<SG\ is\ acyclic> \tag{1}$$

$$<\neg\exists e, j(Last\_op_i(e, j) = lock\ \wedge\ Last\_op_i(e, j+1) = lock)> \tag{2}$$

For an arc denoted by (T_{kl}, T_{st}) to be a member of E, it must be the case that all the operations of T_{kl} are completed before all the operations of T_{st}. This is represented as follows:

$$<(T_{kl}, T_{st}) \in E_i \rightarrow \exists\ e, j, i\ (Last\_op_i(e, j) = lock \wedge Last\_op_i(e, j+1) = unlock \wedge \\ Last\_tran_i(e, j) = T_{st} \wedge Last\_tran(e, j+1) = T_{kl})> \tag{3}$$

It is also necessary to ensure that that if an entity is unlocked then the last scheduled operation for it is an unlock operation. This is represented as follows:

$$<lu_e\ is\ unlocked \rightarrow Last\_op_i(e, 0) = unlock> \tag{4}$$

These four assertions form the invariant, I.

Initially, I is true since, for each *SG*=(*T*, *E*), *E* is empty and $schedule_i$ is the empty schedule. It is then necessary to show that for each incoming operation that *SG* remains acyclic. It is trivial to show that if the loop terminates then the postcondition is implied from I.

The following theorem is useful in understanding the proof outline:

Theorem 3.1: [LuSM92b] At a process ρ_i, if *SG* is acyclic and the assertion expressed in equation (3) is true then if an arc is placed between T_{kl} and T_{st} then the graph $(T, E \bigcup (T_{kl}, T_{st}))$ is acyclic.

There are two types of messages that a lock manager may receive.

(1) The first type occurs when the incoming message's operation is a Read or Write

(2) The second type occurs when the incoming message's operation is a Commit or Abort.

For a Read or Write message, there are two possible situations. In the first situation, the entity of the Read or Write is unlocked. Therefore, the requesting transaction is able to lock the requested entity. This causes a new entry to be added to the schedule. This is denoted by

$$schedule_i = schedule_i\ \hat{}\ (T_{kl}, lock, e)$$

Since, I is true before the above assignment, then we can conclude that after the assignment $Last\_op_i(e, 1) = unlock$ and $Last\_op_i(e, 0) = lock$. All other variables are the same. Hence, we can immediately conclude that the truth of I is preserved.

After the above assignment, another assignment is made that adds an arc to *SG* between $Last\_trans_i(e, 1)$ and the requesting transaction. All other variables remain the same. From Theorem 3.1, we have that *SG* remains acyclic. Since, $Last\_op_i(e,1) = unlock$ and $Last\_op_i(e,0) = lock$, then it is easy to conclude that the truth of I is preserved.

The second situation occurs when the entity of the Read or Write is locked. This implies that the Read/Write operation is to be scheduled later. In either case, no change is made to *SG* or $schedule_i$. Therefore, the truth of I is still preserved.

In case (2), we have a Commit or an Abort. Both of these operations will release all the locks that are currently held by the transaction doing the Commit or Abort. If there are no transactions waiting to lock the released entities, then *SG* does not change, therefore, it still remains acyclic. On the other hand, there may be other transactions waiting to lock the entity. One is taken off the queue and is allowed to lock the entity. If T_{kl} is the releasing transaction and if T_{st} is the new locking transaction then the edge (T_{kl},T_{st}) is added to the set of edges. The graph denoted by $(T, E \cup (T_{kl}, T_{st}))$ is acyclic by Theorem 3.1. Since, $Last\_op_i(e,1) = unlock$ and $Last\_op_i(e,0) = lock$, then $Last\_tran_i(e,1) = T_{kl}$ and $Last\_tran_i(e,0) = T_{st}$, then it is easy to conclude that the truth of I is preserved.

4. CONVERSION TO HAA AND TRANSLATION TO EXECUTABLE ASSERTIONS

In *Changeling*, conversion from the verification proof outline in GAA to HAA requires instrumenting the proof outline with augmented communication of global auxiliary variables. Each assertion of interest in GAA is an assertion in HAA (here, only the invariant, I, is used), since every verification proof outline in GAA is a verification proof outline in the HAA system [LuSM92a]. Each augmented communication requires a g_{ij}. Here, g_{ij} for any two processes ρ_i and ρ_j is

$$g_{ij} = \{\{schedule_r \mid 0 \le r \le n-1\}, SG\}$$

Since coordinating transaction managers will request use of entities managed by many lock managers, then any lock manager LM_i may receive values of $schedule_j$, where $i \ne j$. Consistency is checked by showing that the last update of $schedule_r$, where $0 \le r \le n-1$, coming from TM_s is a prefix of the last update of $schedule_r$ coming from TM_t or vice-versa. Thus, the $\circ$ operator for consistency is simply the prefix operation on two schedules: $schedule_r \in g_{si}$ and $schedule_r \in g_{ti}$ of lengths L_1 and L_2, respectively and $L_1 \le L_2$ is defined as follows:

$$\circ \equiv (\exists k(0 \le k \le L_1 - 1 \wedge select(k, schedule_r \in g_{si}) \ne select(k, schedule_r \in g_{ti})))$$

The function select is used for determining the kth element of the schedule sequence.

Both the transaction and lock managers are being used to test the executable assertions. By the noninterference property of the HAA system, these managers only test available data; no extra communication is forced to obtain a global picture of the entire system; when auxiliary variables arrive, they are checked by the executable assertions. Consistency ensures that other lock managers are correctly sending g_{ij}'s to a particular transaction/lock manager.

To better understand the transmission of auxiliary variables, consider the following sequence of transactions starting at some time k with Transaction T_{00} holding a lock on entity e_1.

k) TM_1 executes
T_{10} requests LM_1 lock entity e_1
and is queued

k+1) TM_0 executes
T_{00} requests LM_1 unlock entity e_1

k+2) LM_1 responds
acknowledgement, $[LM_1, (g_{LM_1TM_0}, (T_{00}, unlock, e_1);), k+2, 1]$

k+3) LM_1 responds
acknowledgement, $[LM_1, (g_{LM_1TM_1},$
$(T_{00}, unlock, e_1)(T_{10}, lock, e_1); (T_{00}, T_{10})), k+3, 1]$

k+4) TM_0 executes
T_{01} requests LM_1 lock entity e_2

k+5) LM_1 responds
acknowledgement, $[LM_1, (g_{LM_1TM_0},$
$(T_{10}, lock, e_1)(T_{01}, lock, e_2); (T_{00}, T_{10})), k+5, 1]$

k+6) TM_0 executes
T_{01} requests LM_n lock entity e_3,
$[TM_0, (g_{TM_0LM_n},$
$(T_{00}, unlock, e_1)(T_{10}, lock, e_1)(T_{01}, lock, e_2); (T_{00}, T_{10})), k+6, 0]$
and is granted access

k+7) TM_1 executes
T_{10} requests LM_n lock entity e_3,
$[TM_1, (g_{TM_1LM_n}, (T_{00}, unlock, e_1)(T_{10}, lock, e_1); (T_{00}, T_{10})), k+7, 1]$
and is queued

Since T_{00} holds a lock on e_1, TM_1's transaction's request is queued. In step k+1, transaction T_{00} requests an unlock which is granted. The current contents of *schedule*$_1$'s changes since TM_0 last communicated with LM_1 is piggybacked as a tuple on the return message. T_{10}'s lock request is now granted. Since we now have a dependency on entity e_1, GAV, SG's corresponding arc is updated. Since TM_1 does not yet have a copy of the changes made to *schedule*$_1$ by TM_0, this entity is piggybacked in the return tuple as well. Finally, TM_0 makes another request to LM_1. On the return, since $(T_{00}, e_1, unlock)$ was already communicated in step k+1, it does not form part of the return tuple by virtue of the variable updating in step k+1. In step k+6, TM_0 requests access to an entity being managed by LM_n. TM_0 piggybacks the contents of *schedule*$_1$'s changes since TM_0 last communicated with LM_1. In step k+7, T_1 requests access to an entity being managed by LM_n. TM_1 piggybacks the contents of *schedule*$_1$'s changes since TM_1 last communicated with LM_1. By looking at steps k+2 and k+3, we see that there should be one operation in common; $(T_{00}, unlock, e_1)$; this allows lock manager LM_n to apply the consistency test. Figure 4.1 depicts the message flow.

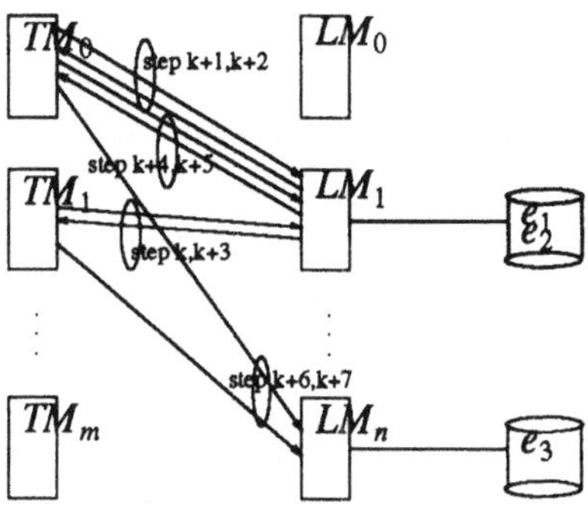

Figure 4.1: System Structure and Example Transaction Scenario

5. FAULT COVERAGE and RUN-TIME EFFICIENCY ASSESSMENT

5.1 Fault Tolerance Analysis of the expected error coverage is important for any error-detecting algorithm. Since we evaluate the invariant, I, at run-time in all managers we can detect any schedule that violates I under a consistent system. Thus, we need to derive bounds on the number of faulty processors such that consistency is maintained.

It is assumed that a transaction manager uniformly requests access to data at all sites. This implies that a transaction communicates with all lock managers. It is assumed that faulty resources have Byzantine [LaSP82] behavior. Theorem 5.1 shows that if we have n lock managers and m transaction managers the system can handle $n-1$ faults occurring on the processors with the lock managers. Theorem 5.2 shows a similar results for faults occurring on the transaction managers' processors.

Theorem 5.1: [LuSM92b] Assume that a transaction manager uniformly requests access to data at all sites. If there are n lock managers and m transaction managers, where $n > 1$, then the system can tolerate $n-1$ lock manager failures and zero transaction manager failures.

Theorem 5.2: [LuSM92b] Assume that a transaction manager uniformly requests access to data at all sites. If there are n lock managers and m transaction managers, where $n > 1$, then the system can tolerate $n-1$ transaction manager failures and zero lock manager failures.

Corollary 5.1: [LuSM92b] Assume that a transaction manager uniformly requests access to data at all sites. In the worst case fault distribution between lock managers and transaction managers, the system can tolerate one fault if there are n lock managers and m transaction managers, where $n > 1$.

5.2 Run-Time Assessment In the implementation, each *schedule*$_i$ is a GAV maintained by LM_i. Each SG entry is simply: (T_{ij}, T_{kl})

From the construction of g_{ij}, the actual set of auxiliary variables sent during a communication between LM_i and TM_j is the differences in *schedule*$_r$, where $0 \leq r \leq n-1$, and SG sent in the previous communication between LM_i and TM_j and the current communication. Since a transaction manager uniformly requests access to data at all sites, then the average difference between old and new schedules is m entries.

SG is updated when there are two transactions that request access to a common entity. In the worst case, there are m entities in common. Therefore, SG increases by at most m elements. This yields the following theorem.

Theorem 5.3: The overhead of the error detecting algorithm is $O(nm)$.

Proof: In the basic algorithm there is one action communicated between LM_i and TM_j. Let this message length be denoted by B. In the transformed algorithm the average message length is (nm+m+1)B giving an overhead of (nm+m+1) = O(nm). □

On the surface, this may seem like an a great deal of overhead. However, the major part of the cost of communication is in the connection set up time between the two communicating processes and not in the actual transfer. Therefore, the additional cost of piggybacking auxiliary variables is minimal.

The derived executable assertions can be implemented in linear time on the length of *schedule* and the size of SG. However, as time progresses, schedules from the lock managers will continue to grow. It is not always necessary to scan the entire SG and schedule data structures each time. This corresponds to not checking old schedule information repeatedly. In terms of the GAA/HAA proof outline, instead of checking the postcondition $Post_i$ and the total order all at once at the end of the scheduler, it can be checked incrementally. The following two theorems give sufficient conditions on when further checking of entries in SG and *schedule*, respectively, is no longer needed. The theory is similar to that used in establishing a consistent recovery line [KuRe86].

Theorem 5.4: If, for a transaction T_{ij} in SG, all entities locked by T_{ij} have been unlocked by T_{ij} and locks successfully acquired by succeeding transactions on all entities, then arcs incident to T_{ij}'s actions no longer need to be checked for participation in a cycle.

Proof: Consider some transaction T_{ij} which accesses a set of entities. We need to show that no cycle can occur involving T_{ij}'s entities after they have all been locked by other transactions under the conditions of the invariant I. Suppose such a cycle exists involving a transaction T_{st} on entity e. By the proof of Theorem 4.1, T_{ij} must have locked some entity unlocked by T_{st}. Thus, by the two-phase protocol, T_{ij} must not have started its unlocking phase. Thus, we have a contradiction; if locks have been acquired on all T_{ij}'s entities, no such transaction T_{st} can exist. Thus, we need not check T_{ij}'s actions further. □

Figure 5.1 shows SG after all T_{ij}'s entities have been locked by other transactions.

For auxiliary variable $schedule_i$ similar reasoning holds.

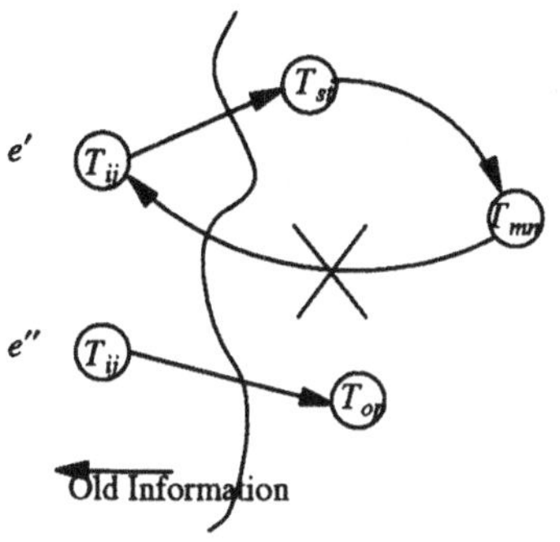

Figure 5.1: Lock Dependencies

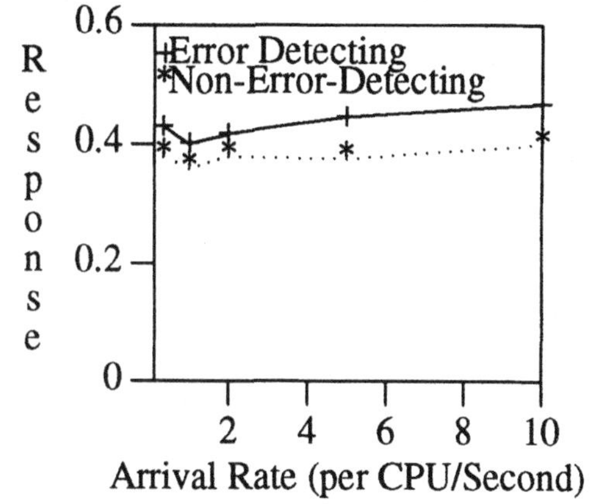

Figure 5.2: Performance

Theorem 5.5: If, for a transaction T_{ij} in $schedule_i$, all entities locked by T_{ij} have been unlocked by T_{ij} and locks successfully acquired by succeeding transactions on all entities, then T_{ij}'s actions are no longer needed in checking I.

Proof: Consider some transaction T_{st} which locks an entity unlocked by T_{ij}. Assertions of I involve only the current operation (from T_{st}) and the last operation (from T_{ij}) on each entity, e, once parts (2) and (3) of I have been checked, tuples involving T_{ij} of $schedule_i$ need never be checked again under I. □

5.3 Experimental Results

We implemented the code with embedding executable assertions resulting from *Changeling* in "C" on a collection of 10 NeXT workstations running the MACH OS interconnected by a 10Mbps CSMA/CD network. We measured the average response time for each transaction at the transaction managers with the results depicted in Figure 5.2. As we expected, the additional message length and computation (from Theorem 5.3 and it's following discussion) have a small effect on the system performance for these small number of processors.

6. Summary

We applied a fundamental technique (*Changeling*) which translates a concurrent verification proof outline into a error-detecting concurrent program on a sound, theoretical basis. It is obvious that the cost of everhead time cannot be avoided, but a good fault-tolerant algorithm will provide minimal overhead. systems.

Little of our work so far has been concerned with locating a fault when an error occurs or in recovery. Further research will also focus on using formal methods to incorporate fault location and recovery into the application-oriented fault tolerance paradigm.

Further research will focus on eventuality (or liveness) assertions. A temporal logic-based proof system [OwLa82] provides a convenient, expressive, way of reasoning in time. The problem of relating temporal assertions to executable assertions, however, is not straightforward, as in the axiomatic techniques. The complexity of the general problem of relating temporal assertions to those in first order logic is disheartening. Thus, the need to explore further the notion of temporal proof systems.

7. BIBLIOGRAPHY

[BeGo82] Bernstein, P. and Goodman, N., "Concurrency Control in Distributed Database Systems," *Computing Surveys*, 13, 2, 1981, pp. 185-221.

[EGLT76] Eswan, K.P., Gray, J.N., Lorie, R.A., and Traiger I.L., "The Notions of Consistency and Predicate Locks in a Database System," *Communications of the ACM*, 19, 11, 1976, pp. 624-633.

[Hoar69] Hoare, C., "An Axiomatic Basis for Computer Programming," *Communications of the ACM*, 12, 10, 1969, 576-583.

[KuRe86] Kuhl, J. and Reddy, S., "Fault Tolerance Considerations in Large, Multiple Pro cessor Systems," *IEEE Computer*, March 1986, pp. 56-67.

[LaSP82] Lamport, L., Shostack, R. and Pease, M., "The Byzantine General's Problem," *ACM Transaction on Programming Language Systems*, vol. 4, July 1982, pp. 38 2-401.

[LeGr81] Levin, G.M and Gries, D., "A Proof Technique for Communicating Sequential Process," *Acta Information*, 15, 1981, 281-302.

[LuMc91] Lutfiyya, H. and McMillin, B., "Comparison of Three Axiomatic Proof Systems," *UMR Department of Computer Science Technical Report Number CSC 91-13*, (Submitted to *Information Processing Letters*)

[LuSM92a] Lutfiyya, H., Schollmeyer, M., and McMillin, B., "Fault-Tolerant Distributed Sort Generated from a Verification Proof Outline," *Second International Workshop on Responsive Computer Systems, 1992* (To Appear)

[LuSM92b] Lutfiyya, H., Su, A., and McMillin, B., "Formal Derivation of an Error-Detecting Distributed Data Scheduler Using Changeling," UMR Technical Report CSc. 92-014.

[McNi92] McMillin, B. and Ni, L., "Reliable Distributed Sorting Through The Application-oriented Fault Tolerance Paradigm," *IEEE Trans. On Parallel and Distributed Computing*, Volume 3, Number 4, July 1992, pp. 411-420.

[OwLa82] Owicki, S. and Lamport, L., "Proving Liveness Properties of Concurrent Programs," *ACM TOPLAS*, Vol. 4, No. 3, July 1982, pp. 455-495.

REAL92: A Combined Specification Language for Real-Time Concurrent Systems and Properties

Nepomniaschy V.A., Shilov N.V
Institute of Informatics Systems
Siberian Division of the Russian Academy of Sciences
630090, Novosibirsk, Russia

Abstract. Real 92 is a new combined specification language for concurrent processes based on nondeterministic dialect of Specification and Design Language (SDL) and on dynamic version of Computation Tree Logic (CTL), both — SDL and CTL — with a real time. The paper includes a survey of syntax and sketch of operational semantics of Real 92, specification examples of systems and properties of concurrent communicating real-time processes.

Introduction

A progress of real-time distributed systems and computer networks stimulates a development of specification languages of both kinds - executional and logical. A specification language is a high level programming language with a great variety of consistent implementations. An executional specification language is oriented to a computer-aided design of a specified system in an imperative style. A logical specification language is oriented to a description of system properties in a declarative style and to a synthesis of a specified system. Examples of an executional specification language and logical specification language are Specification and Description Language SDL [1], [2] and Real-Time Temporal Logic [3], respectively.

A combined specification language consists of two sublanguages for executional and logical specifications. So, a combined specification language is oriented to design, development and synthesis of specified systems and to documentation, testing and verification of specified properties. The simplest example of the combined specification language is the language of annotated programs with PASCAL as the executional part and Firt-Order Logic as the logical part, but Hoare's definition of annotated program was the computer science discovery. In general, an unification of executional and logical specification languages in a joint specification language is a nontrivial problem.

REAL92 is a new version of the combined specification language REAL [4] for distributed systems. It is based on a dialect of SDL and a real-time variant of Dynamic Logic. The paper includes a survey of syntax and a sketch of operational semantics for REAL92. There exist another approaches to formal semantics of SDL-like languages of executional specifications [5], [6], [7]. The main difference is in our consideration of hierarchical structure for specifications using timers and multiple clock concept. Our approach to time is inspirited by [3] and consists in global multiple clocks as a vector of spontaneous events.

This work is supported by the Russian foundation for fundamental research under grant number 93-012-986.

1 A survey of REAL92 syntax

In this part an informal survey of a kernel of REAL92 syntax is given, for a context-free grammar of the kernel readers are referred to Appendix 1. We would also like to give some intuitive semantics of REAL92 constructions. An analogy with *SDL* and *CTL* would be useful too.

A REAL92 specification can be either executional or logical: the first one is intended for design of communicating real-time systems in an imperative style, and the second one is used for expression of their properties in a declarative style.

A specification (executional as well as logical) has a hierarchical features and consists of

- a head,
- a scale,
- a context,
- a scheme,
- a finite set of subspecifications.

A head of a specification defines a name and a kind of the specification: an executional specification can be either a process or a block, and a logical one is either a predicate or a formula. Processes and predicates are elementary specifications and have no subspecifications. Blocks and formulae are composite specifications and should have subblocks and subformulae, respectively.

A scale of a specification is a finite set of linear equalities/ inequalities between uninterpretated time units (zero and infinity may occur too). Each time the unit which occurs in the scale is the tick of an independent clock. So, the scale defines the multiple clocks whose speed must meet all equalities/inequalities from the scale. A duration is a finite set of time intervals. A continuation is the length of a time interval. Only time units of the multiple clocks, zero and infinity can be used as boundaries of durations in a scheme of a specification and as continuations.

A context of a specification is a finite set of type definitions and object declarations (variables, channels, interruptions, states and names). For simplicity, we suppose that a type definition can use finite oriented graphs as pre-defined finite types, natural numbers and strings as pre-defined basic types and the unique constructor for arrays. Each object declared in a context has the unique declaration. A declaration of an object defines attributes of the object and its location among them: a location can be either an own name of a specification or a name of a subspecification. A location of an object declared in a context of a specification is an own name of the specification iff the object occurs in a scheme of the specification; in the other case the location is a name of a subspecification and the object is declared in a context of the subspecification. All declared objects and only declared objects are visible for an external observer.

A declaration of a variable defines the following attributes: an appointment (program or quantor) and a type. Program variables are intended for computations in executional specifications and quantor ones - for quantification in logical specifications.

A declaration of a channel defines the following attributes: a channel role (input, output or inner), a data structure (queue, store or bag), a capacity (unbounded or bounded by a natural number), a finite set of possible signals with characteristics of life-time (till the first request or a continuation) and vectors of parameter names and types. Channels are intended to be a tool for a local communication between two processes or between a process and environment.

A declaration of an interruption defines the following attributes: a characteristics of life-time (till the first request or a continuation) and a vector of parameter names and types. Interruptions are an analog of signals but are intended for broadcasting.

A declaration of a state defines the following attribute: sorts of possible actions (signal acception or delivery, generation or treatment of interruption, task for deadlock avoidance).

A declaration of a name defines the following attribute: a kind of a subspecification with this name (process, block, predicate or formula).

A scheme of an executional specification consists of a diagram and a finite set of fairness conditions. A fairness condition is a name of a logical subspecification (subformula). The sense of fairness conditions is as follows: a behaviour of the executional specification is said to be fair iff each fairness condition is properly valid infinitely often along the behaviour.

A scheme of a logical specification consists of a diagram and a finite set of systems. A system is a name of an executional subspecification (subblock). The sense of systems is as follows: a logical specification is said to be properly valid in a configuration iff it is valid in this configuration and the configuration belongs to a fair behaviour of a system.

A subspecification is a specification itself: a subblock is an executional specification, a subformula is a logical one. A name of each subspecification of a specification is declared in a context of the specification and occurs in a scheme of the specification. A scale of a subspecification is an extension of a scale of a specification. A set of type definitions of a subspecification is an extension of a similar set of specifications. If an object is declared in a context of a specification with a name of a subspecification as a location, then the object is similarly declared in a context of the subspecification but with another location.

At last we have to describe a variety of diagrams and related topics.

A diagram of a process is a finite set of transitions. A transition is marked by a state, consists of a body and duration, is finished by a jump to next states.

For simplicity let us fix a scale and a context. If a possible action of a declared state is

(1) a generation/treatment of an interruption,

(2) an acception/delivery of a signal,

(3) a task for deadlock avoidance,

then respectively a body of a transition marked by the state defines

(1) an interruption management action (set up or treat), a declared interruption and an adequate vector of declared program variables for parameters,

(2) a local communication action (read or write), a declared channel, a possible signal and an adequate vector of declared program variables for parameters,

(3) a task to execute a non-deterministic program which is constructed from assignment to declared program variables and boolean conditions as guards by the sequential composition, deterministic/non-deterministic choices and loops.

An intuitive semantics of transitions is obvious, but we have to remark that an interruption management has the highest priority, a local communication has the middle priority, and a task has the least priority.

A diagram of a block is a finite set of routes. A route consists of a source, a channel or an interruption, and a target. For simplicity, let us fix a context. If for a declared channel there exists a route with this channel in a diagram of a block, then it is the unique route and it is said to be the route of the channel. A source/target of a route is a declared name of a subblock or an environment. A source/target of a declared channel route is an environment iff the channel is declared as an input/output channel; in other case the channel is declared as an inner channel. If a source/target of a declared channel route is a name of a subblock then the channel is similarly declared in a context of the subblock but with an output/input role. If a source/target of a declared interruption route is a name of a subblock, then the channel is similarly declared in the context of the subblock.

A diagram of a predicate can be

(1) a relation between variables and parameters of interruptions and signals,

(2) a locator of control states with durations,

(3) a tester for activity, deadlock or stability of subblocks,

(4) a controller for emptiness, overflow or concurrent access to channels,

(5) a checker for available signals of channels or interruptions with continuations.

A diagram of a formula is constructed from names of subspecifications by

(1) propositional combinations,

(2) universal and existantional quantification over a quantor variable,

(3) dynamic/temporal expressions.

A dynamic/temporal expression consists of a modality over fair behaviours of a system (which is presented by a name of the system) and a modality over moments of time from a duration. Both modalities of a dynamic/temporal expression are chosen among the box and the diamond modalities with an intuitive semantics "for all" and "for some", respectively. So semantics of the modalities is similar to quantification, but deals with behaviours and moments rather than with the values of variables.

2 An Example: Passenger and Slot-Machine

Let us consider the following example: Passenger is buying a railway ticket in automatic booking-office (Slot-Machine). Passenger can:
- see an electronic turned on/off Table with a sum of dropped coins,
- receive returned coins from a special window for a Change,
- get a ticket with a station name from another special Window,
- drop coins into Slot,
- press a Button with a station name or with a return command.

Symmetrically, Slot-Machine is an electro-mechanical automatic machine which can:
- show a sum of dropped coins on a table,
- return all dropped coins back through a Change window,
- print a station name on a ticket and release it through a Window,
- get a station name or return command from Button,
- receive coins through a Slot.

At the same time a Slot-Machine gets energy from an Environment and a gain sum is extracted via a Cashier to an Environment.

It is known from psychological experiments that a Passenger as an average man can make 1-20 solutions and actions per minute. We suppose that Slot-Machine can execute 30-100 operations per second. We consider a case when energy is turned on and a Cashier extracts a gain eventually.

Let us embed the above informal example in REAL92 as an executional specification. We would like to specify Passenger and Slot-Machine as a block with two subblocks, both are processes. So, the head of the specification is:

Passenger and Slot-Machine: block.

The scale of the specification deals with minutes (min), seconds (sec), Passenger's ingenuity (ing) and Machine operations (opr):

1 min = 60 sec;

1 ing $\leq$ 1 min $\leq$ 20 ing;

30 opr $\leq$ 1 sec $\leq$ 100 opr;

For simplicity, we omit some details of alive Passenger and real Slot- Machine: in the example we suppose that there exist coins with nominal 1, 2, 3 and 5 rubles only, the unique wrong coin and only 3 stations - A, B and C. So we omit type definitions because of triviality.

First of all the context of the specification includes the declaration of two interruptions - energy on and energy off - with the location in the diagram of the specification: own interruptions energy on, energy off with life-time till the first request without parameters.

Then the context includes declarations of six channels - five of which are inner and one is an output channel, but all are located in the diagram of the

specification:
own output 1 - element bag channel Cashier

for signal gain with life-time till the first request
with parameter sum $\in$ natural;

own inner 1 - element bag channel Table

for signal turn on with life-time till the first request
with parameter sum $\in$ natural;

for signal turn off with life-time 1 opr
without parameters;

own inner unbounded bag channel Change

for signal coin with life-time till the first request
with parameter nominal $\in \{1, 2, 3, 5\}$,

for signal wrong coin with life-time till the first request
without parameters;

own inner 1 - element bag channel Window

for signal ticket with life-time till the first request
with parameter station $\in \{A, B, C\}$;

own inner 1 - element bag channel Slot

for signal coin with life-time till the first request
without parameter nominal $\in \{1, 2, 3, 5\}$,

for signal wrong coin with life-time till the first request
without parameters;

own inner 1 - element bag channel Buttons

for signal enter with life-time till the first request
with parameter station $\in \{A, B, C\}$,

for signal return with life-time till the first request
without parameters;

At last, the context of the specification includes two declarations of names for four subspecifications - one for two subformulae for fairness conditions and one for two subblocks:
own names Energy, Money for formulae;
own names Passenger, Slot-Machine for processes;
Fairness conditions in the specification consist of two names of subformulae:
Energy; Money;

The diagram in a graphical syntax is represented on the picture below:

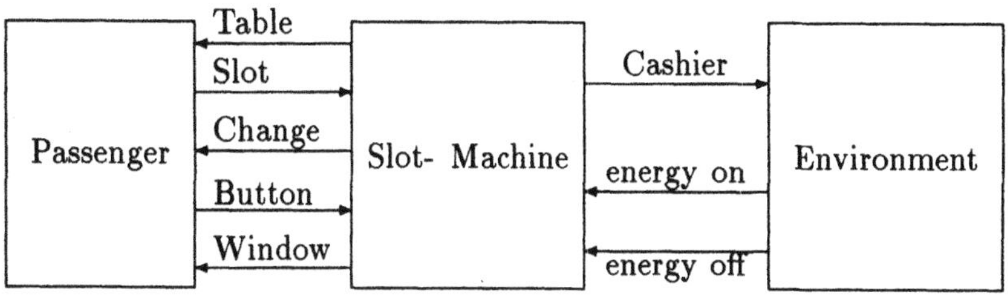

Now we have to give subspecifications - two predicates with names Energy and Money, and two processes with names Passenger and Slot-Machine. As we illustrate expressiveness of REAL92, we give one subformula and one subblock only:

Money: predicate

gain in Cashier.

We specify the process Passenger. The head of the process is:

Passenger: process.

The scale of the process is the same as the scale of the block. The process context includes declarations of variables for the Passenger purse, the nominal of a coin, a desirable station, pre-known ticket price and sum of dropped coins:

own program var purse $\in$ ($\{1,2,3,5\}$ array of natural);

own program var nominal $\in \{1,2,3,5\}$;

own quantor var station $\in \{A, B, C\}$;

own quantor var price $\in$ natural;

own program var sum $\in$ natural;

Of course, the process context includes declarations of channels Table, Change, Window, Slot and Button which are similar to those declarations from the context of the block Passenger-Slot Machine, but a role of channesl Table, Change, Window is input, and a role of channels Slot; Buttons is output. And the process context includes four declarations for nine states:

own state look on Table for acceptance:

own states wait ticket, receive coins for acceptance and deadlock avoidance;

own state press Button, return command, drop to Slot for delivery;

own state compare price, put to purse, pick out coin for deadlock avoidance.

There are no fairness conditions in the process. The diagram of the process is represented in Appendix 2.

At last, let us consider the following property of Passenger process: for any price if Passenger sees that the Table is turned one with zero sum and he has enough money for the price in his purse, then always after some time he will press the Button with the desirable station name. We embed this informal property in REAL92 as a logical specification. The head of this specification is:

Passenger property: formula.

The scale and the context of the logical specification are the same as the scale and the context of the process Passenger. The system of the logical specification consists of the unique name of the process Passenger. The diagram of the specification is represented below:

$\forall$ price
Table is turned on &
Table shows zero &
Passenger sees Table &
enough money $\Rightarrow$
$\Box$Passenger $\Diamond$ from now till infinity
button is pressed & desirable station.

All subformulae of the specification are predicates with the same scales, contexts and systems as the specification. So, we give heads and diagrams for subformulae only:

Table is turned on: predicate
Turn on in Table from now upto 1 ign.
Table shows zero: predicate
Table.Sum = 0
Passenger sees Table: predicate
at look on Table along 1 ign.
Enough money: predicate
$1 \times$ purse [1] $+ 2 \times$ purse [2] $+ 3 \times$ purse [3] $+ 5 \times$ purse [5] $\geq$ price.
Button is pressed: predicate
enter in Button from now upto 1 ign.
desirable station: predicate
button.station = station.

3 A Sketch of REAL93 Semantics

The formal semantics of specifications (executional as well as logical) is defined by means of the space of configurations. Each executional specification is assigned a semantical set of fair behaviours - countable sequences of configurations. Each logical specification is assigned a semantical set of systems configurations where the specification is valid.

Let us fix a specification. A configuration CNF is six-tuple (T, V, C, E, S, N) where

- T is multiple clocks i.e. an integer solution of the scale as a system of linear equalities/inequalities,
- V is variables values,

- $C = (R, B, P)$ with
  - R are numbers of processes which try to read from or to write into channels,
  - B are data structures (bags et c.t.) of signals for channels,
  - P are data structures of parameters values for signals,
- $E = (G, H)$ with
  - G is boolean characteristics of generation for interruptions,
  - H are parameter values for interruptions,
- $S = (M, D)$ with
  - M is boolean characteristics of control activity for states,
  - D is multiple clock counter of delays for states,
- $N = (A, F, L)$ with
  - A is boolean characteristics of activity for executional specification names,
  - F is boolean characteristics of finiteness for executional specification names,
  - L is boolean characteristics of deadlocs for executional specification names.

Note this definition is informal but can be formalized adequately. Also we use standard notations $\perp$ for undefinite value and $\emptyset$ for the empty set.

Let us fix an executional specification. Semantics is defined in terms of events and step rules in Plotkin style [6], [8].

There exist five kinds of events:

- broadcast interruption with parameters,
- consume interruption with parameters,
- put signal with parameters into channel,
- get signal with parameters from channel,
- invisible event.

Note that in visible events all objects are declared.

A step rule has the form

$CND \vdash CNF1\langle EVN\rangle CNF2$,

where CND is the applicability condition, $CNF1$, $CNF2$ are configurations and EVN is an event. An intuitive semantics of the step rule is as follows: if the condition CND holds, then the executional specification can be transformed by the event EVN from the configuration $CNF1$ to the configuration $CNF2$.

A countable sequence of configurations is said to be a behaviour iff for each configuration $CNF1$ and the next one $CNF2$ from this sequence there exists an event EVN and a condition CND such that

$CND \vdash CNF1\langle EVN\rangle CNF2$ is an instance of step rules and the condition CND holds.

The first step rule is the unique rule for blocks: the composition rule. Informally, a step of a block is a projection of simultaneous steps of subblocks. Let us suppose that a block A has subblocks $B1 - Bk$ only. Then

for all $I = 1, ..., k$

$CNF1.I\langle EVN/BI\rangle CNF2.I$,

$CNF1.I/A = CNF1/BI, CNF2.I/A = CNF1/BI$
$\vdash CNF1\langle EVN/A\rangle CNF2$,

where $CNF1$, $CNF2$ are configurations of A, $CNF1.I$ and $CNF2.I$ are configurations of $BI(I = 1, ..., k)$, EVN is an event of some subblocks of A, CNF/C is a projection of a configuration CNF on a specification C, and EVN/C is either the event EVN, if all objects of the event are declared in a specefication C, or invisible event in the other case.

All other step rules deal with processes. So, let us fix a process and a pair of configurations $CNF1 = (T1, ...)$, $CNF2 = (T2, ...)$.

The first rule for the process is the stutter rule. Informally, this rule represents the case when in the active process nothing changes except time. Let us suppose that there exists a transition in the process diagram of the form:

state: duration body jump next states.

Then

$M1(state) = true, T1 \leq T2$,
$D1(state) < sup(duration)$,
$T2 - T1 = D2(state) - D1(state)$,
$\vdash CNF1\langle INVISIBLE\rangle CNF2$,

where sup(duration) is the greatest upper bound of the duration, $CNF2$ differs from $CNF1$ by multiple clocks and by multiple clocks counters for the state.

There exists the group of rules for interruption management. Let us demonstrate only one of them: the step rule for the process generation of an interruption Int with live-time till the first request. Let us suppose that there exists a transition in the diagram of the process of the form:

state: duration
set up Int with Var for par Par
jump next states.

Then

$M1(state) = true, D1(state) \in duration$
$\vdash CNF1\langle broadcast \quad Int \quad with \quad Par\rangle CNF2$,

where differences of $CNF1$ and $CNF2$ are below:

$M2(state) = false, D2(state) = \perp$,
$M2(nextstate) = true, D2(nextstate) = 0$,
$G2(Int) = true, H2(Par) = V1(Var), R2 = R1 - R(state) + R(nextstate)$.

There exists the group of rules for signals management. Let us demonstrate one of those rules too-the step rule for the process writing of a signal $Sign$ with life-time till the first request into a 1-element bag channel Chn. Let us suppose that there exists a transition in the diagram of the process of the form:

state: duration
write Sgn into Chn with var Var for par Par
jump next states,

and in addition rules for interruptions management are not applicable. Then,

$M1(state) = true, D1(state) \in duration$,
$R1(Chn) \geq 1, S(Chn) = \emptyset$
$\vdash CNF1\langle$ put Sgn with Par into Chn $\rangle CNF2$,

where differences of $CNF1$ and $CNF2$ are below:

$M2(state) = false, D2(state) = \bot,$

$M2(nextstate) = true, D2(nextstate) = 0,$

$S(Chn) = \{Sgn\}, P(Chn) = \{V(Var)\}, R2 = R1 - R(State) + R(nextState)$

The last three rules deal with program executions, stabilisation and deadlocks.

The step rule is intended for program execution. Informally, this step consists of program variables transformation by means of a program which is considered as an indivisible operation. Let us suppose that there exists a transition in the diagram of the process of the form:

state: duration execute program jump next states, and rules for interruptions and signal management are not applicable. Then

$M1(State) = true, D1(State) \in duration,$

$V1\langle$ program $\rangle V2$

$\vdash CNF1\langle INVISIBLE\rangle CNF2,$

where differences of $CNF1$ and $CNF2$ are below:

$M2(state) = false, D2(state) = \bot,$

$M2(nextstate) = true, D2(nextstate) = 0,$

$R2 = R1 - R(state) + R(nextstate).$

Here the binary relation $\langle$ program $\rangle$ on variable values is defined as usual.

Let us consider the step rule for deadlock. Informally, a deadlock happens iff temporal restrictions for all possible actions are abrupted: all desired communications have not taken a place in an admitted time and no deadlock avoidence instruction will be applicable. Let us suppose that there exists a transition in the diagram of the process of the form:

state: duration body jump next state

and the stutter rule is not applicable. Then

$M1(state) = true, T1 \leq T2$

$\vdash CNF1\langle INVISIBLE\rangle CNF2,$

where $CNF2$ differs from $CNF1$ by multiple clocks only.

Let us consider step rule for stabilization. Informally, the process is stable iff it have no transitions for executions. Let us suppose that a state does not mark any transition in the diagram of the process. Then

$M1(state) = true, T1 \leq T2$

$\vdash CNF1\langle INVISIBLE\rangle CNF2,$

where $CNF2$ differs from $CNF1$ by multiple clocks only.

Now let us consider a sketch of the semantics for predicates and formulae.

A relation is valid in a configuration iff values of left and right parts of a diagram of the relation evaluated in the configuration satisfy the relation. A locator is valid in a configuration iff the boolean characteristics of activity for a state from a diagram of the locator is true and the delay of this state belongs to a duration. A test for activity, deadlock or halting is valid in a configuration iff the boolean characteristics A, L, or F of a name from the diagram of the test are true, respectively. A checker is valid in a configuration iff a signal contained in the buffer of the channel from a diagram of the checker is equal to the signal from the scheme of the checker along the continuation.

If the semantics of subspecifications of a propositional combination is defined, then the semantics of the propositional combination is defined as usual. For example, if a propositional combination is the conjunction of subformulae names, then the formula is valid in a configuration iff all subformulae are valid in the configuration.

A quantified formula is valid in a configuration iff the subformula of the formula is valid in each (in the case of the universal quantifier) or some (in the case of the existential quantifier) configuration which may differ from the initial configuration in the value of the variable connected by the quantifier only.

A dynamic/temporal formula is valid in a configuration iff

for each (in the case of the first modality is the box) or some (in the case of the first modality is the diamond) fair behaviour of the system which begins from the configuration,

for each (in the case of the second modality is the box) or some (in the case of the second modality is the diamond) configuration from the behaviour and with multiple clocks belongs to the duration

the subformula (followed the modalities) is valid in the configuration.

4 Conclusion: Towards REAL93

The language REAL is an experimental developing language. The REAL92 version is oriented to hand design and documentation of distributed systems and their properties. We are going to extend this current version REAL92 as follows:

1. to permit copies of processes and blocks;
2. to use fixed points as new logical operations.

Then we are going to develop a REAL-system for computer-aided design, documentation, testing and verification of REAL specification. There exists a hope for sound (and complete in some sense) axiomatizations for equivalent transformations of executable specifications and for the validity of logical specifications. At the same time our group has implemented prototyping model-checker for the propositional program logic with fixed points. So, an expected REAL-system will be based upon equivalent transformations, automatic inference and model-checking. There exist model-checking systems for temporal properties of *SDL*-like executable specifications [9], but we are going to develop a model-checker for logical specifications of REAL.

References

1. Specification and Description Language // CCITT, Recomendation Z.100, 1988.
2. Barzdin J.M. et al. Specification Language SDL/PLUS and its applications // Computer Center of Latvian State University, Riga, 1988.

3. Ostroff J.S. Automated Verification of Timed Transition Models // Lecture Notes in Computer Science, v. 407, 1990, p. 247-256.
4. Nepomniaschy V.A., Shilov N.V. A Language for Specifing Systems and Properties of Real-Time Communicating Processes // Methods of Theoretical and Systems Programming // Novosibirsk, 1991, p. 32-45 (in Russian).
5. Broy M. Towards a Formal Foundation of the Specification and Description Language SDL // Formal Aspects of Computing, v. 3, 1991, n. 1, p. 21-57.
6. Orava F. Formal Semantics of SDL Specifications // Proc. Conf. on Protocol Specification, Testing, and Verification - VIII, North-Holland, 1988, p.143-157.
7. Mery D., Mokkedem A. CROCOS: An Integrated Environment for Interactive Verification of SDL Specifications // Lecture Notes in Computer Science, v. 663, 1993, p. 343-356.
8. Plotkin G.D. A Structure Approach to Operational Semantics // Technical report FN-19, Aarhus University, DAIMI, Denmark, 1981.
9. Covalli A.R., Horn F. Proof of Specification properties by using finite state machines and temporal logic // Proc. IFIP Conf. on Protocol Specifications, Testing and Verification - VII, 1987, p. 221-233.

Appendix 1

A Kernel of the Formal REAL92 Syntax

specification :: executional_specification | logical_specification
executional_specification :: process | block
logical_specification :: predicate | formula

process :: process_head scale context fairness_conditions
process_diagram subformulae
block :: block_head scale context fairness_conditions
block_diagram subblocks subformulae

predicate :: predicate_head scale context systems
predicate_diagram subblocks
formula :: formula_head scale context systems
formula_diagram subblocks subformulae

process_ head :: name : PROCESS
block_ head :: name : BLOCK
predicate_ head :: name : PREDICATE
formula_ head :: name : FORMULA

scale :: { linear_equality_of_time_units |
linear_inequality_of_time_units }
continuation :: ALONG time
durution :: { interval }
interval :: left time right time
left :: FROM | AFTER
right :: UPTO | TILL
time :: linear_expression_of_time_units | NOW | INFINITY

context :: { type_definition | object_declaration }

fairness_conditions :: { name }
systems :: { name }

subblocks :: { executional_specification }
subformulae :: { logical_specification }
object_declaration :: variable_declaration | channel_declaration —
interruption_declaration |
state_declaration | name_declaration
variable_declaration :: location appointment VAR variable ∈ type ;
channel_declaration :: location role capacity structure CHN channel
{ FOR signal { WITH PARAMETER name ∈ type }
WITH LIFE-TIME life_time } ;

interruption_declaration :: location INT interruption
{ WITH PARAMETER name ∈ type }
WITH LIFE-TIME life_time ;
state_declaration :: location state FOR action_sort ;
name_declaration :: location name FOR kind ;

location :: OWN | name
appointment :: PROGRAM | QUANTOR
role :: INPUT | OUTPUT | INNER
capacity :: number-ELM | INFINITE
structure :: QUEUE | STORE | BAG
life_time :: TILL THE FIRST REQUEST | continuation
action_sort :: { ACCEPTANCE | DELIVERY | TASK | GENERATION | TREAT }
kind :: PROCESS | BLOCK | PREDICATE | FORMULA

process_diagram :: { transition }
transition :: state : duration body jump ;
jump :: JUMP { state }

block_diagram :: { route }
route :: source via target
via :: channel | interruption
source :: name | ENVIRONMENT
target :: name | ENVIRONMENT

predicate_diagram :: relation | locator | controller | checker | test

formula_diagram :: name | (propositional_combination) |
(quantifier variable formula_diagram) |
(modality name modality duration
formula_diagram)
quantifier :: $\forall$ | $\exists$
modality :: $\Box$ | $\Diamond$

Appendix 2

The Diagram for Passenger in Example (Section 2)

look on Table:
from now upto 1 ign
read with var sum for par sum turn on from Table
jump compare price;

look on Table:
from now upto 1 ign

```
        read without parameters turn off from Table
jump unsatisfaction;

compare price:
        from now upto 1 ign
        execute if sum ≥ price then skip else abort
jump press Button;

compare price:
        from now upto 1 ign
        execute if sum ¡ price then skip else abort
jump pick out coin;

drop to slot:
        from now upto 1 ign
        write coin into slot
                with var nominal for par nominal
jump look on Table;

press Button:
        from now upto 1 ign
        write enter into Button
                with var station for par station
jump wait ticket;

wait ticket:
        from now upto 1 ign
        read ticket from Window
                with var station for par station
jump satisfaction;

wait ticket:
        from 1 ign upto 2 ign
        execute skip
jump return command;

return command:
        from now upto 1 ign
        write return into Button
                without parameters
jump receive coins;

receive coins:
        from now upto 1 ign
        read coin from change
```

```
            with var nominal for par nominal
jump put to purse;

receive coins:
        from 1 ign upto 2 ign
        execute skip
jump unsatisfaction;

put to purse:
        from now upto 1 ign
        execute purse[nominal] := purse[nominal] + 1
jump receive coins;

pick out coin:
        from now upto 1 ign
        execute
                if sum + 5 ≤ price and purse[5] > 0
                then nominal := 5
                else if sum + 3 < price and purse[3] > 0
                        then nominal := 3
                        else if sum + 2 ≤ price and purse[2] > 0
                                then nominal := 2
                                else nominal := 1;
                purse[nominal] := purse[nominal] - 1
jump drop to slot;

pick out coin:
        from 1 ign upto 2 ign
        execute skip
jump return command;
```

Algebraic Calculation of Graph and Sorting Algorithms

Bernhard Möller

Institut für Mathematik, Universität Augsburg, D-86135 Augsburg, Germany,
e-mail: moeller@uni-augsburg.de

Abstract. We introduce operators and laws of an algebra of formal languages, a subalgebra of which corresponds to the algebra of (multiary) relations. This algebra is then used in the formal specification and derivation of some graph and sorting algorithms. This study is part of an attempt to single out a framework for program development at a very high level of discourse, close to informal reasoning but still with full formal precision.

1 Introduction

The transformational or calculational approach to program development has by now a long tradition (see e.g. [3, 7, 1, 13]. There one starts from a (possibly non-executable) specification and transforms it into a (hopefully efficient) program using semantics-preserving rules. Many specifications and derivations, however, suffer from the use of lengthy, obscure and unstructured expressions involving formulas from predicate calculus. This makes writing tedious and error-prone and reading difficult. Moreover, the lack of structure leads to large search spaces for supporting tools.

The aim of modern algebraic approaches (see e.g. [10, 2]) is to make program specification and calculation more compact and perspicuous. They attempt to identify frequently occurring patterns and to express them by operators satisfying strong algebraic properties. This way the formulas involved become smaller and contain less repetition, which also makes their writing safer and speeds up reading (once one is used to the operators). The intention is to raise the level of discourse in formal specification and derivation as closely as possible to that of informal reasoning, so that both formality and understandability are obtained at the same time. In addition, the search spaces for machine assistance become smaller, since the search can be directed by the combinations of occurring operators.

If one succeeds in finding equational laws, proof chains rather than complicated proof trees result as another advantage. Moreover, frequently aggregations of quantifiers can be packaged into operators; an equational law involving them usually combines a series of inference steps in pure predicate calculus into a single one.

We illustrate these ideas by treating some graph and sorting problems within a suitable algebra of formal languages [12]. This differs from the approach of [10, 2] in that we concentrate on properties of the underlying problem domain rather than on those of standard recursions over inductively defined data types.

2 The Algebra of Formal Languages

A formal language is a set of words over letters from some alphabet. Usually these letters are considered as "atomic". We shall take a more liberal view and allow arbitrary objects as letters. Then words over these "letters" can be viewed as representations for tuples or sequences of objects. In particular, if the alphabet consists of (names for) nodes of a directed graph, words can be considered as sequences of nodes and can thus model paths in the graph.

2.1 Words, Languages and Relations

We denote by $A^{(*)}$ the set of all finite words over an alphabet A. A **(formal) language** V over A is a subset $V \subseteq A^{(*)}$. As is customary in formal language theory, to save parentheses a singleton language is identified with its only word and a word consisting just of one letter is identified with that letter. By ε we denote the empty word over A.

A **relation of arity** n is a language R such that all words in R have length n. In particular, the empty language $\emptyset$ is a relation of any arity. There are only two 0-ary relations, viz. $\emptyset$ and ε.

2.2 Pointwise Extension

We define our operations on languages first for single words and extend them pointwise to languages. We explain this mechanism for a unary operation; the extension to multiary ones is straightforward. Since we also need partial operations, we choose $\mathcal{P}(A^{(*)})$, the powerset of $A^{(*)}$, as the codomain for such an operation. The operation will then return a singleton language consisting of the result word, if this is defined, and the empty language $\emptyset$ otherwise. Thus $\emptyset$ plays the role of the error "value" $\perp$ in denotational semantics. Consider now such an operation $f : A^{(*)} \to \mathcal{P}(A^{(*)})$. Then the **pointwise extension** of f is denoted by the same symbol, has the functionality $f : \mathcal{P}(A^{(*)}) \to \mathcal{P}(A^{(*)})$ and is defined by

$$f(U) \stackrel{\text{def}}{=} \bigcup_{x \in U} f(x)$$

for $U \subseteq A^{(*)}$. By this definition, the extended operation distributes through union:

$$f(\cup \mathcal{V}) = \cup \{f(V) : V \in \mathcal{V}\}$$

for $\mathcal{V} \subseteq \mathcal{P}(A^{(*)})$. By taking $\mathcal{V} = \emptyset$ we obtain strictness of the pointwise extension with respect to $\emptyset$:

$$f(\emptyset) = \emptyset \ .$$

Moreover, taking $\mathcal{V} = \{U, V\}$ and using the equivalence $U \subseteq V \Leftrightarrow U \cup V = V$, we also obtain monotonicity with respect to $\subseteq$:

$$U \subseteq V \ \Rightarrow \ f(U) \subseteq f(V) \ .$$

Finally, bilinear equational laws for f, i.e., laws in which each side has at most one occurrence of every variable, are inherited by the pointwise extension (see e.g. [8]).

2.3 Concatenation and Auxiliaries

We now apply this mechanism to the operation of concatenation. It is denoted by $\bullet$ and is associative, with ε as its neutral element:

$$u \bullet (v \bullet w) = (u \bullet v) \bullet w \ ,$$
$$\varepsilon \bullet u = u = u \bullet \varepsilon \ .$$

Since associativity of concatenation and neutrality of ε are expressed by bilinear equational laws, they also hold for the pointwise extension of concatenation to languages:

$$U \bullet (V \bullet W) = (U \bullet V) \bullet W \ ,$$
$$\varepsilon \bullet U = U = U \bullet \varepsilon \ .$$

The **identity** I_a over a letter $a \in A$ is defined by

$$I_a \stackrel{\mathsf{def}}{=} a \bullet a$$

and extended pointwise to sets of letters. In particular, I_A is the binary identity relation (or diagonal) on A.

The operation set calculates the set of letters occurring in a word. It is defined inductively by

$$\begin{aligned} \mathsf{set}\,\varepsilon &= \emptyset \ , & \\ \mathsf{set}\,a &= a & (a \in A) \ , \\ \mathsf{set}\,(u \bullet v) &= \mathsf{set}\,u \ \cup\ \mathsf{set}\,v & (u, v \in A^{(*)}) \ , \end{aligned}$$

and, again, extended pointwise to languages.

Finally, the first and last letters of a word (if any) are given by the operations fst and lst defined by

$$\mathsf{fst}\,\varepsilon = \emptyset = \mathsf{lst}\,\varepsilon \ ,$$
$$\mathsf{fst}\,(a \bullet u) = a \qquad \mathsf{lst}\,(v \bullet b) = b$$

for $a, b \in A$ and $u, v \in A^{(*)}$. Again, these operations are extended pointwise to languages. For a binary relation $R \subseteq A \bullet A$ they have special importance: $\mathsf{fst}\,R$ is the domain of R, whereas $\mathsf{lst}\,R$ is the codomain of R.

As unary operators, fst, lst and set bind strongest.

2.4 Shuffle

An operation on languages that is well-known from the trace theory of parallel processes is the **shuffle**, viz. the set of arbitrary interleavings of words from these languages. It is defined inductively ($a, b \in A, s, t \in A^{(*)}$):

$$\varepsilon \,|||\, t \stackrel{\mathsf{def}}{=} t \ ,$$
$$s \,|||\, \varepsilon \stackrel{\mathsf{def}}{=} s \ ,$$
$$(a \bullet s) \,|||\, (b \bullet t) \stackrel{\mathsf{def}}{=} a \bullet (s \,|||\, (b \bullet t)) \ \cup\ b \bullet ((a \bullet s) \,|||\, t) \ .$$

As usual, the operation is extended pointwise to sets. It is commutative and associative and satisfies

$$\mathsf{set}\,(S \,|||\, T) = \mathsf{set}\,S \ \cup\ \mathsf{set}\,T \ . \tag{1}$$

2.5 Permutations

An important requirement for sorting is that the number of occurrences of each element in the word to be sorted must not be changed by the sorting process. An equivalent requirement is that the result word must be a permutation of the input word. To capture that, we give a definition of permutation in terms of our operators. For a word s we denote the set of its permutations by $permw(s)$. Formally,

$$\begin{aligned} permw(\varepsilon) &\stackrel{\text{def}}{=} \varepsilon \,, \\ permw(a) &\stackrel{\text{def}}{=} a \,, \\ permw(s \bullet t) &\stackrel{\text{def}}{=} permw(s) \,|||\, permw(t) \,. \end{aligned} \tag{2}$$

A useful property is

$$a \,|||\, permw(s) = \bigcup_{u \bullet v \in permw(s)} permw(u) \bullet a \bullet permw(v) \,. \tag{3}$$

We shall also need the set of all permutations of a finite set of letters. This is given by

$$\begin{aligned} perms(\emptyset) &\stackrel{\text{def}}{=} \varepsilon \,, \\ perms(a \cup T) &\stackrel{\text{def}}{=} a \,|||\, perms(T \backslash a) \,, \end{aligned} \tag{4}$$

for $a \in A$ and finite $T \subseteq A$. Note that in general $permw(s) \neq perms(\text{set}\, s)$, since multiple occurrences of a letter are preserved by $permw(s)$ but not by $perms(\text{set}\, s)$. Hence each element of $perms(T)$ is repetition-free.

2.6 Join and Composition

For words s and t over alphabet A we define their **join** $s \bowtie t$ and their **composition** $s \,;\, t$ by

$$\varepsilon \bowtie s = \emptyset = s \bowtie \varepsilon \,, \qquad \varepsilon \,;\, s = \emptyset = s \,;\, \varepsilon \,,$$

and, for $s, t \in A^{(*)}$ and $a, b \in A$, by

$$(s \bullet a) \bowtie (b \bullet t) \stackrel{\text{def}}{=} \begin{cases} s \bullet a \bullet t & \text{if } a = b \,, \\ \emptyset & \text{otherwise} \,, \end{cases} \qquad (s \bullet a) \,;\, (b \bullet t) \stackrel{\text{def}}{=} \begin{cases} s \bullet t & \text{if } a = b \,, \\ \emptyset & \text{otherwise} \,. \end{cases}$$

These operations provide two different ways of "glueing" two words together upon a one-letter overlap: join preserves one copy of the overlap, whereas composition erases it. Again, they are extended pointwise to languages. On relations, the join is a special case of the one used in data base theory. On binary relations, composition coincides with usual relational composition (see e.g. [19]). To save parentheses we use the convention that $\bullet$, $\bowtie$ and ; bind stronger than all set-theoretic operations.

To exemplify the close connection between join and composition further, we consider a binary relation $R \subseteq A \bullet A$ modelling the edges of a directed graph with node set A. Then

$$\begin{aligned} R \bowtie R &= \{a \bullet b \bullet c \,:\, a \bullet b \in R \wedge b \bullet c \in R\} \,, \\ R ; R &= \{a \bullet c \,:\, a \bullet b \in R \wedge b \bullet c \in R\} \,. \end{aligned}$$

Thus, the relation $R \bowtie R$ consists of exactly those paths $a \bullet b \bullet c$ which result from glueing two edges together at a common intermediate node. The composition $R\,;R$ is an abstraction of this; it just states whether there is a path from a to c via some intermediate node without making that node explicit. Iterating this observation shows that the relations

$$R,\quad R \bowtie R,\quad R \bowtie (R \bowtie R),\ \ldots$$

consist of the paths with exactly $1, 2, 3, \ldots$ edges in the directed graph associated with R, whereas the relations

$$R,\quad R\,;R,\quad R\,;(R\,;R),\ \ldots$$

just state existence of these paths between pairs of nodes.

The operations associate nicely with each other and with concatenation:

$$\left.\begin{array}{rcll}
U \bullet (V \bullet W) &=& (U \bullet V) \bullet W \ , & \\
U \bowtie (V \bowtie W) &=& (U \bowtie V) \bowtie W \ , & \\
U\,;(V\,;W) &=& (U\,;V)\,;W & \Leftarrow V \cap A = \emptyset \ , \\
U\,;(V \bowtie W) &=& (U\,;V) \bowtie W & \Leftarrow V \cap A = \emptyset \ , \\
(U \bowtie V)\,;W &=& U \bowtie (V\,;W) & \Leftarrow V \cap A = \emptyset \ , \\
U \bullet (V \bowtie W) &=& (U \bullet V) \bowtie W & \Leftarrow V \cap \varepsilon = \emptyset \ , \\
U \bowtie (V \bullet W) &=& (U \bowtie V) \bullet W & \Leftarrow V \cap \varepsilon = \emptyset \ , \\
U \bullet (V\,;W) &=& (U \bullet V)\,;W & \Leftarrow V \cap \varepsilon = \emptyset \ , \\
U\,;(V \bullet W) &=& (U\,;V) \bullet W & \Leftarrow V \cap \varepsilon = \emptyset \ .
\end{array}\right\} \qquad (5)$$

We shall omit parentheses whenever one of these laws applies.

Interesting special cases arise when one of the operands of join or composition is a relation of arity 1. Suppose $R \subseteq A$. Then

$$R \bowtie S = \{a \bullet u : a \in R \wedge a \bullet u \in S\} \ , \qquad R\,;S = \{u : a \in R \wedge a \bullet u \in S\} \ .$$

In other words, $R \bowtie S$ selects all words in S that start with a letter in R, whereas $R;S$ not only selects all those words but also removes their first letters. Therefore, if S is binary, $R \bowtie S$ is the restriction of S to R, whereas $R;S$ is the image of R under S. Likewise, if $T \subseteq A$ then $S \bowtie T$ selects all words in S that end with a letter in T, whereas $S\,;T$ not only selects all those words but also removes their last letters. Therefore, if S is binary, $S \bowtie T$ is the corestriction of S to T, whereas $S\,;T$ is the inverse image of T under S.

For binary $R \subseteq A \bullet A$ and $S, T \subseteq A$ we have, moreover,

$$S \bowtie R \bowtie T = S \bullet T \cap R \ , \qquad S\,;R\,;T = \begin{cases} \varepsilon & \text{if } S \bullet T \cap R \neq \emptyset \ , \\ \emptyset & \text{otherwise} \ . \end{cases} \qquad (6)$$

If both $R \subseteq A$ and $S \subseteq A$ we have

$$R \bowtie S = R \cap S \ , \qquad R\,;S = \begin{cases} \varepsilon & \text{if } R \cap S \neq \emptyset \ , \\ \emptyset & \text{if } R \cap S = \emptyset \ . \end{cases} \qquad (7)$$

We also have neutral elements for join and composition. Assume $A \supseteq P \supseteq \mathsf{fst}\,V$ and $A \supseteq Q \supseteq \mathsf{lst}\,V$ and $V \cap \varepsilon = \emptyset$. Then

$$P \bowtie V = V = V \bowtie Q \ , \qquad I_P\,;V = V = V\,;I_Q \ . \qquad (8)$$

In special cases join and composition can be transformed into each other: assume $P, Q \subseteq A$ and let R be an arbitrary language. Then

$$P \bowtie R = I_P ; R \,, \qquad R \bowtie Q = R ; I_Q \,, \tag{9}$$

$$P ; (Q \bowtie R) = (P \bowtie Q) ; R \,, \qquad (R \bowtie P) ; Q = R ; (P \bowtie Q) \,. \tag{10}$$

2.7 Assertions, Guards and Conditional

As we have seen in (6) and (7), the nullary relations ε and $\emptyset$ behave like the outcomes of certain tests. Therefore they can be used instead of Boolean values, and we call relational expressions yielding nullary relations **assertions**. Note that in this view "false" and "undefined" both are represented by $\emptyset$. Negation is defined by

$$\overline{\emptyset} \stackrel{\text{def}}{=} \varepsilon \,, \qquad \overline{\varepsilon} \stackrel{\text{def}}{=} \emptyset \,.$$

Conjunction and disjunction of assertions are represented by their intersection and union. To improve readability, we write $B \wedge C$ for $B \cap C = B \bullet C$ and $B \vee C$ for $B \cup C$.

For assertion B and language U we have

$$B \bullet U = U \bullet B = \begin{cases} U & \text{if } B = \varepsilon \,, \\ \emptyset & \text{if } B = \emptyset \,. \end{cases}$$

Hence $B \bullet U$ (and $U \bullet B$) behaves like the expression

$$B \;\triangleright\; U = \mathsf{if}\ B\ \mathsf{then}\ U\ \mathsf{else\ error\ fi}$$

in [11] and can be used for propagating assertions through recursions.

Using assertions we can also define a guarded expression and a conditional by

$$\mathsf{if}\ B_1\ \mathsf{then}\ U_1\ [\!]\ \cdots\ [\!]\ B_n\ \mathsf{then}\ U_n\ \mathsf{fi} \stackrel{\text{def}}{=} \bigcup_{i=1}^{n} B_i \bullet U_i \,,$$

$$\mathsf{if}\ B\ \mathsf{then}\ U\ \mathsf{else}\ V\ \mathsf{fi} \stackrel{\text{def}}{=} \mathsf{if}\ B\ \mathsf{then}\ U\ [\!]\ \overline{B}\ \mathsf{then}\ V\ \mathsf{fi} \,,$$

for assertions B, B_i and languages U, U_i, V. Although the conditional is not monotonic in B, it is monotonic in U and V. So we can still use it in recursions provided recursion occurs only in the branches and not in the condition. Note that the guarded expression has angelic semantics: whenever one of the branches is $\neq \emptyset$, so is the whole expression.

3 Closure Operations

We now study in more detail iterated join and composition of a binary relation with itself, which were already seen to be important for path problems.

3.1 Closures

Consider a binary relation $R \subseteq A \bullet A$. We define the **(reflexive and transitive) closure** R^* and the **transitive closure** R^+ of R by

$$R^* \stackrel{\text{def}}{=} \bigcup_{i \in \mathbb{N}} R^i , \qquad R^+ \stackrel{\text{def}}{=} \bigcup_{i \in \mathbb{N} \setminus 0} R^i ,$$

where, as usual, $R^0 \stackrel{\text{def}}{=} I_A$ and $R^{i+1} \stackrel{\text{def}}{=} R; R^i$. It is well-known that R^* is the least fixpoint of the recursion equations

$$R^* = I_A \cup R ; R^* = I_A \cup R^* ; R .$$

This is important since it allows proofs about R^* using fixpoint induction (see e.g. [9]).

Let G be the directed graph associated with R. We have

$$a ; R^i ; a = \begin{cases} \varepsilon \text{ if there is a path with } i \text{ edges from } a \text{ to } b \text{ in } G , \\ \emptyset \text{ otherwise} . \end{cases}$$

For $S \subseteq A$, the set $S ; R^*$ gives all nodes in A reachable from nodes in S via paths in G, whereas $R^*; S$ gives all nodes in A from which some node in S can be reached.

Analogously to R^* we introduce the **path closure** $R^{\bowtie}$ and the **proper path closure** $R^{\Rightarrow}$ of R by

$$R^{\bowtie} \stackrel{\text{def}}{=} \bigcup_{i \in \mathbb{N}} {}^iR , \qquad R^{\Rightarrow} \stackrel{\text{def}}{=} \bigcup_{i \in \mathbb{N} \setminus 0} {}^iR ,$$

where ${}^0R \stackrel{\text{def}}{=} A$ and ${}^{i+1}R \stackrel{\text{def}}{=} R \bowtie {}^iR$. $R^{\bowtie}$ is the least fixpoint of the recursion equations

$$R^{\bowtie} = A \cup R \bowtie R^{\bowtie} = A \cup R^{\bowtie} \bowtie R .$$

It consists of all finite paths in G including the trivial ones with just one node, whereas $R^{\Rightarrow}$ consists of paths with at least one edge. Hence

$$a \bowtie R^{\bowtie} \bowtie b$$

is the language of all paths between a and b in G.

A uniform treatment of these closure operations within the framework of Kleene algebras (see [5]) can be found in [15].

Moving away from the graph view, the path closure also is useful for general binary relations. Let e.g. $\leq$ be a partial order on A. Then $\leq^{\bowtie}$ is the language of all $\leq$-non-decreasing sequences. If $\leq$ is even a total order, then $\leq^{\bowtie}$ is the language of all sequences which are sorted with respect to $\leq$.

To close this section we present a property that allows localising graph traversals in computing the nodes reachable from some set $T \subseteq A$. Suppose $R \subseteq A \bullet A$. By the fixpoint property of R^* and distributivity we have, for $R \subseteq A \bullet A$,

$$T ; R = T \cup T ; R ; R^* .$$

However, since on the right hand side T is already covered by the first summand, we can restrict our attention in the second one to paths outside T. This is stated in

Lemma 1. *Assume* $R \subseteq A \bullet A$ *and* $T \subseteq A$. *Then, with* $\overline{T} \stackrel{\text{def}}{=} A \backslash T$,

$$T ; R^* = T \cup T ; R ; (\overline{T} \bowtie R)^* .$$

The proof uses (a variant of) fixpoint induction and can be found in [15].

3.2 Applications to Sorting

We have already briefly mentioned the relation between the path closure and ordered sequences. However, the path closure consists of nonempty words only, whereas sorting algorithms usually also work for the empty word. Therefore we define, for binary relation $\prec \subseteq A \bullet A$, the **improper path closure** $\prec^{\leadsto}$ of $\prec$ by

$$\prec^{\leadsto} \stackrel{\text{def}}{=} \varepsilon \cup \prec^{\bowtie} .$$

In the sequel s and t range over $A^{(*)}$ and a over A. Moreover, we abbreviate the assertion $S \bullet T \subseteq \prec$ for $S, T \subseteq A$ by $S \prec T$.

Intersection with the improper path closure distributes through join:

$$(s \bowtie t) \cap \prec^{\leadsto} = (s \cap \prec^{\leadsto}) \bowtie (t \cap \prec^{\leadsto}) .$$

This implies

$$s \bullet a \bullet t \cap \prec^{\leadsto} = (s \bullet a \cap \prec^{\leadsto}) \bowtie (a \bullet t \cap \prec^{\leadsto}) . \tag{11}$$

Moreover, we have the following important property:

$$(s \bullet t) \cap \prec^{\leadsto} = (s \cap \prec^{\leadsto}) \bullet (\mathsf{lst}\, s \prec \mathsf{fst}\, t) \bullet (t \cap \prec^{\leadsto}) . \tag{12}$$

In particular (using also that assertions commute with all languages w.r.t. $\bullet$),

$$(a \bullet t) \cap \prec^{\leadsto} = (a \prec \mathsf{fst}\, t) \bullet a \bullet (t \cap \prec^{\leadsto}) .$$

From this we obtain

Corollary 2. *If* $\prec$ *is transitive then*

$$(a \bullet t) \cap \prec^{\leadsto} = (a \prec \mathsf{set}\, t) \bullet a \bullet (t \cap \prec^{\leadsto}) .$$

Moreover, if for all $t \in T \subseteq A$ *we have* $\mathsf{set}\, t = U$ *then*

$$(a \bullet T) \cap \prec^{\leadsto} = (a \prec U) \bullet a \bullet (T \cap \prec^{\leadsto}) .$$

The additional condition in the extension to languages is necessary, since the property on words is not bilinear.

For transitive relations we have a weak distributivity property w.r.t. shuffle, more precisely:

Lemma 3 (Merging Lemma). $\prec$ *is transitive iff*

$$\forall\, S \,.\, \forall\, T \,.\, (S \,|||\, T) \cap \prec^{\leadsto} = ((S \cap \prec^{\leadsto}) \,|||\, (T \cap \prec^{\leadsto})) \cap \prec^{\leadsto} .$$

Proof. ($\Leftarrow$) Assume $a \prec b$ and $b \prec c$. Then

$$
\begin{array}{ll}
& \emptyset \\
\neq & \{\![\text{ by } a \bullet b \bullet c \in \prec^{\sim}]\!\} \\
& (b \bullet a \bullet c \ \cup \ a \bullet b \bullet c \ \cup \ a \bullet c \bullet b) \cap \prec^{\sim} \\
= & \{\![\text{ definition }]\!\} \\
& ((a \bullet c) \,|||\, b) \cap \prec^{\sim} \\
= & \{\![\text{ assumption }]\!\} \\
& (((a \bullet c) \cap \prec^{\sim}) \,|||\, (b \cap \prec^{\sim})) \cap \prec^{\sim} \\
= & \{\![\text{ definition of } \prec^{\sim}]\!\} \\
& (((a \bullet c) \cap \prec) \,|||\, b) \cap \prec^{\sim} \ .
\end{array}
$$

Now strictness shows $(a \bullet c) \cap \prec \neq \emptyset$, i.e. $a \prec c$.

($\Rightarrow$) We show the property for single words; by bilinearity it propagates to languages. The property is immediate if $s \in \prec^{\sim}$ and $t \in \prec^{\sim}$. So assume now $s \notin \prec^{\sim}$. By strictness the right hand side reduces to $\emptyset$. Moreover, by (12) there are u, a, b, w such that $s = u \bullet a \bullet b \bullet v$ and $a \not\prec b$. Consider now some $w \in s \,|||\, t$. It has the form $w = p \bullet a \bullet q \bullet b \bullet r$ for certain $p, q, r \in A^{(*)}$, since shuffling preserves the relative order of the elements of each argument word. Now the assumption $w \in \prec^{\sim}$ by transitivity implies $a \prec b$, a contradiction. □

The formula in the above lemma states that a word is ordered iff all its scattered subwords are. This property will be crucial for the derivation of sorting algorithms.

4 Graph Algorithms

We now want to use our framework to derive three simple graph algorithms, viz. a reachability algorithm, cycle detection and topological sorting. As further applications, [16] calculates an algorithm computing the length of a shortest connecting path between two graph nodes, whereas [18] deals with an algorithm for finding Hamiltonian cycles and relates it to the selection sort algorithm.

4.1 A Simple Reachability Algorithm

We consider the following problem:

Given a directed graph, represented by a binary relation $R \subseteq A \bullet A$ over a finite set A of nodes, and a subset $S \subseteq A$, compute the set of nodes reachable by paths starting in S.

Hence we define

$$reach(S) \stackrel{\text{def}}{=} S \,;\, R^* \ .$$

The aim is to derive a recursive variant of *reach* from this specification. A termination case is given by $reach(\emptyset) = \emptyset \,;\, R^* = \emptyset$. Moreover, we can exploit Lemma 1:

$$S \,;\, R^* = S \ \cup \ S \,;\, R \,;\, (\overline{S} \bowtie R)^* \ ,$$

where $\overline{S} \stackrel{\text{def}}{=} A\backslash S$. However, since on the right hand side we have $(\overline{S} \bowtie R)^*$ rather than R^*, we cannot fold this into a recursive call to *reach*. To gain flexibility we use the technique of generalisation (see e.g. [17]): we introduce a second parameter T for the set that restricts R by defining

$$re(S,T) = S\,;(\overline{T} \bowtie R)^* \ .$$

By specialising this additional parameter we obtain an embedding of the original problem into the generalised one by $reach(S) = re(S,\emptyset)$. The termination case remains unchanged: $re(\emptyset,T) = \emptyset$. Moreover, we calculate:

$$
\begin{array}{ll}
 & re(S,T) \\
= & \{\!\!\{ \text{ definition of } re \ \}\!\!\} \\
 & S\,;(\overline{T} \bowtie R)^* \\
= & \{\!\!\{ \text{ by Lemma 1 } \}\!\!\} \\
 & S \cup S\,;(\overline{T} \bowtie R)\,;(\overline{S} \bowtie \overline{T} \bowtie R)^* \\
= & \{\!\!\{ \text{ by (10) } \}\!\!\} \\
 & S \cup (S \bowtie \overline{T})\,;R\,;(\overline{S} \bowtie \overline{T} \bowtie R)^* \\
= & \{\!\!\{ \text{ by (7) and Boolean algebra } \}\!\!\} \\
 & S \cup (S\backslash T)\,;R\,;(\overline{S \cup T} \bowtie R)^* \\
= & \{\!\!\{ \text{ definition of } re \ \}\!\!\} \\
 & S \cup re((S\backslash T)\,;R,\ S \cup T) \ .
\end{array}
$$

Altogether,

$$re(S,T) = \textsf{if } S = \emptyset \textsf{ then } \emptyset \textsf{ else } S \cup re((S\backslash T)\,;R,\ S \cup T) \textsf{ fi} \ .$$

Note that by (7) the test $S = \emptyset$ can be expressed by the assertion $S\,;S$. We see that T keeps track of the nodes "already visited", while S is the set of nodes the successors of which still have to be visited.

To see whether this can be used as a recursive routine, we need to analyse the termination behaviour. An obvious idea is to inspect the cardinalities of the sets involved. Whereas the first parameter of *re* can shrink and grow according to the varying outdegrees of nodes, the second parameter never shrinks and is bounded from above by $|A|$. The cardinality actually increases unless $S \subseteq T$. However, in that latter case we have $S\backslash T = \emptyset$, so that the recursion moves into the termination case anyway. So the cardinality of the second parameter can indeed be used as a termination function. By standard techniques using an accumulator and associativity of $\cup$ (see e.g. [17]) one can finally transform this into a tail recursion and from there into loop form.

4.2 Cycle Detection

Formal Specification. Consider again a finite set A of nodes and a binary relation $R \subseteq A \bullet A$. The problem is now:

Determine whether R contains a **cyclic path**, *i.e., a proper path in which some node occurs twice.*

The set of all proper paths is given by the proper path closure $R^{\Rightarrow}$. The set of all proper paths that begin and end in the same node is

$$cyc(R) \stackrel{\text{def}}{=} \bigcup_{a \in A} a \bowtie R^{\Rightarrow} \bowtie a \ .$$

Obviously, R contains a cyclic path iff $cyc(R) \neq \emptyset$. However, $cyc(R)$ will be infinite in case R actually contains a cycle, and so this test cannot be evaluated directly. Rather we have to find equivalent characterisations of the problem.

Lemma 4. *The following statements are equivalent:*
(a) $cyc(R) \neq \emptyset$.
(b) $R^{+} \cap I_A \neq \emptyset$.
(c) $R^{|A|} \neq \emptyset$.
(d) $R^{|A|} ; A \neq \emptyset$.
(e) $A ; R^{|A|} \neq \emptyset$.

For the proof see [15]. Among these equivalent formulations, (d) and (e) seem computationally most promising, since they deal with unary relations which in general are much smaller objects than binary ones. We choose (e) as our starting point and specify our problem as

$$hascycle \stackrel{\text{def}}{=} (A ; R^{|A|} \neq \emptyset) \ .$$

An Iteration Principle. To compute $A ; R^{|A|}$ we define $A_i \stackrel{\text{def}}{=} A ; R^i$ and use the properties of the powers of R:

$$\begin{aligned} A_0 &= A ; R^0 &&= A ; I_A &&= A \ , \\ A_{i+1} &= A ; R^{i+1} &&= A ; (R^i ; R) &&= (A ; R^i) ; R = A_i ; R \ . \end{aligned}$$

The associated function $f : X \mapsto X ; R$ is monotonic. We now state a general theorem about monotonic functions on noetherian partial orders. A partial order $(M, \leq)$ is called **noetherian** if each of its nonempty subsets has a minimal element with respect to $\leq$. An element $x \in S \subseteq M$ is **minimal** in S if $y \in S$ and $y \leq x$ imply $y = x$. Viewing a function $f : M \rightarrow M$ as a binary relation, we can form its closure f^*. Then, for $x \in M$, we have $x ; f^* = \{f^i(x) : i \in \mathbb{N}\}$.

Theorem 5. *Let $(M, \leq)$ be a noetherian partial order and $f : M \rightarrow M$ a monotonic total function.*
(a) *If for $x \in M$ we have $f(x) \leq x$ then $x_\infty = \mathrm{glb}\ (x ; f^*)$ exists and is a fixpoint of f. Moreover, it is the only fixpoint of f in $x ; f^*$.*
(b) *Assume x as in (a) and $y \in M$ with $x_\infty \leq y \leq x$. Then also $y_\infty = \mathrm{glb}\ (y ; f^*)$ exists and $y_\infty = x_\infty$.*
(c) *If M has a greatest element $\top$, then $\top_\infty$ exists and is the greatest fixpoint of f.*

For the proof see [16]. A similar theorem has been stated in [4].

To actually calculate x_∞ we define a function *inf* by

$$inf(y) \stackrel{\text{def}}{=} (x_\infty \le y \le x) \bullet x_\infty \ ,$$

which for fixed x determines x_∞ using an upper bound y. We have the embedding $x_\infty = inf(x)$. Now from the above theorem and the fixpoint property of f^* the following recursion is immediate:

$$inf(y) = (x_\infty \le y \le x) \bullet \text{ if } y = f(y) \text{ then } y \text{ else } inf(f(y)) \text{ fi } .$$

Since M is noetherian, this recursion terminates for every y satisfying $f(y) \le y$, because monotonicity then also shows $f(f(y)) \le f(y)$, so that in each recursive call the parameter decreases properly. In particular, the call $inf(x)$ terminates. This algorithm is an abstraction of many iteration methods on finite sets.

A Recursive Solution. We now return to the special case of cycle detection. By finiteness of A the partial order $(\mathcal{P}(A), \subseteq)$ is noetherian with greatest element A. Therefore A_∞ exists. Moreover, we have

Corollary 6. $A_{|A|} = A_\infty$.

Proof. The length of any properly descending chain in $\mathcal{P}(A)$ is at most $|A|+1$. Hence we have $A_{|A|+1} = A_{|A|}$ and thus $A_{|A|} = A_\infty$. □

So we have reduced our task to checking whether $A_\infty \neq \emptyset$, i.e., whether $inf(A) \neq \emptyset$. For our special case the recursion for *inf* reads (omitting the trivial part $S \subseteq A$)

$$inf(S) = (A_\infty \subseteq S) \bullet \text{ if } S = S\,;R \text{ then } S \text{ else } inf(S\,;R) \text{ fi } .$$

We want to improve this by avoiding the computation of $S\,;R$. By the above considerations we may strengthen the assertion of *inf* by adding the conjunct $S;R \subseteq S$. Thus we only need to worry about the difference between S and $S\,;R$. We define

$$src(S,R) \stackrel{\text{def}}{=} S \backslash (S\,;R) \ .$$

Since $S\,;R$ is the set of successors of S under R, this is the set of **sources** of S, i.e., the set of nodes in S which do not have a predecessor in S.

Now, assuming $S\,;R \subseteq S$, we have $S = S\,;R \Leftrightarrow src(S,R) = \emptyset$ and $S\,;R = S \backslash src(S,R)$, so that we can rewrite *inf* into

$$\begin{aligned} inf(S) = {} & (A_\infty \subseteq S \wedge S\,;R \subseteq S) \bullet \\ & \text{if } src(S,R) = \emptyset \text{ then } S \text{ else } inf(S \backslash src(S,R)) \text{ fi } . \end{aligned}$$

This is an improvement in that $src(S,R)$ usually will be small compared with S. Moreover, the computation of $src(S,R)$ can be facilitated by a suitable representation of R. Plugging *inf* into our original problem of cycle recognition we obtain

$$\begin{aligned} hascycle &= hcy(A) \ , \\ hcy(S) &= (A_\infty \subseteq S \wedge S\,;R \subseteq S) \bullet \\ &\quad \text{if } src(S,R) = \emptyset \text{ then } S \neq \emptyset \text{ else } hcy(S \backslash src(S,R)) \text{ fi } , \end{aligned} \tag{13}$$

which is one of the classical algorithms and works by successive removal of sources. Note that Lemma 4(d) suggests a dual specification to the one we have used; replaying our development for it would lead to an algorithm that works by successive removal of sinks. From the algorithm above we derive in [16] a more efficient one, in which the source sets are computed from an incrementally adjusted vector of in-degrees of the graph nodes. The transition from the tail-recursive functional versions to imperative ones with loops and variables which administer the data structures in place is standard transformational knowledge [17].

4.3 Topological Sorting

The problem of topological sorting can be given as follows:

Given an acyclic directed graph, find a total strict-order $<$ on the nodes such that if there is an edge from node a to node b then $a < b$ holds as well.

A finite total order on the nodes can be conveniently described by a repetition-free word comprising all the nodes, taking $<$ as the relation "occurs before". We can give an inductive definition of this relation as

$$\begin{aligned} bef(\varepsilon) &\stackrel{\text{def}}{=} \emptyset \,, \\ bef(a \bullet s) &\stackrel{\text{def}}{=} a \bullet \mathsf{set}\, s \cup bef(s) \,. \end{aligned}$$

Now we can specify the topological sortings of a directed graph with node set A and acyclic edge relation $R \subseteq A \bullet A$ by

$$topsort(R) \stackrel{\text{def}}{=} \{s : s \in perms(A) \wedge R \subseteq bef(s)\} \,.$$

Since we require a permutation of the set A of all nodes, also isolated nodes are covered.

One possibility to obtain a recursive solution is to exhaust in some fashion the set A of nodes. We therefore generalise the problem to deal with arbitrary subsets $S \subseteq A$:

$$tops(S, R) \stackrel{\text{def}}{=} (R \subseteq S \bullet S) \bullet \{s : s \in perms(S) \wedge R \subseteq bef(s)\} \,,$$

with the embedding $topsort(R) = tops(A, R)$.

Suppose now $R \subseteq S \bullet S$. If $S = \emptyset$, then $tops(\emptyset, R) = \varepsilon$. If $S \neq \emptyset$, choose an arbitrary $s \in tops(S, R)$. Since $s \in perms(S)$ we have $s \neq \varepsilon$, and $s = a \bullet t$ for some $a \in S$ and $t \in perms(S \backslash a)$, so that $\mathsf{set}\, t = S \backslash a$. We want to characterise a and t. First, from (7) it follows that

$$a \bowtie \mathsf{set}\, t = \emptyset = \mathsf{set}\, t \bowtie a \;\wedge\; \overline{a} \bowtie \mathsf{set}\, t = \mathsf{set}\, t = \mathsf{set}\, t \bowtie \overline{a} \,. \tag{14}$$

Let us now investigate the relation between a and R. An easy induction shows $bef(t) \subseteq \mathsf{set}\, t \bullet \mathsf{set}\, t$, which implies, by the definition of bef, that $R \subseteq a \bullet \mathsf{set}\, t \cup \mathsf{set}\, t \bullet \mathsf{set}\, t$. Using monotonicity, distributivity and (7, 14) we obtain from this

$$a \bowtie R \subseteq a \bullet \mathsf{set}\, t \;\wedge\; \overline{a} \bowtie R \subseteq \mathsf{set}\, t \bullet \mathsf{set}\, t \;\wedge\; R \bowtie a \subseteq \emptyset \tag{15}$$

with $\overline{a} \stackrel{\text{def}}{=} A\backslash a$. This, in turn, implies

$$R \bowtie \overline{a} = R \,. \tag{16}$$

Hence

$$\begin{array}{cl} & R \subseteq bef(a \bullet t) \\ \Leftrightarrow & \{\!\![\ \text{definition of } bef\ ]\!\!\} \\ & R \subseteq a \bullet \mathbf{set}\, t \cup bef(t) \\ \Leftrightarrow & \{\!\![\ \text{monotonicity, (14, 15) and Boolean algebra}\ ]\!\!\} \\ & a \bowtie R \subseteq a \bullet \mathbf{set}\, t \wedge \overline{a} \bowtie R \subseteq bef(t) \\ \Leftrightarrow & \{\!\![\ \text{by (15)}\ ]\!\!\} \\ & \overline{a} \bowtie R \subseteq bef(t) \,. \end{array}$$

Moreover, by the definition of *perms* we have $a \bullet t \in perms(S) \Leftrightarrow t \in perms(S\backslash a)$; also, it is easily checked that $\overline{a} \bowtie R \subseteq (S\backslash a) \bullet (S\backslash a)$. Altogether,

$$a \bullet t \in tops(S, R) \;\Rightarrow\; R \bowtie a = \emptyset \;\wedge\; t \in tops(S\backslash a, \overline{a} \bowtie R) \,.$$

Hence we are close to a recursion for *topsort*. We only need to check whether the necessary condition also is sufficient. So suppose now $R \bowtie a = \emptyset$ and $t \in tops(S\backslash a, \overline{a} \bowtie R)$. We need to show $R \subseteq bef(a \bullet t)$.

$$\begin{array}{cl} & R \\ = & \{\!\![\ \text{Boolean algebra and distributivity}\ ]\!\!\} \\ & a \bowtie R \cup \overline{a} \bowtie R \\ = & \{\!\![\ \text{by (16)}\ ]\!\!\} \\ & a \bowtie R \bowtie \overline{a} \cup \overline{a} \bowtie R \\ \subseteq & \{\!\![\ \text{by } R \subseteq S \bullet S \text{ and monotonicity}\ ]\!\!\} \\ & a \bowtie S \bullet S \bowtie \overline{a} \cup \overline{a} \bowtie R \\ = & \{\!\![\ \text{by } a \in S\text{, (7) and Boolean algebra}\ ]\!\!\} \\ & a \bullet (S\backslash a) \cup \overline{a} \bowtie R \\ \subseteq & \{\!\![\ \text{by } t \in tops(S\backslash a, \overline{a} \bowtie R)\ ]\!\!\} \\ & a \bullet \mathbf{set}\, t \cup bef(t) \,. \end{array}$$

Summing up, we have shown the recursion relation

$$a \bullet t \in tops(S, R) \;\Leftrightarrow\; a \in S \wedge R \bowtie a = \emptyset \wedge t \in tops(S\backslash a, \overline{a} \bowtie R) \,.$$

To relate this to the previous section, we observe that

$$a \in S \wedge R \bowtie a = \emptyset \;\Leftrightarrow\; a \in S\backslash(S\,;R) \;\Leftrightarrow\; a \in src(S, R) \,.$$

Since R is assumed to be acyclic, we know from (13) that $S \neq \emptyset \Rightarrow src(S,R) \neq \emptyset$. Moreover, Lemma 4(b) and monotonicity show that the assumption of cyclefreeness also holds for $\overline{a} \bowtie R \subseteq R$. Altogether we have the (obviously terminating) recursion

$$tops(S,R) = \text{if } S = \emptyset \text{ then } \varepsilon \text{ else } \bigcup_{a \in src(S,R)} a \bullet tops(S \backslash a, \overline{a} \bowtie R) \text{ fi} .$$

Again, the repeated *src*-computations are inefficient. The remedy, as in the previous algorithm, lies in carrying along an incrementally updated vector of in-degrees.

5 Sorting Algorithms

We now want to derive several sorting algorithms from a common specification.

5.1 Formal Specification of the Sorting Problem

The task of sorting can be formulated as follows:

Given an alphabet A with a total order $\leq$ on it and a word $s \in A^{(*)}$, find a permutation of s which is sorted in ascending order with respect to $\leq$.

Using our operations, we can formalize this by defining

$$sort(s) \stackrel{\text{def}}{=} permw(s) \cap \leq^{\leadsto} .$$

In other words, $sort(s)$ is specified as the set of all ordered permutations of s.

5.2 Mergesort

We use the definition of the permutation set to calculate $sort(\varepsilon) = \varepsilon$, $sort(a) = a$ and

$$\begin{aligned}
& sort(s \bullet t) \\
= \;& \{\!\![\text{ definition }]\!\!\} \\
& permw(s \bullet t) \cap \leq^{\leadsto} \\
= \;& \{\!\![\text{ by (2) }]\!\!\} \\
& (permw(s) \,|||\, permw(t)) \cap \leq^{\leadsto} \\
= \;& \{\!\![\text{ by Lemma 3 }]\!\!\} \\
& ((permw(s) \cap \leq^{\leadsto}) \,|||\, (permw(t) \cap \leq^{\leadsto})) \cap \leq^{\leadsto} \\
= \;& \{\!\![\text{ definition }]\!\!\} \\
& (sort(s) \,|||\, sort(t)) \cap \leq^{\leadsto} \\
= \;& \{\!\![\text{ abbreviation }]\!\!\} \\
& merge(sort(s), sort(t)) ,
\end{aligned}$$

where, for $s, t \in \leq^{\leadsto}$,

$$merge(s,t) \stackrel{\text{def}}{=} (s \cup t \subseteq \leq^{\leadsto}) \bullet ((s \,|||\, t) \cap \leq^{\leadsto}) \; .$$

From the definition it is immediate that *merge* is commutative and associative with neutral element ε. This gives us the base cases for the recursion. Moreover, we calculate, assuming $a \bullet s, b \bullet t \in \leq^{\leadsto}$,

$$\begin{array}{ll}
& merge(a \bullet s, b \bullet t) \\
= & \{\!\![\text{ definition }]\!\!\} \\
& ((a \bullet s) \,|||\, (b \bullet t)) \cap \leq^{\leadsto} \\
= & \{\!\![\text{ definition }]\!\!\} \\
& (a \bullet (s \,|||\, (b \bullet t)) \;\cup\; b \bullet ((a \bullet s) \,|||\, t)) \cap \leq^{\leadsto} \\
= & \{\!\![\text{ distributivity }]\!\!\} \\
& (a \bullet (s \,|||\, (b \bullet t)) \cap \leq^{\leadsto}) \;\cup\; (b \bullet ((a \bullet s) \,|||\, t) \cap \leq^{\leadsto}) \; .
\end{array}$$

Now we treat the first summand, the second one being symmetric. Again we abbreviate $S \bullet T \subseteq \leq$ by $S \leq T$ (for $S, T \subseteq A$).

$$\begin{array}{ll}
& a \bullet (s \,|||\, (b \bullet t)) \cap \leq^{\leadsto} \\
= & \{\!\![\text{ by Corollary 2 }]\!\!\} \\
& (a \leq \mathsf{set}\,(s \,|||\, (b \bullet t))) \bullet a \bullet ((s \,|||\, (b \bullet t)) \cap \leq^{\leadsto}) \\
= & \{\!\![\text{ definition of } merge \;]\!\!\} \\
& (a \leq \mathsf{set}\,(s \,|||\, (b \bullet t))) \bullet a \bullet merge(s, b \bullet t) \\
= & \{\!\![\text{ by (1) and distributivity }]\!\!\} \\
& (a \leq \mathsf{set}\, s \wedge a \leq \mathsf{set}\,(b \bullet t)) \bullet a \bullet merge(s, b \bullet t) \\
= & \{\!\![\text{ by } a \bullet s, b \bullet t \in \leq^{\leadsto} \text{ and transitivity }]\!\!\} \\
& (a \leq b) \bullet a \bullet merge(s, b \bullet t) \\
= & \{\!\![\text{ definition }]\!\!\} \\
& (a \leq b) \bullet a \bullet merge(s, b \bullet t) \; .
\end{array}$$

Combining the two summands we therefore obtain

$$\begin{array}{rl}
merge(a \bullet s, b \bullet t) = & \mathsf{if}\; a \leq b \;\mathsf{then}\; a \bullet merge(s, b \bullet t) \\
& [\!]\; b \leq a \;\mathsf{then}\; b \bullet merge(a \bullet s, t) \;\mathsf{fi} \; .
\end{array}$$

5.3 Quicksort

This time we use a different way of splitting nonempty words and calculate

$$
\begin{aligned}
& sort(s \bullet a \bullet t) \\
= & \quad \{\!\![\text{ definition }]\!\!\} \\
& permw(s \bullet a \bullet t) \cap \leq^{\leadsto} \\
= & \quad \{\!\![\text{ by (2) }]\!\!\} \\
& (permw(s) \mathbin{|||} a \mathbin{|||} permw(t)) \cap \leq^{\leadsto} \\
= & \quad \{\!\![\text{ commutativity }]\!\!\} \\
& (a \mathbin{|||} permw(s) \mathbin{|||} permw(t)) \cap \leq^{\leadsto} \\
= & \quad \{\!\![\text{ by (2) }]\!\!\} \\
& (a \mathbin{|||} permw(s \bullet t)) \cap \leq^{\leadsto} \\
= & \quad \{\!\![\text{ abbreviation }]\!\!\} \\
& insert(a, s \bullet t) \ ,
\end{aligned}
$$

where

$$insert(a, s) \stackrel{\text{def}}{=} (a \mathbin{|||} permw(s)) \cap \leq^{\leadsto} \ .$$

Now we obtain

$$
\begin{aligned}
& insert(a, s) \\
= & \quad \{\!\![\text{ definition }]\!\!\} \\
& (a \mathbin{|||} permw(s)) \cap \leq^{\leadsto} \\
= & \quad \{\!\![\text{ by (3) }]\!\!\} \\
& \Big(\bigcup_{u \bullet v \in permw(s)} permw(u) \bullet a \bullet permw(v)\Big) \cap \leq^{\leadsto} \\
= & \quad \{\!\![\text{ distributivity }]\!\!\} \\
& \bigcup_{u \bullet v \in permw(s)} permw(u) \bullet a \bullet permw(v) \cap \leq^{\leadsto} \\
= & \quad \{\!\![\text{ by (11) }]\!\!\} \\
& \bigcup_{u \bullet v \in permw(s)} (permw(u) \bullet a \cap \leq^{\leadsto}) \bowtie (a \bullet permw(v) \cap \leq^{\leadsto}) \\
= & \quad \{\!\![\text{ by Corollary 2 and its dual }]\!\!\} \\
& \bigcup_{u \bullet v \in permw(s)} ((permw(u) \cap \leq^{\leadsto}) \bullet a \bullet (\text{set}\, u \leq a)) \bowtie \\
& \qquad\qquad ((a \leq \text{set}\, v) \bullet a \bullet (permw(v) \cap \leq^{\leadsto})) \\
= & \quad \{\!\![\text{ definition of } sort\text{, shifting assertions }]\!\!\} \\
& \bigcup_{u \bullet v \in permw(s)} (\text{set}\, u \leq a) \bullet (a \leq \text{set}\, v) \bullet ((sort(u) \bullet a) \bowtie (a \bullet sort(v))) \\
= & \quad \{\!\![\text{ definition of join }]\!\!\}
\end{aligned}
$$

$$\bigcup_{u \bullet v \in permw(s)} (\text{set } u \leq a) \bullet (a \leq \text{set } v) \bullet (sort(u) \bullet a \bullet sort(v)) \ .$$

Altogether, defining

$$split(s,a) \stackrel{\text{def}}{=} \bigcup_{u \bullet v \in permw(s)} (\text{set } u \leq a) \bullet (a \leq \text{set } v) \bullet \langle u, v \rangle \ ,$$

where $\langle u, v \rangle$ denotes the pair consisting of u and v, we have

$$\begin{aligned} sort(\varepsilon) &= \varepsilon \ , \\ sort(s \bullet a \bullet t) &= \bigcup_{\langle u,v \rangle \in split(s \bullet t, a)} sort(u) \bullet a \bullet sort(v) \ . \end{aligned} \tag{17}$$

This is the quicksort algorithm.

5.4 Treesort

We now also consider binary trees. The empty binary tree is denoted by $\Box$ and the composition of two binary trees l, r and a node value a by the triple $\langle l, a, r \rangle$. The inorder traversal of a binary tree is defined by

$$\begin{aligned} inord(\Box) &\stackrel{\text{def}}{=} \varepsilon \\ inord(\langle l, a, r \rangle) &\stackrel{\text{def}}{=} inord(l) \bullet a \bullet inord(r) \ . \end{aligned}$$

Now we specify the sorted trees representing a word s as

$$trees(s) \stackrel{\text{def}}{=} \{b : inord(b) \in sort(s)\} \ .$$

From this it is easily calculated that $trees(\varepsilon) = \Box$ and

$$s \neq \varepsilon \Rightarrow \Box \notin trees(s) \ . \tag{18}$$

Moreover,

$$\begin{aligned} & trees(s \bullet a \bullet t) \\ = \quad & \{\!\![\text{ definition, (18), (17) }]\!\!\} \\ & \{\langle l, b, r \rangle : inord(\langle l, b, r \rangle) \in \bigcup_{\langle u,v \rangle \in split(s \bullet t, a)} sort(u) \bullet a \bullet sort(v)\} \\ = \quad & \{\!\![\text{ commuting quantifiers, definition }]\!\!\} \\ & \bigcup_{\langle u,v \rangle \in split(s \bullet t, a)} \{\langle l, b, r \rangle : inord(l) \bullet b \bullet inord(r) \in sort(u) \bullet a \bullet sort(v)\} \\ \supseteq \quad & \{\!\![\text{ specializing } b \text{ to } a]\!\!\} \\ & \bigcup_{\langle u,v \rangle \in split(s \bullet t, a)} \{\langle l, a, r \rangle : inord(l) \bullet a \bullet inord(r) \in sort(u) \bullet a \bullet sort(v)\} \\ \supseteq \quad & \{\!\![\text{ specializing } l \text{ and } r]\!\!\} \end{aligned}$$

$$\bigcup_{(u,v)\in split(s\bullet t,a)} \{\langle l,a,r\rangle : inord(l) \in sort(u) \land inord(r) \in sort(v)\}$$

$$= \quad \{\!\!\{ \text{ folding } \}\!\!\}$$

$$\bigcup_{(u,v)\in split(s\bullet t,a)} \langle trees(u), a, trees(v)\rangle \ .$$

Note that proper descendant steps are involved. Altogether,

$$trees(\varepsilon) = \Box \ ,$$

$$trees(s \bullet a \bullet t) \supseteq \bigcup_{(u,v)\in split(s\bullet t,a)} \langle trees(u), a, trees(v)\rangle \ .$$

In a similar manner algorithms for inserting into and deleting from sorted trees can be derived. This is, on a less algebraic basis, discussed in [6].

6 Conclusion

We have shown with several examples how to derive graph and sorting algorithms from formal specifications using standard transformation techniques in connection with a powerful algebra over special operators for that particular problem domain.

Different sets of operators have been used to derive algorithms on pointer structures (see [15, 14]) and binary search trees (see [6]). A long term goal is the construction of a database of such operators, enhanced by indexing and external representation using informal language referring to the intuitive meaning of the operators. Such a component would serve as a "specifier's workbench" as a front end to a formal development tool.

Acknowledgements

Many individuals have helped me with their advice. I gratefully acknowledge helpful remarks from F.L. Bauer, R. Berghammer, R. Bird, J. Desharnais, W. Dosch, H. Ehler, M. Lichtmannegger, O. de Moor, H. Partsch, P. Pepper, M. Russling, G. Schmidt and M. Sintzoff.

References

1. F.L. Bauer, B. Möller, H. Partsch, P. Pepper: Formal program construction by transformations — Computer-aided, Intuition-guided Programming. IEEE Transactions on Software Engineering **15**, 165–180 (1989)
2. R. Bird: Lectures on constructive functional programming. In: M. Broy (ed.): Constructive methods in computing science. NATO ASI Series. Series F: Computer and systems sciences **55**. Berlin: Springer 1989, 151–216
3. R.M. Burstall, J. Darlington: A transformation system for developing recursive programs. J. ACM **24**, 44–67 (1977)

4. J. Cai, R. Paige: Program derivation by fixed point computation. Science of Computer Programming **11**, 197–261 (1989)
5. J.H. Conway: Regular algebra and finite machines. London: Chapman and Hall 1971
6. W. Dosch, B. Möller: Calculating a module for binary search trees. GI-Jahrestagung 1993 (to appear)
7. M.S. Feather: A survey and classification of some program transformation approaches and techniques. In L.G.L.T. Meertens (ed.): Proc. IFIP TC2 Working Conference on Program Specification and Transformation, Bad Tölz, April 14–17, 1986. Amsterdam: North-Holland1987, 165–195
8. P. Lescanne: Modèles non déterministes de types abstraits. R.A.I.R.O. Informatique théorique **16**, 225–244 (1982)
9. Z. Manna: Mathematical theory of computation. New York: McGraw-Hill 1974
10. L.G.L.T. Meertens: Algorithmics — Towards programming as a mathematical activity. In: J. W. de Bakker et al. (eds.): Proc. CWI Symposium on Mathematics and Computer Science. CWI Monographs Vol **1**. Amsterdam: North-Holland 1986, 289–334
11. B. Möller: Applicative assertions. In: J.L.A. van de Snepscheut (ed.): Mathematics of Program Construction. Lecture Notes in Computer Science **375**. Berlin: Springer 1989, 348–362
12. B. Möller: Relations as a program development language. In [13], 373–397
13. B. Möller (ed.): Constructing programs from specifications. Proc. IFIP TC2/WG 2.1 Working Conference on Constructing Programs from Specifications, Pacific Grove, CA, USA, 13–16 May 1991. Amsterdam: North-Holland 1991, 373–397
14. B. Möller: Towards pointer algebra. Institut für Mathematik der Universität Augsburg, Report No. 279, 1993. Also to appear in Science of Computer Programming
15. B. Möller: Derivation of graph and pointer algorithms. Institut für Mathematik der Universität Augsburg, Report No. 280, 1993. Also to appear in B. Möller, H.A. Partsch, S.A. Schuman (eds.): Formal program development. Proc. of an IFIP TC2/WG 2.1 State of the Art Seminar. Lecture Notes in Computer Science. Berlin: Springer (to appear)
16. B. Möller, M. Russling: Shorter paths to graph algorithms. Proc. 1992 International Conference on Mathematics of Program Construction (to appear). Extended version: Institut für Mathematik der Universität Augsburg, Report Nr. 272, 1992. Also to appear in Science of Computer Programming
17. H.A. Partsch: Specification and transformation of programs — A formal approach to software development. Berlin: Springer 1990
18. M. Russling: Hamiltonian sorting. Institut für Mathematik der Universität Augsburg, Report Nr. 270, 1992
19. G. Schmidt, T. Ströhlein: Relations and graphs. Discrete Mathematics for Computer Scientists. EATCS Monographs on Theoretical Computer Science. Berlin: Springer 1993

Automatical Synthesis of Programs with Recursions

Anatolij P. Beltiukov

Udmurt state university, Software department
71, Krasnogerojskaya, Izhevsk, Udmurtia, 426034, RUSSIA
e-mail: belt@matsim.udmurtia.su

Abstract. A method of automatic creation of some programs by non-procedural specifications in a simple logical language is proposed. The language is similar to the data description subsets of some programming. languages (such as pascal, Modula-2, Ada) and is the language without lower level (i.e. programs of all elementary actions is to be contained in the programming system libraries). Programs synthesized by a computer using this method can contain loops and recursions.

A notions of the DEPTH OF CONSTRUCTIONS and the BRANCHING INDEX of the programs is introduced. A complete synthesis procedure exists for creating programs with bounded depth of constructions. The time of the procedure execution is polynomial of the input length when the depth of branching in the created programs and the numbers of arguments of predicates and functions used in the task specification are bounded by constants.

The method of the deductive program synthesis is used. A modified first order predicate calculus is used as the theory required for the method.

1 Introduction

The paper is devoted to the deductive program synthesis. The deductive program synthesis is the program generation using constructive proofs of existence of the objects to be computed. Sucy the approach is founded on realizational semantics of logical propositions by Heyting - Kolmogorov - Kleene - Goedel [14, 17, 16, 10] and continued by a lot of publications [1, 2, 3, 4, 5, 6, 7, 8, 9, 11, 12, 13, 15, 18, 19, 20, 21, 22, 24, 25, 26, 27, 28, 29].

Such systems like PRIZ and similar [30, 31, 23] can be regarded as a practical implementation of this approach. Unfortunately, to obtain effectiity in automatic program generation, similar systems use only propositional calculus to formulate tasks to be solved. This feature essentially restricts the application field of the systems.

Author tries to show a restriction on the task under consideration using predicate calculus that has also an effective solution, but with wider application field owing to more powerful language.

2 Basic notions

Let the following be fulfilled:

1) a set of constructive objects UNIVERSUM is given; the constructive objects can be considered as words in a finite alphabet of SYMBOLS, that does not contain the following three signs

$$(\, , \,)$$

UNIVERSUM elements will be called merely as OBJECTS;

2) a finite list of DATA TYPE NAMES is fixed; the names can be regarded as words in a finite alphabet of LETTERS, that does not contain the following six symbols:

$$> \; | \; : \; (\; , \;)$$

3) a set of objects is put into correspondence to each data type name; this set is called the CONTENTS OF THE DATA TYPE;

4) a set of objects is put into correspondence to any sequence $P(L)$, where P is a data type name, L is a chain of objects (called PARAMETERS), separated with commas; this set is called the CONTENTS OF THE GENERIC DATA TYPE $P(L)$.

Type R is called a FOUNDING data type iff there is no an infinite object sequence

$$o_1, \; o_2, \; o_3, \; \dots \; ,$$

that the contents of the data types

$$R(o_1, o_2), \; R(o_2, o_3), \; \dots$$

are not empty. Let R be the name of a fixed founding data type.

The elements of data types contents BELONG to these data types. Expression $o : T$ is TRUE iff the object o belongs to the data type T.

VARIABLES are words in the letter alphabet which are different from the data types names. Variables will be used to denote objects. VARIABLE LIST is a sequence

$$x_1, x_2, \dots, x_n \; ,$$

where $x1$, ..., xn are variables, $n > 0$. Variable lists will be used to denote corresponding object lists.

FORMULA is a data type name or a sequence

$$P(L) \; ,$$

where P is a data type, L is a variable list. Formulas will be used to denote contents of the corresponding data types and generic data types.

FORM is a sequence

$$x : F \; ,$$

where x is a variable, F is a formula. The variable x is the LEADING VARIABLE of this formula. Forms will be used to denote the facts that objects belong to data types.

PARAMETER DESCRIPTION is a sequence

$$f_1; f_2; \dots; f_n \ ,$$

where f_1, ..., f_n are forms; all the leading variables of the forms required to be different.

A FINITE SEQUENCE OF OBJECTS SATISFIES a given parameter description iff the length of the sequence equals to the number of the forms in the description and substitution of all the occurences of the leading variables in this description by the corresponding objects from the sequence transforms the description into a sequence of true propositions (without variables). In a similar way we can define the sentence: "A finite sequence satisfies a given parameter description AFTER REPLACING given (non-leading in the description) VARIABLES by given OBJECTS".

SPECIFICATION is a sequence

$$s_0 > s_1|s_2|\dots|s_n \ ,$$

where s_0, s_1, ..., s_n - parameters description, $n > 0$. All the leading variables of the description s_0 are required to be not a leading variable of any description s_i at $i > 0$. All the leading variables of the description s_i for $i > 0$ must not occur in s_0. All the variables occuring in s_0 must occur as a leading somewhere in s_0. All the variables, occuring in s_i for $i > 0$ must occur as leading somewhere in s_0 or in s_1.

An algorithm is a REALIZATION of a given specification iff when the algorithm processes any finite object sequence b satisfying the description s_0, it puts out a sequence

$$i \ o \ ,$$

where i is a natural number called THE RETURN CODE, o is an object sequence that satisfies the parameter description s_i after replacing in this definition leading variables of the forms of the description s_0 by corresponding objects of the sequence b.

A PROGRAM SYNTHESIS TASK is a nonempty finite sequence of specifications. The last specification of the sequence is the CONCLUSION of the task. All the previous specifications are its PREMISES. A SOLUTION of this task is an algorithm that builds a realization of the conclusion by realizations of the premises. A program SYNTHESIZER is an algorithm that builds solutions of some program synthesis tasks by these tasks.

Note that the proposed specification language can be considered as a subset of the first order predicate calculus if the leading variables v of forms $v : P(L)$ do not occur in nonleading position. Indeed in that case we can replace commas by conjunctions, | signs by disjunctions, > sign by implication, add appropriate quantifiers by leading variables and brackets, join the premises of the task by conjunction and the conclusion of the task by implication. After these transformations the sense of the task will coincide with the realisational semantics of the obtained formula.

3 The result

In the next sections a method to construct program synthesizers is proposed based on deductive principle. A programming language without lower level is fixed. Synthesized programs can contain several kinds non-nested of loops and recursions. A notions CONSTRUCTION DEPTH, and BRANCHING INDEX are introduced for these programs. The depths of constructions are very small for many practical tasks: from 0 to 3. The branching index is bounded too. It is proven that a problem of the program synthesis is algorithmically solvable when the depths of constructions is bounded by constants. A polynomial algorithm can be constructed provided the number of generic types argument and the branching index are bounded by constants too. The degree of the polynomial is bounded by the maximal number of type arguments.

4 Programs

The proposed programming language for deductively synthesized programs (*D*-language) have the following syntax:

```
<program> ::= return <natural> ( <arguments> )
            | <function name> ( <arguments> ) <cases>

<cases> ::= case ( <arguments> ) <program> endcase
          | case ( <arguments> ) <program> endcase <cases>

<arguments> ::= <variable> | <variable> , <arguments>
```

The language contains no variable definitions because the programs will describe realizations of given specifications. In fact a program in this language defines a solution of some program synthesis task when a FUNCTION NAMES are put into correspondence to the all premises and to the conclusion of the program synthesis task.

Let us define the semantics of the programs. The program

$$return\ i\ (a_1, ..., a_n)$$

puts out the sequence

$$i\ A_1, ..., A_n\ ,$$

where A_j is the value of the variable a_j for any $j = 1, ..., n$.

The program

$$f(a_1, ..., a_n)$$

$$case(r_{1,1}, ..., r_{1,k_1})c_1 endcase$$

$$...$$

$$case(r_{m,1}, ..., r_{m,k_m})c_m endcase$$

executes the algorithm f (the realization of the premise or the conclusion to which the name f put into correspondence) the result of the execution is the sequence i o, where the natural number i points to the case that must be chosen. If the case exists ($i < m + 1$), then the sequence o is split into k_i objects which become values of the variables $r_{i,1}$, ... r_{i,k_i} correspondingly. Then the program c_i is executed. If one of these actions cannot be fulfilled, then the whole program must be aborted.

Note that the execution of the conclusion realization is recursive (as the defined function call) and therefore has a specific feature to avoid an endless recursion. We suppose that recurively called algorithm has additionary argument: a value of the founding data type $R(i, i_0)$, where i is the first argument of the defined function, i_0 is the first argument of the recursive call. Here we suppose that the recursion goes by the first argument to simplify the language.

Though the syntax does not contain a loop notions some cases of recursion can be obviously regarded as loops.

5 An example

Let the following data types be fixed:

Item is a set of all names of produced items (in some industrial plant),

$Cost(i)$ is a set of all possible costs of the item i,

$C(i, j, k)$ is a set of all possible ways to build the item i from the items j and k;

$R(i, j)$ is the set to all possible ways to reduce the item i to the item j (this is a founding type).

A program synthesis task has two premises:

1) a program that computes a given item cost or puts out a way to produce the item from some more simple items:

$$i : Item > c : Cost(i) | j : Item; k : Item; d : C(i, j, k); r : R(i, j); s : R(i, k)$$

2) a program that defines an item cost from costs of its possible components and from a possible way of its production of them:

$$i : Item; j : Item; k : Item; d : C(i, j, k); f : Cost(j); e : Cost(k) > c : Cost(i)$$

The task conclusion is the following specification:

$$i : Item > c : Cost(i)$$

Let us denote a realizations of the premises by *Split* and *Count* correspondingly. A program to be built will be denoted by *Total*. Then a solution of the task can be written as follows:

```
Split(i)
case(c) return 1 (c) endcase
case(j,k,d,r,s)
```

```
            Total(j,r)
            case(f) Total(k,s)
                    case(e) Count(i,j,k,d,f,e)
                            case(c) return 1 (c) endcase
                    endcase
            endcase
    endcase
```

6 Program restrictions

Let a program synthesis task be given. Data type name Q is a CONSTRUCTION type name iff there is a formula $P(L)$ in the task that contains a variable of the type Q or Q(M) among the list L. E.g. the program synthesis task in the previous section has only one constructive data type name: *Item*.

A specification

$$s_0 > s_1|s_2|...|s_n \quad ,$$

is a CONTSRUCTION specification if some formula of some s_i for $i > 0$ contains a construction type name. E.g. the program synthesis task in the previous section has only one construction specification: the specification of *Split*.

A program

$$f(a_1, ..., a_n)\ c_1\ ...\ c_m$$

is a CONSTRUCTION BASED routine if the specification corresponding to f is a construction specification. DEPTH OF CONSTRUCTIONS in a program is the maximal number of nested construction based routines in the program. E.g. the program in the previous section has the construction depth 1.

A program

$$return\ i\ (a_1, ..., a_n)$$

has BRANCHING INDEX 1.

The branching index of a program

$$f(a_1, ..., a_n)$$

$$case(r_1)c_1 endcase$$

$$...$$

$$case(r_m)c_m endcase$$

is the sum of the branching indexes of the programs c_1, ..., c_m. E.g. the branching index of the program in the previous section is 2.

7 The theorems

Dimension of the program synthesis task is the maximal number of arguments of generic data types and functions in the task.

A c-bounded program synthesis task is the task of finding solutions for program synthesis task in the D-language with the construction depth not greater than c.

A c-p-b-bounded program synthesis task is the task of finding solutions for program synthesis task with dimension not greater than p, in the D-language with the construction depth not greater than c and the branching index not greater than b.

Theorem 1. *For any natural number c the c-bounded program synthesis task is algorithmically solvable.*

Theorem 2. *For any natural numbers c, p, b the c-p-b-bounded synthesis task is solvable in polynomial time with polynomial degree p.*

The idea of the proof of the theorem 1 is based on the fact that there is only finite number of essentially different formulas that can be derived as data types in the c-bounded program synthesis task.

The idea of the proof of the theorem 2 is based on the fact that the common number of the above mentioned formulas can be bounded by a polynomial of the required degree.

References

1. Arhangelsky D.A., Taitslin M.A.: A logic for data description. Lecture Notes in Computer Science, **363**, (1989), 2–11.
2. Archangelsky D.A., Taitslin M.A.: A modal linear logic. Lect. Notes in Comp. Sci., (1992).
3. Constable R.L.: Constructive mathematics and automatic program writers. Information Processing 71, North Holland, (1972), **1**, 229–233.
4. Constable R.L.: Constructive mathematics as a programming logic I: some principles of theory. International Foundations of Computation Theory Conference, Lecture notes in Computer Science, bf 158, (1983), 64–77.
5. Constable R.L.: Programs and types. IEEE Symp. Found. Computer science, (1980), 112–128.
6. Galmiche D.: Constructive system for automatic program synthesis. Ifornatika 88, Nice, France, (1988).
7. Galmiche D.: The power of constructive framework for programming synthesis. COLOG-88, Papers presented at the International Comference in Computer Logic, Part I, Tallinn, December 12–16, (1988), 165–176.
8. Girard J.-Y.: Linear Logic. Theoretical computer science. (1987). No 50. 1–102.
9. Goad C.A. Proofs as descriptions of computation. Lecture Notes in Computer Science **87**, (1980), 39–52.
10. Goedel K.: Uber eine bisher noch nicht benutze Erweiteung des finiten Standpunktes. Dialectica 12, (1958), No 3/4, 280–287.

11. Goto S.: Program synthesis through Godel's interpretation. Proc. International Conference on Mathematical Studies of Information Processing, (1978), Kyoto, 287–306.

12. Hayashi S.: Extracting Lisp programs from constructive proofs. A formal theory of constructive mathematics based on Lisp. Publ. RIMS, Kyoto Univ., 19 (1983), 169–191.

13. Hayashi S.: PX - a system extracting programs from proofs. IFIP Conference of Formal Description of Programming Concepts, (1986).

14. Heyting: Matematische Grundlagenforschung, Intuitionismus, Beweisstheorie. Ergebnisse der Math. etc. III, 4, Berlin, (1934).

15. Howard W.A.: The formulae-as-types notion of construction. To H.B.Curry: Essays on Combinatory Logic, - Calculus and Formalism, Academic Press, (1980), 479–490.

16. Kleene S.: Introduction to Metamathematics. Princeton - Toronto, 1952.

17. Kolmogorov. Zur Deutung der Intuitionistischen Logik. Math Zeitschrift, (1932), **25**, 58–65.

18. Kreisel G.: Some uses of proof theory for finding computer programs. Colloc. Intern. Log., Clermont-Ferrant, 1975, p.151.

19. Krivine J.L., Parigot M.: Programming with Proofs. 6th Symposium on Computation Theory, Wendisch-Reitz, November 1987.

20. Manna Z., Waldinger R.: A deductive approach to program synthesis. Proc. 6-th Intern. Conf. on AI, Tokyo, (1979), p.542.

21. Manna Z., Waldinger R.: Towards authomatic program synthesis. CACM **3**, (1971), 151–165.

22. Martin-Loef P.: Constructive mathematics and computer programming. 6th Congress for Logic, Methodology and Phylosophy of Science, North Holland, (1982), 153–175.

23. Matrosov V.M., Oparin G.A., Feoktistov D.G., Zhuravlev A.E., Kiselev S.I.: A software tool and the technology of its application (SATURN-technology) for development and explotation of applied intelligent system for mathimatical modeliing and investigation of control system of complex objects notion. IMASS/IFAC International Workshop "Methods and software for automatic control system" (Irkutsk, USSR, September, 3-5,1991). Irkutsk, Computing Centre, (1991), 92–93.

24. Nepejvoda N.N.: A proof theoretical comparison of program synthesis and program verification. 6-th Intern. Congr. of Logic, Methodology and Philosophy of Science, Hannover, (1979)

25. Nepejvoda N.N.: The connection between the proof theory and computer programming. 6-th Intern. Congr. of Logic, Methodology and Philosophy of Science, Hannover, (1979), **1**, p.7-11.

26. Nepejvoda N.N. The logical approach to programming. Lect. Notes Comp. Sci., (1981), **118**, 261–290.

27. Nordstrom B.: Programming in constructive set theory: some examples. ACM Conference of Functional Programming Languasges and Computer Architecture, (1984), 141–154.

28. Parigot M.: Programming with Proofs: a second order type theory. ESOP'88, Nancy, March (1988).

29. Sato M.: Towards a mathematical theory of program synthesis. Proc. IJCAI **6**, (1979), 757–762.

30. Tyugu E.: A programming system with automatic program synthesis. Lecture Notes in Computer Science **47**: Methods of Algorithmic Language Impleventation, Berlin, (1977), 251–267.
31. Tyugu E.: Using a problem Solver in CAD. Artificial Intelligence and Pattern Recognition in CAD, North-Holland, (1978), 197–210.

Parsing in ISBES

Eerke A. Boiten*

Department of Mathematics and Computing Science
Eindhoven University of Technology
P.O Box 513, 5600 MB Eindhoven, The Netherlands
email: eerke@win.tue.nl

Abstract. It is shown how parsing can be described as a problem in the class ISBES, Intersections of Sets and Bags of Extended Substructures, defined in an earlier paper, by viewing parsing as a generalization of pattern matching in several ways. The resulting description is shown to be a good starting point for the transformational derivation of the Cocke-Kasami-Younger tabular parsing algorithm that follows. This derivation is carried out at the level of bag comprehensions.

Keywords. transformational programming, formal specification, substructures, bags, parsing, Cocke-Kasami-Younger

1 Introduction

Transformational programming is a methodology for the derivation of efficient programs from formal specifications by applying semantics preserving transformations, thus guaranteeing correctness of the final result. For a survey of the transformational method, cf. [Par90]. Current research is no longer concentrated on particular derivations or transformation steps, but on higher level knowledge: data types with their characteristic algorithms and properties (*theories of data*, e.g. [Bir87]); transformation strategies so well understood that they can be automated (*theories of programs*, e.g. [SL90]); derivations and their general shapes (*theories of derivations*, e.g. [PB91]).

The present paper falls in the first category: formal specifications in terms of particular data types, and the algorithms that can be derived from those.

An earlier paper [Boi91b] proposed a general problem, called ISBES (Intersection of a Set and a Bag of Extended Substructures), mainly as a generalization of pattern matching problems. Several instances of this problem were given, most of them pattern matching examples. The claim that *parsing* could also be treated in this way was removed from a preliminary version of the paper, since it was not clear how this could be substantiated.

* This research has been carried out at the University of Nijmegen, Department of Computing Science, sponsored by the Netherlands Organization for Scientific Research (NWO), under grant NF 63/62-518 (the STOP — Specification and Transformation Of Programs — project).

This paper describes how, from an almost trivial specification of parsing as an ISBES problem, by stepwise generalization an ISBES instance is obtained that naturally leads to a derivation of the Cocke-Kasami-Younger parsing algorithm.

Various characteristics of the final solution arise during *both* phases of the development: the formulation of parsing as an ISBES instance, and the derivation of the algorithm. This substantiates the claim that formal specification in ISBES-style has relevance for different kinds of problems besides pattern matching.

2 Types and Operations

Types are denoted as sets of introduction rules with laws on terms. The important type *Struct* of binary structures is defined by introduction rules

$$\frac{}{\varepsilon : \mathit{Struct}(\mathbf{t})} \qquad \frac{x : \mathbf{t}}{\tau(x) : \mathit{Struct}(\mathbf{t})} \qquad \frac{a, b : \mathit{Struct}(\mathbf{t})}{a \sqcup b : \mathit{Struct}(\mathbf{t})}$$

and law

$$x \sqcup \varepsilon = \varepsilon \sqcup x = x,$$

i.e. ε is the unit of $\sqcup$. When $\sqcup$ is associative, we have the type *List* of finite lists (usually with $+\!\!+$ for $\sqcup$). If it is also commutative, we get the type *Bag* of finite bags (with $\uplus$). When, additionally, $\sqcup$ is idempotent, the type *Set* of finite sets is obtained, i.e. $\sqcup$ can be denoted by $\cup$. This family of structured types is commonly named the Boom-hierarchy [Mee89].

A number of useful operators [Bir87] on these types, together colloquially called "Squiggol", are defined below (a, b are of type $\mathit{Struct}(\mathbf{t}_1)$ and c, d have type $\mathit{Struct}(\mathbf{t}_2)$):

- "filter" $\triangleleft$ that takes a predicate and a binary structure, and returns the structure containing those of its elements that satisfy the predicate:

$$\begin{aligned} p \triangleleft \varepsilon &= \varepsilon \\ p \triangleleft (\tau(x) \sqcup a) &= \mathbf{if}\ p(x)\ \mathbf{then}\ \tau(x) \sqcup (p \triangleleft a) \\ &\qquad \mathbf{else}\ p \triangleleft a\ \mathbf{fi} \end{aligned}$$

- "map" $*$ that takes a function and a binary structure, and applies the function elementwise:

$$\begin{aligned} f * \varepsilon &= \varepsilon \\ f * \tau(x) &= \tau(f\ x) \\ f * (a \sqcup b) &= (f * a) \sqcup (f * b) \end{aligned}$$

- "reduce" $/$ that takes a binary operator $\oplus$ and a binary structure, that it reduces by putting $\oplus$ between all elements ($1_\oplus$ denotes the unit of $\oplus$):

$$\begin{aligned} \oplus/\varepsilon &= 1_\oplus \\ \oplus/\tau(y) &= y \\ \oplus/(a \sqcup b) &= (\oplus/a) \oplus (\oplus/b) \text{ for nonempty } a, b \end{aligned}$$

- "cross" X that takes two binary structures and a binary operation, generates all possible combinations of elements from one of each of the structures and combines them using the operation:

$$\mathsf{X}_{\oplus}\ \varepsilon = \varepsilon$$
$$a\,\mathsf{X}_{\oplus}\ \tau(x) = (\oplus x) * a$$
$$a\,\mathsf{X}_{\oplus}\ (c \sqcup d) = (a\,\mathsf{X}_{\oplus}\ c) \sqcup (a\,\mathsf{X}_{\oplus}\ d)$$

Tuples are denoted with <>. Let π_i denote the projection to the i-th component of a tuple; compositions of projections will be denoted with lists of indices, e.g. $\pi_1(\pi_2(\pi_2(x))) = \pi_{1,2,2}(x)$. Lists of two elements will occur frequently; $[a,b]$ will be used for the list $\tau(a) +\!\!+ \tau(b)$. Partial parametrization of prefix functions is denoted using underscores, i.e. $f(a, \_, c)(b) = f(a, b, c)$.

3 ISBES Revisited

The problem class ISBES (Intersection of a Set and a Bag of Extended Substructures) can be specified as follows:

Given
- a structured type $S(\mathbf{t})$,
- an object O of the structured type $S(\mathbf{t})$,
- a set P (patterns) of objects of the type $S^+(\mathbf{t})$, where $S^+(\mathbf{t})$ is a type of tuples, the first component of which is in $S(\mathbf{t})$ and the second is in another type of "labels".
- a function D^+ (decompose) which yields the bag of extended substructures (of type $S^+(\mathbf{t})$) of an object of type $S(\mathbf{t})$,
- a function R (result) which takes a bag of elements of type $S^+(\mathbf{t})$ and returns the "answer" to the problem,

compute

$$R(P \Cap D^+(O))$$

where $\Cap$ is defined by

$$S \Cap B = (\in S) \lhd B.$$

Formal definitions of the terms used in this description can be found in [Boi91b]; informally

- a *structured type* $S(\mathbf{t})$ is a type of objects that consist of *structure*, and *basic elements* of type $\mathbf{t}$. These can be seen as together forming the *information* of the structured object. Examples of structured types are lists, bags, sets, and trees of all possible kinds.
- a *substructure* of a structured object is an object of the same type, that contains a subset of the information present in that object. For each structured type, various useful notions of substructures exists – lists, e.g., have prefixes, suffixes, substrings, and segments.

- an *extended substructure* of a structured object is a substructure, labeled with information about the "position" of the substructure in the original object. An example form, for lists, segments with indices. The bag of all substructures of S that have label E is given by $\phi(S, E)$, where ϕ is a function characterizing the type of substructure chosen.

Some ideas on deriving programs from ISBES-style specifications are given in [Boi91b]. Important is also the following:

Observation 1 *When each label component of an extended substructure uniquely identifies the substructure within the original object, a data type transformation can take place, viz. the substructure can be represented by just its label when the original object is known (in context).*

In the case that a certain bag contains at most one occurrence of each value, it could be said that the bag "is actually a set". There is a subtle difference between the property mentioned above and the property that the bag of all substructures "is actually a set". The data type transformation can only be done when ϕ returns at most singleton bags (e.g. for lists with indices), whereas the bag of all substructures "is a set" in *all* cases where ϕ returns bags containing all different elements (e.g. in the case of segments with offsets).

4 Parsing in ISBES

The problem class ISBES describes generalizations of pattern matching; intuitively *parsing* appears to be a related problem and so maybe also something that could be described as an ISBES problem.

Pattern matching is the problem of deciding whether any element of a certain set of patterns occurs in a structure; parsing is the problem of determining if and how a certain sentence is generated by a certain grammar (to fix terminology for now). A borderline case of pattern matching occurs when the structure is one of the patterns. Parsing can be seen as a pattern matching generalization in at least two ways. In one view, the set of strings generated by the language is considered as the pattern set, and we ask whether the sentence is one of the patterns. This is worked out in detail in Section 4.2. An alternative view is to take a complicated (infinite) structure representing *all* possible sentences generated by the grammar, and as the pattern set (implicitly) only the structures that represent derivations of the sentence. This alternative view will be discussed in Section 4.1.

In order to be able to formally describe these ideas, we now first give a formal description of the parsing problem (what follows now is a Squiggol-ish description of traditional formal language theory).

Definition 2 *A* context free grammar *is a 4-tuple* $< S, V_N, V_T, PR >$ *such that* $S \in V_N$ *and* $PR : V_N \rightarrow Set(List(V))$ *where* $V = V_N + V_T$.

In such a grammar $< S, V_N, V_T, PR >$, S is called the *starting symbol*; elements of V_N are called *nonterminals* and elements of V_T *terminals.* If $r \in PR(X)$, then r is called an *alternative* or *right hand side* of (the *left hand side*) X.

Definition 3 *The* language *generated by a grammar* $< S, V_N, V_T, PR >$ *is the set* $\mathcal{L}(\tau(S))$*, where* $\mathcal{L} : List(V) \rightarrow Set(List(V_T))$ *is defined by:*

$$\begin{aligned} \mathcal{L}(\varepsilon) &= \{\varepsilon\} \\ \mathcal{L}(\tau(x)) &= \{\tau(x)\} \quad (x \in V_T) \\ \mathcal{L}(\tau(x)) &= \cup/\mathcal{L} * PR(x) \quad (x \in V_N) \\ \mathcal{L}(x +\!\!+ y) &= \mathcal{L}(x) \times_{+\!\!+} \mathcal{L}(y) \end{aligned}$$

Definition 4 *A grammar* $< S, V_N, V_T, PR >$ *is in* Chomsky Normal Form *(CNF) iff all right hand sides consist of exactly one terminal or exactly two nonterminals, i.e.*

$$\forall X \in V_N : \forall s \in PR(X) : (\exists x \in V_T : s = \tau(x)) \vee (\exists Y, Z \in V_N : s = [Y, Z]).$$

Note that no nonterminal in a CNF grammar generates the empty string. This holds in particular for the starting symbol, so the empty string cannot be in the language generated.

The following subsections describe completely different ways of describing parsing as an ISBES problem. For completeness, we also note that in principle everything can be described as an ISBES problem: if one takes a trivial notion of substructure, viz. every object is only a substructure of itself, and one takes the pattern set to be some kind of universe, the ISBES problem reduces to computing $R(\{O\})$ for some object O and function R. This implies that "anything" is an ISBES problem, and thus so is parsing. An important point is, indeed, that ISBES is a *way of describing* problems, not a limited class of itself.

4.1 Substructures of Parse Forests

Consider the grammar (incidentally, one in CNF) in Van Wijngaarden notation

S:S, S;
  "a".

i.e. $< S, \{S\}, \{a\}, \lambda S.\{[S, S], [a]\} >$, then the type of parse trees for this grammar is

$$\frac{}{P_2(\text{“}a\text{”}) : S} \qquad \frac{x, y : S}{P_1(x, y) : S}$$

This gives a type describing all different parse trees. Sets of parse trees are also called parse *forests*. Parse forests can be represented in a condensed form as trees, where each node contains a (nonempty) *set* of children[1]. Although descriptions of types via introduction rules are usually understood to describe finite objects only, one can imagine infinite objects as well. The list of infinitely many a's can be viewed as a "supremum" of the type $List(\{a\})$, and likewise the "supremum"

[1] This may be called the *mixed type* of parse trees and sets; we intend to further study mixed types and their properties.

of all parse forests of the grammar above is the parse forest S^{∞}, characterized by

$$S^{\infty} = U(P_1(S^{\infty}, S^{\infty}), P_2(\text{“}a\text{”}))$$

where U (“union”) is a constructor function with no additional properties. Now all parse trees of the grammar can be seen as (finite) subtrees of S^{∞}, viz. those where each node is a ***singleton*** set.

For arbitrary grammars a similar characterization of S^{∞} can be given by simultaneously defining A^{∞} for all $A \in V_N$ by $A^{\infty} = U_A(i * *PR(A))$, where $i(x) = x$ for $x \in V_T$, $i(X) = X^{\infty}$ for $X \in V_N$, and U_A is a unique constructor without additional properties.

For parse trees, define the ***yield*** to be the concatenated list of all its (terminal) leaves. Parsing is then an ISBES problem: the pattern set contains all parse trees that have the string to be parsed as their yield, the structured object is S^{∞}, and substructures are finite subforests with singleton nodes. The infinite parse forest S^{∞} represents all possible choices to be made during parsing, whereas the pattern set represents in an implicit way all possible outcomes of the parsing process. The set of substructures of S^{∞} can be decomposed in several ways. The decomposition that is chosen in a derivation that starts from this specification largely determines what parsing algorithm is obtained. (Note that $\Cap$ distributes through set union, and thus each subset of $D^{+}(S^{\infty})$ can be independently intersected with the pattern set). Parsing algorithms derived from this specification have as an advantage that parse trees are already present in the solution, whereas usually the construction of parse trees has to be grafted upon a recognition algorithm [AU72].

4.2 Stepwise Generalization via the Pattern Set

Using the experience of describing string matching as an ISBES problem [Boi91b], we now give another description of parsing as an ISBES problem. From now on, we consider ***non-empty*** strings and (thus) non-empty segments only. (Formally, this is done by discarding the introduction rule for ε from the type *List*).

First, define the bag of all segments of a string by

$$\mathit{segs}\,(s) = \{\!\![\, p \mid \exists x, y : s = x \mathbin{+\!\!+} p \mathbin{+\!\!+} y \,]\!\!\}$$

where $\{\!\![\;]\!\!\}$ denotes bag comprehension (cf. the appendix). Since $w \in \mathit{segs}\,(w)$, a trivial ISBES description of parsing is given by

$$\begin{aligned} O &\triangleq w \\ P &\triangleq \mathcal{L}(S) \\ D^{+} &\triangleq \mathit{segs} \\ R &\triangleq (w \in) \end{aligned}$$

i.e., compute

$$w \in (\mathcal{L}(S) \Cap \mathit{segs}\,(w)).$$

Although this seems a trivial generalization, it suggests the (in the context of parsing relevant) computation of $\{w' \in segs\,(w) \mid w' \in \mathcal{L}(S)\}$.

The substructures used above are *segments*, and thus it is obvious that choosing *indexed* segments, defined by

$$segsi\,(s) = \{\!\{ <p, <i,j>> \mid p = s[i..j] \}\!\}$$

as *extended* substructures may be helpful in this case.[2] This leads to the following ISBES instance, where the pattern set has been extended to contain elements of the same type as *segsi*:

$$\begin{aligned} D^+ &\triangleq segsi \\ P &\triangleq \{<x, <m, m+|x|>> \mid x \in \mathcal{L}(S) \wedge m \leq m+|x| \leq |w|\,\} \\ R &\triangleq (<w, <0, |w|>> \in). \end{aligned}$$

Note that the predicate for P allows an easy restriction to elements of $\mathcal{L}(S)$ not longer than $|w|$.

For $V_N = \{S\}$ and PR in CNF, this is already quite close to the Cocke-Kasami-Younger (CKY) parsing algorithm. In order to allow for multiple nonterminals in V_N, the pattern set and substructures are extended with nonterminals:

$$\begin{aligned} D^+(s) &\triangleq \{\!\{ <p, <i,j,X>> \mid p = s[i..j] \wedge X \in V_N \}\!\} \\ P &\triangleq \{<x, <m, m+|x|, X>> \mid x \in \mathcal{L}(X) \wedge m \leq m+|x| \leq |w|\,\} \\ R &\triangleq (<w, <0, |w|, S>> \in). \end{aligned}$$

A recursive version of $P \Cap D^+(O)$ can be seen as a parsing algorithm that corresponds to the CKY algorithm when the grammar is in CNF. This will be shown in a derivation below.

5 A Derivation of the CKY Parsing Algorithm

The goal is to compute $R(P \Cap D^+(w))$. First the subexpression $P \Cap D^+(w)$ is reduced to a union of bag comprehensions.

$P \Cap D^+(w)$

$=\{$ definition P, definition D^+ $\}$

$\{<x, <m, m+|x|, X>> \mid x \in \mathcal{L}(X) \wedge m \leq m+|x| \leq |w|\} \Cap$
$\{\!\{ <p, <i,j,X>> \mid p = w[i..j] \wedge X \in V_N \}\!\}$

$=\{$ definition $\Cap$, simplifications $\}$

$\{\!\{ <w[i..i+k], <i, i+k, X>> \mid w[i..i+k] \in \mathcal{L}(X) \}\!\}$

$=\{$ comprehension dummies made explicit $\}$

$\biguplus_{0 \leq i < i+k \leq |w|} \{\!\{ <w[i..i+k], <i, i+k, X>> \mid w[i..i+k] \in \mathcal{L}(X) \}\!\}$

[2] Note that we use the convention of not numbering the elements themselves, but the "borders" between them. E.g. $w[1..2]$ denotes the list consisting of the second element of w, and $w[0..|w|] = w$.

This expression is abstracted by the following *embedding*. From this point on, the string w is assumed to be known in context. Let

$$E(i,k,X) = < w[i..i{+}k], < i, i{+}k, X >> .$$

Define a function *cky* by:

$$cky(i,k : 0 \leq i < i{+}k \leq |w|) = \{\!\!\{ E(i,k,X) \mid w[i..i{+}k] \in \mathcal{L}(X) \}\!\!\} .$$

Thus,

$$P \Cap D^{+}(w) = \biguplus_{0 \leq i < i+k \leq |w|} cky(i,k).$$

Also we have that

$$R(P \Cap D^{+}(w)) = (((= S) \circ \pi_{3,2}) \lhd cky(0,|w|) \neq \emptyset).$$

We derive a recursive definition for *cky*. Effective parsing algorithms are characterized by the fact that they are phrased in terms of the grammar only, without resorting to reference to the generated language. This should also be the driving force here: occurrences of $\mathcal{L}$ need to be replaced by occurrences of PR. This can be done easily when the grammar is in Chomsky Normal Form, as can be seen in the derivation below.

Some of the more complicated calculation rules used below are given in the appendix, as indicated by a number in the hint. The first step is to be a case introduction on k, viz. $k = 1 \vee k > 1$. In the case that $k = 1$ we have:

$cky(i,1)$
$\quad = \{$ definition *cky* $\}$
$\{\!\!\{ E(i,1,X) \mid w[i..i{+}1] \in \mathcal{L}(X) \}\!\!\}$
$\quad = \{$ CNF: strings of length 1 are produced in 1 step$\}$
$\{\!\!\{ E(i,1,X) \mid w[i..i{+}1] \in PR(X) \}\!\!\}$,

which is of the required form: it refers to PR but no longer to $\mathcal{L}$.

In the case that $k > 1$ we have:

$cky(i,k)$
$\quad = \{$ definition *cky* $\}$
$\{\!\!\{ E(i,k,X) \mid w[i..i{+}k] \in \mathcal{L}(X) \}\!\!\}$
$\quad = \{$ CNF: strings of length ≥ 2 are produced in more than 1 step $\}$
$\{\!\!\{ E(i,k,X) \mid \exists\, m,Y,Z : 0 < m < k \wedge Y \in V_N \wedge Z \in V_N$
$\qquad \wedge w[i..i{+}m] \in \mathcal{L}(Y) \wedge w[i{+}m..i{+}k] \in \mathcal{L}(Z)$
$\qquad \wedge [Y,Z] \in PR(X) \}\!\!\}$
$\quad = \{$ bag comprehension over linear index domain, Rule 10 $\}$
$\biguplus_{m=1}^{k-1} \{\!\!\{ E(i,k,X) \mid \exists\, Y,Z : Y \in V_N \wedge Z \in V_N$
$\qquad \wedge w[i..i{+}m] \in \mathcal{L}(Y) \wedge w[i{+}m..i{+}k] \in \mathcal{L}(Z)$

$$\wedge [Y, Z] \in PR(X) \}\!\}$$

$= \{$ generalization of domain of comprehension variable, Rule 12, twice$\}$

$$\biguplus_{m=1}^{k-1} \{\!\{ E(i,k,X) \mid \exists Y', Z' : Y' \in \{\!\{ E(i,m,Y) \mid w[i..i+m] \in \mathcal{L}(Y) \}\!\}$$
$$\wedge Z' \in \{\!\{ E(i+m, k-m, Z) \mid w[i+m..i+k] \in \mathcal{L}(Z) \}\!\}$$
$$\wedge [\pi_{3,2}(Y'), \pi_{3,2}(Z')] \in PR(X) \}\!\}$$

$= \{$ fold cky $\}$

$$\biguplus_{m=1}^{k-1} \{\!\{ E(i,k,X) \mid \exists Y', Z' : Y' \in cky(i,m) \wedge Z' \in cky(i+m, k-m)$$
$$\wedge [\pi_{3,2}(Y'), \pi_{3,2}(Z')] \in PR(X) \}\!\}$$

$= \{$ bag comprehension over product domain, Rule 11 $\}$

$$\biguplus_{m=1}^{k-1} E(i,k,\_) * (\biguplus / (cky(i,m) \ \mathsf{X}_{\oplus} \ cky(i+m, k-m)))$$
where $Y' \oplus Z' = \{\!\{ X \mid [\pi_{3,2}(Y'), \pi_{3,2}(Z')] \in PR(X) \}\!\}$.

Altogether, we have

$cky(i,k : 0 \leq i < i+k \leq |w|)$
$=$ **if** $k = 1$
  then $\{\!\{ E(i,1,X) \mid w[i..i+1] \in PR(X) \}\!\}$
  else $\biguplus_{m=1}^{k-1} E(i,k,\_) * (\biguplus / (cky(i,m) \ \mathsf{X}_{\oplus} \ cky(i+m, k-m)))$
    where $Y' \oplus Z' = \{\!\{ X \mid [\pi_{3,2}(Y'), \pi_{3,2}(Z')] \in PR(X) \}\!\}$ **fi**
  where $E(i,k,X) = < w[i..i+k], < i, i+k, X >>$.

By Observation 1, this can be simplified by representing all intermediate results $E(i,k,X)$ by triples $< i, k, X >$. Formally this is done by a somewhat tedious unfold-fold derivation, starting from the definition $cky'(i,k) = E^{-1} * cky(i,j)$, resulting in

$cky(i,k : 0 \leq i < i+k \leq |w|)$
$= E * cky'(i,k)$ **where**
$cky'(i,k : 0 \leq i < i+k \leq |w|)$
$=$ **if** $k = 1$
  then $\{\!\{ < i, 1, X > \mid w[i..i+1] \in PR(X) \}\!\}$
  else $\biguplus_{m=1}^{k-1} < i, k, \_ > *(\biguplus / (cky'(i,m) \ \mathsf{X}_{\oplus} \ cky'(i+m, k-m)))$
    where $Y' \oplus Z' = \{\!\{ X \mid [\pi_3(Y'), \pi_3(Z')] \in PR(X) \}\!\}$ **fi**,
$E(i,k,X) = < w[i..i+k], < i, i+k, X >>$.

By applying *tabulation* to this version of cky, computing $cky(i,k)$ for increasing values of k, the familiar tabular parsing algorithm is obtained. Effectively, this can be seen as a transition from an inefficient top-down parsing algorithm (which computes $cky(i,k)$ several times for most values of i and k) to an efficient bottom-up parsing algorithm.

6 Final Remarks

Describing parsing as an ISBES style specification proved to be an interesting exercise. The resulting specification could still be called "descriptive" (as opposed to operational), but it did to some extent already "suggest" the Cocke-Kasami-Younger parsing algorithm. It contained already complete information about the intermediate values to be computed for obtaining the final result (which is not necessary for the specification as such), but not in which order they need to be computed (which is necessary for the operational version).

The derivation relies mostly on basic properties of bag comprehensions. It is somewhat unfortunate that the calculation does not proceed on the (higher) level of set-bag intersections, using rules like the ones given in [Boi91b]. This seems due to the fact that, for *efficient* parsing, some of the set-bag intersections that occur in the ISBES parsing specification have to be computed elementwise from previous intersections. The fact that *extended* substructures are carried around, however, makes the calculation seem more complicated than it actually is.

Acknowledgements

Helmut Partsch, Paul Frederiks and Johan Jeuring are thanked for many remarks that improved the clarity and readability of this note.

References

[AU72] A.V. Aho and J.D. Ullman. *The Theory of Parsing, Translation and Compiling, Vol. 1: Parsing.* Prentice-Hall, Englewood Cliffs, New Jersey, 1972.

[Bir87] R.S. Bird. An introduction to the theory of lists. In M. Broy, editor, *Logic of Programming and Calculi of Discrete Design. NATO ASI Series Vol. F36*, pages 5–42. Springer-Verlag, Berlin, 1987.

[Boi91a] E.A. Boiten. Can bag comprehension be used at all? Technical Report 91-21, Dept. of Informatics, K.U. Nijmegen, September 1991.

[Boi91b] E.A. Boiten. Intersections of bags and sets of extended substructures - a class of problems. In [Möl91], pages 33–48.

[Mee89] L.G.L.T. Meertens. Lecture notes on the generic theory of binary structures. In *STOP International Summer School on Constructive Algorithmics, Ameland.* September 1989.

[Möl91] B. Möller, editor. *Proceedings of the IFIP TC2 Working Conference on Constructing Programs from Specifications* North-Holland Publishing Company, Amsterdam, 1991.

[Par90] H. Partsch. *Specification and Transformation of Programs - a Formal Approach to Software Development.* Springer-Verlag, Berlin, 1990.

[PB91] H.A. Partsch and E.A. Boiten. A note on similarity of specifications and reusability of transformational developments. In [Möl91], pages 71–89.

[SL90] D.R. Smith and M.R. Lowry. Algorithm theories and design tactics. *Science of Computer Programming*, 14:305–321, 1990.

A Bag Comprehension and Calculation Rules

As described in [Boi91a], bag comprehension is an operation that can only be soundly used in a limited number of cases. For set comprehensions, one can safely write $\{E(x) \mid P(x)\}$; for list comprehensions one is forced to give a (list) domain for the comprehension variable (e.g., $[E(x) \mid x \leftarrow l \wedge P(x)]$); implicitly, such a domain must be present for bag comprehensions as well. "Sound" bag comprehensions are defined by the following rules (B and B_i denote bags):

$$\{\![\, x \mid x \in B \wedge P(x)]\!\} = P \lhd B \tag{5}$$

$$\{\![\, x \mid \mathbf{false}]\!\} = \emptyset \tag{6}$$

$$\begin{aligned} \{\![\, x \mid \exists y : Q(y,x)]\!\} = \; & \biguplus_y (\mathbf{if}\ \exists x : Q(y,x) \\ & \quad \mathbf{then}\ \{\![\, \mathbf{that}\ x : Q(y,x)]\!\} \\ & \quad \mathbf{else}\ \emptyset\ \mathbf{fi}) \\ & \text{if} \quad \forall x,y,z : (Q(y,x) \wedge Q(y,z)) \Rightarrow x = z \end{aligned} \tag{7}$$

$$\{\![\, f(x) \mid x \in B]\!\} = f * B \tag{8}$$

$$\{\![\, x \oplus y \mid x \in B_1 \wedge y \in B_2]\!\} = B_1 \mathsf{X}_{\oplus} B_2. \tag{9}$$

Most of these rules will be used tacitly in derivations, in particular Rule 8 is often used in combination with another one. Some additional rules to be used are the following.

Bag comprehension over linear index domain. Provided m is of type $\mathbf{m}$, and $\mathbf{m}$ is linearly ordered by $\leq$, and $\{\![\, x \mid Q]\!\}$ is a sound bag comprehension, then

$$\{\![\, x \mid \exists m : a \leq m \leq b \wedge Q]\!\} = \biguplus_{m=a}^{b} \{\![\, x \mid Q]\!\}. \tag{10}$$

Bag comprehension over product domain. If $\{\![\, x \mid P(x,y,z)]\!\}$ is a sound bag comprehension for fixed y and z, then

$$\begin{aligned} & \{\![\, x \mid \exists y,z : y \in Y \wedge z \in Z \wedge P(x,y,z)]\!\} \\ & = \uplus/(Y \mathsf{X}_{\oplus} Z) \\ & \quad \mathbf{where}\ y \oplus z = \{\![\, x \mid P(x,y,z)]\!\}. \end{aligned} \tag{11}$$

Generalization of domain of comprehension variable. This rule is used for introducing bag comprehensions, in order to allow folding later on. For an injective function f,

$$\begin{aligned} & \{\![\, E \mid \exists x : x \in X \wedge P(x) \wedge Q(x)]\!\} \\ & = \{\![\, E \mid \exists x' : x' \in \{\![\, f(x) \mid x \in X \wedge P(x)]\!\} \wedge Q(f^{-1}(x'))]\!\}. \end{aligned} \tag{12}$$

Author Index

Springer-Verlag and the Environment

We at Springer-Verlag firmly believe that an international science publisher has a special obligation to the environment, and our corporate policies consistently reflect this conviction.

We also expect our business partners – paper mills, printers, packaging manufacturers, etc. – to commit themselves to using environmentally friendly materials and production processes.

The paper in this book is made from low- or no-chlorine pulp and is acid free, in conformance with international standards for paper permanency.

Lecture Notes in Computer Science

For information about Vols. 1–660
please contact your bookseller or Springer-Verlag

Vol. 661: S. J. Hanson, W. Remmele, R. L. Rivest (Eds.), Machine Learning: From Theory to Applications. VIII, 271 pages. 1993.

Vol. 662: M. Nitzberg, D. Mumford, T. Shiota, Filtering, Segmentation and Depth. VIII, 143 pages. 1993.

Vol. 663: G. v. Bochmann, D. K. Probst (Eds.), Computer Aided Verification. Proceedings, 1992. IX, 422 pages. 1993.

Vol. 664: M. Bezem, J. F. Groote (Eds.), Typed Lambda Calculi and Applications. Proceedings, 1993. VIII, 433 pages. 1993.

Vol. 665: P. Enjalbert, A. Finkel, K. W. Wagner (Eds.), STACS 93. Proceedings, 1993. XIV, 724 pages. 1993.

Vol. 666: J. W. de Bakker, W.-P. de Roever, G. Rozenberg (Eds.), Semantics: Foundations and Applications. Proceedings, 1992. VIII, 659 pages. 1993.

Vol. 667: P. B. Brazdil (Ed.), Machine Learning: ECML – 93. Proceedings, 1993. XII, 471 pages. 1993. (Subseries LNAI).

Vol. 668: M.-C. Gaudel, J.-P. Jouannaud (Eds.), TAPSOFT '93: Theory and Practice of Software Development. Proceedings, 1993. XII, 762 pages. 1993.

Vol. 669: R. S. Bird, C. C. Morgan, J. C. P. Woodcock (Eds.), Mathematics of Program Construction. Proceedings, 1992. VIII, 378 pages. 1993.

Vol. 670: J. C. P. Woodcock, P. G. Larsen (Eds.), FME '93: Industrial-Strength Formal Methods. Proceedings, 1993. XI, 689 pages. 1993.

Vol. 671: H. J. Ohlbach (Ed.), GWAI-92: Advances in Artificial Intelligence. Proceedings, 1992. XI, 397 pages. 1993. (Subseries LNAI).

Vol. 672: A. Barak, S. Guday, R. G. Wheeler, The MOSIX Distributed Operating System. X, 221 pages. 1993.

Vol. 673: G. Cohen, T. Mora, O. Moreno (Eds.), Applied Algebra, Algebraic Algorithms and Error-Correcting Codes. Proceedings, 1993. X, 355 pages 1993.

Vol. 674: G. Rozenberg (Ed.), Advances in Petri Nets 1993. VII, 457 pages. 1993.

Vol. 675: A. Mulkers, Live Data Structures in Logic Programs. VIII, 220 pages. 1993.

Vol. 676: Th. H. Reiss, Recognizing Planar Objects Using Invariant Image Features. X, 180 pages. 1993.

Vol. 677: H. Abdulrab, J.-P. Pécuchet (Eds.), Word Equations and Related Topics. Proceedings, 1991. VII, 214 pages. 1993.

Vol. 678: F. Meyer auf der Heide, B. Monien, A. L. Rosenberg (Eds.), Parallel Architectures and Their Efficient Use. Proceedings, 1992. XII, 227 pages. 1993.

Vol. 679: C. Fermüller, A. Leitsch, T. Tammet, N. Zamov, Resolution Methods for the Decision Problem. VIII, 205 pages. 1993. (Subseries LNAI).

Vol. 680: B. Hoffmann, B. Krieg-Brückner (Eds.), Program Development by Specification and Transformation. XV, 623 pages. 1993.

Vol. 681: H. Wansing, The Logic of Information Structures. IX, 163 pages. 1993. (Subseries LNAI).

Vol. 682: B. Bouchon-Meunier, L. Valverde, R. R. Yager (Eds.), IPMU '92 – Advanced Methods in Artificial Intelligence. Proceedings, 1992. IX, 367 pages. 1993.

Vol. 683: G.J. Milne, L. Pierre (Eds.), Correct Hardware Design and Verification Methods. Proceedings, 1993. VIII, 270 Pages. 1993.

Vol. 684: A. Apostolico, M. Crochemore, Z. Galil, U. Manber (Eds.), Combinatorial Pattern Matching. Proceedings, 1993. VIII, 265 pages. 1993.

Vol. 685: C. Rolland, F. Bodart, C. Cauvet (Eds.), Advanced Information Systems Engineering. Proceedings, 1993. XI, 650 pages. 1993.

Vol. 686: J. Mira, J. Cabestany, A. Prieto (Eds.), New Trends in Neural Computation. Proceedings, 1993. XVII, 746 pages. 1993.

Vol. 687: H. H. Barrett, A. F. Gmitro (Eds.), Information Processing in Medical Imaging. Proceedings, 1993. XVI, 567 pages. 1993.

Vol. 688: M. Gauthier (Ed.), Ada-Europe '93. Proceedings, 1993. VIII, 353 pages. 1993.

Vol. 689: J. Komorowski, Z. W. Ras (Eds.), Methodologies for Intelligent Systems. Proceedings, 1993. XI, 653 pages. 1993. (Subseries LNAI).

Vol. 690: C. Kirchner (Ed.), Rewriting Techniques and Applications. Proceedings, 1993. XI, 488 pages. 1993.

Vol. 691: M. Ajmone Marsan (Ed.), Application and Theory of Petri Nets 1993. Proceedings, 1993. IX, 591 pages. 1993.

Vol. 692: D. Abel, B.C. Ooi (Eds.), Advances in Spatial Databases. Proceedings, 1993. XIII, 529 pages. 1993.

Vol. 693: P. E. Lauer (Ed.), Functional Programming, Concurrency, Simulation and Automated Reasoning. Proceedings, 1991/1992. XI, 398 pages. 1993.

Vol. 694: A. Bode, M. Reeve, G. Wolf (Eds.), PARLE '93. Parallel Architectures and Languages Europe. Proceedings, 1993. XVII, 770 pages. 1993.

Vol. 695: E. P. Klement, W. Slany (Eds.), Fuzzy Logic in Artificial Intelligence. Proceedings, 1993. VIII, 192 pages. 1993. (Subseries LNAI).

Vol. 696: M. Worboys, A. F. Grundy (Eds.), Advances in Databases. Proceedings, 1993. X, 276 pages. 1993.

Vol. 697: C. Courcoubetis (Ed.), Computer Aided Verification. Proceedings, 1993. IX, 504 pages. 1993.

Vol. 698: A. Voronkov (Ed.), Logic Programming and Automated Reasoning. Proceedings, 1993. XIII, 386 pages. 1993. (Subseries LNAI).

Vol. 699: G. W. Mineau, B. Moulin, J. F. Sowa (Eds.), Conceptual Graphs for Knowledge Representation. Proceedings, 1993. IX, 451 pages. 1993. (Subseries LNAI).

Vol. 700: A. Lingas, R. Karlsson, S. Carlsson (Eds.), Automata, Languages and Programming. Proceedings, 1993. XII, 697 pages. 1993.

Vol. 701: P. Atzeni (Ed.), LOGIDATA+: Deductive Databases with Complex Objects. VIII, 273 pages. 1993.

Vol. 702: E. Börger, G. Jäger, H. Kleine Büning, S. Martini, M. M. Richter (Eds.), Computer Science Logic. Proceedings, 1992. VIII, 439 pages. 1993.

Vol. 703: M. de Berg, Ray Shooting, Depth Orders and Hidden Surface Removal. X, 201 pages. 1993.

Vol. 704: F. N. Paulisch, The Design of an Extendible Graph Editor. XV, 184 pages. 1993.

Vol. 705: H. Grünbacher, R. W. Hartenstein (Eds.), Field-Programmable Gate Arrays. Proceedings, 1992. VIII, 218 pages. 1993.

Vol. 706: H. D. Rombach, V. R. Basili, R. W. Selby (Eds.), Experimental Software Engineering Issues. Proceedings, 1992. XVIII, 261 pages. 1993.

Vol. 707: O. M. Nierstrasz (Ed.), ECOOP '93 – Object-Oriented Programming. Proceedings, 1993. XI, 531 pages. 1993.

Vol. 708: C. Laugier (Ed.), Geometric Reasoning for Perception and Action. Proceedings, 1991. VIII, 281 pages. 1993.

Vol. 709: F. Dehne, J.-R. Sack, N. Santoro, S. Whitesides (Eds.), Algorithms and Data Structures. Proceedings, 1993. XII, 634 pages. 1993.

Vol. 710: Z. Ésik (Ed.), Fundamentals of Computation Theory. Proceedings, 1993. IX, 471 pages. 1993.

Vol. 711: A. M. Borzyszkowski, S. Sokołowski (Eds.), Mathematical Foundations of Computer Science 1993. Proceedings, 1993. XIII, 782 pages. 1993.

Vol. 712: P. V. Rangan (Ed.), Network and Operating System Support for Digital Audio and Video. Proceedings, 1992. X, 416 pages. 1993.

Vol. 713: G. Gottlob, A. Leitsch, D. Mundici (Eds.), Computational Logic and Proof Theory. Proceedings, 1993. XI, 348 pages. 1993.

Vol. 714: M. Bruynooghe, J. Penjam (Eds.), Programming Language Implementation and Logic Programming. Proceedings, 1993. XI, 421 pages. 1993.

Vol. 715: E. Best (Ed.), CONCUR'93. Proceedings, 1993. IX, 541 pages. 1993.

Vol. 716: A. U. Frank, I. Campari (Eds.), Spatial Information Theory. Proceedings, 1993. XI, 478 pages. 1993.

Vol. 717: I. Sommerville, M. Paul (Eds.), Software Engineering – ESEC '93. Proceedings, 1993. XII, 516 pages. 1993.

Vol. 718: J. Seberry, Y. Zheng (Eds.), Advances in Cryptology – AUSCRYPT '92. Proceedings, 1992. XIII, 543 pages. 1993.

Vol. 719: D. Chetverikov, W.G. Kropatsch (Eds.), Computer Analysis of Images and Patterns. Proceedings, 1993. XVI, 857 pages. 1993.

Vol. 720: V.Mařík, J. Lažanský, R.R. Wagner (Eds.), Database and Expert Systems Applications. Proceedings, 1993. XV, 768 pages. 1993.

Vol. 721: J. Fitch (Ed.), Design and Implementation of Symbolic Computation Systems. Proceedings, 1992. VIII, 215 pages. 1993.

Vol. 722: A. Miola (Ed.), Design and Implementation of Symbolic Computation Systems. Proceedings, 1993. XII, 384 pages. 1993.

Vol. 723: N. Aussenac, G. Boy, B. Gaines, M. Linster, J.-G. Ganascia, Y. Kodratoff (Eds.), Knowledge Acquisition for Knowledge-Based Systems. Proceedings, 1993. XIII, 446 pages. 1993. (Subseries LNAI).

Vol. 724: P. Cousot, M. Falaschi, G. Filè, A. Rauzy (Eds.), Static Analysis. Proceedings, 1993. IX, 283 pages. 1993.

Vol. 725: A. Schiper (Ed.), Distributed Algorithms. Proceedings, 1993. VIII, 325 pages. 1993.

Vol. 726: T. Lengauer (Ed.), Algorithms – ESA '93. Proceedings, 1993. IX, 419 pages. 1993

Vol. 727: M. Filgueiras, L. Damas (Eds.), Progress in Artificial Intelligence. Proceedings, 1993. X, 362 pages. 1993. (Subseries LNAI).

Vol. 728: P. Torasso (Ed.), Advances in Artificial Intelligence. Proceedings, 1993. XI, 336 pages. 1993. (Subseries LNAI).

Vol. 729: L. Donatiello, R. Nelson (Eds.), Performance Evaluation of Computer and Communication Systems. Proceedings, 1993. VIII, 675 pages. 1993.

Vol. 730: D. B. Lomet (Ed.), Foundations of Data Organization and Algorithms. Proceedings, 1993. XII, 412 pages. 1993.

Vol. 731: A. Schill (Ed.), DCE – The OSF Distributed Computing Environment. Proceedings, 1993. VIII, 285 pages. 1993.

Vol. 732: A. Bode, M. Dal Cin (Eds.), Parallel Computer Architectures. IX, 311 pages. 1993.

Vol. 733: Th. Grechenig, M. Tscheligi (Eds.), Human Computer Interaction. Proceedings, 1993. XIV, 450 pages. 1993.

Vol. 734: J. Volkert Ed.), Parallel Computation. Proceedings, 1993. VIII, 248 pages. 1993.

Vol. 735: D. Bjørner, M. Broy, I. V. Pottosin (Eds.), Formal Methods in Programming and Their Applications. Proceedings, 1993. IX, 434 pages. 1993.

Vol. 736: R. L. Grossman, A. Nerode, A. P. Ravn, H. Rischel (Eds.), Hybrid Systems. VIII, 474 pages. 1993.

Vol. 737: J. Calmet, J. A. Campbell (Eds.), Artificial Intelligence and Symbolic Mathematical Computing. Proceedings, 1992. VIII, 305 pages. 1993.

Vol. 739: H. Imai, R. L. Rivest, T. Matsumoto (Eds.), Advances in Cryptology – ASIACRYPT '91. X, 499 pages. 1993.